T0189078

Lecture Notes in Artificial Intelligence 12714

Subseries of Lecture Notes in Computer Science

More information about this subseries at http://www.springer.com/series/1244

Kamal Karlapalem · Hong Cheng ·
Naren Ramakrishnan · R. K. Agrawal ·
P. Krishna Reddy · Jaideep Srivastava ·
Tanmoy Chakraborty (Eds.)

Advances in Knowledge Discovery and Data Mining

25th Pacific-Asia Conference, PAKDD 2021
Virtual Event, May 11–14, 2021
Proceedings, Part III

Springer

Editors
Kamal Karlapalem (iD)
IIIT, Hyderabad
Hyderabad, India

Naren Ramakrishnan
Virginia Tech
Arlington, VA, USA

P. Krishna Reddy (iD)
IIIT Hyderabad
Hyderabad, India

Tanmoy Chakraborty (iD)
IIIT Delhi
New Delhi, India

Hong Cheng
Chinese University of Hong Kong
Shatin, Hong Kong

R. K. Agrawal
Jawaharlal Nehru University
New Delhi, India

Jaideep Srivastava
University of Minnesota
Minneapolis, MN, USA

ISSN 0302-9743   ISSN 1611-3349   (electronic)
Lecture Notes in Artificial Intelligence
ISBN 978-3-030-75767-0   ISBN 978-3-030-75768-7   (eBook)
https://doi.org/10.1007/978-3-030-75768-7

LNCS Sublibrary: SL7 – Artificial Intelligence

This Springer imprint is published by the registered company Springer Nature Switzerland AG
The registered company address is: Gewerbestrasse 11, 6330 Cham, Switzerland

General Chairs' Preface

On behalf of the Organizing Committee, it is our great pleasure to welcome you to the 25th Pacific-Asia Conference on Knowledge Discovery and Data Mining (PAKDD 2021). Starting in 1997, PAKDD has long established itself as one of the leading international conferences in data mining and knowledge discovery. Held during May 11–14, 2021, PAKDD returned to India for the second time, after a gap of 11 years, moving from Hyderabad in 2010 to New Delhi in 2021. Due to the unexpected COVID-19 epidemic, the conference was held fully online, and we made all the conference sessions accessible online to participants around the world.

Our gratitude goes first and foremost to the researchers, who submitted their work to the PAKDD 2021 main conference, workshops, and data mining contest. We thank them for the efforts in research, as well as in preparing high-quality online presentations videos. It is our distinct honor that five eminent keynote speakers graced the conference: Professor Anil Jain of the Michigan State University, USA, Professor Masaru Kitsuregawa of the Tokyo University, and also the National Institute of Informatics, Japan, Dr. Lada Adamic of Facebook, Prof. Fabrizio Sebastiani of ISTI-CNR, Italy, and Professor Sunita Sarawagi of IIT-Mumbai, India. Each of them is a leader of international renown in their respective areas, and we look forward to their participation.

Given the importance of data science, not just to academia but also to industry, we are pleased to have two distinguished industry speakers. The conference program was further enriched with three high-quality tutorials, eight workshops on cutting-edge topics, and one data mining contest on the prediction of memory failures.

We would like to express our sincere gratitude to the contributions of the Senior Program Committee (SPC) members, Program Committee (PC) members, and anonymous reviewers, led by the PC co-chairs, Kamal Karlapalem (IIIT, Hyderabad), Hong Cheng (CUHK), Naren Ramakrishnan (Virginia Tech). It is through their untiring efforts that the conference have an excellent technical program. We are also thankful to the other Organizing Committee members: industry co-chairs, Gautam Shroff (TCS) and Srikanta Bedathur (IIT Delhi); workshop co-chairs, Ganesh Ramakrishnan (IIT Mumbai) and Manish Gupta (Microsoft); tutorial co-chairs, B. Ravindran (IIT Chennai) and Naresh Manwani (IIIT Hyderabad); Publicity Co-Chairs, Sonali Agrawal (IIIT Allahabad), R. Uday Kiran (University of Aizu), and Jerry C-W Lin (WNU of Applied Sciences); competitions chair, Mengling Feng (NUS); Proceedings Chair, Tanmoy Chakraborthy (IIIT Delhi); and registration/local arrangement co-chairs, Vasudha Bhatnagar (University of Delhi), Vikram Goel (IIIT Delhi), Naveen Kumar (University of Delhi), Rajiv Ratn Shah (IIIT Delhi), Arvind Agarwal (IBM), Aditi Sharan (JNU), Mukesh Giluka (JNU) and Dhirendra Kumar (DTU).

We appreciate the hosting organizations IIIT Hyderabad and the JNU, Delhi, and all our sponsors for their institutional and financial support of PAKDD 2021. We also appreciate Alibaba for sponsoring the data mining contest. We feel indebted to the

PAKDD Steering Committee for its continuing guidance and sponsorship of the paper and student travel awards.

Finally, our sincere thanks go to all the participants and volunteers. There would be no conference without you. We hope all of you enjoy PAKDD 2021.

May 2021

R. K. Agrawal
P. Krishna Reddy
Jaideep Srivastava

PC Chairs' Preface

It is our great pleasure to present the 25th Pacific-Asia Conference on Knowledge Discovery and Data Mining (PAKDD 2021). PAKDD is a premier international forum for exchanging original research results and practical developments in the space of KDD-related areas, including data science, machine learning, and emerging applications.

We received 768 submissions from across the world. We performed an initial screening of all submissions, leading to the desk rejection of 89 submissions due to violations of double-blind and page limit guidelines. Six papers were also withdrawn by authors during the review period. For submissions entering the double-blind review process, each paper received at least three reviews from PC members. Further, an assigned SPC member also led a discussion of the paper and reviews with the PC members. The PC co-chairs then considered the recommendations and meta-reviews from SPC members in making the final decision. As a result, 157 papers were accepted, yielding an acceptance rate of 20.4%. The COVID-19 pandemic caused several challenges to the reviewing process, and we appreciate the diligence of all reviewers, PC members, and SPC members to ensure a quality PAKDD 2021 program.

The conference was conducted in an online environment, with accepted papers presented via a pre-recorded video presentation with a live Q/A session. The conference program also featured five keynotes from distinguished researchers in the community, one most influential paper talk, two invited industrial talks, eight cutting-edge workshops, three comprehensive tutorials, and one dedicated data mining competition session.

We wish to sincerely thank all SPC members, PC members, and external reviewers for their invaluable efforts in ensuring a timely, fair, and highly effective PAKDD 2021 program.

May 2021

Hong Cheng
Kamal Karlapalem
Naren Ramakrishnan

Organization

Organization Committee

General Co-chairs

R. K. Agrawal	Jawaharlal Nehru University, India
P. Krishna Reddy	IIIT Hyderabad, India
Jaideep Srivastava	University of Minnesota, USA

Program Co-chairs

Kamal Karlapalem	IIIT Hyderabad, India
Hong Cheng	The Chinese University of Hong Kong, China
Naren Ramakrishnan	Virginia Tech, USA

Industry Co-chairs

Gautam Shroff	TCS Research, India
Srikanta Bedathur	IIT Delhi, India

Workshop Co-chairs

Ganesh Ramakrishnan	IIT Bombay, India
Manish Gupta	Microsoft Research, India

Tutorial Co-chairs

B. Ravindran	IIT Madras, India
Naresh Manwani	IIIT Hyderabad, India

Publicity Co-chairs

Sonali Agarwal	IIIT Allahabad, India
R. Uday Kiran	The University of Aizu, Japan ·
Jerry Chau-Wei Lin	Western Norway University of Applied Sciences, Norway

Sponsorship Chair

P. Krishna Reddy	IIIT Hyderabad, India

Competitions Chair

Mengling Feng	National University of Singapore, Singapore

Proceedings Chair

Tanmoy Chakraborty	IIIT Delhi, India

Registration/Local Arrangement Co-chairs

Vasudha Bhatnagar	University of Delhi, India
Vikram Goyal	IIIT Delhi, India
Naveen Kumar	University of Delhi, India
Arvind Agarwal	IBM Research, India
Rajiv Ratn Shah	IIIT Delhi, India
Aditi Sharan	Jawaharlal Nehru University, India
Mukesh Kumar Giluka	Jawaharlal Nehru University, India
Dhirendra Kumar	Delhi Technological University, India

Steering Committee

Longbing Cao	University of Technology Sydney, Australia
Ming-Syan Chen	National Taiwan University, Taiwan, ROC
David Cheung	University of Hong Kong, China
Gill Dobbie	The University of Auckland, New Zealand
Joao Gama	University of Porto, Portugal
Zhiguo Gong	University of Macau, Macau
Tu Bao Ho	Japan Advanced Institute of Science and Technology, Japan
Joshua Z. Huang	Shenzhen Institutes of Advanced Technology, Chinese Academy of Sciences, China
Masaru Kitsuregawa	Tokyo University, Japan
Rao Kotagiri	University of Melbourne, Australia
Jae-Gil Lee	Korea Advanced Institute of Science and Technology, South Korea
Ee-Peng Lim	Singapore Management University, Singapore
Huan Liu	Arizona State University, USA
Hiroshi Motoda	AFOSR/AOARD and Osaka University, Japan
Jian Pei	Simon Fraser University, Canada
Dinh Phung	Monash University, Australia
P. Krishna Reddy	International Institute of Information Technology, Hyderabad (IIIT-H), India
Kyuseok Shim	Seoul National University, South Korea
Jaideep Srivastava	University of Minnesota, USA
Thanaruk Theeramunkong	Thammasat University, Thailand
Vincent S. Tseng	National Chiao Tung University, Taiwan, ROC
Takashi Washio	Osaka University, Japan
Geoff Webb	Monash University, Australia
Kyu-Young Whang	Korea Advanced Institute of Science and Technology, South Korea
Graham Williams	Australian National University, Australia
Min-Ling Zhang	Southeast University, China
Chengqi Zhang	University of Technology Sydney, Australia

Ning Zhong   Maebashi Institute of Technology, Japan
Zhi-Hua Zhou   Nanjing University, China

Senior Program Committee

Fei Wang   Cornell University, USA
Albert Bifet   Universite Paris-Saclay, France
Alexandros Ntoulas   University of Athens, Greece
Anirban Dasgupta   IIT Gandhinagar, India
Arnab Bhattacharya   IIT Kanpur, India
B. Aditya Prakash   Georgia Institute of Technology, USA
Bart Goethals   Universiteit Antwerpen, Belgium
Benjamin C. M. Fung   McGill University, Canada
Bin Cui   Peking University, China
Byung Suk Lee   University of Vermont, USA
Chandan K. Reddy   Virginia Tech, USA
Chang-Tien Lu   Virginia Tech, USA
Fuzhen Zhuang   Institute of Computing Technology, Chinese Academy
                          of Sciences, China
Gang Li   Deakin University, Australia
Gao Cong   Nanyang Technological University, Singapore
Guozhu Dong   Wright State University, USA
Hady Lauw   Singapore Management University, Singapore
Hanghang Tong   University of Illinois at Urbana-Champaign, USA
Hongyan Liu   Tsinghua University, China
Hui Xiong   Rutgers University, USA
Huzefa Rangwala   George Mason University, USA
Jae-Gil Lee   KAIST, South Korea
Jaideep Srivastava   University of Minnesota, USA
Jia Wu   Macquarie University, Australia
Jian Pei   Simon Fraser University, Canada
Jianyong Wang   Tsinghua University, China
Jiuyong Li   University of South Australia, Australia
Kai Ming Ting   Federation University, Australia
Kamalakar Karlapalem   IIIT Hyderabad, India
Krishna Reddy P.   International Institute of Information Technology,
                          Hyderabad, India
Lei Chen   Hong Kong University of Science and Technology,
                          China
Longbing Cao   University of Technology Sydney, Australia
Manish Marwah   Micro Focus, USA
Masashi Sugiyama   RIKEN, The University of Tokyo, Japan
Ming Li   Nanjing University, China
Nikos Mamoulis   University of Ioannina, Greece
Peter Christen   The Australian National University, Australia
Qinghua Hu   Tianjin University, China

Rajeev Raman	University of Leicester, UK
Raymond Chi-Wing Wong	Hong Kong University of Science and Technology, China
Sang-Wook Kim	Hanyang University, South Korea
Sheng-Jun Huang	Nanjing University of Aeronautics and Astronautics, China
Shou-De Lin	Nanyang Technological University, Singapore
Shuigeng Zhou	Fudan University, China
Shuiwang Ji	Texas A&M University, USA
Takashi Washio	The Institute of Scientific and Industrial Research, Osaka University, Japan
Tru Hoang Cao	UTHealth, USA
Victor S. Sheng	Texas Tech University, USA
Vincent Tseng	National Chiao Tung University, Taiwan, ROC
Wee Keong Ng	Nanyang Technological University, Singapore
Weiwei Liu	Wuhan University, China
Wu Xindong	Mininglamp Academy of Sciences, China
Xia Hu	Texas A&M University, USA
Xiaofang Zhou	University of Queensland, Australia
Xing Xie	Microsoft Research Asia, China
Xintao Wu	University of Arkansas, USA
Yanchun Zhang	Victoria University, Australia
Ying Li	ACM SIGKDD Seattle, USA
Yue Xu	Queensland University of Technology, Australia
Yu-Feng Li	Nanjing University, China
Zhao Zhang	Hefei University of Technology, China

Program Committee

Akihiro Inokuchi	Kwansei Gakuin University, Japan
Alex Memory	Leidos, USA
Andreas Züfle	George Mason University, USA
Andrzej Skowron	University of Warsaw, Poland
Animesh Mukherjee	IIT Kharagpur, India
Anirban Mondal	Ashoka University, India
Arnaud Soulet	University of Tours, France
Arun Reddy	Arizona State University, USA
Biao Qin	Renmin University of China, China
Bing Xue	Victoria University of Wellington, New Zealand
Bo Jin	Dalian University of Technology, China
Bo Tang	Southern University of Science and Technology, China
Bolin Ding	Data Analytics and Intelligence Lab, Alibaba Group, USA
Brendon J. Woodford	University of Otago, New Zealand
Bruno Cremilleux	Université de Caen Normandie, France
Byron Choi	Hong Kong Baptist University, Hong Kong, China

Cam-Tu Nguyen	Nanjing University, China
Canh Hao Nguyen	Kyoto University, Japan
Carson K. Leung	University of Manitoba, Canada
Chao Huang	University of Notre Dame, USA
Chao Lan	University of Wyoming, USA
Chedy Raissi	Inria, France
Cheng Long	Nanyang Technological University, Singapore
Chengzhang Zhu	University of Technology Sydney, Australia
Chi-Yin Chow	City University of Hong Kong, China
Chuan Shi	Beijing University of Posts and Telecommunications, China
Chunbin Lin	Amazon AWS, USA
Da Yan	University of Alabama at Birmingham, USA
David C Anastasiu	Santa Clara University, USA
David Taniar	Monash University, Australia
David Tse Jung Huang	The University of Auckland, New Zealand
Deepak P.	Queen's University Belfast, UK
De-Nian Yang	Academia Sinica, Taiwan, ROC
Dhaval Patel	IBM TJ Watson Research Center, USA
Dik Lee	HKUST, China
Dinesh Garg	IIT Gandhinagar, India
Dinusha Vatsalan	Data61, CSIRO, Australia
Divyesh Jadav	IBM Research, USA
Dong-Wan Choi	Inha University, South Korea
Dongxiang Zhang	University of Electronic Science and Technology of China, China
Duc-Trong Le	University of Engineering and Technology, Vietnam National University, Hanoi, Vietnam
Dung D. Le	Singapore Management University, Singapore
Durga Toshniwal	IIT Roorkee, India
Ernestina Menasalvas	Universidad Politécnica de Madrid, Spain
Fangzhao Wu	Microsoft Research Asia, China
Fanhua Shang	Xidian University, China
Feng Chen	UT Dallas, USA
Florent Masseglia	Inria, France
Fusheng Wang	Stony Brook University, USA
Gillian Dobbie	The University of Auckland, New Zealand
Girish Palshikar	Tata Research Development and Design Centre, India
Giuseppe Manco	ICAR-CNR, Italy
Guandong Xu	University of Technology Sydney, Australia
Guangyan Huang	Deakin University, Australia
Guangzhong Sun	School of Computer Science and Technology, University of Science and Technology of China, China
Guansong Pang	University of Adelaide, Australia
Guolei Yang	Facebook, USA

Guoxian Yu	Shandong University, China
Guruprasad Nayak	University of Minnesota, USA
Haibo Hu	Hong Kong Polytechnic University, China
Heitor M Gomes	Télécom ParisTech, France
Hiroaki Shiokawa	University of Tsukuba, Japan
Hong Shen	Adelaide University, Australia
Honghua Dai	Zhengzhu University, China
Hongtao Wang	North China Electric Power University, China
Hongzhi Yin	The University of Queensland, Australia
Huasong Shan	JD.com, USA
Hui Xue	Southeast University, China
Huifang Ma	Northwest Normal University, China
Huiyuan Chen	Case Western Reserve University, USA
Hung-Yu Kao	National Cheng Kung University, Taiwan, ROC
Ickjai J. Lee	James Cook University, Australia
Jaegul Choo	KAIST, South Korea
Jean Paul Barddal	PUCPR, Brazil
Jeffrey Ullman	Stanford University, USA
Jen-Wei Huang	National Cheng Kung University, Taiwan, ROC
Jeremiah Deng	University of Otago, New Zealand
Jerry Chun-Wei Lin	Western Norway University of Applied Sciences, Norway
Ji Zhang	University of Southern Queensland, Australia
Jiajie Xu	Soochow University, China
Jiamou Liu	The University of Auckland, New Zealand
Jianhua Yin	Shandong University, China
Jianmin Li	Tsinghua University, China
Jianxin Li	Deakin University, Australia
Jianzhong Qi	University of Melbourne, Australia
Jie Liu	Nankai University, China
Jiefeng Cheng	Tencent, China
Jieming Shi	The Hong Kong Polytechnic University, China
Jing Zhang	Nanjing University of Science and Technology, China
Jingwei Xu	Nanjing University, China
João Vinagre	LIAAD, INESC TEC, Portugal
Jörg Wicker	The University of Auckland, New Zealand
Jun Luo	Machine Intelligence Lab, Lenovo Group Limited, China
Jundong Li	Arizona State University, USA
Jungeun Kim	ETRI, South Korea
Jun-Ki Min	Korea University of Technology and Education, South Korea
K. Selçuk Candan	Arizona State University, USA
Kai Zheng	University of Electronic Science and Technology of China, China
Kaiqi Zhao	The University of Auckland, New Zealand

Kaiyu Feng	Nanyang Technological University, Singapore
Kangfei Zhao	The Chinese University of Hong Kong, China
Karan Aggarwal	University of Minnesota, USA
Ken-ichi Fukui	Osaka University, Japan
Khoat Than	Hanoi University of Science and Technology, Vietnam
Ki Yong Lee	Sookmyung Women's University, South Korea
Ki-Hoon Lee	Kwangwoon University, South Korea
Kok-Leong Ong	La Trobe University, Australia
Kouzou Ohara	Aoyama Gakuin University, Japan
Krisztian Buza	Budapest University of Technology and Economics, Hungary
Kui Yu	School of Computer and Information, Hefei University of Technology, China
Kun-Ta Chuang	National Cheng Kung University, China
Kyoung-Sook Kim	Artificial Intelligence Research Center, Japan
L Venkata Subramaniam	IBM Research, India
Lan Du	Monash University, Canada
Lazhar Labiod	LIPADE, France
Leandro Minku	University of Birmingham, UK
Lei Chen	Nanjing University of Posts and Telecommunications, China
Lei Duan	Sichuan University, China
Lei Gu	Nanjing University of Posts and Telecommunications, China
Leong Hou U	University of Macau, Macau
Leopoldo Bertossi	Universidad Adolfo Ibañez, Chile
Liang Hu	University of Technology Sydney, Australia
Liang Wu	Airbnb, USA
Lin Liu	University of South Australia, Australia
Lina Yao	University of New South Wales, Australia
Lini Thomas	IIIT Hyderabad, India
Liu Yang	Beijing Jiaotong University, China
Long Lan	National University of Defense Technology, China
Long Yuan	Nanjing University of Science and Technology, China
Lu Chen	Aalborg University, Denmark
Maciej Grzenda	Warsaw University of Technology, Poland
Maguelonne Teisseire	Irstea, France
Maksim Tkachenko	Singapore Management University, Singapore
Marco Maggini	University of Siena, Italy
Marzena Kryszkiewicz	Warsaw University of Technology, Poland
Maya Ramanath	IIT Delhi, India
Mengjie Zhang	Victoria University of Wellington, New Zealand
Miao Xu	RIKEN, Japan
Minghao Yin	Northeast Normal University, China
Mirco Nanni	ISTI-CNR Pisa, Italy
Motoki Shiga	Gifu University, Japan

Shoujin Wang	Macquarie University, Australia
Shu Wu	NLPR, China
Shuhan Yuan	Utah State University, USA
Sibo Wang	The Chinese University of Hong Kong, China
Silvia Chiusano	Politecnico di Torino, Italy
Songcan Chen	Nanjing University of Aeronautics and Astronautics, China
Steven H. H. Ding	Queen's University, Canada
Suhang Wang	Pennsylvania State University, USA
Sungsu Lim	Chungnam National University, South Korea
Sunil Aryal	Deakin University, Australia
Tadashi Nomoto	National Institute of Japanese Literature, Japan
Tanmoy Chakraborty	IIIT Delhi, India
Tetsuya Yoshida	Nara Women's University, Japan
Thanh-Son Nguyen	Agency for Science, Technology and Research, Singapore
Thilina N. Ranbaduge	The Australian National University, Australia
Tho Quan	John Von Neumann Institute, Germany
Tianlin Zhang	University of Chinese Academy of Sciences, China
Tianqing Zhu	University of Technology Sydney, Australia
Toshihiro Kamishima	National Institute of Advanced Industrial Science and Technology, Japan
Trong Dinh Thac Do	University of Technology Sydney, Australia
Tuan Le	Oakland University, USA
Tuan-Anh Hoang	L3S Research Center, Leibniz University of Hanover, Germany
Turki Turki	King Abdulaziz University, Saudi Arabia
Tzung-Pei Hong	National University of Kaohsiung, Taiwan, ROC
Uday Kiran Rage	University of Tokyo, Japan
Vahid Taslimitehrani	PhysioSigns Inc., USA
Victor Junqiu Wei	Huawei Technologies, China
Vladimir Estivill-Castro	Griffith University, Australia
Wang Lizhen	Yunnan University, China
Wang-Chien Lee	Pennsylvania State University, USA
Wang-Zhou Dai	Imperial College London, UK
Wei Liu	University of Western Australia, Australia
Wei Luo	Deakin University, Australia
Wei Shen	Nankai University, China
Wei Wang	University of New South Wales, Australia
Wei Zhang	East China Normal University, China
Wei Emma Zhang	The University of Adelaide, Australia
Weiguo Zheng	Fudan University, China
Wendy Hui Wang	Stevens Institute of Technology, USA
Wenjie Zhang	University of New South Wales, Australia
Wenpeng Lu	Qilu University of Technology (Shandong Academy of Sciences), China

Wenyuan Li   University of California, Los Angeles, USA
Wilfred Ng   HKUST, China
Xiang Ao   Institute of Computing Technology, CAS, China
Xiangliang Zhang   King Abdullah University of Science and Technology,
                              Saudi Arabia
Xiangmin Zhou   RMIT University, Australia
Xiangyu Ke   Nanyang Technological University, Singapore
Xiao Wang   Beijing University of Posts and Telecommunications,
                              China
Xiaodong Yue   Shanghai University, China
Xiaohui (Daniel) Tao   The University of Southern Queensland, Australia
Xiaojie Jin   National University of Singapore, Singapore
Xiaoyang Wang   Zhejiang Gongshang University, China
Xiaoying Gao   Victoria University of Wellington, New Zealand
Xin Huang   Hong Kong Baptist University, China
Xin Wang   University of Calgary, Canada
Xingquan Zhu   Florida Atlantic University, USA
Xiucheng Li   Nanyang Technological University, Singapore
Xiuzhen Zhang   RMIT University, Australia
Xuan-Hong Dang   IBM T.J Watson Research Center, USA
Yanchang Zhao   CSIRO, Australia
Yang Wang   Dalian University of Technology, China
Yang Yu   Nanjing University, China
Yang-Sae Moon   Kangwon National University, South Korea
Yanhao Wang   University of Helsinki, Finland
Yanjie Fu   Missouri University of Science and Technology, USA
Yao Zhou   UIUC, USA
Yashaswi Verma   IIT Jodhpur, India
Ye Zhu   Deakin University, Australia
Yiding Liu   Nanyang Technological University, Singapore
Yidong Li   Beijing Jiaotong University, China
Yifeng Zeng   Northumbria University, UK
Yingfan Liu   Xidian University, China
Yingyi Bu   Google, USA
Yi-Shin Chen   National Tsing Hua University, Taiwan, ROC
Yiyang Yang   Guangdong University of Technology, China
Yong Guan   Iowa State University, USA
Yu Rong   Tencent AI Lab, China
Yu Yang   City University of Hong Kong, China
Yuan Yao   Nanjing University, China
Yuanyuan Zhu   Wuhan University, China
Yudong Zhang   University of Leicester, UK
Yue Ning   Stevens Institute of Technology, USA
Yue Ning   Stevens Institute of Technology, USA
Yue-Shi Lee   Ming Chuan University, China
Yun Sing Koh   The University of Auckland, New Zealand

Yunjun Gao	Zhejiang University, China
Yuqing Sun	Shandong University, China
Yurong Cheng	Beijing Institute of Technology, China
Yuxiang Wang	Hangzhou Dianzi University, China
Zemin Liu	Singapore Management University, Singapore
Zhang Lei	Anhui University, China
Zhaohong Deng	Jiangnan University, China
Zheng Liu	Nanjing University of Posts and Telecommunications, China
Zheng Zhang	Harbin Institute of Technology, China
Zhengyang Wang	Texas A&M University, USA
Zhewei Wei	Renmin University of China, China
Zhiwei Zhang	Beijing Institute of Technology, China
Zhiyuan Chen	University of Maryland Baltimore County, USA
Zhongying Zhao	Shandong University of Science and Technology, China
Zhou Zhao	Zhejiang University, China
Zili Zhang	Southwest University, China

Competition Sponsor

Alibaba Cloud

Host Institutes

Jawaharlal Nehru University

INTERNATIONAL INSTITUTE OF
INFORMATION TECHNOLOGY
H Y D E R A B A D

Contents – Part III

Representation Learning and Embedding

Learning from Data

Representation Learning
and Embedding

Episode Adaptive Embedding Networks
for Few-Shot Learning

Fangbing Liu$^{(\boxtimes)}$ and Qing Wang

Australian National University, Canberra, Australia
{fangbing.liu,qing.wang}@anu.edu.au

Abstract. Few-shot learning aims to learn a classifier using a few labelled instances for each class. Metric-learning approaches for few-shot learning embed instances into a high-dimensional space and conduct classification based on distances among instance embeddings. However, such instance embeddings are usually shared across all episodes and thus lack the discriminative power to generalize classifiers according to episode-specific features. In this paper, we propose a novel approach, namely *Episode Adaptive Embedding Network* (EAEN), to learn episode-specific embeddings of instances. By leveraging the probability distributions of all instances in an episode at each channel-pixel embedding dimension, EAEN can not only alleviate the overfitting issue encountered in few-shot learning tasks, but also capture discriminative features specific to an episode. To empirically verify the effectiveness and robustness of EAEN, we have conducted extensive experiments on three widely used benchmark datasets, under various combinations of different generic embedding backbones and different classifiers. The results show that EAEN significantly improves classification accuracy about 10–20% in different settings over the state-of-the-art methods.

Keywords: Few-shot learning · Episode adaptive embedding

1 Introduction

Few-shot learning has attracted attention recently due to its potential to bridge the gap between the cognition ability of humans and the generalization ability of machine learning models [1, 5, 13, 19]. At its core, few-shot learning aims to learn a classifier using a few labelled instances for each class. This however poses significant challenges to traditional machine learning algorithms which are designed to learn from a large amount of labelled instances. They easily overfit when trained on a small training set, and thus fail to generalize to new classes.

Driven by a simple learning principle: "test and train conditions must match", *episode training* was proposed to deal with the few-shot learning problem [19]. In the episode training setting, each episode contains only a few labelled instances per class (i.e., *support set*) and a number of unlabelled instances (i.e., *query set*) whose classes are to be predicted. Thus, an episode mimics a classification task in few-shot learning scenarios, and a learning model can be trained by conducting a series of classification tasks moving from episode to episode. As reported in [6, 19], compared with traditional supervised training in which labelled instances are from one classification task, episode training leads to a better generalization ability on small training data.

© Springer Nature Switzerland AG 2021
K. Karlapalem et al. (Eds.): PAKDD 2021, LNAI 12714, pp. 3–15, 2021.
https://doi.org/10.1007/978-3-030-75768-7_1

Inspired by [19], episode training has been adopted in many later studies for few-shot learning [4,5,9]. One promising research stream focuses on developing metric-learning-based approaches with episode training [1,16,18,19]. The key idea is to map instances into a high-dimensional embedding space such that their embeddings capture discriminative features for classification. Then, distances between instance embeddings are measured, and unlabelled instances in an episode are classified according to their distances with labelled instances. Although achieving reasonably good performance, most approaches do not consider features specific to classification tasks when embedding instances, i.e., episode-specific features. For example, instances of three classes "dog" (circle), "cat" (cross) and "wolf" (triangle) can be mapped into generic embeddings shown in Fig. 1(a), without considering their episode-specific features. However, it is hard to classify these instances based on their generic embeddings. By embedding instances into an episode-specific embedding space that capture episode-specific features, such as features distinguishing "dog" from "wolf", or "dog" from "cat", as shown in Fig. 1(b) and (c), it is easier to learn classification boundaries within an episode.

Recently, some works [7, 11,20,24] began to explore instance embeddings specific to classification tasks in few-shot learning. They have generally followed two directions: (a) tailoring the embeddings of support instances (i.e., instances in a support set) by learning their inter-class discriminative features within an episode [11,20,24]; (b) adjusting the embeddings of query instances (i.e., instances in a query set) according to their characteristics [7]. For example, support instances were used to refine their generic embeddings via a

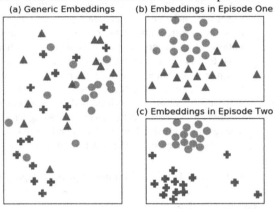

Fig. 1. Instance embeddings (a) in a generic embedding space, and (b) and (c) in an episode-specific embedding space.

set-to-set function in [24]. A task-aware feature embedding network was introduced in [20] to adjust instance embeddings for specific tasks in a meta-learning framework. Nevertheless, none of these methods have fully captured episode-specific features into instance embeddings. They focused on extracting features specific to classes and to instances, whereas neglecting to account for features that align query instances with support instances in a specific episode. Instance embeddings thus lack the discriminative ability to generalize classifiers across episodes with new classes. Moreover, since only a few instances are available in a support set in few-shot learning, the low-data problem also hinders classification performance of these methods.

To circumvent these limitations, we propose *Episode Adaptive Embedding Networks* (EAENs), which leverage the probability distributions of *all instances* in an episode, including instances from both a support set and a query set, to extract representative episode-specific features. Particularly, EAENs consider the probability distributions of all instances in an episode at *each channel-pixel embedding dimension*. This leads to an effective adaptation that transforms generic embeddings into episode-specific embeddings for improved generalisation. Thus, unlike prior works, EAENs have two distinct advantages. First, it alleviates the overfitting issue since it learns based on embeddings of both support and query instances, in contrast to just a few support instances per class used in existing works. Second, it captures features that align query instances with support instances in each specific episode into embeddings. This is important for improving classification performance because metric-learning approaches for few-shot learning mostly rely on measuring distances among instance embeddings. In summary, our main contributions are as follows:

- We propose a novel approach (EAENs) for few-shot learning, which maps instances into an episode-specific embedding space, capturing episode-specific features.
- We derive formulae to exhibit the probability distributions of all instances in an episode with respect to each channel-pixel embedding dimension. This improves the generalization ability of classifiers.
- We conduct experiments to verify the effectiveness and robustness of our approach. Compared with the state-of-the-art models, our approach achieves about 20% accuracy improvement in 5-way 1-shot and about 10% improvement in 5-way 5-shot on both miniImageNet and tieredImageNet datasets, as well as competitive performance on CIFAR-FS dataset.

2   Related Work

Few-shot learning has been extensively studied in recent years [4, 19]. Our work in this paper is broadly related to three streams of research in few-shot learning.

Metric-Learning Approaches. The key idea behind metric-learning approaches is to learn instance embeddings such that discriminative features of instances can be captured by their embeddings in a high-dimensional space [1, 11, 12, 16, 18, 19]. Then, a distance-based classifier is employed to classify instances based on distances between instances in their embedding space. To avoid the overfitting problem in few-shot learning, these approaches often use simple non-parametric classifiers, such as nearest neighbor classifiers [1, 16, 19]. Distances between instance embeddings are typically measured by simple L1 and cosine distances [19]. A recent work proposed to learn such a distance metric for comparing instances within episodes [18].

Meta-learning Approaches. Lots of meta-learning approaches have been proposed for few-shot learning tasks [2, 14, 15, 22]. These approaches aim to minimize generalization error across different tasks and expect that a classifier performs well on unseen tasks [2, 5, 14, 15]. However, they mostly only learn generic embeddings that are same for all tasks. Some recent works have studied task-related embeddings [10, 20]. Since only a few labelled instances are available for each unseen class in a target task, learning discriminative task-related embeddings is hard and these works implicitly relied on the alignment of data distributions between seen and unseen classes. Several works also used data hallucination methods to synthesize instances to help classification [8, 21].

Transductive Approaches. Depending on whether instances in a query set (i.e., query instances) are taken into account when designing a learning model, approaches for few-shot learning can be categorized as being transductive and non-transductive. Several works used query instances and their structure in episodes to conduct a classification task in a transductive way [6,9,13,23]. A label propagation method was proposed in [13], where label information was propagated from instances in a support set to instances in a query set. Graph neural networks were employed to diffuse information from neighbor instances for better embeddings [6]. Assuming that all instances are fully connected with each other, [9] proposed an iterative edge-labeling algorithm to predict edge labels, i.e., whether two instances connected by an edge belong to the same class.

3   Episode Adaptive Embedding Networks

We formulate the few-shot classification problem using *episode training* [19]. Let $\mathcal{D}$ be a set of classes which consists of two disjoint subsets $\mathcal{D}_{train}$ and $\mathcal{D}_{test}$. In a N-way K-shot setting, we randomly sample N classes from $\mathcal{D}_{train}$, and then randomly sample K instances for each class to form a *support set* $S = \{(\mathbf{x}_i, y_i)\}_{i=1}^{N \times K}$ and T instances for each class to form a *query set* $Q = \{(\mathbf{x}_j, y_j)\}_{j=1}^{N \times T}$ in an episode, where y_i is the class of an instance $\mathbf{x}_i$. A classifier is trained to predict the classes of instances in the query set Q, which are compared with their true classes to calculate losses in training.

We propose *Episode Adaptive Embedding Networks* (EAENs) for few-shot classification, which consists of three components: a generic embedding module, an episode adaptive module and a classifier, as illustrated in Fig. 2.

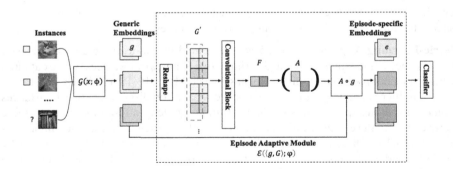

Fig. 2. The framework of Episode Adaptive Embedding Networks.

3.1   Generic Embedding Module

We define a generic embedding module $\mathcal{G}(\mathbf{x}; \phi)$ to be a convolutional block $\mathcal{G}$ with learnable parameters ϕ. Given an instance $\mathbf{x} \in \mathbb{R}^{w \times h \times c}$ where w and h are the width and height of an instance, respectively, and c refers to the number of its channels, a generic embedding module $\mathcal{G}$ takes $\mathbf{x}$ as input and embeds it to a three-dimensional tensor $\mathbf{g} \in \mathbb{R}^{w' \times h' \times c'}$, where w', h' and c' represent the width, height, and number of channels of instance embeddings in a generic embedding space, respectively.

Let $E = (S, Q)$ denote an episode consisting of a support set S and a query set Q. By applying a generic embedding module $\mathcal{G}(\mathbf{x}; \phi)$ on E, we obtain the generic embeddings of all instances in S and Q. For simplicity, we use a *generic embedding matrix* $\mathbf{G} \in \mathbb{R}^{m \times n}$ to represent the generic embeddings of all instances from the episode E, where $m = w' \times h' \times c'$ and $n = N \times (K + T)$.

An instance may appear in one or more episodes. However, given an instance $\mathbf{x}$, the generic embeddings of $\mathbf{x}$ are always same for all episodes. In other words, a generic embedding module $\mathcal{G}(\mathbf{x}; \phi)$ embeds instances into a generic embedding space without taking into account episodes to which instances belong.

3.2 Episode Adaptive Module

An episode adaptive module $\mathcal{E}(\langle \mathbf{g}, \mathbf{G} \rangle; \varphi)$ is defined as a neural network $\mathcal{E}$ with parameters φ. It takes $\langle \mathbf{g}, \mathbf{G} \rangle$ as input, where $\mathbf{g}$ is the generic embeddings of an instance and $\mathbf{G}$ is the generic embedding matrix of an episode E that the instance belongs to, and produces an episode-specific embeddings for the instance w.r.t. the episode E.

Specifically, for each episode E, we first reshape its generic embedding matrix $\mathbf{G}$, which contains the generic embeddings of all instances from E, into a three-dimensional tensor $\mathbf{G}' \in \mathbb{R}^{m \times n \times 1}$. Then, we feed $\mathbf{G}'$ as input to three convolutional layers in order to extract episode-specific features from generic embeddings based on a channel-pixel adaptive mechanism. This process yields episode-specific adaptive vector, each of its element corresponds to a channel-pixel value, to transform instance embeddings from a generic embedding space to an episode-specific embedding space.

Let $\mathbf{G}'(uvk, :, :) \in \mathbb{R}^{n \times 1}$ denote a matrix of instance embeddings at a fixed channel-pixel uvk, i.e., generic embeddings at the location (u, v) of the k-th channel in $\mathbf{G}'$, where $u \in [0, w')$, $v \in [0, h')$ and $k \in [0, c')$. Then, we extract episode-specific features from $\mathbf{G}'(uvk, :, :)$ with a convolutional block which successively applying three convolutional layers with decreasing numbers of kernels (e.g., 64 kernels for the first layer, 32 kernels for the second layer, and 1 kernel for the third layer):

$$\mathbf{P}(uvk, :, i) = \sigma(\mathbf{W}_i^p \circ \mathbf{G}'(uvk, :, :)) \qquad \text{for } i = 1, \ldots, d; \tag{1}$$

$$\mathbf{Z}(uvk, :, j) = \sigma(\mathbf{W}_j^z \circ \mathbf{P}(uvk, :, :)) \qquad \text{for } j = 1, \ldots, f; \tag{2}$$

$$\mathbf{F}(uvk, :, :) = \sigma(\mathbf{W}^a \circ \mathbf{Z}(uvk, :, :)). \tag{3}$$

where $\mathbf{W}_i^p \in \mathbb{R}^{1 \times n}$, $\mathbf{W}_j^z \in \mathbb{R}^{1 \times d}$ and $\mathbf{W}^a \in \mathbb{R}^{1 \times f}$ are the parameters of the i-th kernel of the first convolutional layer, the j-th kernel of the second convolutional layer and the only kernel of the third convolutional layer, respectively, $\circ$ denotes a matrix multiplication, and σ is a non-linear activation function. After extracting episode-specific features from every channel-pixel uvk, we obtain three feature tensors: $\mathbf{P} \in \mathbb{R}^{m \times 1 \times d}$, $\mathbf{Z} \in \mathbb{R}^{m \times 1 \times f}$ and $\mathbf{F} \in \mathbb{R}^{m \times 1 \times 1}$ as outputs of these convolutional layers respectively.

By the feature tensor $\mathbf{F}$, we construct a diagonal matrix $\mathbf{A} = diag(a_i) \in \mathbb{R}^{m \times m}$ with $a_i = \mathbf{F}(uvk, 0, 0)$ on the diagonal. Then, we assign an adaptive value to each channel-pixel of a generic embedding $\mathbf{g}$ to obtain an episode-specific embedding $\mathbf{e}$, through the following linear mapping:

$$\mathbf{e} = \mathbf{A} \circ \mathbf{g}. \tag{4}$$

Intuitively, each diagonal element a_i represents an adaptive value for a generic embedding $\mathbf{g} \in \mathbb{R}^m$ at the location (u, v) of the k-th channel. It is computed according to the distribution of generic embeddings of all instances within an episode $E = (S, Q)$ at the channel-pixel uvk, including support instances in S and query instances in Q.

3.3 Classification

Let $\mathbf{E}_S \in \mathbb{R}^{m \times n_s}$ and $\mathbf{E}_Q \in \mathbb{R}^{m \times n_q}$ denote episode-adaptive embeddings of all instances from the support set S and the query set Q in an episode $E = (S, Q)$, respectively, where $n_s = N \times K$ and $n_q = N \times T$. A classifier predicts classes of query instances in Q based on $\mathbf{E}_S$ and $\mathbf{E}_Q$, as well as the classes of support instances in S.

We use a prototypical network [16] for classification. A *prototype* $\mathbf{e}^t$ is calculated for each class t according to the episode-specific embeddings of all instances in S of class t, where $\mathbf{e}_i$ stands for the episode-specific embedding of the i-th instance in S for the class t.

$$\mathbf{e}^t = \frac{1}{K} \sum_{i=1}^{K} \mathbf{e}_i \tag{5}$$

Let $d(\cdot, \cdot)$ denote a distance between two instance embeddings and $\mathbf{e}_i$ be an episode-specific embedding of a query instance $\mathbf{x}_i$ in Q. Then, the probability that $\mathbf{x}_i$ belongs to a class t is calculated as:

$$p(y = t | \mathbf{e}_i) = \frac{\exp(-d(\mathbf{e}_i, \mathbf{e}^t))}{\sum_{j=1}^{N} \exp(-d(\mathbf{e}_i, \mathbf{e}^j))} \tag{6}$$

The choice of $d(\cdot, \cdot)$ depends on assumptions about data distribution in the episode-specific embedding space. We use the Euclidean distance to define $d(\mathbf{e}_i, \mathbf{e}_j) = \|\mathbf{e}_i - \mathbf{e}_j\|_2$, where $\|\ \|_2$ is the l^2 norm. We thus predict the class $\hat{y}_i$ of $\mathbf{x}_i$ by assigning it to the same class as its nearest prototype:

$$\hat{y}_i = \arg\max_t p(y = t | \mathbf{e}_i). \tag{7}$$

The classifier is optimized by minimizing a cross-entropy loss which averages over the losses of all query instances $\mathbf{x}_i$ in Q w.r.t. their true class y_i:

$$\mathcal{L} = -\frac{1}{n_q} \sum_{i=1}^{n_q} \log p(y = y_i | \mathbf{e}_i) \tag{8}$$

4 Experiments

We evaluate our method to answer the following research questions: [**Q1.**] How does our method perform against the state-of-the-art models for few-shot classification tasks? [**Q2.**] How does our method perform against the state-of-the-art models for semi-supervised classification tasks? [**Q3.**] Is our method robust to different generic embedding networks and different classifiers? [**Q4.**] How effectively our method can leverage instances from a query set for improving performance? We also conduct a case study to visualize how instance embeddings are changed from a general embedding space to an episode-specific embedding space.

Table 1. Few-shot classification accuracies on miniImageNet.

Model	Backbone	5-way 1-shot	5-way 5-shot	10-way 1-shot	10-way 5-shot
MatchingNets [19]	ConvNet-4	43.60	55.30	–	–
MAML [5]	ConvNet-4	48.70	63.11	31.27	46.92
Reptile [14]	ConvNet-4	47.07	62.74	31.10	44.66
PROTO [16]	ConvNet-4	46.14	65.77	32.88	49.29
RelationNet [18]	ConvNet-4	51.38	67.07	34.86	47.94
Label propagation [13]	ConvNet-4	52.31	68.18	35.23	51.24
TPN [13]	ConvNet-4	53.75	69.43	36.62	52.32
GNN [6]	ConvNet-4	50.33	66.41	–	–
EGNN [9]	ConvNet-4	59.18	76.37	–	–
DPGN [23]	ConvNet-4	66.01	82.83	–	–
EA-PROTO (ours)	**ConvNet-4**	**92.95**	**96.55**	**67.66**	**77.64**
MetaGAN [25]	ResNet-12	52.71	68.63	–	–
TADAM [15]	ResNet-12	58.50	76.70	–	–
MetaOptNet [10]	ResNet-12	62.64	78.63	–	–
FEAT [24]	ResNet-12	66.79	82.05	–	–
DPGN [23]	ResNet-12	67.77	84.60	–	–
EA-PROTO (ours)	**ResNet-12**	**93.67**	**96.87**	**70.08**	**77.78**

4.1 Datasets

We conduct experiments on three benchmark datasets: *miniImageNet*, *tieredImageNet* and *CIFAR-FS*. The first two datasets are subsets of ImageNet in different scales, containing RGB images of 84×84 [9, 19]. Besides, *CIFAR-FS* is a subset of CIFAR-100, containing images of 32×32 [3].

4.2 Experimental Setup

Generic Embedding Networks. Experiments are conducted on two widely-used backbones for generic embeddings: ConvNet-4 and ResNet-12 [5, 10, 15–17]. The ConvNet-4 network has four convolutional blocks. Each convolutional block begins with a 3×3 2D convolutional layer, followed by a batch normalization (BN) layer, a 2×2 max-pooling layer and a ReLU activation layer. The ResNet-12 network has four residual blocks with channels of 64, 128, 256, and 64. Each residual block contains three convolutional blocks, which uses a 3×3 convolutional kernel, followed by a BN layer and a LeakyReLU activation layer.

Classifiers. We consider two types of classifiers in experiments: prototypical network [16] and transductive propagation network [13]. Thus, we have two variants of EAEN: (1) *Episode Adaptive Prototypical Networks* (EA-PROTO) uses prototypical network as the classifier, and (2) *Episode Adaptive Transductive Propagation Networks* (EA-TPN) uses transductive propagation network as the classifier.

Evaluation. We follow the episode training strategy for few-shot learning [13, 19]. A N-way K-shot setting is adopted for both training and testing. Following previous

Table 2. Few-shot classification accuracies on tieredImageNet.

Model	Backbone	5-way 1-shot	5-way 5-shot	10-way 1-shot	10-way 5-shot
MAML [5]	ConvNet-4	51.67	70.30	34.44	53.32
Reptile [14]	ConvNet-4	48.97	66.47	33.67	48.04
PROTO [16]	ConvNet-4	48.58	69.57	37.35	57.839
IMP [1]	ConvNet-4	49.60	48.10	–	–
RelationNet [18]	ConvNet-4	54.48	71.31	36.32	58.05
CovaMNET [12]	ConvNet-4	51.19	67.65	–	–
Label propagation [13]	ConvNet-4	55.23	70.43	39.39	57.89
TPN [13]	ConvNet-4	57.53	72.85	40.93	59.17
EGNN [9]	ConvNet-4	63.52	80.24	–	–
DPGN [23]	ConvNet-4	69.43	85.92	–	–
EA-PROTO (ours)	**ConvNet-4**	**92.65**	**96.69**	**70.16**	**82.59**
MetaOptNet [10]	ResNet-12	65.81	81.75	–	–
FEAT [24]	ResNet-12	70.80	84.79	–	–
DPGN [23]	ResNet-12	72.45	87.24	–	–
EA-PROTO (ours)	**ResNet-12**	**91.56**	**97.02**	**74.50**	**83.34**

settings [13, 16], the query number is set to 15 and the performance is measured using classification accuracy over 600 episodes on testing data.

Parameters. The init learning rate is $1e^{-3}$ for ConvNet-4 and $1e^{-4}$ for ResNet-12. In addition, the learning rate of Adam-optimizer decays by half every 10, 000 iterations.

4.3 Few-Shot Learning

To evaluate the effectiveness of our method for few-shot learning, we compare EA-PROTO against the state-of-the-art methods. As CIFAR-FS is a small dataset, we follow [3, 23] to consider 5-way 1-shot and 5-way 5-shot. The results are shown in Tables 1–3.

From Tables 1 and 2, we see that EA-PROTO significantly outperform all baselines on both miniImageNet and tieredImageNet, regardless of using ConvNet-4 or ResNet-12 as the generic embedding net-

Table 3. Few-shot classification accuracies on CIFAR-FS, where † indicates that the results are from [23].

Model	Backbone	5-way 1-shot	5-way 5-shot
MAML† [5]	ConvNet-4	58.90	71.50
PROTO† [16]	ConvNet-4	55.50	72.00
RelationNet† [18]	ConvNet-4	55.00	69.30
DPGN [23]	**ConvNet-4**	**76.40**	**88.40**
EA-PROTO (ours)	ConvNet-4	74.01	80.02

work. Specifically, 1) on miniImageNet, EA-PROTO improves upon the best results of the baselines by a margin 25.9% in 5-way 1-shot and 12.27% in 5-way 5-shot; 2) on tieredImageNet, EA-PROTO improves upon the best results of the baselines by a margin 19.11% in 5-way 1-shot and 9.78% in 5-way 5-shot.

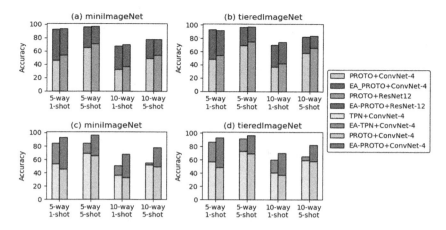

Fig. 3. Classification accuracies with classifiers and generic embedding networks.

Table 4. Semi-supervised classification accuracies on miniImageNet. X-Semi stands for a model X which uses unlabeled instances in a support set. While X stands for a model that only use labeled instances in a support set.

Model	Training strategy	Labeled ratio (5-way 5-shot)				
		20%	40%	60%	80%	100%
GNN [6]	Supervised	50.33	56.91	–	–	66.41
GNN-Semi [6]	Semi-supervised	52.45	58.76	–	–	66.41
EGNN [9]	Supervised	59.18	–	–	–	76.37
EGNN-Semi [9]	Semi-supervised	63.62	64.32	66.37	–	76.37
EA-PROTO (ours)	**Supervised**	**92.95**	**95.03**	**95.89**	**96.24**	**96.55**
EA-PROTO-Semi (ours)	**Semi-supervised**	**93.01**	**95.14**	**96.05**	**96.43**	**96.55**

Table 3 shows that EA-PROTO performs better than all other models except DPGN. In 5-way 1-shot, EA-PROTO improves about 16% on average than the other three models, but performs 2% slightly worse than DPGN. The reason why DPGN has a better performance than EA-PROTO is that the low resolution images (32×32) in CIFAR-FS make generic embeddings of instances contain less useful information compared with those from miniImageNet and tieredImageNet (84×84). This limits the expressiveness of episode specific embeddings learned from CIFAR-FS and accordingly hinders the performance of EA-PROTO. DPGN concatenates the output of the last two layers of a generic embedding network as generic embeddings. Hence, DPGN performs better than all the other models on CIFAR-FS.

4.4 Semi-supervised Learning

For semi-supervised learning, we conduct experiments on miniImageNet in the 5-way 5-shot setting. Following [6,9], we partially label the same number of instances for each

Table 5. Results for an ablation study, where EA-PROTO-S and EA-TPN-S refer to a variant of the methods EA-PROTO and EA-TPN, respectively, which use only instances in a support set to learn their episode-specific embeddings.

Model	Backbone	Dataset	5-way 1-shot	5-way 5-shot	10-way 1-shot	10-way 5-shot
TPN [13]	ConvNet-4		53.75	69.43	36.62	52.32
EA-TPN-S	ConvNet-4	miniImageNet	50.30	68.41	36.15	52.11
EA-TPN	ConvNet-4		84.01	84.43	50.73	54.85
PROTO [16]	ConvNet-4		46.14	65.77	32.88	49.29
EA-PROTO-S	ConvNet-4	miniImageNet	49.64	67.42	34.08	48.94
EA-PROTO	ConvNet-4		92.95	96.55	68.08	78.99

class in a support set, and consider two training strategies: (1) *supervised* – training with only labeled instances in a support set; (2) *semi-supervised* – training with all instances in a support set. These two strategies only differ in whether or not they use unlabeled instances in a support set.

The results are shown in Table 4. We find that: 1) Semi-supervised models achieve better performance compared with their corresponding supervised models. This is because unlabeled instances in a support set help in classification. 2) EA-PROTO-Semi consistently achieves the best performance under all different labeled ratios {20%, 40%, 60%, 80%, 100%}. EA-PROTO-Semi outperforms EGNN-Semi and GNN-Semi significantly. The margin between EA-PROTO-Semi and EGNN-Semi is about 30% when the labeled ratio is 20%, and decreases to 20% when the labeled ratio is 100%. 3) EA-PROTO-Semi has a smaller performance gap between the labeled ratios from 20% to 100% than the other models. This is due to the fact that episode specific embeddings in EA-PROTO are learned from all instances in an episode, regardless whether they are labeled or not, while the other models rely only on labeled instances.

4.5 Robustness Analysis

To evaluate the robustness of our method, we conduct experiments under different combinations of generic embedding networks and classifiers. The results on miniImageNet and tieredImageNet are presented in Fig. 3.

We observe that: (1) Our method is robust to different generic embedding networks. We compare performance of PROTO and EA-PROTO when using ConvNet-4 and ResNet-12 as the generic embedding network separately on miniImageNet and tieredImageNet. Figure 3(a) and (b) shows that our method consistently yields improvement, no matter which generic embedding network or dataset is used. (2) Our method is robust to different classifiers. We compare the performance of PROTO against EA-PROTO, as well as TPN against EA-TPN, when using convNet-4 as the generic embedding network. In Fig. 3(c) and (d), both EA-PROTO and EA-TPN perform better than PROTO and TPN, respectively, on both miniImageNet and tieredImageNet datasets.

4.6   Ablation Analysis

To study how effectively our method can use instances from a query set for improving performance, we conduct an ablation analysis that compares EA-PROTO and EA-TPN (using instances from both support and query sets) against EA-PROTO-S and EA-TPN-S (using only instances in a support set). The results are shown in Table 5.

We observe that: 1) A large performance gap exists between EA-PROTO and EA-PROTO-S, and similarly between EA-TPN and EA-TPN-S. This is due to the fact that there are more instances in a query set than instances in a support set. In 5-way 1-shot setting, the size of a query set is 75 while the size of a support set is 5. Thus, by utilizing 80 instances from both support and query sets, EA-PROTO and EA-TPN can generate better episode specific embeddings than EA-PROTO-S and EA-TPN-S which only use 5 instances from a support set. 2) EA-PROTO-S performs slightly better than PROTO, whereas EA-TPN-S performs slightly worse than TPN. This is because episode adaptive embeddings cannot be effectively computed from instances of a support set. When the number of instances in a support set is limited, computing episode adaptive embeddings only from instances of a support set may even harm performance.

4.7   Case Study

To explore how effectively our method maps instances into an episode-specific embedding space, we conduct a case study using images from miniImageNet dataset. We compare generic embeddings learned from PROTO and TPN with episode-specific embeddings learned from EA-PROTO and EA-TPN in the 5-way 1-shot setting, where ConvNet-4 is used as the generic embedding network. We use t-SNE[1] to visualize embeddings.

Figure 4(a) and (b) shows the t-SNE maps of image

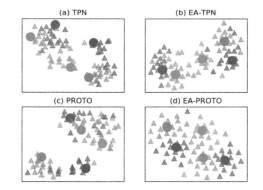

Fig. 4. t-SNE for image embeddings on miniImageNet under the 5-way 1-shot setting. Circles and triangles in each subfigure stand for image embeddings in the support and query sets of an episode, respectively. Different colors indicate different classes.

embeddings in an episode being produced by TPN and EA-TPN, respectively, while Fig. 4(c) and (d) shows the t-SNE maps of image embeddings in an episode being produced by PROTO and EA-PROTO, respectively. Each triangle represents an image embedding in a query set, and each circle represents an image embedding in a support set. From Fig. 4(a) and (b), we can see that the location of each circle is closer to the center of triangles for the same class in Fig. 4(b) than in Fig. 4(a). It indicates that

---

[1] https://lvdmaaten.github.io/tsne/.

the instance embeddings learnt by EA-TPN provide more discriminative information for classifying these instances accurately than the instance embeddings learnt by TPN. There is a similar trend in Fig. 4(c) and (d), indicating EA-PROTO captures more discriminative features (i.e., episode-specific features) in its embeddings than PROTO for classification. These show that episode adaptive embeddings are more discriminative than generic embeddings, which helps improve classification.

5 Conclusion

In this work, we have proposed EAEN[2], a novel approach for learning episode-specific instance embeddings in few-shot learning, EAEN maps generic embeddings to episode-specific embeddings using an episode adaptive module which is learnt from the probability distribution of generic embeddings at each channel-pixel of all instances within an episode. Such episode-specific embeddings are discriminative, and can thus help classify instances in episodes, even when only a few labelled instances are available. Our experimental results on three benchmark datasets have empirically verified the effectiveness and robustness of EAEN. It is shown that EAEN significantly improves classification accuracy compared with the-state-of-the-art methods.

References

1. Allen, K.R., Shelhamer, E., Shin, H., Tenenbaum, J.B.: Infinite mixture prototypes for few-shot learning. arXiv preprint arXiv:1902.04552 (2019)
2. Antoniou, A., Edwards, H., Storkey, A.: How to train your MAML. In: ICLR (2018)
3. Bertinetto, L., Henriques, J.F., Torr, P.H., Vedaldi, A.: Meta-learning with differentiable closed-form solvers. arXiv preprint arXiv:1805.08136 (2018)
4. Chen, W.Y., Liu, Y.C., Kira, Z., Wang, Y.C.F., Huang, J.B.: A closer look at few-shot classification. arXiv preprint arXiv:1904.04232 (2019)
5. Finn, C., Abbeel, P., Levine, S.: Model-agnostic meta-learning for fast adaptation of deep networks. In: ICML, pp. 1126–1135 (2017)
6. Garcia, V., Estrach, J.B.: Few-shot learning with graph neural networks. In: ICLR (2018)
7. Han, C., Shan, S., Kan, M., Wu, S., Chen, X.: Meta-learning with individualized feature space for few-shot classification (2018)
8. Hariharan, B., Girshick, R.: Low-shot visual recognition by shrinking and hallucinating features. In: ICCV, pp. 3018–3027 (2017)
9. Kim, J., Kim, T., Kim, S., Yoo, C.D.: Edge-labeling graph neural network for few-shot learning. In: CVPR, pp. 11–20 (2019)
10. Lee, K., Maji, S., Ravichandran, A., Soatto, S.: Meta-learning with differentiable convex optimization. In: CVPR, pp. 10657–10665 (2019)
11. Li, H., Eigen, D., Dodge, S., Zeiler, M., Wang, X.: Finding task-relevant features for few-shot learning by category traversal. In: CVPR, pp. 1–10 (2019)
12. Li, W., Xu, J., Huo, J., Wang, L., Gao, Y., Luo, J.: Distribution consistency based covariance metric networks for few-shot learning. In: AAAI, vol. 33, pp. 8642–8649 (2019)
13. Liu, Y., et al.: Learning to propagate labels: transductive propagation network for few-shot learning. In: International Conference on Learning Representations (ICLR) (2019)

---

[2] Our code is available at https://www.dropbox.com/s/cll23kem3yswg96/EAEN.zip?dl=0.

14. Nichol, A., Achiam, J., Schulman, J.: On first-order meta-learning algorithms. arXiv preprint arXiv:1803.02999 (2018)

15. Oreshkin, B., López, P.R., Lacoste, A.: TADAM: task dependent adaptive metric for improved few-shot learning. In: NeurIPS, pp. 721–731 (2018)

16. Snell, J., Swersky, K., Zemel, R.: Prototypical networks for few-shot learning. In: NeurIPS, pp. 4077–4087 (2017)

17. Sun, Q., Liu, Y., Chua, T.S., Schiele, B.: Meta-transfer learning for few-shot learning. In: CVPR, pp. 403–412 (2019)

18. Sung, F., Yang, Y., Zhang, L., Xiang, T., Torr, P.H., Hospedales, T.M.: Learning to compare: relation network for few-shot learning. In: CVPR, pp. 1199–1208 (2018)

19. Vinyals, O., Blundell, C., Lillicrap, T., Wierstra, D., et al.: Matching networks for one shot learning. In: NeurIPS, pp. 3630–3638 (2016)

20. Wang, X., Yu, F., Wang, R., Darrell, T., Gonzalez, J.E.: TAFE-Net: task-aware feature embeddings for low shot learning. In: CVPR, pp. 1831–1840 (2019)

21. Wang, Y.X., Girshick, R., Hebert, M., Hariharan, B.: Low-shot learning from imaginary data. In: CVPR, pp. 7278–7286 (2018)

22. Wei, X.S., Wang, P., Liu, L., Shen, C., Wu, J.: Piecewise classifier mappings: learning fine-grained learners for novel categories with few examples. TIP **28**(12), 6116–6125 (2019)

23. Yang, L., Li, L., Zhang, Z., Zhou, X., Zhou, E., Liu, Y.: DPGN: distribution propagation graph network for few-shot learning. In: CVPR, pp. 13390–13399 (2020)

24. Ye, H.J., Hu, H., Zhan, D.C., Sha, F.: Few-shot learning via embedding adaptation with set-to-set functions. In: CVPR, pp. 8808–8817 (2020)

25. Zhang, R., Che, T., Ghahramani, Z., Bengio, Y., Song, Y.: MetaGAN: an adversarial approach to few-shot learning. In: NeurIPS, pp. 2365–2374 (2018)

Universal Representation for Code

Linfeng Liu[1][(✉)], Hoan Nguyen[2], George Karypis[2], and Srinivasan Sengamedu[2]

[1] Tufts University, Medford, MA 02155, USA
`linfeng.liu@tufts.edu`
[2] Amazon Web Services, Seattle, WA 98109, USA
{`hoanamzn,gkarypis,sengamed`}`@amazon.com`

Abstract. Learning from source code usually requires a large amount of labeled data. Despite the possible scarcity of labeled data, the trained model is highly task-specific and lacks transferability to different tasks. In this work, we present effective pre-training strategies on top of a novel graph-based code representation, to produce *universal* representations for code. Specifically, our graph-based representation captures important semantics between code elements (e.g., control flow and data flow). We pre-train graph neural networks on the representation to extract universal code properties. The pre-trained model then enables the possibility of fine-tuning to support various downstream applications. We evaluate our model on two real-world datasets – spanning over 30M Java methods and 770K Python methods. Through visualization, we reveal discriminative properties in our universal code representation. By comparing multiple benchmarks, we demonstrate that the proposed framework achieves state-of-the-art results on method name prediction and code graph link prediction.

Keywords: Code representation · Graph neural network · Pre-training

1 Introduction

Analysis of software using machine learning approaches has several important applications such as identifying code defects [1], improving code search [2], and improving developer productivity [3]. One common aspect of any code-related application is that they learn code representations by following a two-step process. The first step takes code snippets and produces a *symbolic* code representation using program analysis techniques. The second step uses the symbolic code representation to generate *neural* code representations using deep learning techniques.

Symbolic representations need to capture both syntactic and semantic structures in code. Approaches to generating symbolic representations can be categorized as sequence-, tree-, and graph-based. Sequence-based approaches represent code as a sequence of tokens and only capture the shallow and textual structures of the code [4]. Tree-based approaches represent the code via abstract syntax

---

L. Liu—Work done while the author was an intern at Amazon Web Services.

K. Karlapalem et al. (Eds.): PAKDD 2021, LNAI 12714, pp. 16–28, 2021.
https://doi.org/10.1007/978-3-030-75768-7_2

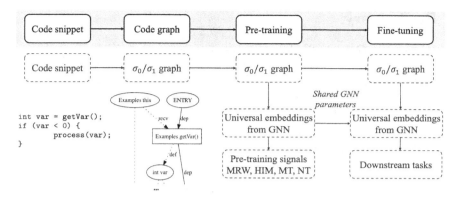

Fig. 1. Model pipeline. The σ_0/σ_1 graph is defined in Sect. 3, and the pre-training signals are defined in Sect. 4.

trees (ASTs) [5] that highlight structural and content-related details in code. However, some critical relations (e.g., control flow and data flow), which often impact machine learning models' success in abstracting code information, are not available in trees. Graph-based approaches augment ASTs with extra edges to partially represent the control flow and the data flow [1,4,6]. Depending on the type of symbolic representation used, the approaches for generating the neural code representations are either sequence-based [3,7] or graph-based [4,6] neural network models. However, these works are generally task-specific, making it hard to transfer the learned representations to other tasks. In addition, the scarcity of labeled data may cause insufficient training in deep learning models.

In this work, we touch upon all the three aspects of ML-based code analysis: symbolic code representation, task-independent neural code representation, and task-specific learning. Figure 1 gives an overview of our approach. For symbolic code representation, we explore two alternatives and show that symbolic code representation (called σ_0/σ_1 graph) which captures richer relations leads to better performance in downstream tasks. For neural code representation, we specialize a recently proposed universal representation for graphs, PanRep [8], to code graphs. And, finally, we explore two tasks to demonstrate the effectiveness of the learned representations: method name prediction (for Python and Java) and link prediction (for Java). Our proposed method consistently improves the prediction accuracy across all experiments.

To summarize, the contributions of this work are as follows:

- We introduce a fine-grained symbolic graph representation for source code, and adapt to 29M Java methods collected from GitHub.
- We present a pre-training framework that leverages the graph-based code representations to produce universal code representations, supporting various downstream tasks via fine-tuning.
- We combine the graph-based representation and the pre-training strategies to go beyond code pre-training with sequence- and tree-based representations.

2   Preliminary

Notation. Let $G = \{\mathcal{V}, \mathcal{E}\}$ denote a *heterogeneous* graph with $|\mathcal{T}|$ node types and $|\mathcal{R}|$ edge types. $\mathcal{V} = \{\{\mathcal{V}^t\}_{t \in \mathcal{T}}\}$ represents the node set, and $\mathcal{E} = \{\{\mathcal{E}^r\}_{r \in \mathcal{R}}\}$ represents the edge set. Each node $v_i^t \in \mathcal{V}^t$ is associated with a feature vector. Throughout the paper, we often use "representation" and "embedding" inter-changeably unless there is any ambiguity.

2.1   Graph Neural Networks

Graph neural networks (GNNs) learn representations of graphs [9]. A GNN typically consists of a sequence of L graph convolutional layers. Each layer updates nodes' representation from their direct neighbors. By stacking multiple layers, each node receives messages from higher-order neighbors. In this work, we utilize the relational graph convolutional network (RGCN) [10] to model our heterogeneous code graphs. RGCN's update rule is given by

$$\mathbf{h}_i^{(l+1)} = \phi \left(\sum_{r \in \mathcal{R}} \sum_{n \in \mathcal{N}_i^r} \frac{1}{c_{i,r}} \mathbf{h}_n^{(l)} \mathbf{W}_r^{(l)} \right),$$

where $\mathcal{N}_i^r$ is the neighbor set of node i under edge type r, $c_{i,r}$ is a normalizer (we use $c_{i,r} = |\mathcal{N}_i^r|$ as suggested in [10]), $\mathbf{h}_i^{(l)}$ is the hidden representation of node i at layer l, $\mathbf{W}_r^{(l)}$ are learnable parameters, and $\phi(\cdot)$ is any nonlinear activation function. Usually, $\mathbf{h}_i^{(0)}$ is initialized as node features, and $\mathbf{h}_i^{(L)}$ (the representation at the last layer) is used as the final representations.

2.2   Pre-training for GNNs and for Source Code

Recently, there is a rising interest in pre-training GNNs to model graph data [11, 12]. To pre-train GNNs, most works encourage GNNs to capture graph structure information (e.g., graph motif) and graph node information (e.g., node feature). PanRep [8] further extends GNN pre-training to heterogeneous graphs.

Pre-training on source code has been studied in [13–15]. However, these models build upon sequence-based code representations and fail to encode code's structural information explicitly. We differ from these works, by pre-training on a novel graph-based code representation to capture code's structural information.

3   Code Graph

Previous Machine Learning (ML) models [4,6,16] are largely based on ASTs to reflect structural code information. Though ASTs are simple to create and use, they have tree-based structures and do not capture control flow and data flow relations. Here, the control flow represents the order of the execution and the data flow represents the flow of data along the computation. For example, to represent a loop snippet, ASTs cannot naturally use an edge pointing from the end of the program statement to the beginning of the loop. In addition, the relation between the definition and uses of a variable is not captured in ASTs. In this work, we represent code as graphs to efficiently capture both control flow and data flow between program elements. We call our code graphs as σ_0 graphs and σ_1 graphs. The σ_0 graphs are related to classical Program Dependence Graphs (PDGs) [17]. The σ_1 graphs build upon the σ_0 graphs and include additional syntactic and semantic information (detailed in Sect. 3.2). Our experiments show that ML models using the σ_1 graphs achieve better prediction accuracy than using the σ_0 graphs.

3.1 The σ_0 Graph

The σ_0 graphs, which relate to PDGs, are used for tasks such as detection of security vulnerabilities and identification of concurrency issues. In σ_0 graphs, nodes represent different kinds of program elements including data and operations; edges represent different kinds of control flow and data flow between program elements. We showcase a σ_0 graph in Fig. 2.

Both nodes and edges in the σ_0 graph are typed. Specifically, we have five node types: entry, exit, data, action, and control nodes. Entry and exit nodes indicate the control flow entering and exiting the graph. Data nodes represent the data in programs such as variables, constants and literals. Action nodes represent the operations on the data such as method calls, constructor calls and arithmetic/logical operations, etc. Control nodes represent control points in the program such as branching, looping, or some special code blocks such as catch clauses and finally blocks. We have two edge types: control and data edges. Control edges represent the order of execution through the programs and data edges represent how data

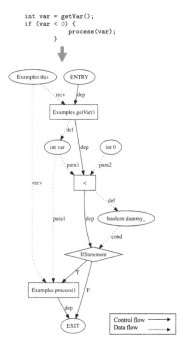

Fig. 2. An example of σ_0 graph.

is created and used in the programs. Examples of edge types include parameter edges which indicate the data flow into operations and throw edges which represent the control flows when exceptions are thrown.

3.2 The σ_1 Graph

One limitation of σ_0 graph is that the downstream modules perform additional analysis such as control dependence and aliasing is not reflected in the σ_0 graph explicitly and requires machine learning models to infer it. We propose σ_1 graph as an augmentation of the σ_0 graph, which is enhanced by additional information. Specifically, σ_1 graph attaches AST node types to nodes in the graph. AST node types capture syntactic information (e.g., InfixOperator) provided by the parser. Higher-order semantic relations such as variable usage (e.g., FirstUse/LastUse), node aliasing, and control dependence are also included as graph edges.

3.3   The Heterogeneous Code Graph

The proposed σ_0 and σ_1 graphs are heterogeneous graphs, i.e. nodes and edges have types and features. Nodes are categorized into five types: `entry`, `exit`, `data`, `action`, and `control`. Node features are attributed by their names. Edge types are the same as edge features, which are defined by their functionalities. Below we first describe node features, followed by edge features.

Entry nodes have feature `ENTRY` and exit nodes have feature `EXIT`. Features of control nodes are their corresponding keywords such as `if`, `while`, and `finally`. Features of variables are types, and the variable names are ignored. For example, `int.x` → `int`; `String.fileName` → `String`. Method names contain the method class, method name, and parameter class. For example, `Request.setConnectionKeepAlive#boolean#`. Here, `Request` is the method class, `setConnectionKeepAlive` is the method name, and `boolean` is the parameter class.

Features of control edges and data edges are defined separately. There are two kinds of control edges: normal control edges and exception control edges. A normal control edge is a directed edge that connects two control or action nodes; defined as `dep`. See Fig. 2. An exception control edge connects an action node to a control node to handle an exception that could be thrown by the action; defined as `throw`. Data edges have five features: `receiver`, `parameter`, `definition`, `condition`, and `qualifier`. Examples include `receiver.call()` as a receiver edge; `call(param)` as a parameter edge, and `foo=bar()` as a definition edge.

3.4   Corpus-Level Graphs

In a typical ML application, the corpus consists of a collection of packages or repositories. Repositories contain multiple files or classes. Classes contain methods. The σ_0 and σ_1 graphs are at *method-level*. Corpus-level graphs are a collection of method-level σ_0/σ_1 graphs. Let σ refer to either σ_0 or σ_1 for notational ease.

4   Method

We propose a new model, Universal Code Representation via GNNs (UniCoRN), to produce universal code representations based on σ graphs. UniCoRN has two components. First, a GNN encoder that takes in σ graphs and generates node embeddings. Second, pre-training signals that train the GNN encoder in an unsupervised manner. By sharing the same GNN encoder across all σ graphs, the learned embeddings reveal universal code properties. Below, we show our design of pre-training signals to help UniCoRN efficiently distill universal code semantics. The instantiation of the GNN encoder is given in the experiment.

4.1 Pre-training Signals

Metapath Random Walk (MRW) Signal. A MRW is a random path that follows a sequence of edge types. We assume node pairs in the same MRW are proximal to each other; accordingly, they should have similar embeddings. For example, for a MRW with nodes [Collection.iterator(), Iterator, Iterator.hasNext()] and edges [definition, receiver], nodes Iterator and Iterator.hasNext() are expected to have similar embeddings. Formally, the signal is defined as

$$\mathcal{L}_{MRW} = \sum_{v,v'\in\mathcal{V}} \log\left(1 + \exp\left(-y \times \mathbf{h}_v^t \mathbf{W}^{t,t'} \mathbf{h}_{v'}^{t'}\right)\right), \tag{1}$$

where $\mathbf{h}_v^t$ and $\mathbf{h}_{v'}^{t'}$ are embeddings for nodes v and v' with node types t and t'. $\mathbf{W}^{t,t'}$ is a diagonal matrix weighing the similarity between different node types. y equals 1 as positive pairs if v and v' are in the same MRW, otherwise y equals -1 as negative pairs. During training, we sample 5 negative pairs per positive pair.

Heterogeneous Information Maximization (HIM) Signal. Nodes of the same type should reside in some shared embedding space, encoding their similarity. On the other hand, nodes of different types, such as control nodes and data nodes, ought to have discriminative embedding space as they are semantically different. However, standard GNNs fail to do so with only local message propagation. Following [8], we use a HIM signal to encode these properties:

$$\mathcal{L}_{HIM} = \sum_{t\in\mathcal{T}} \sum_{v\in\mathcal{V}^t} \left(\log\left(\phi(\mathbf{h}_v^\top \mathbf{W}\mathbf{s}^t)\right) + \log\left(1 - \phi(\tilde{\mathbf{h}}_v^\top \mathbf{W}\mathbf{s}^t)\right)\right). \tag{2}$$

Here, $\mathbf{s}^t = \frac{1}{|\mathcal{V}^t|}\sum_{v\in\mathcal{V}^t} \mathbf{h}_v$ is a global summary of nodes typed t, $\phi(\cdot)$ is a sigmoid function, and $\phi(\mathbf{h}_v^\top \mathbf{W}\mathbf{s}^t)$ quantifies the closeness between a node embedding $\mathbf{h}_v$ and a global summary $\mathbf{s}^t$. Negative samples $\tilde{\mathbf{h}}_v$ are obtained by first row-wise shuffling input node features then propagating through the GNN encoder [8].

Motif (MT) Signal. Code graph has connectivity patterns. For example, node ENTRY has only one outgoing edge; node IfStatement has True and False branch. Such structural patterns can be captured in graph motifs, see Fig. 3. With this observation, we pre-train GNNs with MT signal to generate structure-aware node embeddings for code graphs. Formally, we aim to recover the ground

Fig. 3. Motifs sized 3 and 4.

truth motif around each node, $\boldsymbol{m}_v$, from the node embedding $\mathbf{h}_v$ using an approximator $f_{MT}(\cdot)$ (we use a two-layer MLP),

$$\mathcal{L}_{MT} = \sum_{v \in \mathcal{V}} ||\boldsymbol{m}_v - f_{MT}(\mathbf{h}_v)||_2^2. \tag{3}$$

The ground truth $\boldsymbol{m}_v$ is obtained using a fast motif extraction method [18].

Node Tying (NT) Signal. The corpus-level graph (Cf. Sect. 3.4) contains many *duplicate nodes* that have the same feature (e.g., two ENTRY nodes will be induced from two methods). These duplicate nodes serve as anchors to imply underlying relations among different graphs. We divide duplicate nodes into two categories: strict equality and weak equality. Strict equality refers to duplicate nodes whose semantic meaning should be invariant to their context, including keywords (if, while, do ...), operators (=, *, << ...), entry and exit nodes. Duplicate nodes of strict equality will have the same embedding in all σ graphs. We keep a global embedding matrix to maintain their embeddings. Weak equality refers to other duplicate nodes whose semantic meaning can be affected by their context. For example, two foo() nodes in two methods, or two tied nodes due to the simple qualified types of one or two nodes[1]. We use NT signal to encourage duplicate node of weak equality to have similar embeddings:

$$\mathcal{L}_{NT} = \sum_{k \in \mathcal{K}} \text{ave}(\{||\mathbf{h}_v - \mathbf{g}^k||_2^2\}, v \text{ has feature } k), \tag{4}$$

where $\mathcal{K}$ is the set of distinct node features (exclude strict equality nodes), $\text{ave}(\cdot)$ is an average function, and $\mathbf{g}^k = \text{ave}(\{\mathbf{h}_v\}, v \text{ has label } k)$ is a global summary of nodes featured k. In Eq. (4), we first group nodes featured k, followed by computing the group centers $\mathbf{g}^k$, then minimize the distance between nodes to their group centers.

Pre-training Objective. We combine the four pre-training signals to yield a final objective:

$$\mathcal{L} = \omega_1 \mathcal{L}_{MRW} + \omega_2 \mathcal{L}_{HIM} + \omega_3 \mathcal{L}_{MT} + \omega_4 \mathcal{L}_{NT}, \tag{5}$$

where $\omega_1, \ldots, \omega_4$ balance the importance of different signals. The objective resembles the objective in multi-task learning [19].

4.2 Data Pre-processing and Fine-Tuning

Numeric Node Features. The initial node features are strings (Cf. Sect. 3.3), which need to be cast into numeric forms before feeding into the GNN encoder. To this end, we first split each node's feature into subtokens based on the delimiter ".". Then, language models are used to get subtoken embeddings, in which

---

[1] We use simple types instead of fully qualified types since we create graphs from source files and not builds. In this case, types are not fully resolvable.

Table 1. Dataset statistics. The σ_1 graph doubled the number of edges as the σ_0 graph, providing extra information for code graphs.

Dataset	Repository	Method	Node	Edge
Java (σ_0)	28K	29M	621M	1,887M
Java (σ_1)	28K	29M	529M	3,782M
Python	14K	450K	57M	156M

Table 2. Result for method name prediction on Java dataset. Higher value indicates better performance.

	F1	Precision	Recall
σ_0	21.7	26.1	19.9
σ_1	**23.6**	**27.5**	**22.0**

we use FastText [20]. Finally, we use average subtoken embeddings as the node's numeric feature.

Inverse Edges. We enrich our σ graphs with inverse edges. Recent work has proven improved performance by adding inverse edges to ASTs [6].

Fine-Tuning. After pre-training, we can fine-tune on downstream tasks. Fine-tuning involves adding downstream classifiers on top of the pre-trained node embeddings, and predicting downstream labels. A graph pooling layer [9] might be needed if the downstream tasks are defined on the graph/method level.

5   Experiment

5.1   Dataset

We tested on two real-world datasets, spanning over two programming languages Java and Python. Summary of the datasets is listed in Table 1.

Java. The Java dataset is extracted from 27,581 GitHub packages. In total, these packages contain 29,024,142 Java methods. We convert each Java method into a σ_0 graph and a σ_1 graph. The data split is on package-level, with training (80%), validation (10%), and testing (10%).

Python. The Python dataset is collected from Stanford Open Graph Benchmark (`ogbg-code`) [6]. The total number of Python methods is 452,741, with each method is represented as an AST. These ASTs are further augmented with next-token edges and inverse edges. The data split keeps in line with `ogbg-code`.

5.2   Experimental Setup

We tune hyperparameters on all models based on their validation performances. For Java dataset, we consider a two-layer RGCN with 300 hidden units. For Python dataset, we follow `ogbg-code` and use a five-layer RGCN with 300 hidden units. We use Adam [21] as optimizer, with learning rate ranges from 0.01 to 0.0001. Mini-batch training is adopted to enable training on very large graphs[2].

---

[2] https://github.com/dmlc/dgl/tree/master/examples/pytorch/rgcn-hetero.

We apply dropout at rate 0.2, L2 regularization with parameter 0.0001. The model is first pre-trained on a maximum of 10 epochs, then fine-tuned up to 100 epochs on downstream tasks until convergence. The model is implemented using Deep Graph Library (DGL) [22]. Models have access to 4 T V100 GPUs, 32 CPUs, and 244 GB memory.

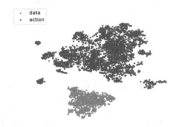
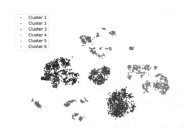

Fig. 4. Visualization of node-level embeddings with t-SNE.

Fig. 5. Visualization of method-level embeddings with t-SNE. Colored with K-means.

5.3 Analysis of Embeddings

We begin by analyzing code embeddings via t-SNE [23] visualization. We study two levels of embeddings: node-level embeddings and method-level embeddings. Here, a method-level embedding summarizes a method, computed by averaging node embeddings in its σ graph. For better visualization, we show results on 10 random Java packages (involving 4,107 Java methods) using σ_1 graphs.

In Fig. 4, we see that data nodes and action nodes are forming separate clusters, indicating our code embeddings preserve important node type information. Figure 5 suggests method embeddings are forming discriminative clusters. By manually annotating each cluster, we discovered that cluster 4 contains 91% (out of all) set functions, cluster 3 contains 78% find functions, and cluster 1 contains 69% functions which end with Exception. This result suggests that our model has the potential to distinguish methods in terms of method functionalities.

5.4 Method Name Prediction

We use pre-trained UniCoRN model to initialize code embeddings. Then following [6,16], we predict method names as downstream tasks. The method name is treated as a sequence of subtokens (e.g. getItemId → [get, item, id]). As in [6], we use independent linear classifiers to predict each subtoken. The task is defined on the method-level: predict one name for one method (code graph). We use attention pooling [24] to generate a single embedding per method. We follow [6,16] to report F1, precision, recall for evaluation. Below we show results on Java and Python separately, as they are used for different testing purposes.

Java. We evaluate the performance gain achieved by switching from the σ_0 graph to the σ_1 graph. We truncate subtoken sequences to a maximal length of 5 to cover 95% of the method names. Vocabulary size is set to 1,000, covering 95% of tokens. Tokens not in the vocabulary are replaced by a special **unknown** token. Similar techniques have been adopted in [6]. We experiment on approximately 774,000 methods from randomly selected 1,000 packages.

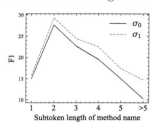

Fig. 6. F1 at name length.

The result is summarized in Table 2. We see that the σ_1 graph outperforms the σ_0 graph, indicating that the extra information provided by the σ_1 graph is beneficial for abstracting code snippets. Figure 6 further supports this observation. The σ_1 graph consistently outperforms the σ_0 graph w.r.t. the F1 score for different method name lengths. Note that the F1 score at method names of length 1 is low. We suspect that some names at this length are not semantically meaningful, such as **a** or **xyz**. Thus, these method names are hard to predict correctly.

Python. We compare the performances of UniCoRN with various baselines on Python. Our experiment setup closely follows `ogbg-code` [6]. For the baseline, `ogbg-code` considers GCN and GIN. Additionally, we introduce two baselines that run GCN and GIN with next-token edges only. We expect these two new baselines to mimic the performance of sequence-based models. In this task, we test three pooling methods: average, virtual node [6], and attention [24].

Results are given in Table 3. We list three observations. First, UniCoRN with attention pooling (UniCoRN[‡]) performs the best, endorsing UniCoRN's superior modeling capacity. Second, UniCoRN with pre-training shows performance gain over UniCoRN without pre-training, verifying the usefulness of our pre-training strategies. Third, GCN(GIN) improves GCN(GIN)-NextTokenOnly, confirming the importance of using structural information.

Example pairs of ground truth and prediction are shown in Fig. 7. The examples of prediction encompass exact matches, such as `get_config` pair, context matches, such as `get_aws_credentials` pair, and mismatches, such as `load_bytes` pair. We showcase a prediction example in Fig. 8.

Table 3. Method name prediction for Python. Pooling: average[†], virtual node[§], and attention[‡]. GCN[†,§] and GIN[†,§]: Reported in [6].

	F1	Precision	Recall
GCN-NextTokenOnly[†]	29.77	31.09	29.18
GIN-NextTokenOnly[†]	29.00	30.98	28.13
GCN[†]	31.63	–	–
GIN[†]	31.63	–	–
UniCoRN w/o pretrain[†]	32.81	35.25	31.71
UniCoRN[†]	33.28	35.28	32.36
GCN[§]	32.63	–	–
GIN[§]	32.04	–	–
UniCoRN[§]	33.80	35.81	32.89
GCN[‡]	32.80	34.72	31.88
GIN[‡]	32.60	34.42	31.77
UniCoRN[‡]	**33.94**	**36.02**	**32.99**

Table 4. MRR and Hit@K(%) results for link prediction. Higher values are better. Superscripts $\mathcal{D}$ and $\mathcal{M}$ denote DistMult and MLP link predictors. Hit@K for random is computed as $K/(1+200)$, where 200 is the number of negative edges per testing edge.

	MRR	Hit@1	Hit@3	Hit@10
Random	–	0.5	1.5	5.0
FastText$^{\mathcal{D}}$	0.01	0.4	1.0	2.3
$\sigma_0^{\mathcal{D}}$	0.26	15.4	28.5	41.3
$\sigma_1^{\mathcal{D}}$	**0.32**	**18.1**	**38.4**	**61.4**
FastText$^{\mathcal{M}}$	0.05	1.9	4.0	8.0
$\sigma_0^{\mathcal{M}}$	0.51	46.1	49.8	58.4
$\sigma_1^{\mathcal{M}}$	**0.53**	**46.2**	**55.0**	**65.2**

Ground truth	Prediction
get_config	get_config
create_collection	create_collectio
get_aws_credentials	get_ec2
wait_for_task_ended	wait_job
add_role	create_permissio
load_bytes	upload_file

Fig. 7. Examples of method name prediction on Python in different degree of consensus. Each pair of results is demonstrated as ground truth name and predicted name.

```
def wait_for_task_ended(self):
    try:
        waiter = self.client.\
            get_waiter('job_execution_complete')
        # timeout is managed by airflow
        waiter.config.max_attempts = sys.maxsize
        waiter.wait(jobs=[self.jobId])
    except ValueError:
        # If waiter not available use expo
        retry ...
```

Prediction: `wait_job`.

Fig. 8. Reasonable prediction based on the code context is observed, though it is inexact match.

5.5 Link Prediction

In this task, we examine how UniCoRN recovers links in code graphs. We follow the same experimental setup as in [8]. We feed two node embeddings of a link to a predictor, a DistMult [25] or a two-layer MLP, to predict the existence of the link. Here, node embeddings are obtained by applying UniCoRN to σ_0/σ_1 graphs, or simply obtained from FastText embeddings. Note that the FastText is the initial embedding of UniCoRN. We freeze UniCoRN after pre-training. When fine-tuning link predictors, we ensure σ_0 and σ_1 have the same set of training edges. During inference, we sample 200 negative edges per testing edge (positive) and evaluate the rank of the testing edge. Evaluations are based on Mean Reciprocal Rank (MRR) and Hit@K (K = 1, 3, 10).

Table 4 shows the results. UniCoRN outperforms FastText. The results suggest that node embeddings from UniCoRN capture neighborhood correlations. We see σ_1 graph again improves σ_0 graph. Figure 9 demonstrates the histogram of scores for 1,000 positive and 1,000 negative edges. The score, which ranges from 0 to 1, indicates the plausibility of the link existence. Positive edges (0.58 ± 0.33) score higher than negative edges (0.13 ± 0.15), with p-value less than 0.00001 using t-test, suggesting that UniCoRN is capable to distinguish positive and negative links in code graphs.

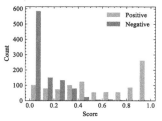

Fig. 9. Scores for positive and negative edges.

6   Conclusion

This paper presents a new model, UniCoRN, to provide a universal representation for code. Building blocks of UniCoRN include a novel σ graph to represent code as graphs, and four effective signals to pre-train GNNs. Our pre-training framework enables fine-tuning on various downstream tasks. Empirically, we show UniCoRN's superior ability to offer high-quality code representations.

There are several possibilities for future works. First, we are looking to enhance UniCoRN with additional code-specific signals. Second, the explainability of the learned code representation deserves further study. The explainability can in turn motivate additional signals to embed desired code properties. Third, more downstream applications are left to be explored, such as bug detection and duplicate code detection upon the availability of labeled data.

References

1. Dinella, E., Dai, H., Li, Z., Naik, M., Song, L., Wang, K.: Hoppity: learning graph transformations to detect and fix bugs in programs. In: ICLR (2019)
2. Cambronero, J., Li, H., Kim, S., Sen, K., Chandra, S.: When deep learning met code search. In: ESEC/FSE, pp. 964–974 (2019)
3. Raychev, V., Vechev, M., Yahav, E.: Code completion with statistical language models. In: PLDI, pp. 419–428 (2014)
4. Allamanis, M., Brockschmidt, M., Khademi, M.: Learning to represent programs with graphs. In: ICLR (2018)
5. Mou, L., Li, G., Zhang, L., Wang, T., Jin, Z.: Convolutional neural networks over tree structures for programming language processing. In: AAAI (2016)
6. Hu, W., et al.: Open graph benchmark: datasets for machine learning on graphs. In: NeurIPS (2020)
7. Hindle, A., Barr, E.T., Su, Z., Gabel, M., Devanbu, P.: On the naturalness of software. In: ICSE, pp. 837–847. IEEE (2012)

8. Ioannidis, V.N., Zheng, D., Karypis, G.: PanRep: universal node embeddings for heterogeneous graphs. In: DLG-KDD (2020)
9. Wu, Z., Pan, S., Chen, F., Long, G., Zhang, C., Philip, S.Y.: A comprehensive survey on graph neural networks. In: TNNLS (2020)
10. Schlichtkrull, M., Kipf, T.N., Bloem, P., van den Berg, R., Titov, I., Welling, M.: Modeling relational data with graph convolutional networks. In: Gangemi, A., et al. (eds.) ESWC 2018. LNCS, vol. 10843, pp. 593–607. Springer, Cham (2018). https://doi.org/10.1007/978-3-319-93417-4_38
11. Hu, W., et al.: Strategies for pre-training graph neural networks. In: ICLR (2020)
12. Jin, W., et al.: Self-supervised learning on graphs: deep insights and new direction. arXiv preprint (2020)
13. Kanade, A., Maniatis, P., Balakrishnan, G., Shi, K.: Learning and evaluating contextual embedding of source code. In: ICML (2020)
14. Feng, Z., et al.: CodeBERT: a pre-trained model for programming and natural languages. In: EMNLP (2020)
15. Svyatkovskiy, A., Deng, S.K., Fu, S., Sundaresan, N.: IntelliCode compose: code generation using transformer. In: ESEC/FSE (2020)
16. Alon, U., Brody, S., Levy, O., Yahav, E.: code2seq: generating sequences from structured representations of code. In: ICLR (2019)
17. Ferrante, J., Ottenstein, K.J., Warren, J.D.: The program dependence graph and its use in optimization. In: TOPLAS (1987)
18. Ahmed, N.K., Neville, J., Rossi, R.A., Duffield, N.G., Willke, T.L.: Graphlet decomposition: framework, algorithms, and applications. KAIS 50(3), 689–722 (2017)
19. Zhang, Y., Yang, Q.: A survey on multi-task learning. CoRR (2017)
20. Bojanowski, P., Grave, E., Joulin, A., Mikolov, T.: Enriching word vectors with subword information. TACL 5, 135–146 (2017)
21. Kingma, D.P., Ba, J.: Adam: a method for stochastic optimization. In: ICLR (2015)
22. Wang, M., et al.: Deep graph library: a graph-centric, highly-performant package for graph neural networks. arXiv (2019)
23. van der Maaten, L., Hinton, G.: Visualizing data using t-SNE. JMLR 9, 2579–2605 (2008)
24. Li, Y., Tarlow, D., Brockschmidt, M., Zemel, R.: Gated graph sequence neural networks. In: ICLR (2016)
25. Yang, B., Yih, W.-T., He, X., Gao, J., Deng, L.: Embedding entities and relations for learning and inference in knowledge bases. In: ICLR (2015)

Self-supervised Adaptive Aggregator Learning on Graph

Bei Lin, Binli Luo, Jiaojiao He, and Ning Gui$^{(\boxtimes)}$

Central South University, Changsha, China
linbei@csu.edu.cn

Abstract. Neighborhood aggregation is a key operation in most of the graph neural network-based embedding solutions. Each type of aggregator typically has its best application domain. The single type of aggregator for aggregation adopted by most existing embedding solutions may inevitably result in information loss. To keep the diversity of information during aggregation, it is mandatory to use the most appropriate different aggregators for specific graphs or subgraphs. However, when and what aggregators to be used remain mostly unsolved. To tackle this problem, we introduce a general contrastive learning framework called Cooker, which supports self-supervised adaptive aggregator learning. Specifically, we design three pretext tasks for self-supervised learning and apply multiple aggregators in our model. By doing so, our algorithm can keep the peculiar features of different aggregators in node embeddings and minimize the information loss. Experiment results on node classification and link prediction tasks show that Cooker outperforms the state of the art baselines in all three compared datasets. A set of ablation experiments also demonstrate that the integration of more types of aggregators generally improves the algorithm's performance and stability.

Keywords: Graph representation learning · Self-supervised learning · Neighborhood aggregation

1 Introduction

The low-dimensional vector embeddings of nodes in large graphs have proved extremely useful as feature inputs for various prediction and graph analysis tasks. The basic idea behind embedding is to use the dimensionality reduction technology to transform nodes on the graph into low-dimensional dense vectors while still preserving the attribute features of nodes and structure features of graphs [4]. In many aggregation techniques, neighborhood aggregations have played a vital role in most of the graph neural network-based embedding algorithms. Since GCN [11] represents a node by aggregating the feature vectors of its neighbors with a weighted sum and a (weighted) element-wise mean, some other aggregators have been proposed, e.g., mean aggregator, LSTM aggregator, pooling aggregator discussed in GraphSage [3], and the sum aggregator used in

© Springer Nature Switzerland AG 2021
K. Karlapalem et al. (Eds.): PAKDD 2021, LNAI 12714, pp. 29–41, 2021.
https://doi.org/10.1007/978-3-030-75768-7_3

GIN [18]. Later, LA-GCN [22] introduced a learnable aggregator and allowed the aggregator to learn to assign different weights to different features within a feature vector with the label information. Each of those aggregators keeps the neighboring information from a single perspective with a simple permutation-invariant function.

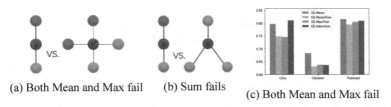

(a) Both Mean and Max fail   (b) Sum fails

(c) Both Mean and Max fail

Fig. 1. (a) and (b) are two examples of graph structures that different aggregators fail to distinguish the two different nodes v and v', Colors indicate different features of nodes. Vector v and v' would get the same embedding with the aggregators, although the graph structures are different; (c) is the node classification result from GraphSAGE with various aggregators.

However, practical networks are complex with different types of features. To keep most of the information during aggregation, sometimes it is mandatory to use different aggregators for different graphs or even different parts of the graph. Figure 1a, b show two examples of graph structures that mean, max and sum aggregators fail to distinguish. Some of them are already discussed in [18]. Figure 1a and b show two possible structures that node v and v' get the same embeddings even though their corresponding graph structures differ. The node classification results from the GraphSAGE variants with different aggregators (Fig. 1c) further illustrate that the selection of aggregators might have significant impacts on the performance of different datasets. All of these examples demonstrate that a single predefined heuristic aggregator might restrict modeling capacity and result in an inevitable loss of information [23].

Thus, it is important to have a solution that can find the most appropriate aggregator for different network structures. Currently limited solutions, e.g., [22] on the learnable aggregator, still demand the labeled data for training. In practice, labeled data are scarce and expensive, sometimes even impossible to acquire in real scenes [24]. It is favorable to have an algorithm that can automatically apply the most suitable aggregator to aggregate node features according to the graph structure without labels. As different aggregators only display peculiar features on certain graph structures, instead of learning to find the best aggregator, we can try to learn the distinguishing feature of different aggregators. Contrastive learning, a method to learn what makes two objects similar or different [5], becomes a natural candidate.

Present Work. To address the above challenging problems, in this paper, we propose a general contrastive learning framework: Cooker, to support self-supervised adaptive aggregator learning. To learn the different distinguishing features of the aggregators at different nodes while still preserving major graph information, we design three pretext tasks [9] and corresponding loss functions for self-supervised learning. The three pretext tasks help to guide our algorithm to capture the differences between different aggregators and keep the peculiar features of different aggregators in node embeddings. Unlike existing embedding approaches that are based on a single aggregation method, this design enables flexible and intelligent aggregator learning by allowing any type of aggregator to be added and evaluated. In this way, our algorithm can retain information from different angles for each node and minimize the information loss.

To verify the effectiveness of the proposed method, we integrate four types of typically used aggregators into Cooker and conduct experiments on a node classification task, where every node is assigned one class label and a link prediction task, where we predict the existence of an edge given a pair of nodes. We contrast the performance of Cooker with state-of-the-art baselines. We experiment with several real-world networks from diverse domains, such as citation networks and protein networks. Experiments results demonstrate that (i) our algorithm outperforms state-of-the-art methods by up to 19.20% on node classification task and up to 24.7% on link prediction task; (ii) the ablation studies show the effect of different pretexts and demonstrate that the integration of more types of aggregators generally indeed improve the algorithm's performance and stability.

2   Related Works

2.1   Aggregator

In order to aggregate information from neighboring nodes, it demands a certain mechanism to aggregate information. In the Graph convolutional network (GCN), Kipf and Welling [11] propose the usage of a weighted sum and an element-wise mean to aggregate the features information of neighboring nodes. Later, GraphSage [3] summarizes the design principle of the aggregator and uses the element-wise mean, a max-pooling neural network, and LSTMs [7] as aggregators to aggregate the information of neighboring nodes. GIN [18] analyzes the expressive power of GNNs and proposes to use the sum aggregator to solve graph isomorphism test. GraphAir [8] proposes a non-linear aggregation of the neighborhood interaction. The mentioned GCN, mean, LSTM, pooling, and sum aggregators are all predefined heuristics that can not adapt to different structures. Limited research has been proposed to make the aggregator learnable. GAT [16] proposes the use of self-attention to evaluate the different influences of neighboring nodes. LA-GCN [22] proposes a learnable aggregator by borrowing the idea from meta-learning. However, this approach is confined in the semi-supervised network embedding as they require information from labels.

2.2   Contrastive Methods

Unsupervised learning aims to learn representations from the data itself without explicit manual supervision. Self-supervised learning is a form of unsupervised learning where the data provides the supervision to train a pretext task, which helps to guide the learning algorithm to capture the underlying patterns of the data. Self-supervised representation learning is highly successful in natural language processing. Today, the self-supervised method has surpassed the supervised method in Pascal VOC detection [5] and also shown excellent results on many other tasks. Behind the rise-up of self-supervised methods is that they all follow contrastive learning. Contrastive methods are central to many popular word-embedding methods [2,12,13], but they are found in many unsupervised algorithms for learning representations of graph-structured input as well. For example, DGI [17] introduces "mutual information", designs a pretext task to classify local-global pairs and negative-sampled counterparts, and uses Jensen-Shannon divergence to maximize the mutual information of the local representations and global representations.

3   Formulating Cooker

Given a network $\mathcal{G} = (\mathcal{V}, E)$ where $\mathcal{V}$ is the node set and E is the edge set, with node feature vectors X_v for $v \in \mathcal{V}$. The task of our algorithm is to learn a low-dimensional representation z_v, $v \in \mathcal{V}$.

3.1   Cooker: A Contrastive Representation Learning Framework

In the network embedding, the major goal is to learn a mapping that embeds nodes, or entire (sub)graphs, as points in a low-dimensional vector space so that the geometric relationships in the embedding space reflect the structure of the original graph [4]. In this process, we want the embedding to take triple responsibilities: 1) keeps node's feature information; 2) keeps the structure information via aggregation, and 3) identifies the most suitable aggregate function for each node. The first two tasks are already well studied but it is challenging to find the most suitable aggregator for any specific node due to the lack of labeling information. Here, instead of finding the most suitable aggregator, we shift the original goal of keeping the peculiar features of different aggregators in node embeddings.

Inspired by recent contrastive learning algorithms, we designed the **Cooker**: a contrastive learning framework that integrates three pretext tasks with a shared embedding layer. This framework can learn a joint embedding space with data from multiple perspectives.

1) **The node restoration task.** This task attempts to restore the original input of the model. By comparing the model's input and the output of the node restoration task, this task can learn the hidden features of the input data and maximum retention of the nodes' feature information.
2) **Classification of node pairs.** The intuition of this task is to keep the structure information in the node representations maximally. We capture the structural properties of nodes via classic random walks. By determining whether

the starting nodes of two random walk paths are the same, this task can capture the structural differences between different nodes and learn the most unique representation for each node.

3) **Classification of aggregators.** In order to keep the peculiar features of different aggregators in node embeddings, multiple aggregators are applied to nodes to aggregate the context information obtained from random walks. This task attempts to determine whether the aggregators applied to two nodes in node pairs are the same. By learning the differences between the information learned by different aggregators, the final node embeddings can contain the unique characteristics of each aggregator.

Since these three tasks are intercorrelated, learning these tasks jointly can improve performance compared to learning them separately

Definition 1. $\{(u_i, v_i)\}_{i=1}^n$ denotes n node pairs, where $u_i, v_i \in \mathcal{V}$ and $(u_i, v_i) \in E$.

Definition 2. $\{l_0^i\}_{i=1}^n$ denotes the set of labels of the second pretext task, classification of node pairs. Specifically, when node u_i and v_i are equal, $l_0^i = 1$, otherwise, $l_0^i = 0$.

Definition 3. $\{l_1^i\}_{i=1}^n$ denotes the set of labels of the third pretext task, classification of aggregators. Specifically, when the aggregators applied upon u_i and v_i are the same, $l_1^i = 1$, otherwise, $l_1^i = 0$.

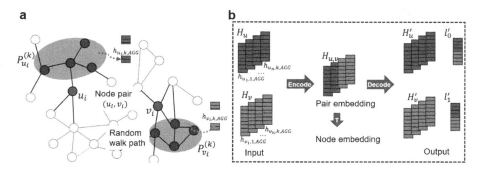

Fig. 2. Cooker architecture. **a.** Date preparation. Here, $P_{u_i}^{(k)}$ denotes the k-th random walk path of node u_i and $h_{u_i,k,AGG}$ denotes the aggregated feature of $P_{u_i}^{(k)}$ and AGG is the aggregate function. **b.** A conceptual diagram for Cooker.

4 Architecture

4.1 Data Preparation

As self-supervised tasks need to generate supervisory signals based on raw data, it is necessary to prepare data for processing. Firstly, a set of node pair (u_i, v_i)

are generated from the edge set E and the self-loop edges. Then as shown in Fig. 2a, for each node u_i, v_i in node pair, we obtain k random walk paths $\{P_{u_i}^{(1)}, ..., P_{u_i}^{(k)}\}$ and $\{P_{v_i}^{(1)}, ..., P_{v_i}^{(k)}\}$ respectively to capture the neighborhood context of nodes. For the neighborhood information, similar to the GNN-based solution, we leverage four most used aggregators, mean, sum, max, and min, to aggregate them, which results in $\{H_u, H_v\}$. Thus, the inputs for Cooker is $H_u = \{H_{u_i}\}_{i=1}^n, H_v = \{H_{v_i}\}_{i=1}^n$.

4.2   Embedding Generation Algorithm

Figure 2b describes the principal structure of Cooker. The inputs of our model are H_u and H_v generated from the data preparation process. Firstly, the inputs are transformed into pair embeddings $H_{u,v}$. We define the function $Encode$ as follows:

$$H_{u,v} \leftarrow \alpha(W^3 \cdot concat(H_u, H_v, \alpha(W^1 \cdot H_u), \alpha(W^2 \cdot H_v))) \tag{1}$$

where W^1, W^2, and W^3 are three layer-wised learnable weight matrices. The aggregated attributes H_u, H_v are fed through two fully connected layer α individually. To prevent gradient disappearance and achieve more stable performance, we adopt the "skip connection", just like ResNet [6] which skips one or more layers and concatenates the output from the front layer to the back layer. The outputs of the two dense layers are concatenated with the two inputs of the model and enter a dense layer, which is the shared embedding layer with a pair-embedding output $H_{u,v}$.

Through the 'Decode' layer, pair representations $H_{u,v}$ are transformed into final four outputs which are used to get the losses of our three pretext tasks:

$$H_u', H_v' \leftarrow \delta(W^4 \cdot H_{u,v}), \delta(W^5 \cdot H_{u,v}) \tag{2}$$

$$l_0', l_1' \leftarrow \theta(W^6 \cdot H_{u,v}), \theta(W^7 \cdot H_{u,v}) \tag{3}$$

H_u' and H_v' are the outputs of two separated dense layers with softmax activation function δ. l_0' and l_1' are obtained from the sigmoid activation function σ.

4.3   Learning the Parameter of Cooker

In order to learn useful, predictive representation in a fully unsupervised setting, we apply four loss functions to the four output representations, H_u', H_u', l_0', and l_1', and tune the weight matrices, $W^k, \forall k \in \{1, ..., 7\}$ via stochastic gradient descent.

For the first pretext task of node restoration, in order to make the learned embedding representation to recover the original input features as comprehensively and effectively as possible, the Kullback-Leibler (KL) divergence is used to measure the asymmetry of the difference between the two the original probability distribution P and the reconstructed probability distribution Q. The original probability distributions of input features H_u, H_v are denoted as P_u and P_v, and the reconstructed probability distributions of $H_u{}'$, $H_v{}'$ are denoted as Q_u and Q_v. The probability distribution of node u_i is calculated by Eq. (4):

$$P_{u_i}(k) = \exp\left(H_{u_i}(k)\right) / \sum_{j=1}^{F} \exp\left(H_{u_i}(j)\right) \tag{4}$$

where, $P_{u_i}(k)$ indicates the kth feature's distribution possibility of node u_i and F is the dimension of features. It is calculated by applying the standard exponential function to each element $H_{u_i}(k)$ and the normalized values are divided by the sum of all these exponents. P_v, Q_u, and Q_v are calculated in the same way.

Therefore, the goal of the node restoration task is to minimize the KL divergence between P_u and P_v, Q_u and Q_v respectively.

$$min \sum_{i=1}^{n} \sum_{j=1}^{F} P_{u_i}(j) \ln\left(\frac{P_{u_i}(j)}{Q_{u_i}^j}\right) \tag{5}$$

$$min \sum_{i=1}^{n} \sum_{j=1}^{F} P_{v_i}(j) \ln\left(\frac{P_{v_i}(j)}{Q_{v_i}^j}\right) \tag{6}$$

For our two binary classification tasks, the binary cross-entropy is used to measure the difference between the model's prediction, l_0', l_1', and a fixed target, l_0, l_1 (Eq. (7), (8)).

$$min\{-\sum_{i=1}^{n} l_0{}^{i'} log l_0{}^i + (1 - l_0{}^{i'})log(1 - l_0{}^{i'})\} \tag{7}$$

$$min\{-\sum_{i=1}^{n} l_1{}^{i'} log l_1{}^i + (1 - l_1{}^{i'})log(1 - l_1{}^{i'})\} \tag{8}$$

4.4 Translator

In order to support node-related downstream tasks, e.g., node classification task, it is essential to convert the embedding of node pairs $H_{u,v}$ to node embedding. As shown in Eq. (9), the embedding of node v is obtained from the pair embedding H_{u_i,v_i}, with u_i and v_i are equal to node v. And similar to the aggregate function, the translator function can be the min, max, mean, and sum.

$$z_v = T(H_{u_i,v_i}), \forall u_i = v_i = v \tag{9}$$

5   Experiments

To evaluate the effectiveness of Cooker, three citation graphs and one protein dataset widely used in graph tasks are selected. A set of experiments have been performed with the node classification and link prediction to check the effectiveness of Cooker. A set of ablation studies are also performed to check the impacts of pretext tasks and the effectiveness in unifying various aggregators.

Table 1. Dataset statistics.

Dataset	#Nodes	#Edges	#Features	#Classes
Cora	2,708	5,429	1,433	7
Citeseer	3,327	4,732	3,703	6
Pubmed	19,717	44,338	500	3
Enzymes	19,580	74,564	18	3

5.1   Dataset Descriptions

In this paper, we use 4 benchmark datasets including three citation networks datasets, one protein datasets and so on. We summarize the statistics of these benchmark datasets in Table 1.

5.2   Experiment Settings

Baseline Methods. To contextualize the empirical results on our algorithm, we compare against various unsupervised baselines, including DeepWalk [14], ProNE [21], MUSAE [15], TADW [19], DGI [17], GraphSAGE [3] and P-GNN [20]. Specifically, P-GNN is used in link prediction task.

For a fair comparison, we set the embedding dimension $d = 128$ and generate an equal number of samples for each method. We set the window size as 10, the walk length as 40, and the number of walks as 30 for DeepWalk. For ProNE, the term number of the Chebyshev expansion k is set to 10, $u = 0.1$, and $\theta = 0.5$. For MUSAE, the window size t is set to 3, walk length $l = 80$, and the number of walks per node $p = 10$. For TADW, we set the weight of regularization term $\lambda = .2$. For DGI, we use the GCN update rule in the encoder and sample 10 and 5 neighbors respectively on each neighborhood sampling depth. For GraphSAGE, we set the Neighborhood sampling depth $K = 1$ with neighborhood sample sizes $S = 5$.

Next we evaluate Cooker model on both node classification and link prediction task settings. For more parameter settings, the batch size is set as 1024 while the training epoch size as 10. For the random walk in our experiments, we set the walk length as 3, 10, and 25. The implementation is based on TensorFlow [1] with the Adam optimizer [10]. We designed our experiments with the goals of (i) verifying the improvement of Cooker over the baseline approaches, (ii) providing a rigorous comparison between the Cooker and different single pretext tasks, and (ii) proving the effectiveness of the combination of multiple aggregators.

Table 2. Node classification results, evaluated with average Micro-F1.

Name	Cora			Citeseer			Pubmed		
Ratio	30%	50%	70%	30%	50%	70%	30%	50%	70%
DeepWalk	.8177	.8146	.8248	.5602	.5809	.5907	.8042	.8016	.8056
ProNE	.7809	.7917	.7979	.5431	.5550	.5730	.8029	.8002	.8012
TADW	.6576	.6646	.6628	.6103	.6139	.6306	.8159	.8233	.8257
MUSAE	.8226	.8132	.8169	.6594	.6404	.6409	.7516	.8025	.7936
GraphSAGE	.8018	.8361	.8290	**.7296**	.7337	.7362	.8382	.8437	.8383
DGI-gcn	.8128	.8257	.8337	.6894	.6963	.7126	.8311	.8328	.8329
Cooker	**.8496**	**.8605**	**.8640**	.7224	**.7381**	**.7440**	**.8477**	**.8468**	**.8478**

5.3 Results on Node Classification Task

This task is a well-adopted node based task. The classifier used in this paper is Logistic Regression with L2 regularization. The embedding of vertices from different solutions are taken as the features to train classifiers with different training ratios, from 30%, 50% to 70%, and classification accuracy is evaluated with remaining data. To measure the classification performance, We repeat this process 10 times.

The classification results are shown in Tables 2. The best results in each case are marked in bold and the second ones are underlined. From these tables, we can find that Cooker generally outperforms the other state-of-the-art methods. Moreover, Cooker achieves very stably performance within all training ratio ranges across different datasets. Specifically, We achieve the biggest improvement over TADW of 19.20% on Cora at 30% training data.

The experiment results affirm the effectiveness of the automatic aggregation learning with the self-supervised learning tasks. Good performance on node classification can only be achieved if the learned representation could encode global topological information and finely discriminate the similarity and differentiation between nodes. We have the reasons to believe that the supervision built from the data itself could capture the inherent characteristics of data.

5.4 Results on Link Prediction Task

The goal of this task is to predict the existence of an edge between two nodes. The classifier is Logistic Regression with L2 regularization. Specifically, we randomly select 20% existing links from the original network as positive samples and an equal number of nonexistent links as negative samples of the testing set, with remaining 80% exiting links as positive samples and equal number of nonexistent links as negative samples of the training set. The average ROC AUC is used for evaluation.

The link prediction accuracy of three citation datasets is shown in Table 3. From this table, we can observe that our proposed method significantly outperform all the other baseline methods in all datasets. Especially, the biggest improvement over TADW on Citeseer is up to 24.6%, which further verifies the effectiveness of our method.

Table 3. Link prediction task, measured in ROC AUC. Standard deviation errors are given.

Method	Cora	Citeseer	Pubmed
DeepWalk	.811 ± .009	.819 ± .006	.799 ± .002
ProNE	.838 ± .003	.838 ± .007	.832 ± .002
TADW	.659 ± .000	.692 ± .005	.705 ± .000
MUSAE	.814 ± .002	.844 ± .001	.817 ± .001
GraphSAGE	.829 ± .011	.834 ± .037	.865 ± .000
DGI-gcn	.670 ± .014	.672 ± .015	.761 ± .005
P-GNN	.809 ± .007	.749 ± .016	.817 ± .000
Cooker	**.887** ± .003	**.919** ± .004	**.952** ± .002

5.5   Ablation Studies

In this section, a set of ablation studies are performed. Due to page limits, the Enzymes dataset is selected due to its sensitivity towards both the types of aggregators and the types of pretext tasks.

Cooker and Single Pretext Tasks. Cooker is a contrastive learning framework that consists of three tasks. To compare Cooker with the methods of a single pretext task, we run a set of ablation study experiments on several datasets. We provide an apples-to-apples comparison of three different pretext tasks and all the settings and hyperparameters are the same as reported in Sect. 5.2. Figure 3a shows the comparison results on a protein dataset (Enzymes). From the ablation study experiments, we have the following observations. First, Cooker is the best combination of pretext tasks we test. Second, we can find that among all the single pretext tasks, the node restoration task has the strongest performance and the other two pretext classification tasks ("uv" and "agg") have a relatively small effect. Third, the results of two pretext classification tasks indicate that they indeed can capture some information although their performance is not competitive enough. The experiment results suggest that representations learned by the various pretext tasks have different strengths and weaknesses, and combining pretext tasks can yield further improvements.

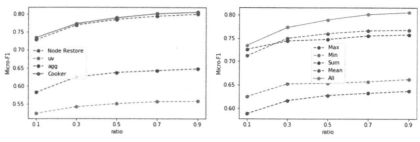

(a) Comparisons among pretext tasks on Enzymes

(b) Comparisons among aggregator on Enzymes

Fig. 3. Node classification results with ablation studies.

Different Aggregators. To prove that Cooker can learn more information by adopting various aggregators, we conduct node classification tasks with different types of aggregators and their combinations. Figure 3b summarizes the performance with different aggregation function on Enzymes. Overall, We found that the combination of four aggregators performed the best, especially when the training data scare. For example, the experiment with the combination of four aggregators outperforms the method with Max aggregator by 16% on 10% training data of Enzymes. Moreover, it is easy to find that different aggregators would result in significant performance divergences. We also find that the Mean aggregator generally achieves the best performance than other types of aggregators in most of the datasets. It matches the findings of existing works [17]. However, for the much more complex dataset, e.g. Enzymes, the Mean operator can not catch certain data. It lags behind cooker with four aggregators by about 3% points in Micro-F1. It clearly indicates the importance of the combination of different aggregators for complex graphs.

6   Conclusion

In this paper, we propose a general contrastive learning framework so-called Cooker, which can support self-supervised adaptive aggregator learning. To learn the different distinguishing features of the aggregators at different nodes while still preserving major graph information, we design three pretext tasks for self-supervised learning. Unlike existing embedding approaches that are based on a single aggregation method, this design enables flexible and intelligent aggregator learning by allowing any type of aggregator to be added and evaluated. In this way, our algorithm can retain information from different angles for each node and minimize the information loss. Extensive experiments on different graph tasks have demonstrated the superior performance of our proposed method. Due to its capabilities integration different pretext tasks, we believe Cooker can be directly applied to other domains with different pretext tasks.

References

1. Abadi, M., et al.: TensorFlow: a system for large-scale machine learning. In: 12th {USENIX} Symposium on Operating Systems Design and Implementation ({OSDI} 16), pp. 265–283 (2016)
2. Collobert, R., Weston, J.: A unified architecture for natural language processing: deep neural networks with multitask learning. In: Proceedings of the 25th International Conference on Machine Learning, pp. 160–167 (2008)
3. Hamilton, W., Ying, Z., Leskovec, J.: Inductive representation learning on large graphs. In: Advances in Neural Information Processing Systems, pp. 1024–1034 (2017)
4. Hamilton, W.L., Ying, R., Leskovec, J.: Representation learning on graphs: methods and applications. arXiv preprint arXiv:1709.05584 (2017)
5. He, K., Fan, H., Wu, Y., Xie, S., Girshick, R.: Momentum contrast for unsupervised visual representation learning. arXiv preprint arXiv:1911.05722 (2019)
6. He, K., Zhang, X., Ren, S., Sun, J.: Deep residual learning for image recognition. In: Proceedings of the IEEE Conference on Computer Vision and Pattern Recognition, pp. 770–778 (2016)
7. Hochreiter, S., Schmidhuber, J.: Long short-term memory. Neural Comput. 9(8), 1735–1780 (1997)
8. Hu, F., Zhu, Y., Wu, S., Huang, W., Wang, L., Tan, T.: GraphAIR: graph representation learning with neighborhood aggregation and interaction. arXiv preprint arXiv:1911.01731 (2019)
9. Jing, L., Tian, Y.: Self-supervised visual feature learning with deep neural networks: a survey. IEEE Trans. Pattern Anal. Mach. Intell. (2020)
10. Kingma, D.P., Ba, J.: Adam: a method for stochastic optimization. arXiv preprint arXiv:1412.6980 (2014)
11. Kipf, T.N., Welling, M.: Semi-supervised classification with graph convolutional networks. arXiv preprint arXiv:1609.02907 (2016)
12. Mikolov, T., Sutskever, I., Chen, K., Corrado, G.S., Dean, J.: Distributed representations of words and phrases and their compositionality. In: Advances in Neural Information Processing Systems, pp. 3111–3119 (2013)
13. Mnih, V., et al.: Playing atari with deep reinforcement learning. arXiv preprint arXiv:1312.5602 (2013)
14. Perozzi, B., Al-Rfou, R., Skiena, S.: DeepWalk: online learning of social representations. In: Proceedings of the 20th ACM SIGKDD International Conference on Knowledge Discovery and Data Mining, pp. 701–710 (2014)
15. Rozemberczki, B., Allen, C., Sarkar, R.: Multi-scale attributed node embedding. arXiv preprint arXiv:1909.13021 (2019)
16. Veličković, P., Cucurull, G., Casanova, A., Romero, A., Lio, P., Bengio, Y.: Graph attention networks. arXiv preprint arXiv:1710.10903 (2017)
17. Veličković, P., Fedus, W., Hamilton, W.L., Liò, P., Bengio, Y., Hjelm, R.D.: Deep graph infomax. arXiv preprint arXiv:1809.10341 (2018)

18. Xu, K., Hu, W., Leskovec, J., Jegelka, S.: How powerful are graph neural networks? arXiv preprint arXiv:1810.00826 (2018)
19. Yang, C., Liu, Z., Zhao, D., Sun, M., Chang, E.: Network representation learning with rich text information. In: Twenty-Fourth International Joint Conference on Artificial Intelligence (2015)
20. You, J., Ying, R., Leskovec, J.: Position-aware graph neural networks. arXiv preprint arXiv:1906.04817 (2019)
21. Zhang, J., Dong, Y., Wang, Y., Tang, J., Ding, M.: ProNE: fast and scalable network representation learning. In: Proceedings of 28th International Joint Conference on Artificial Intelligence (IJCAI), pp. 4278–4284 (2019)
22. Zhang, L., Lu, H.: A feature-importance-aware and robust aggregator for GCN. In: Proceedings of the 29th ACM International Conference on Information & Knowledge Management, pp. 1813–1822 (2020)
23. Zhu, X., Ghahramani, Z., Lafferty, J.D.: Semi-supervised learning using Gaussian fields and harmonic functions. In: Proceedings of the 20th International Conference on Machine Learning (ICML 2003), pp. 912–919 (2003)
24. Zhu, X., Goldberg, A.B.: Introduction to semi-supervised learning. In: Synthesis Lectures on Artificial Intelligence and Machine Learning, vol. 3, no. 1, pp. 1–130 (2009)

A Fast Algorithm for Simultaneous Sparse Approximation

Guihong Wan$^{(\boxtimes)}$ and Haim Schweitzer

Department of Computer Science, The University of Texas at Dallas, Richardson,
TX 75080, USA
{Guihong.Wan,HSchweitzer}@utdallas.edu

Abstract. Simultaneous sparse approximation problems arise in several domains, such as signal processing and machine learning. Given a dictionary matrix X of size $m \times n$ and a target matrix Y of size $m \times N$, we consider the classical problem of selecting k columns from X that can be used to linearly approximate the entire matrix Y. The previous fastest nontrivial algorithms for this problem have a running time of $O(mnN)$. We describe a significantly faster algorithm with a running time of $O(km(n + N))$ with accuracy that compares favorably with the slower algorithms. We also derive bounds on the accuracy of the selections computed by our algorithm. These bounds show that our results are typically within a few percentage points of the optimal solution.

Keywords: Simultaneous sparse approximation · Multiple measurement vectors · Multi-target regression · Spectral pursuit

1 Introduction

Sparse approximation, or alternatively, sparse representation, has attracted significant attention in fields of signal processing, image processing, and machine learning (e.g., [16,27,33]). It originally arose in the study of linear regression in which a target vector is approximated by a linear combination of several selected features to foster interpretability and avoid overfitting.

The problem that we discuss in this paper is an extension of the classical sparse approximation problem to multiple targets. The goal is to identify a small number of "atoms" (columns of the dictionary matrix) that can be used to linearly approximate all the columns of another "target" matrix. This formulation is referred to as simultaneous sparse approximation (e.g., [17,25]). It has various applications, such as supervised feature selection for multi-target regression [34], human action recognition in videos [11], multi-sensor image fusion [31], and hyperspectral image unmixing [28].

We present an efficient selection algorithm with a running time that depends linearly on the number of columns of the relevant matrices. This significantly improves on previous algorithms which have a running time depending on the product of these parameters.

© Springer Nature Switzerland AG 2021
K. Karlapalem et al. (Eds.): PAKDD 2021, LNAI 12714, pp. 42–54, 2021.
https://doi.org/10.1007/978-3-030-75768-7_4

The main idea is to select columns from the dictionary matrix whose span is close to the dominant spectral components of the target matrix. The following two-stage procedure is proposed. In the first stage, a greedy algorithm is used to select k columns from the dictionary matrix. In each iteration, a column that is most related to the first left eigenvector of the residual target matrix is selected. In the second stage, an iterative "bidirectional selection" algorithm is used to improve the results produced in the first stage.

1.1   Problem Formulation

The problem of simultaneous sparse approximation can be stated as follows. Let $Y = (y_1 \ldots y_N)$ be a "target" matrix of m rows and N columns. Let $X = (x_1 \ldots x_n)$ be a "dictionary" matrix of m rows and n columns (n is typically very large). Consider an approximation of Y in terms of X:

$$Y \approx X\tilde{A}, \tag{1}$$

where $\tilde{A}$ is the coefficient matrix of size $n \times N$. The approximation in Eq. (1) is considered a sparse approximation if only a small number of rows of $\tilde{A}$ are nonzero. See, e.g., [4,19,24,25]. Let $k \leq n$ be the number of nonzero rows in $\tilde{A}$, then the approximation in (1) can be written as:

$$Y \approx SA, \tag{2}$$

where $S = (x_{s_1} \ldots x_{s_k})$ is the $m \times k$ selection matrix consisting of the k columns of X corresponding to the k nonzero rows of $\tilde{A}$, and A is the coefficient matrix of size $k \times N$, created from the k nonzero rows of $\tilde{A}$. Thus, the goal of the problem described in (2) is to select k columns from X such that all the columns of Y can be simultaneously approximated by a linear combination of the selected columns in S. The quality of the selected columns in S is measured by the following (Frobenius norm) error:

$$E(S) = \min_A \|Y - SA\|_F^2, \quad \text{where } S \text{ is a subset of } X \text{ columns.} \tag{3}$$

The following cases are special sub-problems:

- When $N = 1$, the problem is called sparse approximation, where Y is a single vector and X is a dictionary. It is also called supervised subset selection when Y corresponds to labels for the data matrix X. See, e.g., [9,20].
- When $X = Y$, the problem is known as "unsupervised feature selection" if each column corresponds to a data feature. It is known as "representative selection" if each column corresponds to a data point. See, e.g., [1,13,30].

Optimal solutions for the optimization in (3), as well as approximate solutions within a constant are known to be NP-hard even when $N = 1$ [8,20].

1.2   Related Work

Extensive studies were directed at the $N = 1$ case (e.g., [9,18,21,24,32]). Applications include signal processing e.g., [18,24] and supervised feature selection in linear regression e.g., [9,21,32]. An optimal algorithm proposed in [9] finds the best solution, but it is not feasible for large k and large datasets. Previously proposed approximate solutions can be roughly divided into three groups: forward selection, backward elimination, and convex relaxation. Forward selection sequentially adds columns that improve the quality the most. Backward elimination starts with the full selection and sequentially removes the columns that affect the quality the least. As shown in [32], these two techniques have limitations due to their greedy behavior. The author of [32] introduced an adaptive algorithm that combines these two ideas to alleviate the flaws. The convex relaxation approaches replace some natural constraints (sometimes defined in terms of the l_0 norm) with convex constraints. An example is the l_1 norm used in the Lasso technique [23]. In [21], the authors treated the subset selection as a bi-objective optimization problem. Their algorithm is optimal for data drawn from Exponential Decay distribution but not for the general case.

The case where $X = Y$ is known as the unsupervised feature selection or unsupervised column subset selection. See, e.g., [1,2,4,10,14,30]. Recently, a fast algorithm [30] was proposed, which greedily selects columns that are closest to the first left eigenvector of a residual matrix. Another recent study in [14] proposes to iteratively improve a selection using a bi-directional stepwise refinement. Both these studies address the $X = Y$ case. Our method is motivated by these two recent studies, generalizing them to the supervised multiple-target case. We show that the speed advantage of this approach over other approaches becomes bigger for large N.

In the multi-target case, the target matrix Y has $N > 1$ columns. We observe that one cannot simply apply an $N = 1$ algorithm separately to each column of Y. The challenge is to find columns in X that can simultaneously approximate all the columns of Y. Previously proposed algorithms for this general case are typically greedy. See, e.g., [3–5,19,25]. Some of these algorithms are generated from the $N = 1$ case. For example, the Simultaneous Orthogonal Matching Pursuit (SOMP) [25] is generated from the Orthogonal Matching Pursuit (OMP) [22,24] to handle a target matrix of N columns. Similarly, the Simultaneous Orthogonal Least Squares (SOLS) is a direct extension of the Orthogonal Least Squares (OLS) [22]. See [4,5] for the analysis of the SOLS algorithm. An algorithm established in [4] improves the SOLS algorithm in terms of speed at the cost of increased memory. In [19], the authors improved the speed of the SOLS algorithm by a recursive formulation. The running time and the memory requirements of some of these algorithms are summarized in Table 1.

1.3   Our Approach

Suppose the matrix S in (3) is not constrained to be a submatrix of X. Then it is known that the k columns of the optimal solution matrix S are the left

Table 1. Complexity of various algorithms. T is the number of iterations.

Algorithms	Time complexity	Memory complexity
SOMP [25]	$O(kmnN)$	$O(m(N+k))$
S-SBR [3]	$O(TkmnN)$	$O(m(N+k))$
SOLS [6]	$O(kmnN)$	$O(m(n+N))$
CM [4]	$O(nN(m+k))$	$O(m(n+N))+nN$
ISOLS [19]	$O(mnN)$	$O(km)+2n$
SPXY (this work)	$O(km(n+N))$	$O(m(n+N))$

eigenvectors of Y corresponding to the k largest singular values (see e.g., [10]). Our algorithmic approach is based on this result.

Motivated by [30] and [14], we propose a two-stage algorithm that we call the Spectral Pursuit for the matrices X and Y (SPXY). It runs significantly faster than previously proposed algorithms, and its accuracy compares very favorably with the current state of the art. We also show how to derive a bound on how far the selection computed by SPXY is from the optimal solution. (Recall that the computation of the optimal solution is NP-hard.)

In the first stage we use a greedy technique to select k columns from the dictionary matrix. The algorithm runs k iterations. In each iteration the column selected is the one most similar to the left eigenvector corresponding to the largest singular value. The two matrices are then projected on the null space of the columns that were already selected. The null space of the selected columns indicates a subspace that the selected columns cannot span. This first stage algorithm is greedy, and gets "stuck" in a local minimum. In the second stage we use a bidirectional stepwise technique to further improve the results.

Similar to many other data analysis techniques, the performance of our SPXY algorithm is data-dependent. Since the general problem is NP-hard, there can be situations where one may be tempted to use other more accurate algorithms, such as exhaustive search, in the hope of significantly improving the accuracy of the result. To this end, we derive bounds on how far the results given by our algorithm are from the optima. As we show, in many practical problems our results are provably within a small percentage of the optima. Thus, in such cases, even the exhaustive search can provide almost no improvement. In summary our main contributions are as follows:

- We introduce the SPXY algorithm that has a linear running time in terms of the numbers of columns of the two matrices X and Y. This running time is significantly faster than the current state of the art that requires running time proportional to the product of the numbers of columns of X and Y.
- The accuracy of the SPXY algorithm compares favorably with the current state of the art.
- We derive bounds on how far the solutions produced by SPXY are from the optimal solutions.

2   The Proposed Algorithm

In this section, a computationally efficient algorithm is proposed for approximating the solution to the NP-hard selection problem of minimizing Eq. (3).

Algorithm 1. Spectral Pursuit XY algorithm

1: **procedure** SPXY(X, Y, k)
2:     $S \leftarrow$ SELECT(X, Y, k).   ▷ select k columns from X to approximate Y.
3:     $S \leftarrow$ IMPROVE(X, Y, k, S).   ▷ improve the selection of previous stage.
4:     **return** S
5: **end procedure**

Algorithm 2. The selection algorithm

1: **procedure** SELECT(X, Y, k)
2:     $S = \{\}$.
3:     **while** $|S| < k$ **do**
4:         $u =$ the first left eigenvector of Y corresponding to its largest singular value.
5:         $i =$ index of the column from X, which most correlates with u.
6:         $S = S \cup \{i\}$; $q_i = x_i / \|x_i\|$.
7:         $Y = Y - q_i q_i^T Y$; $X = X - q_i q_i^T X$.
8:     **return** S
9: **end procedure**

The top view of the proposed algorithm is shown as Algorithm 1, with two stages involved. In the first stage, a greedy algorithm is introduced to select k columns from the dictionary matrix. In the next stage, an efficient non-greedy algorithm is used to improve the results.

2.1   The Selection Algorithm

We first observe that the vectors: $q_1 \ldots q_k$, generated by Algorithm 2, are mutually orthogonal. Define $Q = (q_1 \ldots q_k)$, where $Q \in \mathbb{R}^{m \times k}$. The projection of Y on Q can be written as $Q^T Y$. The reconstruction of Y from this projection can be written as: $\hat{Y} = Q Q^T Y$. Therefore, the optimization problem (3) can be restated as:

$$Q = \operatorname{argmin}_Q \|Y - QQ^T Y\|_F^2, \tag{4}$$

where Q is restricted to be the basis of k columns from X. As mentioned before, this is an NP-hard problem. If the constraint on Q is relaxed (it can be any matrix with size $m \times k$ with orthonormalized columns), then it is known that the minimizer of problem (4) is the k left eigenvectors of Y corresponding to the k largest singular values (e.g., [10]). We call the k eigenvectors corresponding to the k largest singular values as the first k eigenvectors.

Algorithm 3. The improvement algorithm

1: **procedure** IMPROVE(X, Y, k, S)
2:     $S_{\text{opt}} \leftarrow S$; error$_{\min} \leftarrow$ the current error value.
3:     iter $= 0$.
4:     **while** a stopping criterion is not met **do**
5:         $i = \text{mod}(\text{iter}, k)$.
6:         $Q_{\bar{i}} =$ basis of columns in S except the ith column of S.
7:         $R_Y = Y - Q_{\bar{i}}Q_{\bar{i}}^T Y, \quad R_X = X - Q_{\bar{i}}Q_{\bar{i}}^T X.$
8:         $u =$ the first left eigenvector of R_Y.
9:         $j =$ index of the most correlated column in R_X with u.
10:         $q_j =$ the normalized jth column of R_X; error $= \|R_Y\|_F^2 - \|q_j^T R_Y\|_F^2.$
11:         **if** error $<$ error$_{\min}$ **then:**
12:             $S[i] = j$.   ▷ Improvement
13:             $S_{\text{opt}} \leftarrow S$; error$_{\min} =$ error.
14:         iter $\mathrel{+}= 1$.
15:     **end while**
16:     **return** S_{opt}
17: **end procedure**

In the selection algorithm, we modify (4) into two sub-problems. The first one relaxes the constraint that Q must be the basis of k columns from X. This relaxation makes finding the solution tractable at the expense of resulting in a solution that may not correspond to columns in X. To fix this problem, we introduce the second sub-problem that reimposes the underlying constraint that selects the column that has highest correlation with the vector computed in the first sub-problem. These two sub-problems are formulated as follows:

$$
\begin{aligned}
u &= \text{argmin}_u \|Y - uu^T Y\|_F^2, \ s.t. \ \|u\| = 1, \\
i &= \text{argmin}_i |u^T q_i|,
\end{aligned}
\tag{5}
$$

where $q_i = x_i / \|x_i\|$. The resulting i is the index of the selected column. The first sub-problem is equivalent to computing the first left eigenvector of Y (e.g., [10]). After solving for u (which is not necessarily one of the columns in X), we find the column that matches u the most (has the least angle with u).

After selecting the first column x_i, we project all columns onto the orthogonal space of this selected column. This forms the residual matrices: $Y = Y - q_i q_i^T Y$, $X = X - q_i q_i^T X$. In the future iterations, we solve problem (5) on the residual matrices. This process continues until k columns are selected.

Algorithm 2 shows the selection algorithm. To simplify notation, we do not distinguish between the selection matrix and the selection indexes. To evaluate the algorithm complexity we observe that there are several recently proposed efficient algorithms for computing eigenvectors. In particular, the randomized eigendecomposition algorithm [12] can be used to compute u in $O(mN)$ time. With this we conclude that the time complexity of Algorithm 2 is $O(km(n+N))$, and its memory complexity is $O(m(n + N))$.

2.2   The Improvement Algorithm

Since Algorithm 2 is greedy we propose to improve its result by iteratively revising the selection as long as it can be locally improved. Similar to the selection algorithm in Sect. 2.1, we modify the problem into two sub-problems. The first one is built upon the assumption that $k-1$ columns from X are already selected and the objective is to select next best column. The sub-problems are formulated as:

$$u_i = \mathrm{argmin}_u \|Y - Q_{\bar{i}}Q_{\bar{i}}^T Y - uu^T Y\|_F^2, \ s.t. \ \|u\| = 1,$$
$$j = \mathrm{argmin}_j |u_i^T q_j|. \tag{6}$$

Here $Q_{\bar{i}}$ is the basis of $S_{\bar{i}}$ obtained by removing the ith selection in S. The first sub-problem is equivalent to finding the first left eigenvector of the residual matrix of Y: $R_Y = Y - Q_{\bar{i}}Q_{\bar{i}}^T Y$. The residual matrix of X is given by: $R_X = X - Q_{\bar{i}}Q_{\bar{i}}^T X$, and $q_j = r_X^j / \|r_X^j\|$ is the normalized jth column of R_X. The index j found in this way is the new selection which will replace the ith selection in S.

The improvement algorithm is shown as Algorithm 3. The stopping criterion that we use is a pre-defined maximum number of iterations. The algorithm can be efficiently implemented by using the rank-one update for QR factorization [7] with complexity per iteration of $O(km(n + N))$, and memory complexity of $O(m(n + N))$. In our implementation we limit the algorithm to run no more than 30 iterations. The algorithm terminates if there is no improvement in 5 iterations. The convergence behavior is studied in Sect. 5.

3   Fractional Bound

We proceed to show how to obtain nontrivial bound on how close the computed result is to the optimal solution. As described in (4) the algorithms minimize the following error:

$$E(S) = E(Q) = \|Y - QQ^T Y\|_F^2 = \|Y\|_F^2 - \|Q^T Y\|_F^2 = \|Y\|_F^2 - G(S),$$

where Q is an orthogonal basis of S, and $G(S) = \|Q^T Y\|_F^2$.

Since $\|Y\|_F^2$ is independent of S, the minimization of $E(S)$ is equivalent to the maximization of $G(S)$. Let E_{opt} be the smallest possible error, and let G_{opt} be the largest possible value of $G(S)$. They are related by: $E_{\mathrm{opt}} = \|Y\|_F^2 - G_{\mathrm{opt}}$. We define the fractional bound in terms of $G(S)$ as follows:

$$\tilde{b}_f(S) = (G_{\mathrm{opt}} - G(S))/G_{\mathrm{opt}}.$$

A smaller $\tilde{b}_f$ value indicates a better result, and in particular, $\tilde{b}_f = 0$ implies an optimal solution. Unfortunately both G_{opt} and E_{opt} are unknown. However, we observe that $\tilde{b}_f(S)$ is monotonically increasing as a function of G_{opt}. This implies that any nontrivial upper bound of G_{opt} can be used to obtain a nontrivial estimate of $\tilde{b}_f(S)$ as follows: If $G_{\mathrm{upper}} \geq G_{\mathrm{opt}}$ then:

$$\tilde{b}_f(S) = \frac{G_{\mathrm{opt}} - G(S)}{G_{\mathrm{opt}}} = 1 - \frac{G(S)}{G_{\mathrm{opt}}} \leq 1 - \frac{G(S)}{G_{\mathrm{upper}}}.$$

Let U_k be the matrix computed from the first k left eigenvectors of Y. Then for any S we have: $E(U_k) \leq E(S)$, so that for any S, $G(U_k) = \|Y\|_F^2 - E(U_k) \geq G(S)$. This shows that $G(U_k) \geq G_{\text{opt}}$, and gives the following formula for provable fractional bound:

$$b_f = 1 - G(S)/G(U_k). \tag{7}$$

4 Robustness

The sparse approximation algorithms are vulnerable to outliers. The span of outliers usually covers a bigger subspace, which may not be desirable to be represented. In our algorithm, at each iteration, we compute the first left eigenvector of the residual target matrix to guide the selection. If this eigenvector is robust, outliers in the target matrix X will be automatically rejected. We show that this eigenvector is the most robust spectral component against perturbation of matrix the Y. Additionally, we can use robust principal component analysis algorithms (e.g., [15, 26]) to compute this eigenvector.

Consider the second moment matrix of Y: $B = YY^T = \sum_{i=1}^{N} y_i y_i^T$. The first eigenvector of this matrix is same as the first left eigenvector of Y. If an outlier is added, it will perturb this matrix. The following lemma proven in [30] shows the robustness of ith eigenvector of B against the perturbation.

Lemma 1. Let B be a symmetric matrix with rank r, $(\lambda_1, \ldots, \lambda_r)$ be its eigenvalues (in decreasing order) and $(u_1, \ldots, u_r)$ be the corresponding eigenvectors. Then: $\frac{\|\partial u_i\|_2}{\|\partial B\|_F} \leq \sqrt{\sum_{j \neq i} \frac{1}{(\lambda_i - \lambda_j)^2}}$.

Define the sensitivity of the ith eigenvector as: $s_i = \sqrt{\sum_{j \neq i} \frac{1}{(\lambda_i - \lambda_j)^2}}$. It is known that s_1 is smaller than other $s_i, \forall i > 1$ if the gap between consecutive eigenvalues is monotonously decreasing [30].

5 Experimental Results

We describe experiments on various datasets that are publicly available, and compare the proposed algorithm with the following algorithms: SOMP [25]; S-SBR [3]; SOLS [6]; CM [4]; ISOLS (the exact version is used) [19]. Besides, the results of the random selection are shown. The error and bound are defined in (3) and (7), respectively. In experimental results, they are shown in percentage: error $= \frac{E(S)}{\|Y\|_F^2} * 100$; bound $= b_f * 100$. The computational efficiency and the selection accuracy of our algorithm are demonstrated.

5.1   Quantitative Comparison

In the first experiment, we evenly split the datasets into two matrices, serving as X and Y. The results are shown in Table 2. In the second experiment, we randomly split the datasets with the proportion of $X{:}Y = 3{:}1$. The results are shown in Table 3. We choose to split the datasets, since the learning of the dictionary matrix is another big topic and task-dependent (e.g., [29]). As mentioned in [19], the results of SOLS, CM, and ISOLS are exactly same (different in terms of speed). We show the results of ISOLS, as it is the fastest among these three algorithms. The parameter γ for S-SBR is 0, since k is known in our case.

Observe that our algorithm is much faster than other algorithms. The running time of ISOLS almost does not change as the increase of k. However, its initial step can be very expensive for big dense datasets. Taking the Duke breast cancer dataset in Table 2 as an example, its initial step takes 8.94 s. The overall running time for our algorithm is less than 1 s.

Table 2. Comparison when data is split with proportion to 1:1. "–" indicates that the algorithm runs more than 30 min without results.

k	Random	SOMP [25]	S-SBR [3]	ISOLS [19]	SPXY (this work)		
	Error	Error/time (s)	Error/time (s)	Error/time (s)	Error	Bound	Time (s)
Duke breast cancer X:44 × 3, 565 Y:44 × 3, 565							
10	40.3	30.7/0.73	29.6/3.83	29.6/9.13	**29.1**	4.2%	**0.13**
20	22.4	18.5/1.42	16.7/9.97	16.7/9.37	**16.3**	3.1%	**0.16**
30	11.3	9.4/2.15	**7.9**/18.04	**7.9**/9.41	8.0	1.8%	**0.24**
YearPredictionMSD X:91 × 257, 672 Y:91 × 257, 673							
10	9.2	–	–	–	**5.6**	0.9%	**28**
20	5.3	–	–	–	**2.7**	0.6%	**37**
30	2.8	–	–	–	**1.5**	0.5%	**45**
Sift X:128 × 500, 000 Y:128 × 500, 000							
10	36.3	–	–	–	**27.2**	6.1%	**76**
20	24.7	–	–	–	**19.8**	5.5%	**97**
30	21.7	–	–	–	**15.4**	5.2%	**117**
Sift:transpose X:500, 000 × 128 Y:500, 000 × 128							
10	**60.3**	–	–	–	60.5	49%	**61**
20	60.4	–	–	–	**59.7**	53%	159
30	59.4	–	–	–	**59.3**	54%	**150**

The errors given by our algorithm are typically smaller or similar to the errors of ISOLS. Additionally, our errors come with bounds indicating how close the solutions are to the optima. The bounds are usually within 10%. The reason for high bound values on Sift:transpose dataset is that X cannot explain Y well, since the errors are very big and do not change as the increase of k. It is further discussed in Sect. 5.2. The errors given by SOMP are much bigger. One of the reasons is that SOMP does not update the matrix X on the null-space in each iteration. Clearly, other algorithms are not practical for big datasets.

Table 3. Comparison when data is split with proportion to 3:1.

k	Random Error	SOMP [25] Error/time (s)	S-SBR [3] Error/time (s)	ISOLS [19] Error/time (s)	SPXY (this work) Error	Bound	Time (s)
Duke breast cancer X:$44 \times 5,347$ Y:$44 \times 1,783$							
10	44.9	31.4/0.84	30.3/4.85	30.3/6.89	**29.8**	**4.5%**	**0.15**
20	24.3	18.5/1.65	17.0/11.56	17.0/7.15	**16.9**	**3.4%**	**0.22**
30	12.4	9.5/2.46	8.0/20.01	8.0/7.29	**7.9**	**1.6%**	**0.26**
YearPredictionMSD X:$91 \times 386,508$ Y:$91 \times 128,873$							
10	11.0	–	–	–	**5.4**	**0.7%**	**23**
20	5.3	–	–	–	**2.7**	**0.6%**	**31**
30	2.9	–	–	–	**1.4**	**0.4%**	**37**
Sift X:$128 \times 750,000$ Y:$128 \times 250,000$							
10	38.9	–	–	–	**26.8**	**5.7%**	**63**
20	27.3	–	–	–	**19.7**	**5.5%**	**110**
30	19.9	–	–	–	**15.1**	**4.9%**	**134**
Extended Yale Face B X:$32,256 \times 1,488$ Y:$32,256 \times 496$							
10	23.6	18.6/214	18.0/176	18.0/433	**17.6**	**7.4%**	**18**
20	18.2	15.1/432	**13.9**/552	**13.9**/362	14.1	8.2%	**39**
30	15.5	12.8/649	**11.7**/1105	**11.7**/372	12.1	8.1%	**53**

5.2   Correlation Values of Algorithm 2

In the first stage, Algorithm 2 greedily selects the column with highest correlation value in each iteration. Figure 1 shows the correlation values of the selected columns as the run of Algorithm 2 for various datasets (split with proportion to 1:1) with $k = 30$. Generally, as the increase of the iteration, the maximum correlation value between the first left eigenvector of the residual matrix of Y and the normalized residual columns of X decreases. For the transpose of Sift dataset, the correlation value for the first iteration is 0.78, but for later iterations, they are very small (almost 0). It shows that X cannot explain Y well.

5.3   Convergence of Algorithm 3

In this experiment, we investigate the convergence of Algorithm 3 in the second stage. Here we set the maximum number of iterations to be 50 without early termination. The results are shown in Fig. 2 and Fig. 3. The red lines correspond to errors or bounds given in the first stage. As expected, the algorithm improves errors and bounds sharply at the beginning.

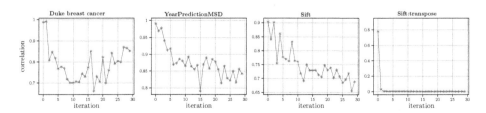

Fig. 1. Correlation values of selected columns for various datasets.

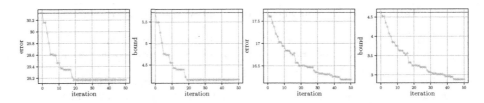

Fig. 2. Convergence of Algorithm 3 on Duke breast cancer dataset (1:1). The left two plots are for $k = 10$. The right two plots are for $k = 20$.

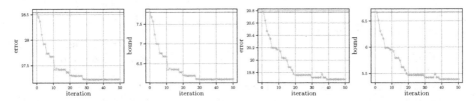

Fig. 3. Convergence of Algorithm 3 on Sift dataset (1:1). The left two plots are for $k = 10$. The right two plots are for $k = 20$.

6   Conclusion

The problem discussed in this paper, simultaneously approximating one entire matrix in terms of a small number of selected columns from another matrix, is well-known, and appears to have many practical applications.

A novel two-stage selection algorithm, referred to as Spectral Pursuit for X and Y (SPXY), is presented, which selects columns capturing the spectral characteristics of the target matrix. What we found surprising is that it is possible to efficiently implement the algorithms with linear complexity w.r.t. the size of the two matrices. We show experimentally that our algorithm can outperform the state-of-the-art methods.

In addition to producing a solution, our algorithm produces a bound on how far the solution is from the optimum. The quality of the bound is data-dependent. In some cases (e.g., when X cannot explain Y well), it is loose, but in other cases, it shows that the computed result is very close to the optimal solution.

References

1. Arai, H., Maung, C., Schweitzer, H.: Optimal column subset selection by A-star search. In: AAAI 2015, pp. 1079–1085. AAAI Press (2015)
2. Arai, H., Maung, C., Xu, K., Schweitzer, H.: Unsupervised feature selection by heuristic search with provable bounds on suboptimality. In: AAAI 2016 (2016)
3. Belmerhnia, L., Djermoune, E.H., Brie, D.: Greedy methods for simultaneous sparse approximation. In: 22nd European Signal Processing Conference (2014)
4. Çivril, A., Magdon-Ismail, M.: Column subset selection via sparse approximation of SVD. Theoret. Comput. Sci. **421**, 1–14 (2012)

5. Chen, J., Huo, X.: Theoretical results of sparse representations of multiple measurement vectors. IEEE Trans. Signal Process. **54**(12), 4634–4643 (2006)
6. Cotter, S.F., Rao, B.D., Engen, K., Kreutz-Delgado, K.: Sparse solutions to linear inverse problems with multiple measurement vectors. ASP **53**(7), 2477–2488 (2005)
7. Daniel, J., Gragg, W., Kaufman, L., Stewart, G.: Reorthogonalization and stable algorithms for updating the Gram-Schmidt QR factorization. Math. Comput. **30**, 772–795 (1976)
8. Davis, G., Mallat, S., Avellaneda, M.: Adaptive greedy approximations. Constr. Approx. **13**(1), 57–98 (1997)
9. Furnival, G., Wilson, R.: Regressions by leaps and bounds. Technometrics **42**, 69–79 (1974)
10. Golub, G.H., Van-Loan, C.F.: Matrix Computations, 4th edn. The Johns Hopkins University Press, Baltimore (2013)
11. Guha, T., Ward, R.K.: Learning sparse representations for human action recognition. IEEE Trans. Pattern Anal. Mach. Intell. **34**, 1576–1588 (2011)
12. Halko, N., Martinsson, P., Tropp, J.: Finding structure with randomness: probabilistic algorithms for constructing approximate matrix decompositions. SIAM Rev. **53**, 217–288 (2011)
13. He, B., Shah, S., Maung, C., Arnold, G., Wan, G., Schweitzer, H.: Heuristic search algorithm for dimensionality reduction optimally combining feature selection and feature extraction. In: AAAI 2019, pp. 2280–2287. AAAI Press, California (2019)
14. Joneidi, M., et al.: Select to better learn: fast and accurate deep learning using data selection from nonlinear manifolds. In: CVPR, pp. 7819–7829 (2020)
15. Lerman, G., Maunu, T.: Fast, robust and non-convex subspace recovery. Inf. Inference J. IMA **7**(2), 277–336 (2018)
16. Mairal, J., Bach, F., Ponce, J.: Sparse Modeling for Image and Vision Processing. Foundations and Trends in Computer Graphics and Vision (2014)
17. Malioutov, D., Cetin, M., Willsky, A.: A sparse signal reconstruction perspective for source localization with sensor arrays. IEEE Trans. Signal Process. **53**, 3010–3022 (2005)
18. Mallat, S.: A Wavelet Tour of Signal Processing. Academic Press, Cambridge (1999)
19. Maung, C., Schweitzer, H.: Improved greedy algorithms for sparse approximation of a matrix in terms of another matrix. IEEE TKDE **27**(3), 769–780 (2015)
20. Natarajan, B.: Sparse approximate solutions to linear systems. SIAM J. Comput. **24**, 227–234 (1995)
21. Qian, C., Yu, Y., Zhou, Z.: Subset selection by pareto optimization. In: Advances in Neural Information Processing Systems. Curran Associates, Inc. (2015)
22. Soussen, C., Gribonval, R., Idier, J., Herzet, C.: Joint k-step analysis of orthogonal matching pursuit and orthogonal least squares. IEEE Trans. Inf. Theory **59**, 3158–3174 (2013)
23. Tibshirani, R.: Regression shrinkage and selection via the Lasso. J. R. Stat. Soc. B **58**, 267–288 (1996)
24. Tropp, J.A.: Greed is good: algorithmic results for sparse approximation. IEEE Trans. Inf. Theory **50**(10), 2231–2242 (2004)
25. Tropp, J.A., Gilbert, A.C., Strauss, M.J.: Algorithms for simultaneous sparse approximation. Part I: greedy pursuit. Signal Process. **86**(3), 572–588 (2006)
26. Wan, G., Schweitzer, H.: A new robust subspace recovery algorithm (student abstract). In: the 35th National Conference on Artificial Intelligence (AAAI) (2021)
27. Wright, J., Ma, Y., Mairal, J., Sapiro, G., Huang, T., Yan, S.: Sparse representation for computer vision and pattern recognition. Proc. IEEE **98**, 1031–1044 (2010)

28. Xu, X., Shi, Z.: Multi-objective based spectral unmixing for hyperspectral images. ISPRS J. Photogramm. Remote Sens. **124**, 54–69 (2017)
29. Xu, Y., Li, Z., Yang, J., Zhang, D.: A survey of dictionary learning algorithms for face recognition. IEEE Access **5**, 8502–8514 (2017)
30. Zaeemzadeh, A., Joneidi, M., Rahnavard, N., Shah, M.: Iterative projection and matching: finding structure-preserving representatives and its application to computer vision. In: CVPR 2019 (2019)
31. Zhang, Q., Liu, Y., Blum, R., Han, J., Tao, D.: Sparse representation based multi-sensor image fusion for multi-focus and multi-modality images: a review. Inf. Fusion **40**, 57–75 (2018)
32. Zhang, T.: Adaptive forward-backward greedy algorithm for sparse learning with linear models. In: Advances in Neural Information Processing Systems (2009)
33. Zhang, Z., Xu, Y., Yang, J., Li, X., Zhang, D.: A survey of sparse representation: algorithms and applications. IEEE Access **3**, 490–530 (2015)
34. Zhu, X., Hu, R., Lei, C., Thung, K.H., Zheng, W., Wang, C.: Low-rank hypergraph feature selection for multi-output regression. World Wide Web **22**(2), 517–531 (2017). https://doi.org/10.1007/s11280-017-0514-5

STEPs-RL: Speech-Text Entanglement for Phonetically Sound Representation Learning

Prakamya Mishra[(⊠)]

Mumbai, India

Abstract. In this paper, we present a novel multi-modal deep neural network architecture that uses speech and text entanglement for learning phonetically sound spoken-word representations. STEPs-RL is trained in a supervised manner to predict the phonetic sequence of a target spoken-word using its contextual spoken word's speech and text, such that the model encodes its meaningful latent representations. Unlike existing work, we have used text along with speech for auditory representation learning to capture semantical and syntactical information along with the acoustic and temporal information. The latent representations produced by our model were not only able to predict the target phonetic sequences with an accuracy of 89.47% but were also able to achieve competitive results to textual word representation models, Word2Vec & FastText (trained on textual transcripts), when evaluated on four widely used word similarity benchmark datasets. In addition, investigation of the generated vector space also demonstrated the capability of the proposed model to capture the phonetic structure of the spoken-words. To the best of our knowledge, none of the existing works use speech and text entanglement for learning spoken-word representation, which makes this work the first of its kind.

Keywords: Speech recognition · Spoken language processing · Representation learning

1 Introduction

Speaking and listening are the most common ways in which humans convey and understand each other in daily conversations. Nowadays, the speech interface has also been widely integrated into many applications/devices like Siri, Google Assistant, and Alexa [13]. These applications use speech recognition-based approaches [3,11] to understand the spoken user queries. Like speech, the text is also a widely used medium in which people converse. Recent advances in language modeling and representation learning using deep learning approaches [2,7,24] have proven to be very promising in understanding the actual meanings of the textual data, by capturing semantical, syntactical, and contextual relationships between the textual words in their corresponding learned fixed-size vector representations.

Such computational language modeling is difficult in the case of speech for spoken language understanding because unlike textual words, (1) spoken words can have different meanings of the same word when spoken in different tones/expressions [9], (2)

P. Mishra—Independent Researcher.

K. Karlapalem et al. (Eds.): PAKDD 2021, LNAI 12714, pp. 55–66, 2021.
https://doi.org/10.1007/978-3-030-75768-7_5

it is difficult to identify sub-word units in speech because of the variable-length spacing and overlapping between the spoke-words [34], and (3) use of stress/emphasis on few syllables of a multi-syllabic word can increase the variability of speech production [27]. Although the textual word representations capture the semantical, syntactical, and contextual properties, they fail to capture the tone/expression. Using only speech/audio data for training spoken-word representations results in semantically and syntactically poor representations.

So in this paper, we propose a novel spoken-word representation learning approach called STEPs-RL that uses speech and text entanglement for learning phonetically sound spoken-word representations, which not only captures the acoustic and contextual features but also are semantically, syntactically, and phonetically sound. STEPs-RL is trained in a supervised manner such that the learned representations can capture the phonetic structure of the spoken-words along with their inter-word semantic, syntactic, and contextual relationships. We validated the proposed model by (1) evaluating semantical and syntactical relationships between the learned spoken-word representations on four widely used word similarity benchmark datasets, and comparing its performance with the textual word representations learned by Word2Vec & FastTexT (obtained using transcriptions), and (2) investigating the phonetical soundness of the generated vector space.

The rest of the paper is organized as follows: Sect. 2 describes the related work; Sect. 3 explains the proposed model architecture; Sect. 4 will describe the datasets used, pre-processing pipeline, and training details for reproducibility. Then experimental results are explained in Sect. 5 and finally we conclude in Sect. 6.

2 Related Work

Earlier, speech processing was done using feature learning-based models like deep neural networks (DNN) [28]. The DNN models were able to capture contextual and temporal information from the speech-based data after the introduction of sequential neural networks like RNNs [16], LSTMs [25], Bi-LSTMs [10,36], and GRUs [29,33]. Recent research by [23] has presented the use of a transformer-based self-supervised speech representation learning approach called TERA that uses multi-target auxiliary tasks. TERA is trained by generating acoustic frame reconstructions; [30] introduced wav2vec which is a CNN based model pre-trained in a unsupervised manner using contrastive loss to learn raw audio representations; [20] explored the use of black-box variational inference for linguistic representation learning of speech using an unsupervised generative model; [26] proposed Contrastive Predictive Coding (CPC) for extracting representations from high dimension data by predicting future in latent space, using autoregressive models; [18] proposed a novel variational autoencoder based model that learns disentangled and interpretable latent representations of sequential data in an unsupervised manner; [22] used BERT encoder for learning phonetically aware contextual speech representation vectors; [4] proposed a Word2Vec type sequence-to-sequence autoencoder model for embedding variable-length audio segments. Other works on learning fixed-length spoken-word vector representations that use multi-task learning include [5,6,19,21,32].

3   Model

In this paper, we propose STEPs-RL: Speech-Text Entanglement for Phonetically Sound Representation Learning. STEPs-RL is a novel spoken-word representation learning approach which entangles speech and text based contextual information for learning phonetically sound spoken-word representations. The model architecture is shown in Fig. 1. Given a target spoken-word represented by S^t, its left and right contextual spoken-words represented by $S_{ctx}^l = \{S^i\}_{t-1-m}^{t-1}$ & $S_{ctx}^r = \{S^i\}_{t+1}^{t+1+m}$ respectively (m represents the context window size), along with the textual word embeddings of the corresponding spoken-words represented by $W_{ctx}^l = \{W^i\}_{t-1-m}^{t-1}$, W^t & $W_{ctx}^r = \{W^i\}_{t+1}^{t+1+m}$, the proposed model tries to learn a vector representation of the target spoken-word that not only captures the semantic-based, syntax-based and acoustic-based information but also captures the phonetic-based information.

Here, a single spoken-word $S^i \in \mathbb{R}^{n \times d_{mfcc}}$ consists of a sequence of acoustic features represented by d_{mfcc}-dimensional Mel-frequency Cepstral Coefficients (MFCCs); $W^i \in \mathbb{R}^{d_w}$ represents the d_w-dimensional pre-trained textual word embedding of the corresponding spoken-word. Each of the spoken-word is padded with silence, so that they all consists of a sequence of n acoustic features.

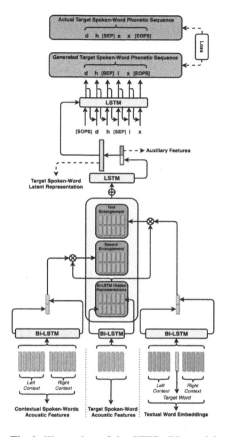

Fig. 1. Illustration of the STEPs-RL model architecture.

Our approach uses Bidirectional-LSTM [31] for capturing the contextual information. Bidirectional-LSTM (also known as Bi-LSTM), uses two LSTM [15] networks $(\overrightarrow{LSTM}, \overleftarrow{LSTM})$ to capture contextual information in opposite directions (forward and backward) of a sequence $(t_1, ...t_T)$. The final hidden representations corresponding to the sequence tokens is generated by concatenating ($\oplus$) the hidden representations $(\overrightarrow{h_i}, \overleftarrow{h_i})$ generated by both the LSTM networks. So the final hidden representation of the ith token can be represented as shown in Eq. 1.

$$\overrightarrow{h_i} = \overrightarrow{LSTM}(t_i, \overrightarrow{h_{i-1}}), \quad \overleftarrow{h_i} = \overleftarrow{LSTM}(t_i, \overleftarrow{h_{i+1}}), \quad h_i = \overrightarrow{h_i} \oplus \overleftarrow{h_i} \tag{1}$$

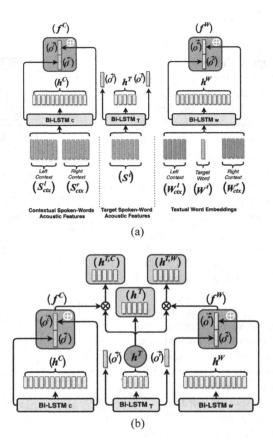

(a)

(b)

Fig. 2. (a) STEPs-RL Phase 1: Each of the individual Bi-LSTM captures contextual information. (b) STEPs-RL Phase 2: Speech & Text entanglement with target spoken word.

STEPs-RL consist of three independent Bi-LSTM networks represented by $BiLSTM_C$, $BiLSTM_T$ and $BiLSTM_W$ to capture contextual information respectively from (1) The acoustic features of the left and right contextual spoken-words represented by S^l_{ctx} & S^r_{ctx}, (2) The acoustic features of the target spoken-word represented by S^t, and (3) The pre-trained textual word embeddings of the corresponding target spoken-word, left contextual spoken-words and right contextual spoken-words represented by W^t, W^l_{ctx} & W^r_{ctx} respectively.

$$h^C, \overrightarrow{o^C}, \overleftarrow{o^C} = BiLSTM_C([S^l_{ctx}, S^r_{ctx}]) \tag{2}$$

$$h^T, \overrightarrow{o^T}, \overleftarrow{o^T} = BiLSTM_T([S^t]) \tag{3}$$

$$h^W, \overrightarrow{o^W}, \overleftarrow{o^W} = BiLSTM_W([W^l_{ctx}, W^t, W^r_{ctx}]) \tag{4}$$

As shown in Eqs. 2, 3, and 4, all the three Bi-LSTM networks generate a final hidden state representation corresponding to each timestamp (h^C, h^T, h^W), a final output

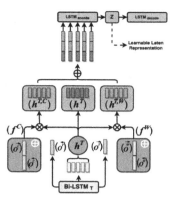

Fig. 3. STEPs-RL Phase 3: Latent representation learning

of the corresponding forward LSTM network ($\overrightarrow{o^C}$, $\overrightarrow{o^T}$, $\overrightarrow{o^W}$), and a final output of the corresponding backward LSTM network ($\overleftarrow{o^C}$, $\overleftarrow{o^T}$, $\overleftarrow{o^W}$). The final forward and backward outputs of $BiLSTM_C$ & $BiLSTM_W$ are concatenated to generate f^C & f^W respectively, which will later act as context vectors during the entanglement of speech and text.

$$f^C = \overrightarrow{o^C} \oplus \overleftarrow{o^C}, \quad f^W = \overrightarrow{o^W} \oplus \overleftarrow{o^W} \tag{5}$$

For intuition (as shown in Fig. 2a), f^C represents the final contextual representation of the spoken-words present in context of the target spoken-word, and f^W represents the final semantical and syntactical contextual representation of all the corresponding textual words. In other words, f^C captures the acoustic/speech-based contextual information whereas f^W captures the text-based contextual information. Both f^C & f^W, are then used to entangle speech and text-based contextual information with the target spoken-word by generating new speech and text entangled bidirectional hidden state representations ($h^{T,C}$ & $h^{T,W}$) of the target spoken-word using the hidden representations generated by $BiLSTM_T$, as shown in Eqs. 6 and 7.

$$h^{T,C} = [h_1^{T,C}, h_2^{T,C}, ..., h_n^{T,C}] = h^T \otimes f^C; \quad h_i^{T,C} = \alpha_i^{T,C} \times h_i^T \tag{6}$$

$$h^{T,W} = [h_1^{T,W}, h_2^{T,W}, ..., h_n^{T,W}] = h^T \otimes f^W; \quad h_i^{T,W} = \alpha_i^{T,W} \times h_i^T \tag{7}$$

In the above equations, ($\otimes$) represents an element wise attention function; $h^{T,C}$ & $h^{T,W}$ represents the newly generated speech-entangled and text-entangled hidden representations respectively; $\alpha_i^{T,C}$ & $\alpha_i^{T,W}$ represents the speech-entangled and text-entangled attention scores respectively, corresponding to the ith timestamp of the hidden representations generated by $BiLSTM_T$. The attention scores $\alpha_i^{T,C}$ & $\alpha_i^{T,W}$ are generated by taking the dot product ($\bullet$) of each of the timestamps of h^T with the context vectors f^C & f^W respectively, as shown in Eq. 8. Same is illustrated in Fig. 2b.

$$\alpha_i^{T,C} = h_i^T \bullet f^C, \quad \alpha_i^{T,W} = h_i^T \bullet f^W \tag{8}$$

Next, the proposed model uses the newly generated speech-entangled and text-entangled hidden representations $h^{T,C}$ & $h^{T,W}$, along with the original bidirectional hidden state representations h^T of the target spoken-word (generated from $BiLSTM_T$), to generate a latent vector representation z of the target spoken-word by stacking (illustrated in Fig. 3) all these three hidden representations on top of each other and passing it through a simple encoder LSTM network $\overrightarrow{LSTM_{encode}}$.

$$z = \overrightarrow{LSTM_{encode}}([h^{T,C} \oplus h^{T,W} \oplus h^T]), \quad z_{new} = zW_1 + z_{aux}W_2 + B \quad (9)$$

In Eq. 9, z represents a fixed size latent vector which is the output of the encoder LSTM network. To add more information about the speaker, the proposed model linearly combines the latent vector with an auxiliary vector z_{aux} to generate a new latent representation z_{new} of the target spoken-word. This new latent representation $z_{new} \in \mathbb{R}^{d_e}$ is a d_e-dimensional vector representation that the proposed model tries to learn. In Eq. 9, $W_1 \in \mathbb{R}^{d \times d_e}$ and $W_2 \in \mathbb{R}^{d_a \times d_e}$ represents the combination weights and B represents the bias. These weights and biases are learnable in nature. The auxiliary vector $z_{aux} \in \mathbb{R}^{d_a}$ is a one-hot vector of size d_a that consists of information related to the speaker's gender/dialect or both. Such an auxiliary vector was introduced because usually, the pronunciation of different words usually depends on the speaker's gender and dialect and hence can help learn phonetically sound spoken-word representations.

Next, the proposed model uses a decoder LSTM network $\overrightarrow{LSTM_{decode}}$ to predict the sequence of phonetic symbols $Y = ([y_1, ..., y_k])$ of the corresponding target spoken-word using the above generated latent representation of the target spoken-word z_{new}, as shown in Eq. 10 and 11.

$$P_\theta(y_i | Y_{<i}, z_{new}) = \Upsilon(h_i^d, y_{i-1}) \quad (10)$$

$$h_i^d = \Psi(h_{i-1}^d, y_{i-1}) \quad (11)$$

Here, Ψ represents a function that generates the hidden vectors h_i^d (hidden state representations of the decoder network), and Υ represents a function that computes the generative probability of the one-hot vector y_i (target phonemic symbol). The hidden vector h_i^d is z_{new}, and y_i is the one-hot vector of "[SOP]" when $i = 0$. Here "[SOP]" represent the start of phoneme token. The proposed model uses cross-entropy as its training loss function as shown in Eq. 12, where cross-entropy loss L is computed using the actual target spoken-word phonetic sequence ($Y = ([y_1, ..., y_k])$) and the predicted target spoken-word phonetic sequence ($\hat{Y} = ([\hat{y}_1, ..., \hat{y}_k])$).

$$L(Y, \hat{Y}) = \sum_i^k y_i \log \frac{1}{\hat{y}_i} \quad (12)$$

4   Dataset and Experimental Setup

For our experiments, we used the DARPA TIMIT acoustic-phonetic Continuous Speech Corpus [8]. This corpus contains 16 kHz audio recordings of 630 speakers of 8 major American English dialects of which approximately 70% were male and 30% were

Table 1. Gender and dialect distribution of the speakers in TIMIT speech corpus.

	Dialect							
Gender	1	2	3	4	5	6	7	8
Male	63%	70%	67%	69%	63%	65%	74%	67%
Female	27%	30%	27%	31%	37%	35%	26%	33%
Total	**8%**	**16%**	**16%**	**16%**	**16%**	**7%**	**16%**	**5%**

female, as shown in Table 1. The corpus consists of 6300 (5.4 h) phonetically rich utterances by different speakers (10 by each speaker) along with their corresponding time-aligned orthographic, phonetic, and word transcriptions.

All the recordings were segmented according to the spoken-word boundaries using the transcriptions and were paired with their left and right context spoken-words and their corresponding textual words along with the phonetic sequence of the target spoken-word. All the spoken-word utterances were represented by their MFCC representations and the textual words were represented by their pre-trained textual word embeddings, where the MFCC representations and the textual word embeddings were of the same size ($d_{mfcc} = d_w$). One-hot encoded dialect (8-dimensional) and gender (2-dimensional) vectors were used as auxiliary information vectors. We used the standard train (462 speakers and 4956 utterances) and test (168 speakers and 1344 utterances) set of the TIMIT speech corpus for training and testing the proposed model. Due to computational resource limitations, a context window size of 3 was used. In all the experiments the MFCC representations and the textual word embeddings were of the same size ($d_{mfcc} = d_w \in \{50, 100, 300\}$). For the textual word embeddings, the proposed model used two different widely used pre-trained word embeddings i.e., (1) Word2Vec [24], which are word-based embeddings, and (2) FastText [2], which are character-based embeddings. For all the experiments, the proposed model was trained for 20 epochs using a mini-batch size of 100. The initial learning rate was set to 0.01 and Adam optimizer was used for optimization. The Bi-LSTM and LSTM nodes were regularised using an L2 regularizer with a penalty of 0.01. Early stopping was used to avoid over-fitting. The size of the target spoken-word latent representation z_{new} was set to 50-, 100- & 300 for comparison. All the spoken-words were represented by a sequence of 50 phonetic symbols using the original unique 27 phonetic symbols present in the corpus along with our four newly introduced symbols ("**[SOPS]**" for the start of each phonetic sequence, "**[SEP]**" for separation/space between phonetic symbols, "**[PAD]**" for padding and "**[EOPS]**" for the end of each phonetic sequence).

5   Results

For evaluation, we first tested the proposed model on the phonetic sequence prediction task with different spoken-word latent representation & textual word embedding sizes, and also tested the performance of the model using different types of textual word embeddings (Word2Vec & FastText). We compared the phonetic sequence prediction accuracy (%) of the base STEPs-RL model (w/o any auxiliary information)

Table 2. Phonetic sequence prediction results on the TIMIT speech corpus. We present here the comparison of testing set accuracy (%) of the STEPs-RL model using different sets of auxiliary information (gender (**G**), dialect (**D**)) with the base STEPs-RL model using no auxiliary information. The comparison is done for different textual word embedding sizes $d_w = \{50, 100, 300\}$, different spoken-word latent representation sizes $d_e = \{50, 100, 300\}$ and different word embeddings like Word2Vec (*w*) and FastText (*f*). The best performance in each configuration is marked in **bold**, row of the best performing model is highlighted in grey, the overall best performance is further marked in red and its configuration is marked in **blue**.

Spoken-Word Latent Rep. Size (d_e) →	$d_e = 50$						$d_e = 100$						$d_e = 300$					
Textual Word Embedding Size (d_w) →	$d_w = 50$		$d_w = 100$		$d_w = 300$		$d_w = 50$		$d_w = 100$		$d_w = 300$		$d_w = 50$		$d_w = 100$		$d_w = 300$	
Textual Word Embeddings Used →	*w*	*f*	*w*	*f*	*w*	*f*	*w*	*f*	*w*	*f*	*w*	*f*	*w*	*f*	*w*	*f*	*w*	*f*
STEPs-RL + No auxiliary information	71.67	73.24	75.22	78.41	84.76	84.98	73.44	76.72	78.36	81.23	86.90	86.92	75.22	79.75	80.59	83.73	87.31	86.90
STEPs-RL + D	83.87	84.01	86.11	86.36	85.89	87.89	86.83	86.85	86.75	87.54	87.23	88.00	81.50	82.05	85.90	86.36	87.40	87.99
STEPs-RL + G	87.93	86.97	87.44	85.60	87.16	88.28	87.15	87.12	87.98	**88.20**	**88.92**	**88.54**	87.16	87.30	88.91	88.45	88.10	88.59
STEPs-RL + D + G	**88.91**	**88.04**	**88.10**	**88.28**	**88.94**	**88.78**	**88.27**	**88.90**	**89.14**	88.08	88.59	89.18	89.47	**89.21**	**89.35**	**89.38**	**88.63**	**89.41**

with its variants that use different sets of auxiliary information like gender/dialect or both. The results are shown in Table 2. It was observed that increasing the spoken-word representation size resulted in better performance but was not so evident in the case of textual word embedding size. It was also observed that in general using Word2Vec textual word embeddings achieved better results compared to using FastText textual word embeddings. The addition of auxiliary information like dialect and gender showed clear improvements in accuracy when compared to the base STEPs-RL model, validating the use of this type of auxiliary information for spoken-word representation learning. It was also found that STEPs-RL was able to perform best when it used both dialect (**D**) and gender (**G**) together in its auxiliary vector (**STEPs-RL+D+G**). So for further evaluations, we will only consider the target spoken-word representations generated from the STEPs-RL+D+G model using the configurations marked blue in Table 2. Table 3a illustrates examples of four different spoke-words along with their actual corresponding phonetic sequences and the phonetic sequences predicted by the STEPs-RL+D+G model. These examples demonstrate the ability of the STEPs-RL+D+G model to encode phonetic-based information in their corresponding latent representations.

To further evaluate the latent representations generated from STEPs-RL+D+G, we use intrinsic methods to test the semantic or syntactic relationships between these generated latent representations of the spoken-words present in the corpus. To do so, we use 4 benchmark word similarity datasets and compare the performance of the spoken-word representations generated from STEPs-RL+D+G with the representations generated by text-based language models (Word2Vec & FastText) trained on the textual transcripts. The word similarity datasets include SimeLex-999 [14], MTurk-771 [12], WS-353 [35] and Verb-143 [1]. These datasets contain pairs of English words and their corresponding human-annotated word similarity ratings. The word similarities between the spoken-words (in case of STEPs-RL+D+G) and the textual-words (in case of Word2Vec and FastText) were obtained by measuring the cosine similarities between their corresponding representation vectors.

Table 3. (a) Examples of the phonetic sequences generated by STEPs-RL+D+G model. (b) Performance of STEPs-RL+D+G compared to Word2Vec & FastText on four benchmark word similarity datasets.

<table>
<tr><th colspan="3" align="center">(a)</th><th colspan="5" align="center">(b)</th></tr>
<tr><th>Ground Truth Word</th><th>Ground Truth Phonetic Sequence</th><th>Generated Phonetic Sequence</th><th>Dataset</th><th># Word Pairs</th><th>STEPs + D + G (ρ)</th><th>Word2Vec (ρ)</th><th>FastText (ρ)</th></tr>
<tr><td>that</td><td>dh ae tcl</td><td>dh ax tc l</td><td>SimLex</td><td>136</td><td>0.2552</td><td>0.2792</td><td>0.3375</td></tr>
<tr><td>shelter</td><td>s sh eh l tcl t axr</td><td>s tsh ah el t er</td><td>MTurk</td><td>55</td><td>0.5312</td><td>0.6536</td><td>0.6771</td></tr>
<tr><td>near</td><td>n ih axr</td><td>n ehh axr</td><td>WS</td><td>17</td><td>0.2726</td><td>0.3248</td><td>0.3082</td></tr>
<tr><td>heck</td><td>hv eh kcl k</td><td>hv ah ncl k</td><td>Verb</td><td>26</td><td>0.3260</td><td>0.3508</td><td>0.3676</td></tr>
</table>

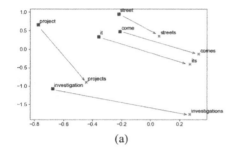
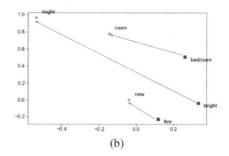

(a)   (b)

Fig. 4. Difference vectors corresponding to (a) Set 1: Word pairs differ in last few phonemes (b) Set 2: Word pairs differ in first few phonemes.

Table 3b reports Spearman's rank correlation coefficient ρ between the human rankings and the ones generated by STEPs-RL+D+G, Word2Vec, and FastText. Since there were many words present in these datasets which were not present in the TIMIT speech corpus, only those word pairs were considered in which both the word were present in the TIMIT speech corpus. Table 3b shows that the performance of the spoken-word representations generated from STEPs-RL+D+G was comparable to the performance of textual word representations generated from Word2Vec and FastText. This demonstrates that our proposed model was also able to capture semantic-based and syntax-based information, although the scores were slightly less compared to Word2Vec and FastText. We believe that the primary reason for this difference is the disparity in the way different speakers speak. The same word can be spoken in different ways and can have different meanings based on the tone and expression which may in return lead to an entirely different representation for the same word. In addition to that, these word similarity datasets are for the textual words, which do not take into account the tone and the expression aspect. Also, to the best of our knowledge, no other such word similarity dataset exists for the spoken-words. So keeping in mind these issues, the performance of the proposed model validates its ability to capture semantical and syntactical information in the representations it generates.

Next, we try to investigate the phonetical soundness of the vector space generated by the proposed model. A vector space can be said to be phonetically sound if the spoken-word representations of the words having similar pronunciations are present close to each other in the vector space. For this investigation we use 2 sets of randomly chosen word pairs:

- **Set 1:** (street, streets), (come, comes), (it, its), (project, projects), (investigation, investigations)
- **Set 2:** (few, new), (bright, night), (bedroom, room)

Here, in Set 1 the word pairs differ in the last few phonemes and in Set 2 the word pairs differ in the first few phonemes. To illustrate the relationship between these word pairs, first, the difference vectors were computed between the average spoken-word vector representation of the words present in the above-mentioned word pairs, and then these high dimensional difference vectors were reduced to 2-dimensional vectors using PCA [17], to interpret these vectors. The difference vectors corresponding to Set 1 & Set 2 are shown in Fig. 4. It can be observed in the figures that the difference vectors are similar in directions and magnitude. In both the figures, phonetic replacements lead to similar transformations, for example (come → comes) is similar to (it → its) in Fig. 4a, and (few → new) is similar to (bright → night) in Fig. 4b. These transformations are not perfectly similar because we are taking an average of the same word spoken by different speakers having different accents and pronunciations, but despite this, the transformations are still very close to each other. All these experiments demonstrate the quality of spoken-word vector representations generated by the proposed model using speech and text entanglement which not only are semantically and syntactically adequate but are also phonetically sound.

6   Conclusion

In this paper, we introduced STEPs-RL for learning phonetically sound spoken-word representations using speech and text entanglement. Our approach achieved an accuracy of 89.47% in predicting phonetic sequences when both gender and dialect of the speaker are used in the auxiliary information. We also compared its performance using different configurations and observed that the performance of the proposed model improved by (1) increasing the spoken word latent representation size, and (2) the addition of auxiliary information like gender and dialect. We were not only able to validate the capability of the learned representations to capture the semantical and syntactical relationships between the spoken-words but were also able to illustrate soundness in the phonetic structure of the generated vector space.

References

1. Baker, S., Reichart, R., Korhonen, A.: An unsupervised model for instance level subcategorization acquisition. In: Proceedings of the 2014 Conference on Empirical Methods in Natural Language Processing (EMNLP), pp. 278–289 (2014)
2. Bojanowski, P., Grave, E., Joulin, A., Mikolov, T.: Enriching word vectors with subword information. Trans. Assoc. Comput. Linguist. **5**, 135–146 (2017)
3. Bourlard, H.A., Morgan, N.: Connectionist Speech Recognition: A Hybrid Approach, vol. 247. Springer, New York (2012). https://doi.org/10.1007/978-1-4615-3210-1
4. Chen, Y., Huang, S., Lee, H., Wang, Y., Shen, C.: Audio Word2Vec: sequence-to-sequence autoencoding for unsupervised learning of audio segmentation and representation. IEEE/ACM Trans. Audio Speech Lang. Process. **27**(9), 1481–1493 (2019)

5. Chorowski, J., Weiss, R.J., Bengio, S., van den Oord, A.: Unsupervised speech representation learning using WaveNet autoencoders. IEEE/ACM Trans. Audio Speech Lang. Proc. **27**(12), 2041–2053 (2019). https://doi.org/10.1109/TASLP.2019.2938863

6. Cui, J., et al.: Multilingual representations for low resource speech recognition and keyword search. In: 2015 IEEE Workshop on Automatic Speech Recognition and Understanding (ASRU), pp. 259–266 (2015)

7. Devlin, J., Chang, M.W., Lee, K., Toutanova, K.: BERT: Pre-training of deep bidirectional transformers for language understanding. In: Proceedings of the 2019 Conference of the North American Chapter of the Association for Computational Linguistics: Human Language Technologies, vol. 1 (Long and Short Papers), Minneapolis, Minnesota, pp. 4171–4186. Association for Computational Linguistics, June 2019. https://doi.org/10.18653/v1/N19-1423, https://www.aclweb.org/anthology/N19-1423

8. Garofolo, J.S., Lamel, L.F., Fisher, W.M., Fiscus, J.G., Pallett, D.S., Dahlgren, N.L.: DARPA TIMIT acoustic-phonetic continuous speech corpus CD-ROM {TIMIT} (1993)

9. Glass, J.: Challenges for spoken dialogue systems. In: Proceedings of the 1999 IEEE ASRU Workshop, vol. 696 (1999)

10. Graves, A., Jaitly, N., Mohamed, A.r.: Hybrid speech recognition with deep bidirectional LSTM. In: 2013 IEEE Workshop on Automatic Speech Recognition and Understanding, pp. 273–278. IEEE (2013)

11. Graves, A., Mohamed, A.r., Hinton, G.: Speech recognition with deep recurrent neural networks. In: 2013 IEEE International Conference on Acoustics, Speech and Signal Processing, pp. 6645–6649. IEEE (2013)

12. Halawi, G., Dror, G., Gabrilovich, E., Koren, Y.: Large-scale learning of word relatedness with constraints. In: Proceedings of the 18th ACM SIGKDD International Conference on Knowledge Discovery and Data Mining, pp. 1406–1414 (2012)

13. Herff, C., Schultz, T.: Automatic speech recognition from neural signals: a focused review. Front. Neurosci. **10**, 429 (2016)

14. Hill, F., Reichart, R., Korhonen, A.: SimLex-999: evaluating semantic models with (genuine) similarity estimation. Comput. Linguist. **41**(4), 665–695 (2015)

15. Hochreiter, S., Schmidhuber, J.: Long short-term memory. Neural Comput. **9**(8), 1735–1780 (1997). https://doi.org/10.1162/neco.1997.9.8.1735

16. Hori, T., Cho, J., Watanabe, S.: End-to-end speech recognition with word-based RNN language models. In: 2018 IEEE Spoken Language Technology Workshop (SLT), pp. 389–396. IEEE (2018)

17. Hotelling, H.: Analysis of a complex of statistical variables into principal components. J. Educ. Psychol. **24**(6), 417 (1933)

18. Hsu, W.N., Zhang, Y., Glass, J.: Unsupervised learning of disentangled and interpretable representations from sequential data. In: Advances in Neural Information Processing Systems (2017)

19. Kamper, H.: Truly unsupervised acoustic word embeddings using weak top-down constraints in encoder-decoder models. In: ICASSP 2019–2019 IEEE International Conference on Acoustics, Speech and Signal Processing (ICASSP), pp. 6535–3539 (2019)

20. Khurana, S., et al.: A convolutional deep Markov model for unsupervised speech representation learning (2020)

21. Li, X., Wu, X.: Modeling speaker variability using long short-term memory networks for speech recognition. In: INTERSPEECH (2015)

22. Ling, S., Salazar, J., Liu, Y., Kirchhoff, K.: BERTphone: phonetically-aware encoder representations for utterance-level speaker and language recognition. In: Proceedings of Odyssey 2020 The Speaker and Language Recognition Workshop, pp. 9–16 (2020). https://doi.org/10.21437/Odyssey.2020-2

23. Liu, A.T., Li, S.W., Yi Lee, H.: TERA: self-supervised learning of transformer encoder representation for speech (2020)
24. Mikolov, T., Sutskever, I., Chen, K., Corrado, G.S., Dean, J.: Distributed representations of words and phrases and their compositionality. In: Advances in Neural Information Processing Systems, pp. 3111–3119 (2013)
25. Moriya, Y., Jones, G.J.: LSTM language model adaptation with images and titles for multimedia automatic speech recognition. In: 2018 IEEE Spoken Language Technology Workshop (SLT), pp. 219–226. IEEE (2018)
26. van den Oord, A., Li, Y., Vinyals, O.: Representation learning with contrastive predictive coding. CoRR abs/1807.03748 (2018). http://arxiv.org/abs/1807.03748
27. Polka, L., Orena, A.J., Sundara, M., Worrall, J.: Segmenting words from fluent speech during infancy–challenges and opportunities in a bilingual context. Dev. Sci. **20**(1), e12419 (2017)
28. Purwins, H., Li, B., Virtanen, T., Schlüter, J., Chang, S., Sainath, T.: Deep learning for audio signal processing. IEEE J. Sel. Top. Sign. Process. **13**(2), 206–219 (2019)
29. Ravanelli, M., Brakel, P., Omologo, M., Bengio, Y.: Light gated recurrent units for speech recognition. IEEE Trans. Emerg. Top. Comput. Intell. **2**(2), 92–102 (2018)
30. Schneider, S., Baevski, A., Collobert, R., Auli, M.: wav2vec: unsupervised pre-training for speech recognition. In: Proceedings of Interspeech 2019, pp. 3465–3469 (2019). https://doi.org/10.21437/Interspeech.2019-1873
31. Schuster, M., Paliwal, K.K.: Bidirectional recurrent neural networks. IEEE Trans. Sign. Process. **45**(11), 2673–2681 (1997)
32. Tan, T., et al.: Speaker-aware training of LSTM-RNNs for acoustic modelling. In: 2016 IEEE International Conference on Acoustics, Speech and Signal Processing (ICASSP), pp. 5280–5284 (2016)
33. Tang, Z., Shi, Y., Wang, D., Feng, Y., Zhang, S.: Memory visualization for gated recurrent neural networks in speech recognition. In: 2017 IEEE International Conference on Acoustics, Speech and Signal Processing (ICASSP), pp. 2736–2740. IEEE (2017)
34. Vincent, E., Barker, J., Watanabe, S., Le Roux, J., Nesta, F., Matassoni, M.: The second 'CHiME' speech separation and recognition challenge: an overview of challenge systems and outcomes. In: 2013 IEEE Workshop on Automatic Speech Recognition and Understanding, pp. 162–167. IEEE (2013)
35. Yang, D., Powers, D.M.: Verb similarity on the taxonomy of WordNet. Masaryk University (2006)
36. Zeyer, A., Doetsch, P., Voigtlaender, P., Schlüter, R., Ney, H.: A comprehensive study of deep bidirectional LSTM RNNs for acoustic modeling in speech recognition. In: 2017 IEEE International Conference on Acoustics, Speech and Signal Processing (ICASSP), pp. 2462–2466. IEEE (2017)

RW-GCN: Training Graph Convolution Networks with Biased Random Walk for Semi-supervised Classification

Yinzhe Li and Zhijie Ban[✉]

College of Computer Science, Inner Mongolia University, Hohhot, China
banzhijie@imu.edu.cn

Abstract. Graph convolution networks (GCN) have recently been one of the most powerful methods in various tasks such as node classification and graph clustering. In the present study, we propose RW-GCN which utilizes biased random walk to assist in feature aggregation and GCN training process. RW-GCN employs biased random walks to generate node pairs. These pairs can be utilized to build a symmetric matrix to replace the adjacent matrix for GCN training. With these pairs generated above, we train the latent representation vectors by skip-gram. Our experiments demonstrate that compared to GCN, our model generates better results on node classification tasks performed on multiple datasets. In this way, both homophily and structural equivalence can be considered. Results of experiments on three datasets are presented to prove the availability of our method.

Keywords: Node classification · Representation learning · Graph learning

1 Introduction

Graph structure, as a suitable carrier of entities and relations, is extensively used in modeling real-world problems. In the field, graph node classification is the main task in graph learning, which uses the information related to the entities to make class predictions for the entities in the network [2]. In recent years, a series of unsupervised learning methods have been proposed. They train node representation vectors to learn abstract features of nodes and then use them for downstream tasks. For example, DeepWalk [13], which conducts a random walk in a graph, then adopts the method of training word embeddings [10, 11] to learn the node embeddings from a random walk sequence. Subsequently, the representation learning technique incorporating the neural network method has received extensive attention. In this kind of method, the features of the entities are transmitted and aggregated along the edges, the error is calculated based on a small number of training samples, and the model training and parameter updating are carried out with the training method of the neural network. For instance, compared with feedforward neural network, the trainable weight matrix is also applied for feature transformation layer by layer in GCN. However, the difference is, to utilize the relationship between entities, a relation matrix is used for feature aggregation of adjacent nodes. In this way, the information of network structure is merged with the learning process of node features.

© Springer Nature Switzerland AG 2021
K. Karlapalem et al. (Eds.): PAKDD 2021, LNAI 12714, pp. 67–76, 2021.
https://doi.org/10.1007/978-3-030-75768-7_6

Nevertheless, we find that not all nodes in the graph can benefit from this pattern of feature aggregation. Merging the neighborhood's attribute features into a central node may improve the representation ability of the features in the hidden layer. Moreover, it may also lead to confusion, as shown in Fig. 1(b). In such case, it is possible that merely using the structural information of the graph will achieve better classification results. Therefore, reasonable speculation is that we need to strike a balance between the aggregation of features and the appropriate representation of features in latent space, aiming to further strengthen the comprehensive utilization of the two types of information.

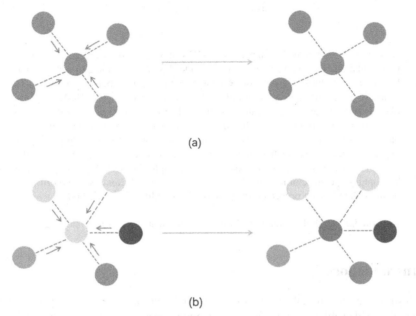

(a)

(b)

Fig. 1. The confusion caused by the characteristics of average aggregate neighborhood nodes. The nodes of each color in the figure represent a different class, and for the case shown in (a), the average feature aggregation may be beneficial for the classification of the central nodes. For the case shown in (b), the average aggregation feature may cause the central node's feature to lose focus, that is, after a round of feature aggregation, the new feature of the central node may mislead its correct classification.

In the current work, we propose RW-GCN, a new graph convolutional neural network model, which integrates the unsupervised graph representation learning into the graph convolutional neural network training process in order to assist the semi-supervised node classification task. In our model, a random walk sequence is obtained from each node in the graph, and then node pairs are generated from the walk sequence, just like the method used in [5] and [13]. We employed the bias random walk proposed in [5] rather than the standard random walk to take advantage of the two properties of homophily and structural equivalence [5]. When input features in the hidden layer are transformed into abstract features, they are passed to the output layer to perform node classification task. Additionally, based on skip-gram [11], we perform context nodes prediction task with these node pairs. Then, the two loss functions are combined, and the parameters of the hidden layer are updated until the model converges. Our experiments indicate that the method proposed in the present study achieves a good balance between feature aggregation and feature representation, thus alleviating the confusion caused by simple feature aggregation in GCN.

In summary, our contributions are as follows:

1. We propose RW-GCN, a new method of graph convolutional neural network, which combines the unsupervised node representation learning and the training process of graph convolutional network to assist the semi-supervised node classification task.
2. We conduct experiments and compare results with other semi-supervised node classification methods on three academic network datasets to demonstrate the effectiveness of our method.

2   Related Work

Graph Representation Learning. Representation learning, also known as feature learning, is a technique that concentrates on abstracting entities' attributes. It automatically learns features from input data that are conducive to machine learning applications, consequently avoiding the complexity and redundancy of manual design features [7]. For example, word embedding in natural language processing is a technology that bases on a neural network to learn vocabulary representation vectors. Graph representation learning is the application of data representing learning in the graph learning field, which aims to learn the feature representations of nodes, edges, communities, and subgraphs in the graph, so as to provide input data carriers with better properties for downstream tasks. In this field, the main methods are as follows: methods based on matrix decomposition, Which use matrix decomposition technology to learn the node representation vector [13, 14] with the aim to approximatethe similarity between nodes. [5, 13] introduce a shallow unsupervised graph representation learning method, which adopts neural network to predict neighborhood nodes and updates the hidden layer weight of the neural network to obtain the representation vector of nodes. VGAE [8] uses Encoder-Decoder architecture to learn low-dimensional latent representation vectors of nodes by reducing reconstruction losses. Compared with the manual design features, the entity representation features obtained based on representation learning can generate better effects in various downstream tasks.

Node Classification. When the relationship between nodes is not considered, the task of node classification is the same as other classification tasks. That is, the classifier is trained according to features and labels of samples and then applied to other samples. In the real world, various types of sample individuals are not completely independent of each other, such as social networks, where two users who follow each other are obviously more likely to have similar attributes or types than two users who are not related to each other [3]. In the citation network, papers that have a citation relationship with each other are more likely to belong to the same research field. Consequently, the ability to comprehensively utilize the attributes of samples and the relationships between samples will be of great significance to the classification task.

Graph Neural Networks. The study of graph neural networks is closely related to graph embedding [3, 4]. Traditional deep learning methods are not well suited for processing non-Euclidean data, prompting researchers to extend deep learning methods on the graph. Based on the ideas of convolutional neural network, recurrent neural network, and depth automatic encoder, the researchers defined and designed GNN, a neural network structure specially used for processing graph data. In the field of GNN, other successful neural network architectures are widely used for reference [6, 9, 16]. These techniques have enabled the graph neural network to obtain remarkable development in recent years.

3 Problem Definition

Firstly, we introduce notations that are used in this paper. Formally, let $G = (V, E)$ represent a graph, where V signifies the vertices of the graph, and E denotes the edges between vertices. In the present study, we discuss undirected graphs; that is, the edge from V_i to V_j is equivalent to the edge from V_j to V_i. For some types of graphs, attribute features are attached to nodes. We denote the attribute features by $X \in \mathbb{R}^{|V| \times n}$, where n is the size of the feature space for attribute vectors. In addition, we also denote the labels of nodes by $Y \in \mathbb{R}^{|V| \times |y|}$, where y is the set of labels.

In this paper, we focus on semi-supervised classification. In this task, a large amount of unlabeled data is added to a small amount of labeled data to train together for parameter learning. For labeled data, both the structure of graph and the features of nodes can be exploited, while for unlabeled data, only the structure of graph is available. The goal of RW-GCN is to train the model and thus it can classify the nodes that are not used in the training set.

4 Method

4.1 Overview

Figure 2 illustrates the proposed model RW-GCN. At the beginning of the model, we have a graph $G = (V, E)$ and an attributes matrix X. In graph convolution network these two components are used to train hidden layers' parameters, while in RW-GCN we begin from doing random walks with the nodes in graph G to generate sequences. Specifically, given two parameters N and L, for each node in graph G, N paths of length L are generated. According to the given size of sliding window, we extract the node pairs used for context prediction from the sequences. These node pairs will be used after GCN encodes the input features to abstract features. However, before that, a sparse matrix M is created and initialized with zeros. For each node pair (v_i, v_j), where $i, j \leq |V|$, it increases the value of $M[i, j]$ and $M[j, i]$ by 1. When all the node pairs are counted the matrix M is going to be normalized to replace the Laplacian matrix which is used to guide feature aggregation. In hidden layers, GCN encoder maps input features to low-dimensional representations. In addition, these representations are passed into output layer and Softmax function to perform the node classification task. Moreover, a context nodes prediction task like DeepWalk or node2vec is performed with the extracted node pairs. In the rest of the current section, we will introduce the details of our method.

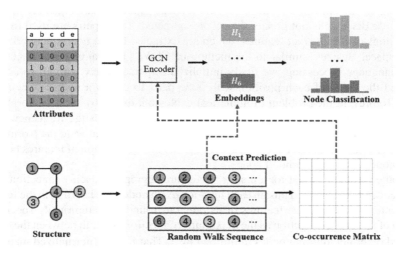

Fig. 2. The framework of RW-GCN. The blue dashed line represents the training of the graph convolutional neural network with the generated matrix, while the red dotted line represents the task of extracting node pairs from the random walk sequence and then making context node prediction. (Color figure online)

4.2 Unsupervised Node Feature Learning

Inspired by word embedding in natural language processing, [13] proposed DeepWalk, a method for learning node embedding based on random walk. For each node u observed in the sequence, DeepWalk optimizes the following objective function:

$$\max_f \sum_{u \in V} \log Pr(N_S(u)|f(u)) \qquad (1)$$

where $f(u)$ represents a representation of node u, S denotes the size of the sliding window to determine the neighborhood and $N_S(u)$ signifies neighborhood of node u. In this way, DeepWalk embeds nodes into the abstract feature space. Therefore, node embeddings have property like word embeddings, that is, nodes in similar contexts have closer representations.

Compared with the simple random walk of DeepWalk, the node2vec proposed in [5] introduces two parameters to determine whether the random walk tends to search locally or to visit nodes deeper in the network, aiming to control the generation of the random walk sequences. Consequently, the two properties of homophily and structural equivalence are used. Homophily emphasizes that nodes close to each other in the network should be embedded in closer positions, while structural equivalence emphasizes that nodes with the same structure roles in the network should be embedded in closer positions.

In our method, we employee this biased random walk to obtain node sequences. However, we do not aim to perform unsupervised representation learning immediately. According to the specified window size S, for each node in the sequence, the node pairs whose distance is not greater than S are extracted. These pairs are used to adjust the positions of the abstract features, which are extracted by the neural network in the feature space. Besides, similar to the method of GloVe [12] training word vectors in natural language processing, we firstly initialized a sparse matrix with all values of 0 and filled the value of each position with node pairs to obtain a node co-occurrence matrix. To prevent the problem of numerical explosion, the matrix will be normalized and thus its value shrinks to between [0, 1]. This matrix records the frequency of co-occurrence between nodes. Therefore, it can be used as an alternative to the Normalized Laplacian Matrix in standard GCN in order to guide the aggregation of features between central nodes and neighborhood nodes.

Another problem is that for a matrix for feature aggregation, self-connection of the central nodes is necessary, Thus, in each iteration, each node can absorb both the features of its context nodes and the features of its previous round. We empirically specify the strength of this self-connection as the maximum connection strength between the central node and other neighboring nodes multiplied by 2. That is, for a normalized matrix $\widehat{M}$, we have:

$$\widehat{M}[i, i] = max\left(\widehat{M}[i]\right) * 2 \tag{2}$$

4.3   Graph Convolution Network

Graph Convolution Network is one of the spectral approaches which works with a spectral representation of the graphs. Given a graph $G = (V, E)$ and the corresponding attribute features X, the architectures of GCN basically have the following form:

$$h^{(l+1)} = \sigma\left(\widehat{A}h^{(l)}W^{(l)}\right) \tag{3}$$

where $\widehat{A}$ is a matrix that carries the structure information of the graph and $W^{(l)}$ is trainable parameters matrix in the l-th hidden layer. Compared with standard feedforward neural network, GCNs not only perform feature transformation, but also aggregate features from the nodes' neighborhoods specified by $\widehat{A}$. For example, if $\widehat{A}$ is the adjacency matrix of graph, every node in every layer aggregates its 1-hop adjacent nodes' features. Nevertheless, in this way, it loses its own features that passed from the previous layer. GCN adopts the following equation:

$$h^{(l+1)} = \sigma\left(\left(I_V + D^{-\frac{1}{2}}AD^{-\frac{1}{2}}\right)h^{(l)}W^{(l)}\right) \tag{4}$$

to update nodes' features where Symmetric Normalized Laplacian is used to replace $\widehat{A}$. In this formula, I_V is identity matrix, A is adjacency matrix and D is degree matrix of graph G. Obviously, repeating this operation could lead to numerical instabilities and exploding/vanishing gradients. [2] introduces renormalization trick to alleviate this problem: $I_V + D^{-\frac{1}{2}}AD^{-\frac{1}{2}}H^{(l)}W^{(l)} \rightarrow \tilde{D}^{-\frac{1}{2}}\tilde{A}\tilde{D}^{-\frac{1}{2}}$, $\tilde{A} = A + I_V$ and $\tilde{D}_{ii} = \sum_j \tilde{A}_{ij}$.

$$h^{(l+1)} = \sigma\left(\tilde{D}^{-\frac{1}{2}}\tilde{A}\tilde{D}^{-\frac{1}{2}}h^{(l)}W^{(l)}\right) \tag{5}$$

The Laplacian matrix and its variants compute the local average of each node as its new representation, and thus each neighborhood node has the same importance. However, in RW-GCN, to fit the unsupervised context node prediction task, we replace the Laplacian matrix with the matrix M_{NOR} that we generated before. Therefore, a single iteration in the model will become the following form:

$$h^{(l+1)} = \sigma\left(M_{NOR}h^{(l)}W^{(l)}\right) \tag{6}$$

where each node uses weighted average to aggregate features from its S-order neighborhoods. The weight of each neighbor node is proportional to the number of times it co-occurs with the central node in a window of length S in all random walk sequences.

4.4 Global Loss Function

At this stage we combine the tasks of node classification and context nodes prediction. For node classification task, input features are transformed with two-layers graph convolution neural network and SoftMax activation function is applied to every node's output features. Therefore, we describe this process as:

$$Z = softmax\left(\widehat{M}\,ReLU\left(\widehat{M}XW^{(0)}\right)W^{(1)}\right) \tag{7}$$

Accordingly, the cross-entropy loss function is used to calculate the error of each iteration:

$$L_{nc} = -\sum_{i \in |V_{train}|} \sum_{k=1}^{K} y_{ik} \log z_{ik} \tag{8}$$

where V_{train} is the set of node indices that have labels.

In hidden layer we use abstract features for the task of context nodes prediction. That is, for every node pair (v_i, v_j) in pair set P, the goal is to maximize the probability of predicting the node v_j with v_i's hidden layer representation $f(v_j)$. As a result, the optimization object is minimizing the following loss function:

$$L_{cp} = -\sum_{(v_i, v_j) \in P} \log Pr\left(v_j | f(v_i)\right) \tag{9}$$

Now we can obtain the global lose function:

$$L = L_{nc} + L_{cp} = -\sum_{i \in |V_{train}|} \sum_{k=1}^{K} y_{ik} \log z_{ik} - \sum_{(v_i, v_j) \in P} \log Pr\left(v_j | f(v_i)\right) \tag{10}$$

To optimize this object function, the weight matrixes of hidden layers can be updated and converged to a reasonable position.

5 Experiments

5.1 Datasets

We conducted our document classification experiment on three citation network data sets [15]. Each dataset was composed of two parts, including a list of links between documents and the input feature in the form of bag-of-words. Each document has a corresponding class label. Table 1 shows the detailed introduction of the data. The label rate in the table denotes the proportion of labeled data used for training.

Table 1. Introduction to the data set used

Dataset	Nodes	Edges	Classes	Features	Label rate
Cora	2708	5429	7	1433	0.052
Citeseer	3327	4372	6	3703	0.036
Pubmed	19717	44338	3	500	0.003

5.2 Experimental Set-Up

In our experiments, we basically followed the same hyperparameters setting of GCN; that is, the two-layer neural network architecture was used, with a dropout rate of 0.5 and a learning rate of 0.01. In the hidden layer, we involved 64 hidden units. We use Adam as our optimizer, and 200 epochs of training were conducted on each data set. For all data sets, we only adjust parameters on Cora and apply the same parameters for the other two data sets.

In the part of the random walk, we kept node2vec's hyperparameters setting. For the three datasets, we set both parameters controlling the random walk to 0.25. For each node, a random walk sequence of length ten is generated from that node. This process is repeated for 5 rounds for each node. Then, a sliding window of length 3 is used to generate node pairs. In the context prediction task, we extract 10% of the node pairs for each iteration for training.

We report experiments conducted under two types of dataset split, one using the same dataset split method as other methods, and the other setting where we maintain the same ratio of the training set, validation set, and test set in each dataset, but with a completely random split method.

5.3 Baselines

We mainly compared with GCN and its comparison methods, including manifold regularization (ManiReg) [1], semi-supervised embedding (SemiEmb) [17], label propagation (LP) [19], Skip gram Based Graph Embeddings (DeepWalk) [13], and Planetoid [18]. In the comparison of GCN, we made a comparison between the two data sets where the randomly divided data sets were annotated later.

5.4   Results

As shown in Table 2, we report the comparison between our experimental results and other methods'. The results of the other methods listed here are taken from [18] and [9]. In addition, we also compared the classification accuracy of randomly split datasets. We kept the size of the training set and the test set unchanged, randomly split each dataset 10 times, and conducted training and evaluation of the model. Moreover, the experimental results of random split are also compared with the results in [9].

Table 2. Accuracy comparisons in node classification tasks

Method	Citeseer	Cora	Pubmed
ManiReg	60.1	59.5	70.7
SemiEmb	59.6	59.0	71.1
LP	45.3	68.0	63.0
DeepWalk	43.2	67.2	65.3
Planetoid	64.7	75.1	73.9
GCN	70.3	81.5	79.0
RW-GCN	**71.2**	**83.4**	**80.42**
GCN (rand.splits)	67.9	80.1	78.9
RW-GCN (rand.splits)	**70.44**	**83.05**	**80.35**

According to experimental results, it can be found that GCN trained with biased random walk has achieved various degrees of improvement in the three datasets. RW-GCN performs better on randomly split datasets than on public split dataset experiment. To some extent, this reflects that RW-GCN is not easily affected by the network structure or attribute features of partial data.

6   Conclusion

In this paper, we propose a graph convolutional neural network (RW-GCN) based on partial random walk training, and use it to carry out experiments on semi-supervised node classification tasks. By comparing our proposed method with other methods in multiple data sets, we can find that our proposed method has achieved a relatively significant improvement. In the future, we will further modify this method and try to reduce its computational complexity.

Acknowledgements. This work is supported by the National Natural Science Foundation of China (No. 61662053) and the Natural Science Foundation of Inner Mongolia in China (Grant nos. 2018BS06001).

References

1. Belkin, M., Niyogi, P., Sindhwani, V.: Manifold regularization: a geometric framework for learning from labeled and unlabeled examples. J. Mach. Learn. Res. **7**, 2399–2434 (2006)
2. Bhagat, S., Cormode, G., Muthukrishnan, S.: Node classification in social networks. In: Aggarwal, C. (ed.) Social Network Data Analytics, pp. 115–148. Springer, Boston (2011). https://doi.org/10.1007/978-1-4419-8462-3_5
3. Cai, H.Y., Zheng, V.W., Chang, C.C.: A comprehensive survey of graph embedding: problems, techniques, and applications. IEEE Trans. Knowl. Data Eng. **30**(9), 1616–1637 (2018)
4. Cui, P., Wang, X., Pei, J., et al.: A survey on network embedding. IEEE Trans. Knowl. Data Eng. **31**(5), 833–852 (2018)
5. Grover, A., Leskovec, J.: node2vec: scalable feature learning for networks. In: Proceedings of the 22nd ACM SIGKDD International Conference on Knowledge Discovery and Data Mining, pp. 855–864. ACM (2016)
6. Hamilton, W.L., Ying, R., Leskovec, J.: Inductive representation learning on large graphs. Adv. Neural Inf. Process. Syst. **30**, 1025–1035 (2017)
7. Hamilton, W.L., Ying, R., Leskovec, J.: Representation learning on graphs: methods and applications. IEEE Data Eng. Bull. **40**(3), 52–74 (2017)
8. Kipf, T.N., Welling, M.: Variational graph auto-encoders. In: NIPS Workshop on Bayesian Deep Learning (2016)
9. Kipf, T.N., Welling, M.: Semi-supervised classification with graph convolutional networks. In: International Conference on Learning Representations (2017)
10. Mikolov, T., Sutskever, I., Chen, K., et al.: Distributed representations of words and phrases and their compositionality. Adv. Neural Inf. Process. Syst. **26**, 3111–3119 (2013)
11. Mikolov, T., Corrado, G., Chen, K., et al.: Efficient estimation of word representations in vector space. In: International Conference on Learning Representations (2013)
12. Pennington, J., Socher, R., Manning, C.D.: GloVe: global vectors for word representation. In: Proceedings of the Empirical Methods in Natural Language Processing, vol. 14, pp. 1532–1543 (2014)
13. Perozzi, B., Al-Rfou, R., Skiena, S.: DeepWalk: online learning of social representations. In: Proceedings of the 20th ACM SIGKDD International Conference on Knowledge Discovery and Data Mining, pp. 701–710. ACM (2016)
14. Qiu, J., Dong, Y., Ma, H., et al.: Network embedding as matrix factorization: unifying DeepWalk, LINE, PTE, and node2vec. In: Proceedings of the Eleventh ACM International Conference on Web Search and Data Mining, pp. 459–467 (2018)
15. Sen, P., Namata, G., Bilgic, M., et al.: Collective classification in network data. AI Mag. **29**(3), 93–106 (2008)
16. Velickovic, P., Cucurull, G., Casanova, A., et al.: Graph attention networks. In: International Conference on Learning Representations (2018)
17. Weston, J., Ratle, F., Mobahi, H., Collobert, R.: Deep learning via semi-supervised embedding. In: Montavon, G., Orr, G.B., Müller, K.-R. (eds.) Neural Networks: Tricks of the Trade. LNCS, vol. 7700, pp. 639–655. Springer, Heidelberg (2012). https://doi.org/10.1007/978-3-642-35289-8_34
18. Yang, Z., Cohen, W.W., Salakhutdinov, R.: Revisiting semi-supervised learning with graph embeddings. arXiv:1603.08861 (2016)
19. Zhu, X., Ghahramani, Z., Lafferty, J., et al.: Semi-supervised learning using Gaussian fields and harmonic functions. In: International Conference on Machine Learning, vol. 3, pp. 912–919 (2003)

Loss-Aware Pattern Inference: A Correction on the Wrongly Claimed Limitations of Embedding Models

Mojtaba Nayyeri[1,2]([✉]), Chengjin Xu[1], Yadollah Yaghoobzadeh[3], Sahar Vahdati[2], Mirza Mohtashim Alam[1,2], Hamed Shariat Yazdi[1], and Jens Lehmann[1,4]

[1] University of Bonn, Bonn, Germany
`nayyeri@cs.uni-bonn.de`
[2] InfAI Lab, Dresden, Germany
{`vahdati,mohtasim`}`@infai.org`
[3] Microsoft, Redmond, USA
`yayaghoo@microsoft.com`
[4] Fraunhofer IAIS, Dresden, Germany
`jens.lehmann@iais.fraunhofer.de`

Abstract. Knowledge graph embedding models (KGEs) are actively utilized in many of the AI-based tasks, especially link prediction. Despite achieving high performances, one of the crucial aspects of KGEs is their capability of inferring relational patterns, such as symmetry, antisymmetry, inversion, and composition. Among the many reasons, the inference capability of embedding models is highly affected by the used loss function. However, most of the existing models failed to consider this aspect in their inference capabilities. In this paper, we show that disregarding loss functions results in inaccurate or even wrong interpretation from the capability of the models. We provide deep theoretical investigations of the already exiting KGE models on the example of the TransE model. To the best of our knowledge, so far, this has not been comprehensively investigated. We show that by a proper selection of the loss function for training a KGE e.g., TransE, the main inference limitations are mitigated. The provided theories together with the experimental results confirm the importance of loss functions for training KGE models and improving their performance.

1 Introduction

Recent years witnessed a great attention on the topic of knowledge graph embedding (KGE) models such that they rapidly became one of the state-of-the-art methods for Link Prediction. One of the primary KGE models is TransE which gained a lot of attention due to its simplicity and high performance. Some follow up models tried to improve TransE in terms of encoding relation types such as 1-many, or certain relation patterns such as reflexive, and symmetric [5,7,15]. While the community got into a paradigm of proposing new embedding models

© Springer Nature Switzerland AG 2021
K. Karlapalem et al. (Eds.): PAKDD 2021, LNAI 12714, pp. 77–89, 2021.
https://doi.org/10.1007/978-3-030-75768-7_7

by (only) focusing on the score function and competing on decimal improvements of the results, the actual cause that was rooted in the loss function, remained overlooked. Although, in a separate track, several loss functions have been proposed [16], its role in studying the capability of KGE models in presence of relational patterns have been majorly ignored. This neglected fact resulted in inaccurate or even wrong interpretations of the model capability.

Our investigations showed that, this problem originates in the initial assumption used for model capability proofs. To formally show this, let (h, r, t) be a positive triple in a KG where h, t are the entities and r is the relation between them which are to be embedded in $(\mathbf{h}, \mathbf{r}, \mathbf{t})$ (vector representation). This fact started from the TransE model, and continued by TransH, TransR, SimplE [6,7,15], where the assumption of $\mathbf{h} + \mathbf{r} \approx \mathbf{t}$ was used in evaluating the limitations of the models in encoding of patterns. This has caused a boom of new KGE models addressing the claimed limitations by proposing only new score functions. On the example of symmetric relations (e.g. brotherOf), this means that by enforcing a relation r to be symmetric ($\mathbf{t} + \mathbf{r} \approx \mathbf{h}$), the relation embedding is then enforced to be $\mathbf{r} = \mathbf{0}$. In this way, all of the corresponding vectors for entities that are related to each other via r relation will be equal. This had the risk of being interpreted as "model disability" in encoding symmetric patterns. While this problem was interpreted as model limitation, it was caused by incompatible equality assumption in the loss functions. We consider this as off-track argument followed by many of the KGE proposals overlooking the root cause. Here, we cover some of these works. In [6], additional limitations of TransE, FTransE [4], STransE [9], TransH and TransR are addressed which are listed here: (i) if the models encode a reflexive relation r, they automatically encode symmetric; (ii) if the models encode a reflexive relation r, they automatically encode transitive and; (iii) if entity h_1 has relation r with every entity in $\Delta \in \mathcal{E}$ and entity h_2 has relation r with one of entities in Δ, then h_2 must have the relation r with every entity in Δ. All of these limitations are justified by the initial assumption which was never fulfilled as they were incompatible with the utilized loss functions. Thus, these proposed approaches are only shedding the light on the underlying score functions.

Even a recent KGE model namely RotatE followed the same problem, which is a highly valuable work, however, it also claims that TransE is not capable of encoding symmetric patterns (Table 2 in [11]), considering the same assumption (i.e. $\mathbf{h} + \mathbf{r} \approx \mathbf{t}$). The equality assumption is satisfied when the loss function enforces $\|\mathbf{h} + \mathbf{r} - \mathbf{t}\| \approx 0$. The claimed disability has been argued to be caused by the score function of TransE. However, none of the existing loss functions (i.e. Margin Ranking Loss [1] and Adversarial Los [11]) hold this assumption during the optimization process, rather such losses take $\|\mathbf{h} + \mathbf{r} - \mathbf{t}\| \leq \gamma_1$ where γ_1 is upper-bound of positive scores. Therefore, most of the identified limitations of the existing KGEs and the proposed solutions for them have been based on an assumption that was not fulfilled by any of the existing loss functions. We re-studied the reported limitations of the TransE model employing the "appropriate" assumption compatible with the used loss function, as we shall see in the body of this paper, TransE is capable of encoding symmetric patterns. Although we highlight that ignoring loss functions caused inaccurate results on studying the limitations of TransE, it is generalziable for other existing KGE models as well as for the models yet to come (if this paradigm continuous). The impact of work is in blocking such a continuous misinterpretation of the inference capability of KGE models influenced by ignored loss functions with a long standing inappropriate assumption. Moreover, our theoretical finding is consistent with the recent experimental studies [10] which highlight that old models perform as well as the recent state-of-the-art models if they are trained with the same setting (using the same boosting techniques).

In summary, our main contributions are the following:

- We show that the different loss functions enforce different upper-bounds and lower-bounds for the scores of positive and negative samples respectively.
- We illustrate that the existing theories corresponding to the limitations of translation-based models are inaccurate since they only consider the score functions. We prove theoretically and later experimentally that the selection of loss functions is critical and can mitigate the main limitations.
- Using symmetric relation patterns, we obtain a proper upper-bound of positive triples score to enable encoding of symmetric patterns.
- We prove that applying translation in the complex space gives a more powerful model while efficiency in memory and time is preserved.

2   Related Works

Most of the previous work, majorly investigate the capability of the translation-based embedding models solely considering the formulation of the score functions. Accordingly, in this section, we review the score functions of TransE and its variants.

The score of **TransE** [1] is initially defined as $f_r(h,t) = \|\mathbf{h} + \mathbf{r} - \mathbf{t}\|$. In order to overcome the problems of TransE in encoding of relational patterns, **TransH** [15] was proposed where the score function was modified as $f_r(h,t) = \|\mathbf{h}_\perp + \mathbf{r} - \mathbf{t}_\perp\|$. In this way, each entity (**e**) is projected to a relation space ($\mathbf{e}_\perp = \mathbf{e} - \mathbf{w}_r \mathbf{e} \mathbf{w}_r^T$). Using this score function, the TransH model reported itself to be capable of encoding reflexive, one-to-many, many-to-one and many-to-many relations. This effort was done while the identified problem of TransE being incapable of encoding relational patterns was not valid. However, other works [6,7] started to build up on top of TransH addressing its problems. For example, encoding reflexive pattern leads to undesired encoding of both symmetric and transitive relations [6].

TransR [7] was then proposed with a new score function that projects each entity (**e**) to the relation space by using a matrix provided for each relation ($\mathbf{e}_\perp = \mathbf{e}\mathbf{M}_r, \mathbf{M}_r \in R^{d_e \times d_r}$). The chain of new score functions has continued with **TransD** [5] which provides two vectors for each individual entities and relations $(\mathbf{h}, \mathbf{h}_p, \mathbf{r}, \mathbf{r}_p, \mathbf{t}, \mathbf{t}_p)$. Head and tail entities are projected using the following matrices: $\mathbf{M}_{rh} = \mathbf{r}_p^T \mathbf{h}_p + \mathbf{I}^{m \times n}$, $\mathbf{M}_{rt} = \mathbf{r}_p^T \mathbf{t}_p + \mathbf{I}^{m \times n}$. The score function of TransD is similar to the score of TransH.

Recently, the **RotatE** [11] model has been proposed to address encoding of relational patterns with a new score function. It rotates the head to the tail entity using relation in the Complex space. Using constraints on the norm of entity vectors, the model is reformed to TransE. The scoring function of RotatE is $f_r(h, t) = \|\mathbf{h} \circ \mathbf{r} - \mathbf{t}\|$, where $\mathbf{h}, \mathbf{r}, \mathbf{t} \in C^d$, and $\circ$ is element-wise product. RotatE obtains the state-of-the-art results using very big embedding dimension (1000) and a lot of negative samples (1000). **TorusE** [3] fixes the problem of regularization in TransE by applying translation on a compact Lie group. The model has several variants including mapping from Torus to Complex space. In this case, the model is regarded as a special case of RotatE [11] applying rotation instead of translation in the Complex space. According to [11], TorusE is not defined on the entire Complex space. Therefore, it has less representation capacity. TorusE needs a very big embedding dimension (10000 as reported in [3]) which is a limitation. All of these state-of-the-art models only focus on proposing a new score function based on an assumption that was not valid for the used losses.

3   Loss-Aware Pattern Inference

In this section, we first introduce the wrongly interpreted limitations of the embedding models especially TransE and its follow ups (Sect. 3.1). Then, the limitations are re-investigated in the light of *score* and *loss* functions where we show that the corresponding theoretical proofs are inaccurate because the effect of loss function is ignored (Sect. 3.2). So, we propose new theories and prove that each of the limitations of TransE is resolvable by revising either the *scoring* function which the community continued with high effort or re-investigating the *loss* with regard to the limitations in the base assumption (our work).

3.1   Wrong Interpretations of KGE Models Presented as Limitations

Here we discuss the wrong interpretations of the KGE models which was reported in several literature that ended up to be presented as their limitations [6,11,14,15]. We focus on the reported limitations (L) of translation-based embedding models and their encoding capabilities for relation patterns (e.g. reflexive, symmetric) as following:

L1 TransE cannot encode reflexive relations when the relation vector is non-zero [15].

L2 TransE cannot encode a relation r which is neither reflexive nor irreflexive. To see that, if TransE encodes the relation r, we have $\mathbf{h}_1 + \mathbf{r} = \mathbf{h}_1$ and $\mathbf{h}_2 + \mathbf{r} \neq \mathbf{h}_2$, resulting $\mathbf{r} = \mathbf{0}$, $\mathbf{r} \neq \mathbf{0}$ which is a contradiction [14].

L3 TransE cannot encode symmetric relation when $\mathbf{r} \neq \mathbf{0}$. If r is symmetric, then: $\mathbf{h} + \mathbf{r} = \mathbf{t}$ and $\mathbf{t} + \mathbf{r} = \mathbf{h}$. Thus, $\mathbf{r} = \mathbf{0}$ and all entities appeared in head or tail of triples will have the same vector [11].

L4 If r is reflexive on $\Delta \in \mathcal{E}$, where $\mathcal{E}$ is the set of all entities in the KG, then r must also be symmetric [4].

L5 If r is reflexive on $\Delta \in \mathcal{E}$, r must also be transitive [9].

L6 If entity h_1 has relation r with every entity in Δ and entity h_2 has relation r with one of entities in Δ, then h_2 must have relation r with every entity in Δ [6].

Limitations 3 to 5 have been reported for TransE, however they are genelaizable for all the follow up models. Limitations 4 to 6 have been reported for TransE, FTransE, STransE, TransH and TransR.

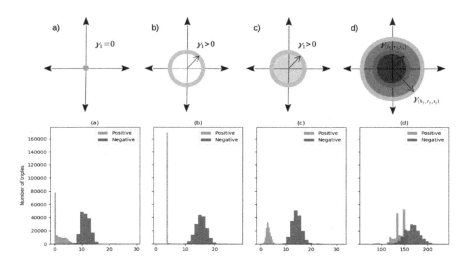

Fig. 1. Top: Visualization of truth region (positive) according to Table 1. The residual vector ϵ, (a) becomes $\mathbf{0}$, (b) lies on the border of a sphere with radius γ_1, (c) lies inside of a sphere with radius γ_1, and (d) $\epsilon_{(h_1,r_1,t_1)}$ lies inside of a sphere with radius $\gamma_{(h_1,r_1,t_1)}$. **Bottom:** The histogram of scores when TransE is trained on WordNet using the losses of Eq. 2 ($\gamma_1 = 0$), 2 ($\gamma_1 = 4$), 4 ($\gamma_1 = 4$) and 6 ($\gamma = 6$) respectively. Each histogram is the approximation of the corresponding conditions (a)–(d).

3.2   Re-investigation of the Reported Limitations

Here, we aim at analyzing the limitations of TransE (in real and complex spaces) by considering the effect of both *score* and *loss* functions. A loss function determines the score boundary within which a triple is positive or negative. A KGE model considers a triple (h, r, t) to be positive if its score is in a region of truth and a triple (h', r, t') to be negative if its score is in a region of falsity.

Such a boundary enforces an assumption through which the capability of embedding models in encoding relation pattern is investigated. For instance, in TransE with score function of $f_r(h,t) = \|\mathbf{h}+\mathbf{r}-\mathbf{t}\|$, the already used assumption for boundary of positive and negative samples are $f_r(h,t) = 0$ and $f_r(h',t') > 0$, respectively. However, this can not be fulfilled (or even approximated) by the considered loss functions of the state-of-the-art models (e.g. margin ranking loss [1] and RotatE loss [11]).

Table 1. Region of truth and falsity.

Condition	Positive	Negative	$\gamma_1, \gamma_2 \in R$
(a)	$f_r(h,t) = \gamma_1,$	$f_r(h',t') \geq \gamma_2$	$\gamma_1 = 0, \quad \gamma_2 > 0$
(b)	$f_r(h,t) = \gamma_1$	$f_r(h',t') \geq \gamma_2$	$\gamma_2 > \gamma_1 > 0$
(c)	$f_r(h,t) \leq \gamma_1$	$f_r(h',t') \geq \gamma_2$	$\gamma_2 > \gamma_1 > 0$
(d)	$f_r(h,t) \leq \gamma_{1(h,r,t)}$	$f_r(h',t') \geq \gamma_{2(h,r,t)}$	$\gamma_{2(h,r,t)} > \gamma_{1(h,r,t)} > 0$

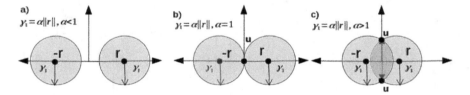

Fig. 2. Condition for encoding symmetric relation: (a) when $\alpha < 1$, the model cannot encode it. (b) when $\alpha = 1$, the intersection of two hyperspheres is a point. $\mathbf{u} = \mathbf{0}$ means embedding vectors of all the entities should be equal. Symmetric cannot be encoded. (c) when $\alpha > 1$, symmetric can be encoded as there are more than one point in the intersection of two hyperspheres.

To re-investigate and address the reported limitations, we propose four new conditions (Table 1) where a triple can be considered positive or negative by the score function. This is done by considering the defined thresholds for upper- and lower-bounds (decision boundary) in the scores of both positive and negative triples. We show that these conditions can be approximated by designing appropriate loss functions. In this regards, we adapt four loss functions based on [8] and propose compatible conditions for each of them (see Table 1). To better comprehend this, we illustrated the conditions in Fig. 1. The condition (a) indicates that a triple is positive if $\mathbf{h} + \mathbf{r} = \mathbf{t}$ holds. It means that the length of *residual vector* i.e. $\epsilon = \mathbf{h} + \mathbf{r} - \mathbf{t}$, is zero. It is the most strict condition that expresses the extent to which a triple is positive. Authors in [6,11] consider this condition to prove their theories as well as the limitation of TransE in the encoding of symmetric relations. However, the employed loss function fails to approximate (a), rather it fulfills condition (c) which results a void limitation in that setting. The condition (b) considers a triple to be positive if its residual vector lies on a hypersphere with radius γ_1. It is less restrictive than (a) which only considers a point in the vector space to express the positiveness of a triple. The optimization problem that approximates the conditions (a) ($\gamma_1 = 0$) and (b) ($\gamma_1 > 0$) is as follows:

$$
\begin{cases}
\min_{\xi_{h,t}} \sum_{(h,r,t)\in S^+} \xi_{h,t}{}^2 \\
s.t.\ f_r(h,t) = \gamma_1,\ (h,r,t) \in S^+ \\
f_r(h',t') \geq \gamma_2 - \xi_{h,t},\ (h',r,t') \in S^- \\
\xi_{h,t} \geq 0
\end{cases}
\tag{1}
$$

where S^+, S^- are the sets of positive and negative samples. $\xi_{h,t}$ are slack variables to reduce the effect of noise in negative samples. One loss function that approximates the conditions (a) and (b) is as follows where for case (a), we set $\gamma_1 = 0$ and for case (b) we set $\gamma_1 > 0$ in the formula.

$$
\mathcal{L}_{a|b} = \sum_{(h,r,t)\in S^+} \left(\lambda_1 \|f_r(h,t) - \gamma_1\| + \sum_{(h',r,t')\in S^-_{(h,r,t)}} \lambda_2 \max(\gamma_2 - f_r(h',t'), 0) \right).
\tag{2}
$$

The condition (c) considers a triple to be positive if its residual vector is inside a hypersphere of radius γ_1. The optimization problem that approximates the condition (c) is:

$$
\begin{cases}
\min_{\xi_{h,t}} \sum_{(h,r,t)\in S^+} \xi_{h,t}{}^2 \\
f_r(h,t) \leq \gamma_1,\ (h,r,t) \in S^+ \\
f_r(h',t') \geq \gamma_2 - \xi_{h,t},\ (h',r,t') \in S^- \\
\xi_{h,t} \geq 0
\end{cases}
\tag{3}
$$

The loss function that approximates the condition (c) is:

$$
\mathcal{L}_c = \sum_{(h,r,t)\in S^+} \left(\lambda_1 \max(f_r(h,t) - \gamma_1, 0) + \sum_{(h',r,t')\in S^-_{(h,r,t)}} \lambda_2 \max(\gamma_2 - f_r(h',t'), 0) \right).
\tag{4}
$$

Remark: The loss function which is defined in [16] is slightly different from the loss in Eq. 4. The former loss slides the margin while the latter fixes the margin by inclusion of a lower-bound for the score of negative triples. Both losses put an upper-bound for scores of positive triples. Apart from the loss 4, the RotatE loss [11] also approximates the condition (c). The formulation of the RotatE loss is as follows:

$$
\mathcal{L}_c^{RotatE} = -\sum_{(h,r,t)\in S^+} \left(\log \sigma(\gamma - f_r(h,t)) + \sum_{(h',r,t')\in S^-_{(h,r,t)}} \log \sigma(f_r(h',t') - \gamma) \right).
\tag{5}
$$

The condition (d) is similar to (c), but it provides different γ_1, γ_2 for each triple. Using (d), there is a triple-specific region of truth for each positive triple (h, r, t) and its corresponding negative triple (h', r, t'). Margin ranking loss [1] approximates (d). Defining $[x]_+ = \max(0, x)$, the loss is:

$$\mathcal{L}_d = \sum \sum [f_r(h, t) + \gamma - f_r(h', t')]_+ . \tag{6}$$

To re-investigate the limitations, we must assume that the relation vectors do not get zero values otherwise we will have the same embedding for head and tail which is undesirable. Here, the conditions (a) to (d) are presented by the following theorems.

Theorem T1. *(Addressing L1)*: TransE (real and complex) cannot infer a reflexive relation pattern with a non-zero relation vector under (a). However, under (b–d), TransE (real and complex) can infer reflexive pattern.

Theorem T2. *(Addressing L2)*: (i) TransE (complex) can infer a relation which is neither reflexive nor irreflexive under (b–d). (ii) TransE (real) cannot infer a relation which is neither reflexive nor irreflexive under (a–d).

Theorem T3. *(Addressing L3)*: (i) TransE (complex) can infer symmetric relations under (a–d). (ii) TransE (real) cannot infer symmetric relations under (a) with non-zero vector for relation. (iii) TransE (real) can infer a symmetric relation under (b–d). *Proof:* Here, due to space problems we only prove (iii) as a representative for other proofs.

Under (b), for TransE we have $\|\mathbf{h} + \mathbf{r} - \mathbf{t}\| = \gamma_1$ and $\|\mathbf{t} + \mathbf{r} - \mathbf{h}\| = \gamma_1$. The necessity condition for encoding symmetric relation is $\|\mathbf{h} + \mathbf{r} - \mathbf{t}\| = \|\mathbf{t} + \mathbf{r} - \mathbf{h}\|$. This implies $\|\mathbf{h}\| \cos(\theta_{h,r}) = \|\mathbf{t}\| \cos(\theta_{t,r})$. Let $\mathbf{h} - \mathbf{t} = \mathbf{u}$, by definition we have $\|\mathbf{u} + \mathbf{r}\| = \gamma_1, \|\mathbf{u} - \mathbf{r}\| = \gamma_1$. Now let $\gamma_1 = \alpha\|r\|$, we have:

$$\begin{cases} \|\mathbf{u}\|^2 + (1 - \alpha^2)\|\mathbf{r}\|^2 = -2\langle \mathbf{u}, \mathbf{r} \rangle \\ \|\mathbf{u}\|^2 + (1 - \alpha^2)\|\mathbf{r}\|^2 = 2\langle \mathbf{u}, \mathbf{r} \rangle \end{cases} \tag{7}$$

Therefore, there is: $\|\mathbf{u}\|^2 + (1 - \alpha^2)\|\mathbf{r}\|^2 = -(\|\mathbf{u}\|^2 + (1 - \alpha^2)\|\mathbf{r}\|^2)$, which can be written as $\|\mathbf{u}\|^2 = (\alpha^2 - 1)\|\mathbf{r}\|^2$. To avoid contradiction, we must have $\alpha > 1$. Once $\alpha > 1$, we have $\cos(\theta_{u,r}) = \pi/2$. Therefore, TransE can encode symmetric relation with condition (b), when $\gamma_1 = \alpha\|r\|$ and $\alpha > 1$. Figure 2 shows different conditions for encoding symmetric relation. Conditions (c–d) are directly resulted from (b), as it is subsumed by (c) and (d). That completes the proof.

Theorem T4. *(Addressing L4)*: For TransE (real and complex) (i) Limitation L4 holds under (a). (ii) Limitation L4 is not valid under (b–d).

Theorem T5. *(Addressing L5)*: For TransE (real and complEx) (i) Limitation L5 holds under (a). (ii) Limitation L5 holds is not valid under (b–d).

Theorem T6. *(Addressing L6)*: For TransE (real and complex), (i) Limitation L6 holds under (a). (ii) Limitation L6 is not valid under (b–d).

4   Experiments and Evaluations

In this section, we evaluate the performance of TransE in real and complex spaces with different loss functions used for theoretical analysis of the limitations. The experiments are done for a link prediction task with the aim of completing the triple $(h, r, ?)$ or $(?, r, t)$ by predicting the missing entities for h or t. Filtered Mean Rank (MR), Mean Reciprocal Rank (MRR) and Hits@10 are the evaluation metrics [7,13]. We used two evaluation dataset namely FB15K-237 [12] and WN18RR [2].

4.1   Experimental Setup

We implement TransE (real and complex) with the losses 2, 4 and 6 in PyTorch. Adagrad is used as an optimizer and 100 mini-batches have been generated in each iteration. The hyperparameter corresponding to the score function is embedding dimension d. We add slack variables to the losses 2 and 4 to have soft margin as in [8]. The loss 4 is rewritten as follows:

$$\min_{\substack{\xi_{h,t}^r \\ (h,r,t) \in S^+}} \sum \left(\lambda_0 \, \xi_{h,t}^r{}^2 + \lambda_1 \max(f_r(h,t) - \gamma_1, 0) + \lambda_2 \sum_{(h,r,t) \in S_{h',r,t'}^-} \max(\gamma_2 - f_r(h',t') - \xi_{h,t}^r, 0) \right).$$

(8)

4.2   Results and Discussion

In this part, we compare TransE (real and complex) and RotatE trained by using the losses 2 (condition (a), (b)), 4 (condition (c)) and the RotatE loss (condition (c)). For FB15K-237, we set the embedding dimension to 300 and the number of negative samples to 256. For WN18RR, we set the embedding dimension and the number of negative samples to 300 and 250 respectively. We additionally use adversarial negative sampling technique from [11] that we have applied for all the models.

Analysis of the Results: Table 2 presents a comparison of TransE (real and complex) and RotatE trained by different losses. TransE in Eq. 2 ($\gamma_1 = 0$) is trained by using the loss in Eq. 2 when $\gamma_1 = 0$. TransE in Eq. 2 ($\gamma_1 > 0$) refers to the TransE model which is trained by using the loss Eq. 2 when γ_1 is a non-zero positive value. The TransE model which is trained by the losses in Eq. 4 and the RotatE loss (i.e., $\mathcal{L}_c^{RotatE}$) are denoted by TransE 4 and TransE$\mathcal{L}_c^{RotatE}$ respectively. Similar notations are considered for TransE (complex) and RotatE when they are trained by using different loss functions. The loss 2 with $\gamma_1 = 0$ approximates the condition (a), and the approximation is done with condition (b) for $\gamma_1 > 0$.

Table 2. Link prediction results. Rows 1–4: TransE trained using condition (a), (b) (with loss 2 (c) (with the loss 4) and (c) (with the RotatE loss) with no injected relation patterns. Rows 5–8 TransE (complex) trained using condition (a), (b), (c) (with the loss 4) and (c) (with the RotatE loss) with no injected relation patterns. Rows 9–10: RotatE trained using condition (c) (with the loss 4) and (c) (with the RotatE loss) with no injected relation patterns.

	FB15K-237			WN18RR		
	MR	MRR	Hits@10	MR	MRR	Hits@10
TransE 2 ($\gamma_1 = 0$)	222	27.4	45.7	<u>3014</u>	19.3	47.4
TransE 2 ($\gamma_1 > 0$)	198	31.3	50.5	3942	21.4	50.3
TransE 4	181	32.3	<u>52.1</u>	3451	<u>23.5</u>	<u>53.9</u>
TransE $\mathcal{L}_c^{RotatE}$	<u>179</u>	<u>32.5</u>	51.9	3594	23.3	53.6
TransE (complex) 2 ($\gamma_1 = 0$)	213	28.5	47.3	<u>3014</u>	31.2	49.5
TransE (complex) 2 ($\gamma_1 > 0$)	194	31.9	50.8	3942	41.3	50.8
TransE (complex) 4	177	<u>32.8</u>	<u>52.1</u>	3435	<u>44.3</u>	<u>55.0</u>
TransE (complex) $\mathcal{L}_c^{RotatE}$	<u>176</u>	32.7	51.9	3537	44.2	54.7
RotatE 4	<u>194</u>	33.0	<u>52.0</u>	<u>3806</u>	<u>47.8</u>	<u>56.9</u>
RotatE $\mathcal{L}_c^{RotatE}$	196	33.0	51.8	3943	47.3	56.5

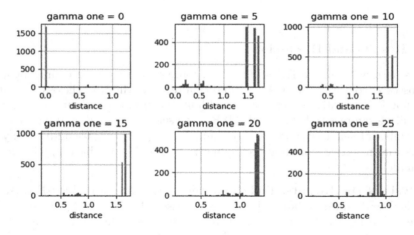

Fig. 3. Histogram of $\|\mathbf{h} + \mathbf{r} - \mathbf{t}\|$ for reflexive triple ($\mathbf{h} = \mathbf{t}$) per different γ_1.

The condition (c) can be approximated by using the loss in Eq. 4 and the RotatE loss (i.e., $\mathcal{L}_c^{RotatE}$). However, the loss Eq. 4 provides a better separation for positive and the negative samples than the RotatE loss. According to the Table 2, the loss 4 obtains a better performance than the other losses in each class of the studied models. It is consistent with our theories indicating that the condition (c) is less restrictive.

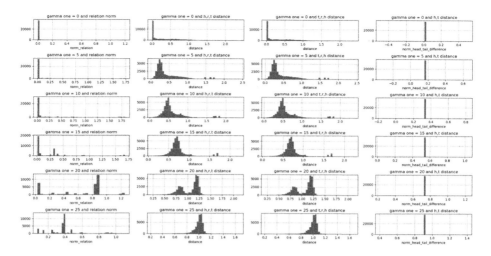

Fig. 4. Scores of symmetric triple for different gamma.

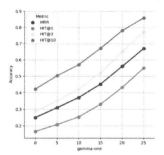

Fig. 5. Accuracy with respect to evaluation metric.

Although we only focus on translation-based KGEs, the theories can be generalized to different models including the RotatE model. We see that the loss in Eq. 4 improves the performance of RotatE. Regarding the Table 2, the loss in Eq. 2 ($\gamma_1 = 0$) gets the worst results. It confirms our theories that with the condition (a), most of the limitations are held. However, with the condition (c), the limitations no longer exist. Previously, there have not been any loss that approximates the condition (a). However, most of the theories presented above corresponding to the main limitations of the translation-based class of embedding models (L1–L6) have been proven using the condition (a) while the used loss didn't approximate the condition. Therefore, in all of the previous works the theories and experimental justifications have not been accurate.

4.3   Further Analysis of Theories

In Theorem Th1-Th6, we proved that most of the claimed limitations about TransE are inaccurate or even incorrect due to wrong assumptions not to be fulfilled by the used losses. More concretely, the claimed limitations were not rooted in the formulation of score function of TransE. Even worse, the claimed limitations were not also rooted in the loss function. The claimed limitations are analytically derived by using wrong assumption which is not fulfilled by the used loss function. Here we visualize the histogram of distance function ($\|\mathbf{h}+\mathbf{r}-\mathbf{t}\|$) to encode symmetric relation using various $\gamma_1 = \{0, 5, 10, 15, 20, 25\}$. While most of proofs corresponding to the limitations of TransE have been done with condition (a) ($\gamma_1 = 0$), the used loss function (margin ranking loss does not fulfill that condition (see Fig. 1)). Figure 3 visualizes the histogram of distance when the relation r is reflexive.

In the case of reflexive relation, distance $\|e + r - e\| = \|r\|$ is the norm of relation. According to this results, when $\gamma_1 = 0$, the norm of relation must be zero to consider the triple as positive. Therefore, with non-zero relation, all triples will be recognized as negative sample. In the case of $\|r\| = 0$, for all other triples in the form of (h, r, t) where $h \neq t$, the embeddings of head and tail must be equal which is undesired. However, from the Fig. 3 (the cases of $\gamma_1 \neq 0$, we can see that most of reflexive relations are non-zero. Moreover, we have $\|e + r - e\| = \|r\| \leq \gamma_1 \neq 0$. Therefore, the triples (e, r, e) are learned as positive by the model while the embeddings of relation is not zero. This confirms that the claim about TransE not being capable of encoding reflexive relation is no longer valid when margin ranking loss is used. Figure 4 shows that only when $\gamma_1 = 0$, the embedding of relation becomes zero, which addresses the limitations of L3. According to the last figure of the first row, when $\gamma_1 = 0$, the embeddings of head and tail becomes equal. Therefore, with non-zero relation vector, the TransE model cannot encode symmetric relation with condition (a). However, from the last row of Fig. 4, with a bigger value for upper-bound of positive triples $\gamma_1 = 25$, the embedding of relation is not zero and when all triples (h, r, t) and their symmetric (t, r, h) are learned as positive (encoding symmetric by TransE) based on the second and third columns of the last row in Fig. 4. Moreover, from the last sub-figure of last row, we see that embedding of head and tail are different. This shows symmetric relation is properly learned by the model. From Fig. 5, we observe that by increasing γ_1, the accuracy of the model increases when learning is done on symmetric patterns. As a conclusion, the mentioned limitations of TransE do not exist because there none of the previous loss functions fulfill the condition (a).

5   Conclusion

In this paper, we re-investigated the main limitations of Translation-based embedding models from two aspects: *score* and *loss*. We showed that different loss functions enforce different boundaries for triple scores, affecting the limitations of embedding models in encoding relation patterns such as symmetric. Therefore, the existing theories corresponding to the limitations of the KGE models are inaccurate because the effect of loss functions has been ignored. Accordingly, we presented new theories about the limitations by consideration of the effect of score and loss functions. The TransE model (in both real and Complex space) is trained by using various loss functions on standard datasets. According to the experiments, TransE in complex space with appropriate loss function significantly outperformed other existing translation-based embedding models. It got competitive performance with the other embedding models while it is more efficient in time and memory. Beside the performance-related improvements, the main impact of our work is the correction it provides between the initial assumption and the used loss function by most of the already existing embedding models. The objective is to influence the future embedding models and shed light on the effect of loss functions.

Acknowledgements. We acknowledge the support of the EU projects TAILOR (GA 952215), Cleopatra (GA 812997), the BmBF project MLwin, the EU Horizon 2020 grant 809965, and ScaDS.AI (01/S18026A-F).

References

1. Bordes, A., Usunier, N., Garcia-Duran, A., Weston, J., Yakhnenko, O.: Translating embeddings for modeling multi-relational data. In: NIPS Conference, pp. 2787–2795 (2013)
2. Dettmers, T., Minervini, P., Stenetorp, P., Riedel, S.: Convolutional 2D knowledge graph embeddings. In: Thirty-Second AAAI Conference (2018)
3. Ebisu, T., Ichise, R.: Toruse: knowledge graph embedding on a lie group. In: Thirty-Second AAAI Conference (2018)
4. Feng, J., Huang, M., Wang, M., Zhou, M., Hao, Y., Zhu, X.: Knowledge graph embedding by flexible translation. In: KR Conference (2016)
5. Ji, G., He, S., Xu, L., Liu, K., Zhao, J.: Knowledge graph embedding via dynamic mapping matrix. In: Proceedings of the 53rd Annual Meeting of the Association for Computational Linguistics and the 7th International Joint Conference on Natural Language Processing (Volume 1: Long Papers), vol. 1, pp. 687–696 (2015)
6. Kazemi, S.M., Poole, D.: Simple embedding for link prediction in knowledge graphs. In: NIPS Conference, pp. 4284–4295 (2018)
7. Lin, Y., Liu, Z., Sun, M., Liu, Y., Zhu, X.: Learning entity and relation embeddings for knowledge graph completion. In: Twenty-Ninth AAAI Conference (2015)
8. Nayyeri, M., Vahdati, S., Lehmann, J., Shariat Yazdi, H.: Soft marginal TransE for scholarly knowledge graph completion. arXiv preprint arXiv:1904.12211 (2019)
9. Nguyen, D.Q., Sirts, K., Qu, L., Johnson, M.: Stranse: a novel embedding model of entities and relationships in knowledge bases. arXiv preprint arXiv:1606.08140 (2016)
10. Ruffinelli, D., Broscheit, S., Gemulla, R.: You can teach an old dog new tricks! on training knowledge graph embeddings. In: International Conference on Learning Representations (2019)
11. Sun, Z., Deng, Z.H., Nie, J.Y., Tang, J.: Rotate: knowledge graph embedding by relational rotation in complex space. arXiv preprint arXiv:1902.10197 (2019)
12. Toutanova, K., Chen, D.: Observed versus latent features for knowledge base and text inference. In: Proceedings of the 3rd Workshop on Continuous Vector Space Models and their Compositionality, pp. 57–66 (2015)
13. Wang, Q., Mao, Z., Wang, B., Guo, L.: Knowledge graph embedding: a survey of approaches and applications. IEEE Trans. Knowl. Data Eng. **29**(12), 2724–2743 (2017)
14. Wang, Y., Gemulla, R., Li, H.: On multi-relational link prediction with bilinear models. In: Thirty-Second AAAI Conference (2018)
15. Wang, Z., Zhang, J., Feng, J., Chen, Z.: Knowledge graph embedding by translating on hyperplanes. In: Twenty-Eighth AAAI Conference (2014)
16. Zhou, X., Zhu, Q., Liu, P., Guo, L.: Learning knowledge embeddings by combining limit-based scoring loss. In: Proceedings of the 2017 ACM CIKM Conference, pp. 1009–1018. ACM (2017)

SST-GNN: Simplified Spatio-Temporal Traffic Forecasting Model Using Graph Neural Network

Amit Roy[✉], Kashob Kumar Roy, Amin Ahsan Ali, M. Ashraful Amin, and A. K. M. Mahbubur Rahman

Artificial Intelligence and Cybernetics Lab, Independent University, Bangladesh, Dhaka, Bangladesh
{2015-616-801,2015-016-762}@student.cse.du.ac.bd
{aminali,aminmdashraful,akmmrahman}@iub.edu.bd
https://www.agencylab.org/

Abstract. To capture spatial relationships and temporal dynamics in traffic data, spatio-temporal models for traffic forecasting have drawn significant attention in recent years. Most of the recent works employed graph neural networks(GNN) with multiple layers to capture the spatial dependency. However, road junctions with different hop-distance can carry distinct traffic information which should be exploited separately but existing multi-layer GNNs are incompetent to discriminate between their impact. Again, to capture the temporal interrelationship, recurrent neural networks are common in state-of-the-art approaches that often fail to capture long-range dependencies. Furthermore, traffic data shows repeated patterns in a daily or weekly period which should be addressed explicitly. To address these limitations, we have designed a **S**implified **S**patio-temporal **T**raffic forecasting **GNN(SST-GNN)** that effectively encodes the spatial dependency by separately aggregating different neighborhood representations rather than with multiple layers and capture the temporal dependency with a simple yet effective weighted spatio-temporal aggregation mechanism. We capture the periodic traffic patterns by using a novel position encoding scheme with historical and current data in two different models. With extensive experimental analysis, we have shown that our model (Code is available at github.com/AmitRoy7781/SST-GNN) has significantly outperformed the state-of-the-art models on three real-world traffic datasets from the Performance Measurement System (PeMS).

Keywords: Traffic forecasting · Spatio-temporal modeling · Graph neural network

1 Introduction

In recent years, future traffic prediction is getting interests among researchers from the area of Intelligent Transportation System(ITS). Generally, the traffic

---

A. Roy and K. K. Roy—Equal Contribution.

© Springer Nature Switzerland AG 2021
K. Karlapalem et al. (Eds.): PAKDD 2021, LNAI 12714, pp. 90–102, 2021.
https://doi.org/10.1007/978-3-030-75768-7_8

intensity of given sensors refers to the speed of people/vehicles passing through those sensors on traffic networks at each timestamp. Accurate forecasting of future traffic speeds has plenty of advantages such as it would help citizens not only to bypass the crowded path but also to schedule an efficient trip in advance. However, the task of traffic forecasting is challenging because the traffic in a busy metropolitan city changes across different locations throughout the different time periods every day. Also, different traffic patterns are observed on weekdays and weekends. Hence, there lies a complex spatio-temporal relationship in traffic data that makes the task of accurate traffic prediction challenging.

As the traffic network of a city can be modeled as a graph with traffic speed of different nodes (road junctions) across different timestamps, most of the recent approaches [1,4,5,7,9,10,12,12] have tried to design the problem of traffic fore-casting as a regression task. In these models, the spatial relationship among dif-ferent nodes are captured using graph neural networks (GNNs) [6,8] and recur-rent neural networks are employed to consider the temporal dependency [13]. To mention a few, STGCN [13] is the first approach to apply graph convolution to capture spatial representation in traffic forecasting along with recurrent units for temporal dependencies. On the other hand, DCRNN [9] employed bi-directional random walk to preserve spatial relation and GRU for temporal dependencies.

In spite of the extensive efforts for future traffic prediction, the challenge is not solved yet due to a couple of reasons. Firstly, state-of-the-art models have a common practice to increase the receptive field by using multi-layer GNNs to capture the spatial traffic information from different-hop neighborhoods. How-ever, the immediate neighboring junctions might have different impacts on the target node's traffic pattern from the distant junctions. Multi-layer GNNs suffer from over-smooth problem [2] while aggregating the information from different hop neighboring junctions in more layers which results in less informative spatial representations. Instead, directly employing the representation of different-hop neighbors towards the fully connected layers will be more effective to encode the impact of different hop neighboring junctions [14]. Secondly, traditional spatio-temporal models apply recurrent neural networks e.g., LSTM, GRU to encode the temporal information. However, recurrent neural networks often fail to per-form well to forecast the traffic in long-range prediction as spatial traffic at dif-ferent timestamps has a varying scale of impact on the target node's pattern. To encode the temporal dependency explicitly, we propose a novel spatio-temporal weighted aggregation scheme that can learn the importance of the spatial rep-resentation from previous timestamps. Also, we stack the representation of dif-ferent timestamps to obtain the final representation that allows our model in handling long-range dependencies effectively.

Finally, traffic data shows repetitive daily patterns across days in a week. To learn these trends in traffic data, an ideal model should consider the current day pattern as well as the daily pattern seen in the traffic data. Here, we define the current day pattern as the traffic situation observed in the last hour on the current day and the daily pattern as the traffic intensities exist in the same time period in the last one week. Most of the researchers put their contribution to

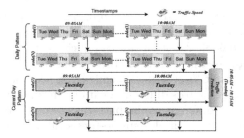

Fig. 1. Predicting the traffic of 10:05 AM–11:00 AM on Tuesday by observing the traffic data of the past hour from the last seven days as well as the present day to capture the daily pattern and current day pattern.

learning the current day pattern. For instance, to predict traffic speed at 10:05 AM–11:00 AM on Tuesday, recent researchers propose frameworks to learn the pattern from 9:05 AM–10:00 AM on the present day (Tuesday) which is depicted as current day pattern in Fig. 1. However, current day pattern information might not be enough to model city traffic. In our work, we learn the traffic pattern effectively with two different models named as the current-day model and historical model where the current-day model analyze the past hour data on the current day and the historical model deals with the past hour traffic intensity in the last seven days (Fig. 1). Lastly, the traffic intensity in a metropolitan city varies throughout different time periods in a day across weekdays and weekends. Therefore, we enhance the generalization capability of our model with a novel position encoding scheme which helps our model to distinguish between traffic data of different periods of the day on both weekdays and weekends. In summary, the key contribution of our work SST-GNN includes:

- We directly utilize the representation of different hop neighbors rather than using multi-layer GNNs to explicitly focus on the spatial dependency of traffic intensity from road junctions at different hop distance.
- We capture the temporal dependency with a simple weighted aggregation of the spatial representations from the different timestamps and finally stacking them to capture inter-timestamp dependency.
- We propose a simple yet effective framework to extract current day and daily information through two different models: current-day model and historical model. The framework uses neighborhood aggregation based graph neural networks to learn the node embeddings.
- We propose a position encoding scheme that can encode the periodic information of days and weeks into traffic data which can be easily extended to months and even for years.
- From the extensive experimental analysis, we show the efficacy of our model. Our model SST-GNN outperforms the state-of-the-art models in predicting the traffic speed of the next 15, 30, 45, and 60 min.

2 Background Study

Related Works: In the early years, various statistical and machine learning techniques such as Auto-Regressive Integrated Moving Average (ARIMA), Historical Average (HA), Support Vector Regression (SVR), and Kalman filters have been widely used for traffic forecasting. However, in recent years, graph neural networks (GNN) have achieved greater success in modeling real-life traffic. GNNs are able to encode the spatial dependency between neighbor nodes in a graph into their hidden representation by employing different feature aggregation scheme. Graph Convolution Networks [3,8] apply spectral convolutions to learn structural dependency as well as feature information. On the other hand, GraphSAGE [6] introduced a neighborhood aggregation strategy to preserve the inter-relationship among proximal nodes. As GNNs succeeds in learning representations for various downstream machine learning tasks, several recent works have employed graph convolution to learn node representations that can extract spatial relations from the traffic network. STGCN [13] has modeled spatial and temporal relations using a convolutional network. The diffusion process is used to model the traffic networks in DCRNN [9] that captures the spatial relations by using the bidirectional random walks and GRU for temporal dependencies. Besides, several recent works [4,10,12] have achieved good performance.To capture the spatio-temporal dependency among nodes in the embedded space, Graph Wavenet [12] learns a self-adaptive dependency matrix where the receptive field increases with the number of layers. Very recent work LSGCN [7] proposes a new graph attention network called cosAtt and incorporates the cosAtt and GCN into the spatial gated block and linear gated block to iteratively predict future traffic intensity. We observe that state-of-the-art models fail to capture the impact of different hop neighborhoods for a targer node in traffic networks explicitly. Also, the RNN-based models are incompetent to learn temporal dependencies in long term prediction. To address the above challenges, we explicitly capture the impact of different-hop neighborhoods on target node's traffic with a simple yet effective spatio-temporal aggregation scheme and stack the embeddings of intermediate timestamps to learn temporal dependencies across different timestamps. Capturing the traffic of different hop neighborhood with simplified spatio-temporal aggregation improves our models performance than the state-of-the-art traffic forecasting models.

Preliminaries and Problem Definition: A traffic network is represented as a graph $G = (V, A)$ where V is the set of nodes that denote road junctions and $A \in \mathbb{R}^{|V| \times |V|}$ is the adjacency matrix of the graph, where $A_{i,j} = 1$ if junction i and j are connected by an road and 0 otherwise. Each node also contains some features of a junction representing traffic flow, speed, occupancy etc. As traffic at different nodes change over time, the traffic features of a node u at timestamp t is denoted as $X_u^{<t>} \in \mathbb{R}^d$ where d denotes the feature dimension and $X^{<t>} \in \mathbb{R}^{|V| \times d}$ represents the traffic features of all nodes at timestamp t. The graph at a timestamp t is denoted as timestamp graph $G^{<t>}$.Note that, all timestamp graphs are structurally identical to each other. However, a traffic forecasting framework takes a sequence of T timestamp graphs with their node features $(X^{<1>}, X^{<2>}, \ldots\ldots, X^{<T>})$ as input and predicts the traffic intensities of nodes at next n timestamps that is $(Y^{<T+1>}, Y^{<T+2>}, \ldots\ldots, Y^{<T+n>})$.

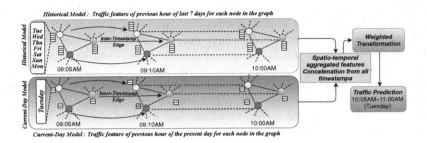

Fig. 2. Inter-timestamp edges are introduced between identical nodes of consecutive timestamps e.g. an edge between a blue node at timestamp 09:05 AM and a blue node at timestamp 09:10 AM where the same color indicates identical nodes. Although both historical and current-day model deals with the same spatio-temporal graph consisting of all timestamp graphs over 5 min interval in the past hour of the prediction window, the historical model considers traffic features from last week to capture the repeated daily patterns while the current-day model uses only current day (e.g. Tuesday) information to find current day patterns in traffic data. Spatial dependency is captured through aggregating features from different neighborhoods on each timestamp graph while temporal dependency is preserved by performing temporal aggregation among the node representations learned from previous timestamps which are depicted in Fig. 3. Finally, concatenation followed by weighted transformation is performed to compute the spatio-temporal embeddings of nodes which are used for traffic prediction. (Color figure online)

3   Proposed Model

In this section, we describe the whole architecture of our proposed framework that can effectively capture spatio-temporal dependencies between road junctions. We discuss spatio-temporal graph and positional encoding scheme for performing spatio-temporal aggregation and capturing periodicity in traffic data respectively. After that, we present spatio-temporal aggregation with two different models namely historical model and current-day model and concluded with the final embedding and training process. A high-level overview has been presented in Fig. 2 and Fig. 3.

Spatio-Temporal Graph: To capture the complex spatio-temporal dependencies between nodes across different timestamp graphs, we introduce inter-timestamp edges between identical nodes of consecutive timestamp graphs as shown in Fig. 2 where the same color indicates identical nodes. Afterward, to learn embeddings of nodes, we perform our proposed spatio-temporal aggregation on a spatio-temporal graph that consists of previous T timestamp graphs from the prediction window with their inter-timestamp edges.

Positional Encoding: To extract informative traffic features from different periods of the day, we need to encode the relative position of the different time periods in our model. Following the relative positioning concept widely used in transformer based attention mechanism in Machine Translation [11], we have used positional encoding with a sinusoidal function to provide position information on different timestamps. We ensure that the sinusoidal function for each day completes a full cycle within a day. Hence, any time duration can be represented as a repetitive portion of the sine curve of each day. For example, the sinusoidal curve will have the same pattern during the time slot (9:05 AM–10:00 AM) daily. Hence, this positional encoding will help the model capture daily pattern indeed. Moreover, there might be a weekly pattern in traffic such as specific days that might have the same kind of traffic. Also, the proposed framework needs to see whether the patterns are coming from weekdays or weekends. To capture this kind of weekly pattern, we also propose another full cycle of a sine wave for each week. Therefore, the final position encoding has been achieved by Eq. 1.

$$\mathcal{P}^{<t>} = sin(\frac{2\pi t}{24 \times hr\_sample}) + sin(\frac{2\pi t}{24 \times 7 \times hr\_sample}) \tag{1}$$

where t denotes a particular timestamp and hr_sample represents the number of observed data samples in an hour. The idea can be extended to capture monthly repetition with another full cycle sine wave that completes in a month.

Spatio-Temporal Aggregation: We develop a spatio-temporal aggregation scheme to encode spatial as well as temporal dependencies into the embeddings of nodes that have been shown in Fig. 3. It has two components as follows:

– **Spatial Aggregation**: In real-life traffic networks, it can be observed that all higher-order neighborhoods are not equally important for a target node. Different hop neighborhood may carry distinct information that should be captured explicitly. Therefore, we perform information aggregation over nodes in different neighborhoods separately in each timestamp graph as follows,

$$X_{(k)}^{<t>} = D_{(k)}^{-1} A_{(k)} X^{<t>}; \qquad S_u^{<t>} = \sum_{k=1}^{K} X_{(k),u}^{<t>} W_{(k)}^{<t>} \tag{2}$$

where, $A_{(k)}$ denotes k^{th}-hop neighborhood - meaning that $|A_{(k)}|_{i,j} = 1$ only if node i and j are exactly k hop away from each other otherwise 0, $D_{(k)}$ is the degree matrix of $A_{(k)}$, $X_{(k)}^{<t>}$ is the degree-normalized mean of kth-hop neighbor-embeddings at timestamp t, Further, we perform weighted aggregation among the mean representations of different-hop neighborhoods up to K hop away from node u to compute the spatial embeddings of node u denoted as $S_u^{<t>}$ where $W_k^{<t>}$ is learnable weight parameters to capture the impact of kth-hop neighborhood at timestamp t. Explicit aggregation of different hop neighborhood embeddings helps to differentiate the impacts of the traffic intensities from different hop neighbor nodes on a target node.

- **Temporal Aggregation**: To capture temporal dynamics among different timestamp graphs, temporal embeddings of nodes, $\tilde{Z}_u^{<t>}$, at timestamp t are computed through aggregating spatio-temporal embeddings from the earlier timestamps as follows,

$$\tilde{Z}_u^{<t>} = ReLU(\sum_{i=1}^{t-1}(W^{<i>}Z_u^{<i>})) \tag{3}$$

where $Z_u^{<i>}$ is the spatio-temporal embedding of u and $W^{<i>}$ is the learnable weight at timestamp i.

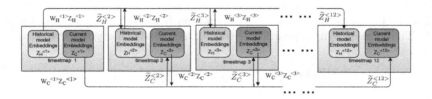

Fig. 3. Spatio-Temporal Aggregation Scheme: To capture complex spatio-temporal dependencies in traffic networks, the historical model concatenates the spatial embeddings from different hop neighborhoods at timestamp t with temporal embedding $\tilde{Z}_H^t$ - the weighted aggregation of $(Z_H^1, \ldots, Z_H^{t-1})$, to learn spatio-temporal embeddings Z_H^t. Similarly, current-day model performs the same process.

After that we concatenate the ego(target node), spatial and temporal embeddings of node u to learn the spatio-temporal embedding of node u at timestamp t, $Z_u^{<t>}$ as following,

$$Z_u^{<t>} = ReLU(W_{sptemp}^{<t>}(X_u^{<t>} \parallel S_u^{<t>} \parallel \tilde{Z}_u^{<t>})) + \mathcal{P}^{<t>} \tag{4}$$

where W_{sptemp} is a learnable parameter at timestamp t and $\parallel$ denotes concatenation operation while $\mathcal{P}^{<t>}$ represents the positional encoding of timestamp t. In Eq. 4, temporal embedding of u at timestamp t, $\tilde{Z}_u^{<t>}$ captures the temporal dependencies of traffic from previous 1 to $t-1$ timestamps while spatial embedding $S_u^{<t>}$ leverages information from different hop neighborhoods of node u. Moreover, our model can achieve its best generalization ability by keeping the ego(target node), spatial and temporal information separate without mixing them. Furthermore, the periodic information of traffic data is also preserved by incorporating the positional encoding value of timestamp t into node embeddings. Therefore, Eq. 4 ensures that our model can learn complex traffic flow information across different hop neighbor road junctions as well as from different timestamps effectively.

Historical Model: To preserve the historical traffic information of previous days, we propose a novel historical model that analyzes the daily patterns. In the historical model, we assign the feature vector of node u, $X_{H_u}^{<t>} \in \mathbb{R}^P$ as the traffic speed at timestamp t of last P days. Therefore, the historical model captures the traffic pattern of the last P days of previous T timestamps from the prediction window. The motivation behind using the historical model is to capture the periodic nature of traffic data from the history of the last $P = 7$ days. On each timestamp t, we perform spatio-temporal aggregation to learn historical spatio-temporal embedding $Z_{H_u}^{<t>}$ for each node u as shown in Fig. 3.

Current-Day Model: The current-day model only considers the traffic speed at timestamp t of current day, $X_{C_u}^{<t>} \in \mathbb{R}$ as the feature vector of each node in the network just like the traditional traffic forecasting frameworks. Hence, the current-day model focuses on the last T timestamps of the present day (prediction day) to capture the traffic pattern on the current day. Similar to the historical model, in our current-day model we also perform spatio-temporal aggregation on each timestamp network to find current day spatio-temporal embedding $Z_{C_u}^{<t>}$ for node u at timestamp t that has been shown in Fig. 3.

Final Embedding: After obtaining the desired embeddings for node u by applying spatio-temporal aggregation for $T = 12$ timestamps in the historical and current-day model, the embeddings from both models are concatenated and combined into final embedding Z_{F_u} for each node u in input traffic network as follows,

$$\widetilde{Z}_{F_u} = Z_{H_u}^{<1>} \parallel \ldots \parallel Z_{H_u}^{<T>} \parallel Z_{C_u}^{<1>} \parallel \ldots \parallel Z_{C_u}^{<T>} \tag{5}$$

$$Z_{F_u} = W_F . \widetilde{Z}_{F_u} \tag{6}$$

where $Z_{H_u}^{<t>}$ and $Z_{C_u}^{<t>}$ represents the spatio-temporal embeddings from historical and current-day models respectively for node u at timestamp t and W_F is the learnable weight parameter. Combining the embeddings from all timestamps in Eq. 5 enables our model to gain more expressiveness, in contrast existing models only focus on the embedding from last timestamp that limits the expressiveness to some extent. Finally, we have used a two-layer neural network to predict the traffic intensities at different nodes and update all the parameters by optimizing supervised mean squared error(MSE) as the loss function.

4 Experimental Analysis

In this section, we describe datasets, dataset preprocessing, and experiment setup followed by the elaborate analysis of observed results.

Dataset Description: To prove the effectiveness of our proposed model, we have conducted experiments on three publicly available real-life traffic datasets PeMSD7, PeMSD4, and PeMSD8 [7] that are widely used for performance comparison in previous works such as STGCN [13], ASTGCN [5], LSGCN [7]. PeMSD7 contains the traffic data of California that consists of the traffic speed of 228 sensors with 832 road segments while the time span is from May, 2012 to June, 2012 (only weekdays). We choose the first month of traffic data as the training set while the rest are split equally into validation and test set. PeMSD4 consists of the traffic data of San Francisco with 307 sensors on 340 roads. The time span of the dataset is January-February in 2018 and we choose the first 47 days as the training set while the rest are used as validation and test set. Lastly, PeMSD8 consists of the traffic data from San Bernardino with 170 detectors on 295 roads, ranging from July to August in 2016. We select the first fifty days as the training and the rest are used as the validation and test set. All three datasets contain traffic feature with an interval of five minutes. In all the experiments, we consider traffic speed as the traffic feature for all three datasets.

Data Preprocessing: Adjacency matrix of the sensor network is constructed using a thresholded Gaussian kernel, $A_{ij} = 1$ only if $i \neq j$ and $exp(-\frac{d_{ij}^2}{\delta}) \geq \epsilon$, otherwise 0 where A_{ij} determines edge between sensor i and j which is related with d_{ij} (the distance between sensor i and j). To control the distribution and sparsity of adjacency matrix A, we set the thresholds $\delta = 0.1$ and $\epsilon = 0.5$

Experimental Settings: The experiments are conducted on a Linux computer (GeForce RTX2080 Ti GPU) where both historical and current-day model adopts 60 min time window i.e. previous 12 timestamps are used to predict traffic of the next 15, 30, 45, and 60 min. In historical model, the input feature vector of each node comprises the traffic speed of the last seven days while the current-day model considers the traffic speed of the current day in the corresponding timestamp. For PeMSD7, we aggregate spatial information from the 2-hop neighborhood while 4-hop neighbors are considered for the other two datasets. We train our model by minimizing Mean Square Error (MSE) as the loss function with ADAM optimizer for 500 epochs. For all the datasets, we set the initial learning rate 0.001 with a decay rate of 0.5 every seven epochs. To report the performance comparison among different models, we opt Mean Absolute Errors (MAE), Root Mean Squared Errors (RMSE) and Mean Absolute Percentage Errors (MAPE) as the evaluation metrics.

Table 1. Performance comparison in traffic prediction (**Best**, 2nd Best)

Datasets	Models	15 min			30 min			45 min			60 min		
		MAE	RMSE	MAPE	MAE	RMSE	MAPE	MAE	RMSE	MAPE	MAE	RMSE	MAPE
PeMSD7	DCRNN (2018)	2.22	4.25	5.16	3.04	6.02	7.46	3.64	7.24	9.00	4.15	8.20	10.82
	STGCN (2018)	2.24	4.01	5.28	3.04	5.74	7.46	3.61	6.85	9.26	4.08	7.69	10.23
	ASTGCN (2019)	2.85	5.15	7.25	3.35	6.12	8.67	3.70	6.77	9.73	3.96	7.20	10.53
	Graph WaveNet (2019)	2.17	3.87	4.85	2.90	5.40	6.86	3.23	6.29	8.06	3.75	7.02	9.58
	LSGCN (2020)	2.22	3.98	5.14	2.96	5.47	7.18	3.43	6.39	8.51	3.81	7.09	9.60
	SST-GNN(ours)	**2.04**	**3.53**	**4.77**	**2.67**	**4.80**	**6.66**	**3.17**	**5.79**	**8.00**	**3.48**	**6.39**	**9.04**
PeMSD4	DCRNN (2018)	1.35	2.94	2.68	1.77	4.06	3.71	2.04	4.77	4.78	2.26	5.28	5.10
	STGCN (2018)	1.47	3.01	2.92	1.93	4.21	3.98	2.26	5.01	4.73	2.55	5.65	5.39
	ASTGCN (2019)	2.12	3.96	4.16	2.42	4.59	4.80	2.60	4.97	5.20	2.73	5.21	5.46
	Graph WaveNet (2019)	1.30	2.68	2.67	1.70	3.82	3.73	1.95	4.16	4.25	2.03	4.65	4.60
	LSGCN (2020)	1.45	2.93	2.90	1.82	3.92	3.84	2.04	4.47	4.42	2.22	4.83	4.85
	SST-GNN(ours)	**1.23**	**2.53**	**2.37**	**1.82**	**3.47**	**3.69**	**2.13**	**3.86**	**3.93**	**2.13**	**4.45**	**4.69**
PeMSD8	DCRNN (2018)	1.17	2.59	2.32	1.49	3.56	3.21	1.71	4.13	3.83	1.87	4.50	4.28
	STGCN (2018)	1.19	2.62	2.34	1.59	3.61	3.24	1.92	4.21	3.91	2.25	4.68	4.54
	ASTGCN (2019)	1.49	3.18	3.16	1.67	3.69	3.59	1.81	3.92	3.98	1.89	4.13	4.22
	LSGCN (2020)	1.16	2.45	2.24	1.46	3.28	3.02	1.66	3.75	3.51	1.81	4.11	3.89
	SST-GNN(ours)	**1.03**	**2.08**	**1.86**	**1.39**	**2.80**	**2.67**	**1.62**	**3.28**	**3.20**	**1.74**	**3.57**	**3.50**

4.1   Experiment Results

Comparison with Baselines: In Table 1, we present the performance comparison of our model named SST-GNN with the state-of-the-art models STGCN, DCRNN, ASTGCN, Graph WaveNet and LSGCN in 15, 30, 45, and 60 min traffic prediction. In Table 1, it is easy to observe that our model outperforms all baseline models in both long and short-term predictions for all three evaluation metrics on PeMSD7, PeMSD4, and PeMSD8. The second-best performance has been observed for the recent work Graph Wavenet in dataset PeMSD7, PeMSD4, and for LSGCN in PeMSD8. Graph Wavenet learns an adaptive adjacency matrix with different granularity whereas LSGCN analyzes long-term and short-term patterns explicitly by employing attention-guided GCN and GLU. It is obvious that our model is able to capture complex spatio-temporal relationship more accurately through the proposed spatio-temporal aggregation scheme to outperform both the Graph Wavenet and LSGCN with reasonable margins. A number of architectural factors facilitate these improvements. Firstly, keeping the representations from different neighboring junctions separate allows the proposed model to learn the impact of different hop neighbors on the target node's traffic. Moreover, our model captures the important historical pattern (daily pattern) by analyzing the data from the last seven days. The historical module helps our proposed framework in both long-term and short-term prediction with significantly better performance than Graph Wavenet, LSGCN, and other models. Thirdly, weighted/attention based aggregation of the representations from the different time stamps facilitates long-term prediction. Careful observations of Table 1 reveals that our model achieves significant performance in long-term predictions (45, and 60 min) for all three datasets. Finally, the position encoding helps our model to distinguish between different patterns that existed in different parts of the day.

Table 2. Performance Comparison of Historical Model, Current-Day Model with SST-GNN(combined model) on PeMSD8

Models	15 min			30 min			45 min			60 min		
	MAE	RMSE	MAPE	MAE	RMSE	MAPE	MAE	RMSE	MAPE	MAE	RMSE	MAPE
Current-day only	1.22	2.62	2.35	1.44	2.87	2.76	1.98	3.77	3.87	2.29	4.06	4.51
Historical only	1.93	3.94	3.85	2.21	4.23	4.36	2.24	4.26	4.41	2.47	4.55	4.70
SST-GNN	**1.03**	**2.08**	**1.86**	**1.39**	**2.80**	**2.67**	**1.62**	**3.28**	**3.20**	**1.74**	**3.57**	**3.50**

Ablation Study on Contributions from Current-Day and Historical Models: We perform ablation analysis to determine which part of the model brings the main performance gain. In Table 2, we present the performance comparison among current-day model, the historical model, and the combined SST-GNN model on PeMSD8. We observe that the performance of current-day model is competitive with the state-of-the-art models showing the effectiveness of spatio-temporal aggregation scheme with current-day traffic data. Though

Table 3. SST-GNN's performance on PeMSD8 while trained on PeMSD7

Models	15 min			30 min			45 min			60 min		
	MAE	RMSE	MAPE	MAE	RMSE	MAPE	MAE	RMSE	MAPE	MAE	RMSE	MAPE
LSGCN (trained with PeMSD8; tested on PeMSD8)	1.16	2.45	2.24	**1.46**	3.28	3.02	1.66	3.75	3.51	**1.81**	4.11	3.89
SST-GNN(ours) (trained with PeMSD7; tested on PeMSD8)	**1.14**	**2.12**	**2.07**	**1.46**	**2.76**	**2.71**	**1.54**	**3.15**	**3.00**	1.94	3.69	3.74

only the current-day model or historical model can not outperform the base-lines, the combined model achieves significant performance gain, demonstrating the significance of both historical and current day traffic data on performance gain.

Generalization Ability: To observe the generalization ability of SST-GNN, we train it with PeMSD7 and test it on PeMSD8. Particularly, the PeMSD7 dataset doesn't include any weekends. However, the PeMSD8 dataset is com-paratively large and it contains both weekdays and weekends. We compare our performance with LSGCN where the LSGCN has been solely trained and tested with PeMSD8. From Table 3, it is easy to notice that our proposed model's per-formance (trained with PeMSD7; tested on PeMSD8) outperforms LSGCN while even the LSGCN is trained and tested with the PeMSD8. The only exceptions are the MAEs for 30 min (equal MAEs) and 60 min. The results demonstrate that though the PeMSD7 does not have weekends, the SST-GNN with posi-tional encoding allows proper attentional weights towards historical weekdays and weekends as well as the current day pattern for the test data.

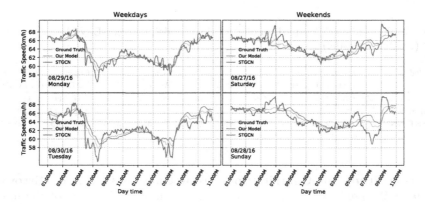

Fig. 4. Different periodic daily patterns on weekdays and weekends on PeMSD8. On the left, we can see speed decreases in morning peak and evening rush hours on weekdays whereas different traffic patterns are present on weekends.

Traffic Periodicity on Weekends and Weekdays: In Fig. 4, we plot the traffic speed of two consecutive weekdays and weekends from the PeMSD8 dataset to show how our model has learned the daily periodicity. The left column of plots demonstrates ground truths and predictions for two consecutive weekdays whereas the right column of plots depicts the ones for two consecutive weekends. In Fig. 4, we can notice that our model can capture the daily periodicity and generalize among different time periods of the weekdays and weekends performing better than STGCN as our models prediction curve is more close to ground truth. In other words, the model can sufficiently distinguish the daily patterns between weekdays and weekends while capturing the historical and current-day patterns. Particularly, the model captures the normal weekend patterns with slower traffic around the afternoon (previous weekend). It can also generalize sufficiently well in morning peaks and evening rush hours for weekdays as it can see the periodicity information from past weekdays through positional encoding.

5   Conclusion

Traffic data include repeated patterns on a daily and weekly basis. To capture the periodicity in traffic data, we design a novel spatial-temporal traffic forecasting framework that includes two different models namely historical and current-day model. The historical patterns are captured by observing the traffic history of the past seven days while the current-day model deals with the current day traffic data. Both of the models capture the spatial interrelation from different hop neighborhoods by separately aggregating different hop neighbor representations while temporal dependency is captured via a weighted spatio-temporal aggregation scheme. Again, we added relative positioning to the node's representation so that our model can distinguish traffic pattern variations from the different periods of a day as well as can discriminate different days in a week. The experimental analysis of real-life datasets verifies the effectiveness of our model in capturing the periodicity of traffic data.

Acknowledgements. This project is supported by ICT Division, Government of Bangladesh, and Independent University, Bangladesh (IUB).

References

1. Chen, C., et al.: Gated residual recurrent graph neural networks for traffic prediction. In: AAAI (2019)
2. Chen, D., Lin, Y., Li, W., Li, P., Zhou, J., Sun, X.: Measuring and relieving the over-smoothing problem for GNNs from the topological view. In: AAAI (2020)
3. Defferrard, M., Bresson, X., Vandergheynst, P.: Convolutional neural networks on graphs with fast localized spectral filtering. In: NIPS (2016)
4. Fang, S., Zhang, Q., Meng, G., Xiang, S., Pan, C.: Gstnet: global spatial-temporal network for traffic flow prediction. In: IJCAI (2019)
5. Guo, S., Lin, Y., Feng, N., Song, C., Wan, H.: Attention based spatial-temporal graph convolutional networks for traffic flow forecasting. In: AAAI (2019)

6. Hamilton, W., Ying, Z., Leskovec, J.: Inductive representation learning on large graphs. In: NIPS (2017)
7. Huang, R., Huang, C., Liu, Y., Dai, G., Kong, W.: LSGCN: long short-term traffic prediction with graph convolutional networks. In: IJCAI (2020)
8. Kipf, T.N., Welling, M.: Semi-supervised classification with graph convolutional networks. In: ICLR (2017)
9. Li, Y., Yu, R., Shahabi, C., Liu, Y.: Diffusion convolutional recurrent neural network: data-driven traffic forecasting. In: ICLR (2018)
10. Park, C., et al.: ST-GRAT: a novel spatio-temporal graph attention networks for accurately forecasting dynamically changing road speed. In: CIKM (2020)
11. Vaswani, A., et al.: Attention is all you need. In: NIPS (2017)
12. Wu, Z., Pan, S., Long, G., Jiang, J., Zhang, C.: Graph wavenet for deep spatial-temporal graph modeling. In: IJCAI (2019)
13. Yu, B., Yin, H., Zhu, Z.: Spatio-temporal graph convolutional networks: a deep learning framework for traffic forecasting. In: IJCAI (2018)
14. Zhu, J., Yan, Y., Zhao, L., Heimann, M., Akoglu, L., Koutra, D.: Beyond homophily in graph neural networks: current limitations and effective designs. In: NIPS (2020)

VIKING: Adversarial Attack on Network Embeddings via Supervised Network Poisoning

Viresh Gupta[✉] and Tanmoy Chakraborty

Indraprastha Institute of Information Technology Delhi, New Delhi, India
{viresh16118,tanmoy}@iiitd.ac.in

Abstract. Learning low-level node embeddings using techniques from network representation learning is useful for solving downstream tasks such as node classification and link prediction. An important consideration in such applications is the robustness of the embedding algorithms against adversarial attacks, which can be examined by performing perturbation on the original network. An efficient perturbation technique can degrade the performance of network embeddings on downstream tasks. In this paper, we study network embedding algorithms from an adversarial point of view and observe the effect of poisoning the network on downstream tasks. We propose VIKING, a supervised network poisoning strategy that outperforms the state-of-the-art poisoning methods by up to 18% on the original network structure. We also extend VIKING to a semi-supervised attack setting and show that it is comparable to its supervised counterpart.

1 Introduction

Several network analysis problems involve prediction over nodes and edges in a network. Traditional network science methods achieve this by analyzing networks using graph properties (such as centrality based techniques) or network factorization [4]. These methods are slow and do not scale very well. Advancements in NLP, especially word vectors [11] have inspired several embedding algorithms for graph data as well. This family of embedding algorithms uses random walks to sample neighboring nodes and optimizes node embeddings to minimize a predefined objective (e.g. Deepwalk [12] and Node2Vec [7]).

Even though network embeddings have recently gained considerable attention, embedding algorithms have not been studied to check their robustness to attacks in the network. Network attacks can generally be classified into *whitebox attacks* (where all information including parameters of the learned classifier and the model used are known) and *blackbox attacks* (where no information about the system is known; however, access to predictions is available). Blackbox attacks on networks can happen in two ways, viz. poisoning and evasion attacks. Poisoning attacks modify the network before the algorithm/model is trained; whereas evasion attacks focus on fooling a model after it has been trained [3].

In this work, we focus on developing a black-box attack method (blind to the embedding algorithm used) and achieve it via network poisoning. Although

© Springer Nature Switzerland AG 2021
K. Karlapalem et al. (Eds.): PAKDD 2021, LNAI 12714, pp. 103–115, 2021.
https://doi.org/10.1007/978-3-030-75768-7_9

adversarial attacks can happen in several ways, as far as undirected, unattributed, and unweighted networks are concerned, the only applicable poisoning attack is either the addition of an edge or the removal of an edge. These attacks are simply referred to as **edge flipping**. A significant portion of studies pertains to developing adversarial attacks on image data. However, several recent efforts [5,15,16] have also shed light on adversarial attacks on semi-supervised network learning models such as Graph Convolutional Networks (GCNs) [8]. However, there has been a lack of studies that explore the effect of perturbations on an unsupervised node embedding method. We are particularly interested in probing how the absence of a supervision signal during embedding affects such attacks.

We focus on homogeneous non-attributed networks, i.e., networks in which nodes do not have any attribute; all nodes represent the same kind of entity; and all edges are undirected and unweighted. We propose adversarial attack on network embeddings via **super̲vised network̲ poison̲ing (VIKING)**, which, unlike previous strategies, incorporates supervision during attack time (in the form of node labels). Our investigation shows that VIKING, even in a semi-supervised setting, is extremely effective. It is able to degrade the performance of the unsupervised embedding methods (and supervised embedding methods like GCN as well), leading to a decrease in the micro F1-score for node classification (up to 19% when applied on synthetic networks and up to 18% when applied on three real-world networks). VIKING also performs efficiently on the link prediction task, decreasing average precision by up to 50%. The semi-supervised counterpart of VIKING performs similar to the other baseline methods, demonstrating the possibility of attacks even with partial knowledge.

In short, our contributions are summarized below:

- We develop VIKING a generic adversarial attacking framework for discrete network features.
- We quantify the assumption of homophily behind random walk based embedding methods using node labels.
- We develop a supervised attacking strategy using the above label-based heuristic.
- We extend VIKING to a semi-supervised setting (VIKING^s) and show that it is equally effective.

Reproducibility: The code and the datasets are public at the following link: https://github.com/virresh/viking.

2   Related Work

We focus on scalable node embedding approaches (mostly based on random walks) due to their flexibility for downstream tasks. For a survey on embedding methods and the advancements, readers are referred to [6]. In the present work, our focus is on analyzing the vulnerability to adversarial attacks of those methods that do not have access to supervision signals for embedding (like Deepwalk [12], Node2Vec [7], LINE [7]) or partial access (like GCN [8]).

Attacks on networks have often focused on exploiting some model parameters. In [15] GCN is linearized for attacking, [5] creates adversarial examples using reinforcement learning approach, [16] exploits meta-gradients for attacking. However, [2] develops a poisoning strategy for attacking networks, they only leverage the information of loss function. It is possible to use supervision to degrade the performance of embedding methods; however to the best of our knowledge, no study has investigated such supervised attack methods for random walk based embedding approaches.

3   Preliminaries

In order to design an adversarial attack, we take the formal problem definition from [2], which is a bi-level optimization problem and break it down to a relatively easier problem. To limit the attacker's activities, we assume a budget restriction. We focus on a variety of embedding approaches such as – random walk based approaches (DeepWalk and Node2Vec), large scale embedding algorithm (LINE) and a semi-supervised embedding method (GCN). We consider networks denoted by $G = (V, E)$ such that they are unweighted, unattributed and undirected, with V denoting the set of vertices and E denoting the set of edges.

Let $A \in \{0, 1\}^{|V| \times |V|}$ be the adjacency matrix associated with network G. The goal of learning network embeddings is to learn a representation $Z = [z_v] \in R^K$ for each node $v \in V$ such that $K << |V|$. Such embeddings (Z) are then used by a downstream function ζ to perform an end task. As an attacker, our goal is to flip some values in A so that the new adjacency matrix A' results in learning embeddings $Z' = [z'_v]$ which in turn results in comparatively worse performance for the downstream function ζ.

In most practical scenarios, there are constraints to the amount of perturbation allowed on the network. Therefore, we fix a budget of b flips in total. Every flip f changes A, such that $||A' - A||_0 = 2$. Thus, this problem boils down to the following bi-level optimization:

$$(A')^* = \underset{A'}{\operatorname{argmax}} \, \Delta(A', Z^*) \qquad (1)$$

$$Z^* = \underset{Z}{\operatorname{argmin}} \, \zeta(A', Z) \qquad (2)$$

subject to $|A' - A|_0 = 2b$. Δ is the loss function that we will have to design specific to the problem and ζ is the embedding algorithm's loss.

Solving this optimization problem is challenging. We shall solve this bi-level optimization problem approximately by converting it to a single optimization problem and use a simple brute-force solution. To achieve this, we will decouple the embeddings Z^* from the attacking function and replace it with a proxy (defined in Sect. 4.1).

4   Network Poisoning Strategy

In this section, we begin by describing the general attacking framework to decide which edges to flip, followed by our approach where we plug in a parameter in the general attacking framework that makes use of supervision in the time of attack, and then propose a simple semi-supervised extension.

4.1   Generic Poisoning Framework

Adding or removing an edge leads to the flipping of two symmetric entries in an adjacency matrix. We form a candidate edge set from which some edges will be chosen for addition or removal as required. The procedure to compute the candidate edge set is as follows:

- **Edges to be removed:** We randomly mark one edge attached with each node as *safe*. These edges will not be removed during the poisoning attack. All the *unsafe* edges now belong to our candidate edge set. This ensures that there are no isolated vertices introduced in the resultant network even if we remove all the edges from the candidate edge set. However, if there were isolated vertices present in the network, then they will remain as is.
- **Edges to be added:** Since all non-edges in the original network can be potentially added, we compute the adjacency matrix of the network complement and then include each edge in the upper triangular portion of the matrix (to avoid repetitions) in our potential edge addition set. Since this set is generally large, we randomly sample a reasonable number of candidate edges from this set.

As mentioned before, we need to constrain our attack with a budget of b edge flips. Accordingly, we assign an importance value imp_e to every edge of the candidate set. This importance value shall determine the top b edges to be flipped and give us the optimized poisoned network A' (i.e., the adjacency matrix).

In the general attack strategy, there can be several ways to determine imp_e; however, we discuss here a simple method to compute the *importance value* for each candidate edge with respect to a particular graph property/feature d. Suppose there is a node feature d that we want to attack (e.g., node centrality or degree). Let us restrict d to be a discrete variable having at most D possible values (in case of continuous values, we can just set this to the dimensionality of the feature). We can then represent a feature matrix as $F \in \{0,1\}^{|V| \times D}$. The key insight into developing this attack is to observe that embeddings aim to preserve similarity in the embedding space. Thus, if all nodes have the same feature value, then that feature becomes useless in the embedding space.

Hence, our objective is to make this feature value of each node as close to each other as possible. For this purpose, we compute imp_e of each candidate edge flip by some function (let's say θ) of the feature/property d with respect to the new attacked graph A'. i.e.:

$$imp_e = C - \theta(F, A') \tag{3}$$

where C is a constant and θ is our defined loss.

In order to compute θ, we can "aggregate" the neighbour features by multiplying the adjacency matrix with feature matrix and weight them elementwise by the original node's feature value $((A \times F) \circ F)$; where $\circ$ represents elementwise multiplication. Since this vector is arbitrary, we will normalize this with the unweighted equivalent $(A \times F)$ and compare the distance (l2 norm) with the normalized importance of the original graph, in order to call it imp_e. Note that even though this imp_e is for determining the importance of a candidate edge, it assigns a value to the whole graph after the edge e was flipped. These operations can be summarized as follows

$$\theta(F, A') = ||(\mathrm{diag}(\sum(A' \times F)))^{-1} \sum((A' \times F) \circ F)||_2 \qquad (4)$$

where $\sum$ represents the summation across first dimension (resulting in a vector of $|V| \times 1$ for $G = (V, E)$). A' is the poisoned graph after flipping exactly one edge e and F is the feature matrix as defined above. Since we are comparing the importance across several flips with the original graph (A), we treat C as the importance of the original graph, i.e.: $C = \theta(F, A)$.

We describe the motivation for choosing such a loss function. The multiplication of two binary matrices $A \in \{0,1\}^{|V| \times |V|}$ and $F \in \{0,1\}^{|V| \times D}$ produces a new matrix $AF \in \mathbb{W}^{|V| \times D}$ with features accumulated from its neighbors. Element-wise multiplication with F again weights out the aggregated values for each node by its own feature value. A row-wise summation $\sum(AF \circ F) \in \mathbb{W}^{|V|}$ provides a numeric characterization with which we can approximate its closeness to a given value. The division by $\sum AF$ is done for normalization and allows computing distance of the numeric characterization using a constant value.

Importance of Using Constant C: We aim to bring the feature values of each node as close to each other as possible. One way of achieving this would be to bring the numeric characterization of feature values for each node as close to a fixed constant as possible. This will automatically imply bringing the values for each node closer to each other. Thus C is a scalar which allows us to choose the value at which we want our feature values in F to converge.

Now we can convert our bi-level optimisation to a single objective – minimizing imp_e. This is the proxy that can be used instead of embeddings Z^* in Eq. 1. As for solving the objective Δ, we can simply use the following brute force approach:

For every edge flip e, we can assign an importance value to the attacked graph (A') with imp_e and then sort the candidate edge set in descending order of the importance values. Now we can pick top b edges from this sorted candidate edge set to obtain the final attacked graph $(A')^*$.

To summarize, the higher the importance of a flip, the more number of node features will be brought closer on flipping that edge.

This method constitutes our generic attack framework. We can use a known node importance measure instead assuming a generic feature d, such as vertex degree or node centrality. This framework can thus be used for both unsupervised, semi-supervised or supervised attacks. To use it in unsupervised fashion, one can use a node-property that can be computed from the graph. To use

---

Algorithm 1. VIKING algorithm

Input: Adjacency matrix A of network, candidate set CS for possible edge flips
Parameter: Constant C, flip budget b
Output: Adjacency matrix A' of the poisoned network
1: **for all** flip f in CS **do**
2:     Compute $imp_f = \theta(F_\eta, A')$
3: **end for**
4: **sort** CS on the basis of Δ_f in decreasing order
5: **choose** top_f = First b flips from sorted CS
6: **apply** top_f to A
7: **return** A'

---

it in supervised or semi-supervised fashion, we just need to use an external information such as node-labels. We will show its utility as a supervised and semi-supervised method in the following sections.

4.2 VIKING: Our Proposed Poisoning Strategy

In this section, we provide a logical choice for the feature matrix F. For this purpose, we first recall why the word vector approach works. As stated in [11], embeddings try to capture similarity i.e., – "birds of a feather flock together". This is an instance of homophily which is exhibited in several real-world networks, and this is the insight that makes network embeddings a useful tool. Thus an ideal candidate for F would be a parameter that quantifies homophily at every node.

For this purpose, let us define for every node an intracommunity-intercommunity ratio $\eta = \frac{n_{same}(x)}{n_{diff}(x)}$, where $n_{diff}(x)$ is the number of neighbors of x that do not belong to the same community as x, and $n_{same}(x)$ is the number of neighbors of x that belong to the same community as x. As shown in Sect. 4.1, the generic loss function makes η converge to a constant value of our choice. Algorithm 1 summarizes VIKING. In order to estimate η, we only need to replace F in Eq. 3 by the community label matrix F_η of the given graph (i.e., a one-hot representation of the community labels of every node in the graph), where $F_\eta \in \mathbb{W}^{|V| \times \alpha}$, α is the number of unique labels/communities, and $\mathbb{W}$ represents the set of whole numbers. This particular computation of η is natively supported by our generic attack framework by simply plugging in the labels. This will result in the following objective:

$$(A')^* = \underset{A'}{\mathrm{argmin}}\, C - \theta(F_\eta, A') \tag{5}$$

Time Complexity: We choose to select $k|E|$ edges for addition (for some constant k) and at most $|E|$ edges for removal. Assuming that the average degree in the network is $\overline{d}$, the time complexity of VIKING is $O(|E||V|\overline{d})$ which is linear in the number of edges.

Table 1. Network statistics. The flip budget shown here is default.

Network	#nodes	#edges	Flip budget (%)	#communities
LFR	1000	11000	1000 (9.09)	3
Cora	2810	7981	1000 (12.25)	7
ForestFire	1000	2721	500 (18.37)	30
PolBlogs	1222	16717	1000 (5.98)	2
CiteSeer	2110	7388	1000 (18.74)	6

4.3 VIKING^s: Semi-supervised Extension

To extend VIKING into a semi-supervised setting, we apply a learned feature matrix F_η' instead of using the ground-truth F_η. This can be done using the node-classification task. First, we generate initial unsupervised embeddings Z_0 using an embedding method (skipgram DeepWalk in this case), and use these embeddings in conjunction with a logistic regression classifier trained with $x\%$ of ground-truth community labels (we take $x = 10$). The logistic regression is then used to predict F_η', the surrogate label matrix to be used in place of L in Eq. 5. The edges selected by VIKING^s can then be flipped and the resultant poisoned graph can be used for different downstream tasks (node classification and link prediction).

5 Datasets

We use two kinds of networks for the evaluation. Table 1 provides the statistics of the networks.

Synthetic Networks: We use two synthetic network generators:

(i) **LFR networks** are commonly used as benchmark for community detection in networks [9]. We treat every community as a class/label to train a node classifier. We perform experiments on different networks by choosing the power law exponent for the degree distribution to be 3, the power law exponent for the community size distribution to be 2, the desired average degree of nodes to be 20, the minimum size of communities to be 200, and varying the fraction of intra-community edges incident to each node (μ).

(ii) **Forest Fire** is another method for generating graphs. Since it doesn't directly provide us community structure, we use Louvain algorithm [1] for creating community partitions. The graph is generated using the forest fire model [10], for 1000 nodes with parameters *forward burning rate* of 0.4 and a *backward burning rate* of 0.2.

Real-World Networks: We used three standard, publicly available datasets for this purpose. These datasets are the same as the ones used in [2]. The first of them is the Cora dataset which is a citation network. The second dataset is PolBlogs. It is a network of political blogs belonging to either rightwing or leftwing, leading

Table 2. Micro F1-scores for node classification (three edge flip strategies: A: Addition, R: Removal, C: Combination) and average precision for link prediction (LP) on real world and LFR networks ($\mu = 0.3$). For link prediction, we only report results for edge deletion which has major impact on the embedding methods (effect of other edge flipping strategies is not shown here due to lack of space). The results of the best method and the second ranked method are highlighted in bold and red, respectively. **Lower value implies more efficiency to attack the embedding methods.**

	Method	SVD DeepWalk				Skipgram DeepWalk				Node2Vec				LINE				GCN			
		A	R	C	LP	A	R	C	LP	A	R	C	LP	A	R	C	LP	A	R	C	LP
(a) Cora	Clean	0.82	0.82	0.82	0.95	0.81	0.80	0.81	0.95	0.77	0.79	0.76	0.95	0.76	0.74	0.73	0.95	0.79	0.78	0.79	0.59
	Random	0.76	0.81	0.78	0.90	0.74	0.80	0.76	0.92	0.70	0.77	0.71	0.91	0.69	0.72	0.72	0.91	0.76	0.78	0.76	0.53
	UNSUP	0.81	0.76	0.76	0.89	0.79	0.73	0.72	0.93	0.75	0.68	0.67	0.90	0.73	0.62	0.61	0.93	0.77	0.75	0.74	0.46
	VIKING*	0.73	0.79	0.74	0.89	0.70	0.77	0.70	0.92	0.64	0.74	0.59	0.89	0.69	0.72	0.69	0.91	0.74	0.78	0.73	0.52
	VIKING	0.72	0.77	0.72	0.89	0.68	0.75	0.68	0.91	0.65	0.71	0.58	0.90	0.67	0.69	0.68	0.90	0.73	0.74	0.72	0.49
(b) PolBlogs	Clean	0.95	0.95	0.95	0.77	0.95	0.95	0.95	0.46	0.95	0.95	0.95	0.37	0.93	0.94	0.93	0.63	0.96	0.96	0.96	0.35
	Random	0.94	0.95	0.95	0.76	0.93	0.95	0.95	0.46	0.94	0.95	0.94	0.41	0.93	0.94	0.93	0.64	0.94	0.95	0.95	0.39
	UNSUP	0.95	0.95	0.94	0.77	0.95	0.95	0.95	0.46	0.95	0.94	0.95	0.32	0.93	0.91	0.91	0.65	0.80	0.93	0.94	0.40
	VIKING*	0.84	0.90	0.84	0.76	0.83	0.89	0.83	0.45	0.85	0.90	0.84	0.36	0.90	0.89	0.90	0.62	0.90	0.90	0.90	0.36
	VIKING	0.83	0.90	0.83	0.76	0.81	0.89	0.81	0.45	0.82	0.89	0.84	0.33	0.90	0.88	0.90	0.63	0.90	0.90	0.90	0.36
(c) CiteSeer	Clean	0.69	0.69	0.69	0.88	0.67	0.66	0.66	0.95	0.66	0.65	0.66	0.92	0.58	0.59	0.57	0.95	0.64	0.63	0.64	0.40
	Random	0.56	0.60	0.55	0.72	0.53	0.57	0.54	0.82	0.51	0.57	0.52	0.80	0.51	0.57	0.52	0.81	0.55	0.57	0.55	0.09
	UNSUP	0.63	0.52	0.52	0.77	0.55	0.50	0.48	0.91	0.53	0.45	0.47	0.85	0.49	0.38	0.38	0.81	0.59	0.52	0.59	0.10
	VIKING*	0.54	0.58	0.54	0.76	0.51	0.60	0.51	0.85	0.49	0.57	0.49	0.77	0.50	0.49	0.51	0.84	0.55	0.58	0.55	0.10
	VIKING	0.50	0.60	0.49	0.74	0.46	0.56	0.46	0.81	0.44	0.56	0.45	0.76	0.45	0.50	0.46	0.79	0.51	0.55	0.51	0.08
(d) LFR	Clean	0.63	0.63	0.63	0.63	0.75	0.74	0.75	0.35	0.59	0.62	0.60	0.34	0.71	0.71	0.71	0.38	0.59	0.58	0.59	0.13
	Random	0.62	0.61	0.62	0.65	0.73	0.73	0.72	0.45	0.60	0.60	0.57	0.34	0.70	0.71	0.70	0.37	0.58	0.58	0.58	0.13
	UNSUP	0.59	0.58	0.58	0.62	0.73	0.70	0.71	0.48	0.59	0.59	0.56	0.32	0.67	0.66	0.67	0.41	0.52	0.50	0.60	0.13
	VIKING*	0.57	0.58	0.58	0.60	0.72	0.68	0.71	0.36	0.59	0.58	0.58	0.34	0.67	0.69	0.67	0.33	0.62	0.57	0.55	0.10
	VIKING	0.56	0.55	0.55	0.59	0.68	0.63	0.67	0.38	0.57	0.59	0.56	0.34	0.67	0.66	0.66	0.32	0.50	0.60	0.53	0.11
(e) FFire	Clean	0.76	0.76	0.76	0.69	0.67	0.65	0.67	0.76	0.51	0.49	0.49	0.55	0.67	0.67	0.66	0.70	0.82	0.82	0.82	0.55
	Random	0.57	0.73	0.60	0.64	0.53	0.65	0.55	0.69	0.33	0.42	0.40	0.46	0.54	0.65	0.58	0.61	0.58	0.65	0.60	0.22
	UNSUP	0.70	0.63	0.64	0.60	0.61	0.56	0.57	0.73	0.43	0.42	0.41	0.49	0.63	0.50	0.50	0.58	0.69	0.68	0.66	0.17
	VIKING*	0.56	0.67	0.57	0.53	0.52	0.58	0.52	0.62	0.37	0.39	0.29	0.41	0.55	0.55	0.54	0.52	0.58	0.59	0.59	0.13
	VIKING	0.57	0.63	0.57	0.57	0.53	0.58	0.54	0.66	0.38	0.42	0.37	0.45	0.57	0.53	0.56	0.54	0.57	0.59	0.58	0.13

to binary classification. The third dataset is the CiteSeer dataset which is also a citation network.

6   Experimental Evaluation

The evaluation of our poisoning strategy is done using two downstream tasks – node classification and link prediction. For node-classification, network embeddings of the poisoned network are generated using an unsupervised embedding method and used as features within a simple logistic regression. The logistic regression is trained on the network embeddings obtained after poisoning the network. Only 10% of the available node labels are used for training logistic regression. Micro F1-scores are reported by averaging over 10 runs. Similarly for link-prediction, we compute cosine similarity of embeddings for node pairs and use it as prediction score for computing average precision (AP).

For evaluation, we use two techniques of generating candidate edges discussed in Sect. 3, viz addition and removal. Additionally, we combine both these in a combined strategy to observe the effect of using both the candidates together in the experiments.

Competing Methods: The following are the competing methods: (i) "Clean" refers to the clean network before poisoning; (ii) "Random" refers to the random

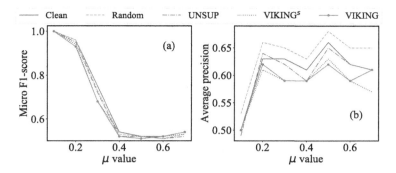

Fig. 1. Variation of performance with μ on LFR networks for (a) node classification, and (b) link prediction. μ varies from 0.1 to 0.7. Note that on extreme μ values, the attack methods fail to significantly reduce the accuracy.

edge flip baseline; (iii) "UNSUP" refers to the unsupervised poisoning baseline used in [2]; (iv) "VIKING" refers to results when the network is poisoned using our proposed method (Sect. 4.2); and (v) "VIKING^s" refers to the semi-supervised version of VIKING (Sect. 4.3).

Embedding Methods: We run all experiments using three unsupervised embedding algorithms – DeepWalk, Node2Vec, and LINE. For DeepWalk, we use two variants – the Skipgram version [12] and the SVD version [13]. The SVD version of DeepWalk approximates the objective of random walks by using matrix factorization. For Node2Vec, we use the author's original implementation provided in the SNAP package[1]. Similarly for LINE, we use the author's original implementation[2].

We also use a semi-supervised method – Graph Convolutional Network (GCN, [8]). For GCN, we use a two level network with the middle layer's size equal to the embedding dimension and final layer's size equal to the number of communities. These are trained with the node labels; hence these embeddings do not perform well on the link-prediction task relative to the node community labels. The GCN network is implemented with the help of DGL library [14].

Performance Over Tasks: Table 2 shows that VIKING performs well over both node classification and link prediction. However, even its semi-supervised setting, VIKING^s performs comparable to the supervised counterpart. On real-world datasets, VIKING performs extremely well, achieving at least 13% score reduction on each dataset using Skipgram DeepWalk. The attack strategy is also effective across various random walk based methods such as Node2Vec (LINE), resulting in decrements of up to 18% (9%) in Cora, 13% (6%) in PolBlogs, and 18% (20%) in Citeseer network. In most tasks across all the networks, VIKING and VIKING^s outperform the baseline attacking strategies.

---

[1] http://snap.stanford.edu/node2vec/.
[2] http://github.com/tangjianpku/LINE.

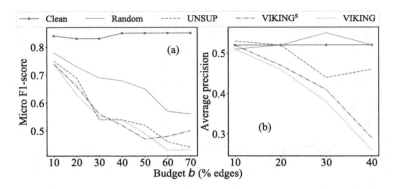

Fig. 2. Performance variation with budget for (a) node classification, and (b) link prediction on PolBlogs network. For both tasks, the performance shows a sharp decrease with VIKING outperforming others.

Since the absolute budget for all the networks is same (except Forest Fire in which the number of edges is already quite low), i.e., 1000 (Table 1), we see that the least performance degradation is in PolBlogs, which is expected because the fraction of edges flipped is small. This is also confirmed by Fig. 2. Also, the number of unique labels does not seem to affect the performance of our attack. VIKING is successful in all three networks where the number of unique labels ranges from 2 to 7.

Effect of Community Mixing (μ) in LFR Networks): Figure 1 shows the performance on the LFR network by varying μ from 0.1 to 0.7. Similar pattern is observed with VIKING outperforming others at every value of μ. We report values on LFR graphs at a non-extreme generator parameter $\mu = 0.3$ in Table 2. Even for synthetic network, VIKING and VIKING^s dominate as the strategies for attacking with a decrease in performance of 4% and 5% on Node2Vec and LINE, respectively. VIKING is clearly a better attacking strategy across the various embedding methods considered.

Side-by-Side Diagnostics: Interestingly, for link prediction, sometimes the results seem counter intuitive. For example, in Skipgram DeepWalk embeddings for LFR graphs, we observe that on all the poisoned graphs, the performance has actually improved (across all poisoning strategies) by a score of $1 - 13\%$. However, the increase is least in case of VIKING and VIKING^s. It seems to be a result of insufficient attack budget and remains a limitation of VIKING along with other strategies.

Effect of Varying Budget b: Since PolBlogs has the maximum number of edges amongst all the networks, we use it to observe how the budget affects the accuracy of the downstream task. Candidate edges have both addition and removal involved. Figure 2 clearly shows that VIKING is better than other alternatives in both node classification (Fig. 2(a)) and link prediction (Fig. 2(b)).

Inspecting Performance of Semi-supervised Approach: So far, we assumed full knowledge of the underlying network (training dataset only). Here, we discuss the observations without using full knowledge of node labels. Table 2 also shows the performance of various embedding methods in the semi-supervised attack setting with VIKING^s. The attack is successful even with partial knowledge. An attacker does not need to know the labels of all the nodes; information of even 10% labels can be effective (refer Sect. 4.3 for details on strategy).

Analysing Adversarial Edges: To investigate if the edges selected by VIKING have any distinct characteristics that may be used to eliminate usage of labels, we analyse node degrees and edge betweenness centrality values of the selected edges from Cora. Figures 3 (a, c) show a logarithmically binned heatmap of fraction of adversarial edges w.r.t. total number of edges for each degree. Nodes across all – low, medium and high – degrees are used in the adversarial edges. Figures 3 (b, d) show the distribution of edge centrality of both adversarial and non-adversarial edges. The distribution of both adversarial (red) and non-adversarial (green) edges is extremely close. Both the distributions peak at the same edge betweenness centrality value (0.0005 and 0.001 after normalisation in Figs. 3(b) and 3(d), respectively) and have a similar spread of distribution. We conclude that developing the attack heuristic isn't trivial and simple measures like degree and centrality are not sufficient to detect edges selected by VIKING.

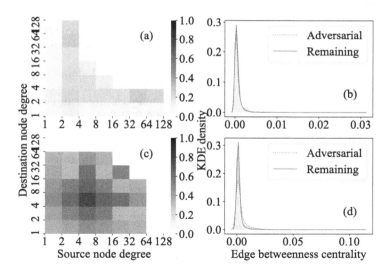

Fig. 3. Analysis of adversarial edge properties: (a, c) degree distribution of nodes on adversarial edges; (b, d) betweenness centrality distribution of adversarial edges. The distribution of properties such as node degrees and edge centralities do not suggest any intelligent heuristic. (a)–(b) correspond to Cora; (c)–(d) correspond to Citeseer. (Color figure online)

7 Conclusions

We studied the robustness of random-walk based embeddings and measured the performance of our heuristic method for supervised poisoning attacks on network data. We presented VIKING, a generic framework for adversarial poisoning attacks on networks. The experiments performed on multiple datasets included comparisons with existing methods to establish the effectiveness of VIKING. Furthermore, a semi-supervised extension of VIKING was also tested which demonstrates the efficacy of poisoning attacks even with partial label knowledge. Based on the current study, we conclude that in network science an important need is to develop robust embedding methods for large-scale networks.

Acknowledgments. The work was partially supported by ECR/2017/00169 (SERB) and the Ramanujan Fellowship.

References

1. Blondel, V.D., Guillaume, J.L., Lambiotte, R., Lefebvre, E.: Fast unfolding of communities in large networks. J. Stat. Mech.: Theory Exp. **2008**, P10008 (2008)
2. Bojchevski, A., Günnemann, S.: Adversarial attacks on node embeddings via graph poisoning. In: ICML. Proceedings of Machine Learning Research, PMLR (2019)
3. Chakraborty, A., Alam, M., Dey, V., Chattopadhyay, A., Mukhopadhyay, D.: Adversarial attacks and defences: a survey. arXiv preprint arXiv:1810.00069 (2018)
4. Chen, H., Perozzi, B., Al-Rfou, R., Skiena, S.: A tutorial on network embeddings. arXiv preprint arXiv:1808.02590 (2018)
5. Dai, H., et al.: Adversarial attack on graph structured data. arXiv preprint arXiv:1806.02371 (2018)
6. Goyal, P., Ferrara, E.: Graph embedding techniques, applications, and performance: a survey. Knowl.-Based Syst. **151**, 78–94 (2018)
7. Grover, A., Leskovec, J.: node2vec: scalable feature learning for networks. In: SIGKDD, pp. 855–864. ACM (2016)
8. Kipf, T.N., Welling, M.: Semi-supervised classification with graph convolutional networks. In: ICLR, pp. 1–14 (2017)
9. Lancichinetti, A., Fortunato, S., Radicchi, F.: Benchmark graphs for testing community detection algorithms. Phys. Rev. E **78**(4), 046110 (2008)
10. Leskovec, J., Kleinberg, J., Faloutsos, C.: Graph evolution: densification and shrinking diameters. ACM TKDD **1**(1), 2-es (2007)
11. Mikolov, T., Chen, K., Corrado, G.S., Dean, J.: Efficient estimation of word representations in vector space. CoRR abs/1301.3781 (2013)
12. Perozzi, B., Al-Rfou, R., Skiena, S.: Deepwalk: online learning of social representations. In: SIGKDD (2014)
13. Qiu, J., Dong, Y., Ma, H., Li, J., Wang, K., Tang, J.: Network embedding as matrix factorization: unifying DeepWalk, LINE, PTE, and node2vec. In: WSDM, pp. 459–467 (2018)

14. Wang, M., et al.: Deep graph library: towards efficient and scalable deep learning on graphs. In: ICLR Workshop on Representation Learning on Graphs and Manifolds, pp. 1–7 (2019)
15. Zügner, D., Akbarnejad, A., Günnemann, S.: Adversarial attacks on neural networks for graph data. In: SIGKDD, pp. 2847–2856 (2018)
16. Zügner, D., Günnemann, S.: Adversarial attacks on graph neural networks via meta learning. In: ICLR, pp. 1–15 (2019)

Self-supervised Graph Representation Learning with Variational Inference

Zihan Liao, Wenxin Liang, Han Liu[⊠], Jie Mu, and Xianchao Zhang

School of Software, Key Laboratory for Ubiquitous Network and Service Software
of Liaoning Province, Dalian University of Technology, Dalian, China
{elvis,jiem}@mail.dlut.edu.cn
{wxliang,hanliu,xczhang}@dlut.edu.cn

Abstract. Graph representation learning aims to convert the graph-structured data into a low dimensional space in which the graph structural information and graph properties are maximumly preserved. Graph Neural Networks (GNN)-based methods have shown to be effective in dealing with the graph representation learning task. However, most GNN-based methods belong to supervised learning, which depends heavily on the data labels that are difficult to access in real-world scenarios. In addition, the inherent incompleteness in data will further degrade the performance of GNN-based models. In this paper, we propose a novel self-supervised graph representation learning model with variational inference. First, we strengthen the semantic relation between node and graph level in a self-supervised manner to alleviate the issue of over-dependence on data labels. Second, we utilize the variational inference technique to capture the general pattern underlying the data, thus guaranteeing the model robustness under some data missing circumstances. Extensive experiments on three widely used citation network datasets show that our proposed method has achieved or matched state-of-the-art results on link prediction and node classification tasks.

Keywords: Graph representation learning · Self-supervised learning · Variational inference

1 Introduction

In recent years, graph representation learning has raised a surge of interest [8,19,22]. It aims to model the graph-structured data with an efficient representation that can generalize well to a wide range of downstream graph mining tasks, such as node classification, link prediction, community detection, etc. Generally, the graph-structured data in real life contains the graph topology and a large amount of high-dimensional attribute information. For example, in the Twitter network, there exists both friend relationships (topology) and user portraits (attribute information). Furthermore, each node in the graph belongs to a specific category, which is difficult to achieve in reality. In the field of computer vision, many works try to use data augmentation strategy [23,24] to get more

K. Karlapalem et al. (Eds.): PAKDD 2021, LNAI 12714, pp. 116–127, 2021.
https://doi.org/10.1007/978-3-030-75768-7_10

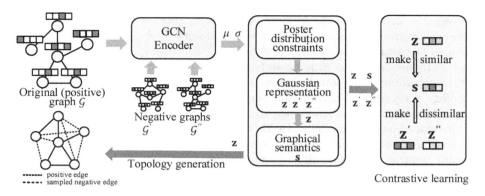

Fig. 1. The overall framework of SGRLVI. The topology and properties of graph $\mathcal{G}$ are first fed into the GCN encoder to obtain the nodes' distribution, which is constrained to approximate the standard Gaussian distribution. We sample the Gaussian representation of each node through the reparameterization trick [11] and then calculate the graphical semantics accordingly. We regard the original graph $\mathcal{G}$ as the positive graph. In addition, we construct two negative graphs and obtain the corresponding Gaussian representations of the negative nodes. All the positive and negative representations are utilized to conduct the contrastive self-supervised learning. Finally, the representations of positive graph nodes are exploited for topology generation.

labeled data to improve the supervision of model training, such as image flipping and image rotating. However, such data augmentation strategies are not suitable for graph-structured data because flipping and rotating operations can neither change the adjacent structure nor the attribute information of the graph. Therefore, an unsupervised graph representation learning method that does not require label information during training shows its flexibility and becomes a challenging problem to be solved urgently.

Thanks to the recent emergence of self-supervised learning methods [14], many works seek to obtain valuable information based on the data itself to strengthen the model training process to achieve better performance. In natural language processing and computer vision, high-quality continuous representations can be trained in a self-supervised manner by predicting context information or solving various pretext tasks [2,20]. However, most graph representation learning algorithms adopt labels as external guidance, which are hard and expensive to access. Many researchers have proposed unsupervised countermeasures [4] in response to such a situation while requiring sophisticated design and rich professional knowledge. Accordingly, in the graph mining domain, whether it is possible to train a representation learning model by relying solely on the abundant unlabeled data itself in a self-supervised way has become a problem that researchers are committed to solving. One important work, namely DGI [26], sets the goal to maximize mutual information to train graph representation.

In this paper, we propose a novel self-supervised graph representation learning model with variational inference (SGRLVI) that gets rid of the dependence on labels. The model does not require manually sophisticated design or rich professional knowledge but supervises the training process by extracting critical information from the abundant unlabeled data itself. As illustrated in Fig. 1, through the graph convolutional network (GCN) encoder, we obtain the Gaussian representations of each node and calculate the graphical semantic representation accordingly. By jointly optimizing the contrastive learning module and topology generation as the training objective, high-quality node representation can be learned. Moreover, the generative model of variational autoencoder (VAE) is introduced into our model to tackle the problem of incompleteness in data.

Extensive experiments are conducted on three widely used datasets. The results confirm the rationality of our proposed model, which can achieve or match state-of-the-art results on link prediction and node classification tasks.

2 Related Work

2.1 Self-supervised Learning

Self-supervised learning has attracted significant research interests due to its ability to use custom pseudo-labels as supervision and avoid over-dependence on data labels [9]. It has successfully exhibited strong performance in many fields, such as computer vision [2], natural language processing [20], and robotics [10]. Currently, mainstream self-supervised learning models can be mainly divided into generative learning and contrastive learning [14]. The former one is mainly to reconstruct the original data through auto-regressive models and autoencoders. Unlike the generative model, contrastive learning is a discriminative method, which aims to group the samples with similar representations in the embedding space closer and vice versa. In computer vision, the commonly used method is to adopt data augmentation techniques (e.g., image rotation) to get more positive samples and then train the model to access the ability to distinguish positive or negative samples. However, such kind of data augmentation techniques are not suitable for graph-structured data. In this work, our approach is to achieve contrastive self-supervised learning on graphs based on the perspective of natural language without any external reliance on labels.

2.2 Graph Representation Learning

Traditional graph representation learning methods mostly utilize random walk [3] or matrix factorization [27] tools to solve the graph representation learning problem. With the rise of Graph Neural Networks (GNNs), many graph representation learning methods based on GNNs have been proposed and achieve better performance [5,13,25]. GCN [13] pioneers a graph convolutional network, in which each graph convolutional layer can successfully merge the node attributes and the topology of the graph. GAT [25] proposes a graph attention layer on the basis of GCN, which can selectively integrate the attribute information of neighboring nodes in each propagation process. Despite their success, the premise for these approaches to achieve excellent performance is the need for data to provide external labels to supervise the models' training, which is usually hard to access in realistic scenarios. In contrast, by manually sophisticated design, unsupervised methods are more flexible and have broader applications. Graph-SAGE [5] successfully leverages node attributes to generate node embeddings via a hierarchical sampling and aggregating framework. More recently, DGI [26] introduced the idea of DIM [7] into graph representation learning, which maximizes the mutual information between patch and graph level. However, DGI only excavates the mesoscopic community structure in the graph and ignore the microscopic structure, such as the link information. In this work, we propose a novel graph representation learning model, which the learned graph representation can not only capture the mesoscopic community structure but also preserve the link information in the graph.

3   The Proposed Method

3.1   Problem Formulation

For an undirected graph $\mathcal{G} = (\mathcal{V}, \mathcal{E}, \mathbf{X})$, where $\mathcal{V}$ is the set of n nodes, $\mathcal{E}$ represents the set of edges between the nodes and $\mathbf{X} \in \mathbb{R}^{n \times m}$ is the graph property matrix where $\mathbf{x}_i$ is the m-dimensional attribute vector of v_i. Besides, each graph is associated with an adjacency matrix $\mathbf{A}$. For graph $\mathcal{G}$, our objective is to learn a continuous latent representation $\mathbf{z}_i \in \mathbb{R}^d$ for each node $v_i \in \mathcal{V}$, where d $(d < m)$ is the final dimension of the representation. In particular, the node representations should be generalized well to downstream graph mining tasks.

3.2   Overall Framework

The overall Framework of SGRLVI is shown in Fig. 1. The GCN encoder accepts graph topology and attributes as input, outputs the mean and variance of the Gaussian distribution of the nodes $q_\phi(\mathbf{Z}|\mathbf{A}, \mathbf{X})$ over the latent space. Following the idea from VAE [11], we approximate the obtained node distribution to standard Gaussian distribution and sample a Gaussian representation $\mathbf{z}$ for each node. Then we calculate the graphical semantic representation $\mathbf{s}$ based on the node representation. We treat the original graph as the positive graph and construct two negative graphs in the initial stage of each epoch and also get the sampled Gaussian representation of the negative nodes $\mathbf{z}', \mathbf{z}''$. All the representations are adopted to conduct the contrastive self-supervised learning module, and the positive node representations are also exploited for topology generation.

3.3   Variational Inference

Regarding the incompleteness of the data, we leverage the VAE generative framework. Specifically, we treat $\mathbf{X}$, $\mathbf{A}$, and $\mathbf{Z}$ as random variables and expect the generative model can learn the patterns underlies the data.

Given a node v_i with its attribute $\mathbf{x}_i$, we aim to generate its topology $\mathbf{a}_i$ from the latent representation $\mathbf{z}_i$. We denote $q_\phi(\mathbf{Z}|\mathbf{A}, \mathbf{X})$ as the inference model with parameters ϕ to be learned to approximate the intractable posterior $p_\theta(\mathbf{Z}|\mathbf{A}, \mathbf{X})$. By minimizing the KL-divergence between them, which is formulated as:

$$D_{KL}(q_\phi(\mathbf{Z}|\mathbf{A},\mathbf{X})||p_\theta(\mathbf{Z}|\mathbf{A},\mathbf{X})) = \int q_\phi(\mathbf{Z}|\mathbf{A},\mathbf{X}) \ln \frac{q_\phi(\mathbf{Z}|\mathbf{A},\mathbf{X})}{p_\theta(\mathbf{Z}|\mathbf{A},\mathbf{X})}\, d\mathbf{Z}$$

$$= \int q_\phi(\mathbf{Z}|\mathbf{A},\mathbf{X})\big(\ln q_\phi(\mathbf{Z}|\mathbf{A},\mathbf{X}) + \ln p_\theta(\mathbf{A}|\mathbf{X}) - \ln p_\theta(\mathbf{Z},\mathbf{A}|\mathbf{X})\big)\, d\mathbf{Z} \qquad (1)$$

$$= \ln p_\theta(\mathbf{A}|\mathbf{X}) + \mathbb{E}_{q_\phi(\mathbf{Z}|\mathbf{A},\mathbf{X})}\big(\ln q_\phi(\mathbf{Z}|\mathbf{A},\mathbf{X}) - \ln p_\theta(\mathbf{Z},\mathbf{A}|\mathbf{X})\big).$$

According to Eq. 1, the log-likelihood of topology given attributes $\ln p_\theta(\mathbf{A}|\mathbf{X})$ can be represented as:

$$\ln p_\theta(\mathbf{A}|\mathbf{X}) = D_{KL} + \mathbb{E}_{q_\phi(\mathbf{Z}|\mathbf{A},\mathbf{X})}\big(\ln p_\theta(\mathbf{Z},\mathbf{A}|\mathbf{X}) - \ln q_\phi(\mathbf{Z}|\mathbf{A},\mathbf{X})\big), \qquad (2)$$

for the value of the KL-divergence term is greater than 0, so maximizing the log-likelihood of the node topology given by its attributes is equivalent to maximizing its Evidence Lower Bound (ELBO) using an inference model [11]:

$$\begin{aligned}\text{ELBO} &= \mathbb{E}_{q_\phi(\mathbf{Z}|\mathbf{A},\mathbf{X})}\big(\ln p_\theta(\mathbf{Z},\mathbf{A}|\mathbf{X}) - \ln q_\phi(\mathbf{Z}|\mathbf{A},\mathbf{X})\big) \\ &= \mathbb{E}_{q_\phi(\mathbf{Z}|\mathbf{A},\mathbf{X})} \ln p_\theta(\mathbf{A}|\mathbf{Z},\mathbf{X}) - D_{KL}(q_\phi(\mathbf{Z}|\mathbf{A},\mathbf{X})||p_\theta(\mathbf{Z})),\end{aligned} \qquad (3)$$

where the approximated posterior $q_\phi(\mathbf{Z}|\mathbf{A},\mathbf{X})$ is a Gaussian distribution $\mathcal{N}(\mu,\sigma^2\mathbf{I})$ with the mean μ and variance σ^2 are learned by GCN encoder which composed of a two-layer GCN network. Note that ELBO consists of two parts. In the first term, given the sampled latent representation $\mathbf{Z}$ obtained by probabilistic encoder q_ϕ and attributes $\mathbf{X}$, the model seeks to generate the adjacency $\mathbf{A}$. During the decoding phase, we adopt the simple inner product decoder proposed by VGAE [12]. We define the loss function L_{recon} as:

$$L_{recon} = -\sum_{(i,j)\in\mathcal{E}} \text{sigmoid}(\mathbf{z}_i^T\mathbf{z}_j) + \sum_{(k,l)\in\mathcal{E}'} \text{sigmoid}(\mathbf{z}_k^T\mathbf{z}_l), \qquad (4)$$

where $\mathcal{E}'$ represents the set of negative edges that we randomly selected with the same number of edges in $\mathcal{E}$. As for the second KL-divergence term in Eq. 3, the posterior Gaussian distribution of the nodes' representations in the latent space $q_\phi(\mathbf{Z}|\mathbf{A},\mathbf{X})$ is expected to approximate the prior distribution $p_\theta(\mathbf{Z})$. Following the setting of VAE, we define the prior distribution of latent representation as standard Gaussian distribution, i.e., $p_\theta(\mathbf{Z}) = \mathcal{N}(0,\mathbf{I})$. The second term in Eq. 3 has an alternate form for calculation, which is employed as loss function L_{poster}.

$$D_{KL}(q_\phi(\mathbf{Z}|\mathbf{A},\mathbf{X})||p_\theta(\mathbf{Z})) = \frac{1}{2n}\sum_{i\in\mathcal{V}}\big((\sigma_i^2+\mu_i^2) - \log(\sigma_i^2) - 1\big), \qquad (5)$$

$$L_{poster} = \frac{1}{2n}\sum_{i\in\mathcal{V}}\big((\sigma_i^2+\mu_i^2) - \log(\sigma_i^2) - 1\big), \qquad (6)$$

thus, the training objective of Eq. 6 enables the learned node distribution to approximate the prior standard Gaussian distribution.

To sum up, given the attributes and topology of the graph, we adopt a variational autoencoder to learn a Gaussian representation $\mathbf{z}$ for each node through variational inference, and $\mathbf{z}$ can be used to generate the node topology.

Algorithm 1. SGRLVI

Input: Graph $\mathcal{G} = (\mathcal{V}, \mathcal{E}, \mathbf{X})$; trade-off parameter λ.
Output: representation vector $\mathbf{z}_i$ for each node $v_i \in \mathcal{V}$.
 1: Generate negative graphs $\mathcal{G}', \mathcal{G}''$.
 2: Initialize parameters of GCN encoder with Xavier Initialization.
 3: **while** L not converge **do**
 4:     Compute μ_i, σ_i by GCN encoder.
 5:     Compute $\mu_i', \mu_i'', \sigma_i', \sigma_i''$ by GCN encoder.
 6:     Sample $(\mathbf{z}_i, \mathbf{z}_i', \mathbf{z}_i'') \sim (q_\phi(\mathbf{z}), q_\phi(\mathbf{z}'), q_\phi(\mathbf{z}''))$ using Eq. 7.
 7:     Compute the graphical semantic representation $\mathbf{s}$ using Eq. 8.
 8:     Compute the overall loss L according to Eq. 10.
 9:     Back propagation and update parameters of SGRLVI.
10: **end while**
11: **return** representation $\mathbf{z}_i$ for each node $v_i \in \mathcal{V}$.

3.4   Contrastive Graphical Semantics Preservation

In this subsection, we adopt a contrastive self-supervised learning approach to preserve the graphical semantics. Specifically, we analogize a graph to a paragraph in natural language. The paragraph has central semantics, which represents its core idea. We believe that a graph should also contain such central semantics. We first sample the Gaussian representation of each node v_i based on the posterior distribution by employing the reparameterization trick [11].

$$\mathbf{z}_i = \mu_i + \sigma_i \cdot \epsilon_i, \epsilon_i \sim U(0, 1), \tag{7}$$

where $\cdot$ is the element-wise product and ϵ_i represents a random noise, one can note that since the node representations are obtained through GCN, they all fuse the local neighborhood information. As for the graphical central semantics $\mathbf{s}$, we believe that the attributes of all graph nodes represent the overall semantics. For example, the occurrence frequency of keywords involved in all papers in the citation network reflects the citation network's theme. In this paper, we leverage the sigmoid average pooling as the readout operation to obtain $\mathbf{s}$.

$$\mathbf{s} = \mathrm{sigmoid}(\frac{1}{n} \sum_{i \in \mathcal{V}} \mathbf{z}_i). \tag{8}$$

We believe that in natural language, a phrase composed of multiple context words must be related to the paragraph's central semantics. Similarly, we treat the original graph as the positive sample and assume each positive graph node representation containing neighborhood information (i.e., $\mathbf{z}$) should be similar to the graphical semantic representation (i.e., $\mathbf{s}$). We maximize the similarity between them. As for negative samples, we also start from the perspective of a paragraph. Suppose the order of the words in the paragraph changes, the semantics of the paragraph will also change. To take a simple example as an illustration, "I will study first, and then play outside" and "I will play outside first, and then study" exactly means different semantics. In addition, if the

Table 1. Statistics of datasets.

Dataset	# Nodes	# Edges	# Attributes	# Labels
Cora	2,708	5,278	1,433	7
Citeseer	3,312	4,552	3,703	6
Pubmed	19,717	44,324	500	3

words involved in the paragraph change directly, the semantics alter directly. We consider the two negative samples of the graph accordingly. The first one corresponds to the former, we randomly shuffle each node's topology under other conditions unchanged. The second one, we randomly transform the column of node attribute vectors without any other changes. These two manual graphs are constructed as negative samples. By sharing the parameters of the GCN encoder with graph $\mathcal{G}$, we also get their node Gaussian representations $\mathbf{z}'$ and $\mathbf{z}''$. Finally, we treat it as a classification problem and use cross-entropy as the loss function:

$$L_{cl} = -\sum_{i\in\mathcal{V}} \Big(\log d(\mathbf{z}_i, \mathbf{s}) * 2 + \log(1 - d(\mathbf{z}'_i, \mathbf{s})) + \log(1 - d(\mathbf{z}''_i, \mathbf{s})) \Big), \qquad (9)$$

where $d(\cdot)$ is the inner product to measure the similarity between node representation and graphical semantics. In general, we adopted a contrastive self-supervised learning method to assist the model in learning more useful information by making use of unlabeled data alone.

Overall, the objective of SGRLVI consists of three parts, namely topology generation, posterior distribution regularization, and graphical semantics preservation. The final loss function can be formulated as:

$$L = L_{recon} + L_{poster} + \lambda L_{cl}, \qquad (10)$$

where λ is a trade-off parameter. We adopt Adam optimizer to minimize the loss based on Eq. 10 until convergence. Finally, we take $\mathbf{z}$ as the final node representation. The learning algorithm is summarized in Algorithm 1.

4   Experiments

4.1   Experimental Settings

Datasets. We select three widely used datasets to verify the performance of SGRLVI. The statistics are shown in Table 1. Cora [1] contains 2,708 papers from 7 categories and 5,278 citation links. Each paper is associated with a 1,433-dimensional bag-of-words vector. Citeseer [21] consists of 3,312 papers from 6 categories and 4,552 links, and a 3,703-dimensional attribute vector accompanies each paper. Pubmed [18] contains 19,717 papers of 3 classes and 44,324 citation links. The attributed vector of each node is a 500-dimensional TF-IDF vector.

Table 2. Link prediction results over Cora, Citeseer and Pubmed. The **bold** represents the best performance among unsupervised methods. ($\pm$) denotes the standard deviation of 10 independent trials, and (l) indicates the algorithm is a semi-supervised method.

Dataset	Cora		Citeseer		Pubmed	
	AUC	AP	AUC	AP	AUC	AP
DeepWalk	0.831 ± 0.13	0.837 ± 0.11	0.795 ± 0.08	0.816 ± 0.12	0.842 ± 0.08	0.831 ± 0.07
GAE	0.913 ± 0.04	0.920 ± 0.05	0.894 ± 0.05	0.899 ± 0.02	0.944 ± 0.02	0.945 ± 0.02
VGAE	0.924 ± 0.01	0.926 ± 0.01	0.918 ± 0.02	0.921 ± 0.02	0.954 ± 0.00	0.947 ± 0.01
Gra.SAGE	0.805 ± 0.03	0.813 ± 0.04	0.832 ± 0.02	0.821 ± 0.03	0.867 ± 0.01	0.849 ± 0.04
CAN	0.923 ± 0.02	0.929 ± 0.02	0.935 ± 0.01	0.937 ± 0.01	0.980 ± 0.02	0.977 ± 0.02
SIG-VAE	0.958 ± 0.03	0.951 ± 0.05	0.953 ± 0.02	0.956 ± 0.03	0.951 ± 0.06	0.942 ± 0.06
DGI	0.946 ± 0.02	0.941 ± 0.03	0.943 ± 0.04	0.941 ± 0.04	0.959 ± 0.04	0.963 ± 0.04
SGRLVI	$\mathbf{0.977 \pm 0.02}$	$\mathbf{0.972 \pm 0.03}$	$\mathbf{0.983 \pm 0.01}$	$\mathbf{0.981 \pm 0.01}$	$\mathbf{0.983 \pm 0.02}$	$\mathbf{0.980 \pm 0.02}$
GCN(l)	0.912 ± 0.00	0.908 ± 0.01	0.915 ± 0.01	0.916 ± 0.02	0.917 ± 0.03	0.912 ± 0.02
GAT(l)	0.933 ± 0.03	0.927 ± 0.03	0.930 ± 0.02	0.933 ± 0.02	0.927 ± 0.03	0.932 ± 0.03

Baselines. We compare SGRLVI with the representative models of the following methods: network embedding (NE)-based, VAE-based, and GNN-based models.

– **NE-based:** DeepWalk [19] is the pioneer of network embedding, which generates the node embeddings by random walks and the Skip-gram model proposed in Word2vec [17].
– **VAE-based:** VGAE [12] is an unsupervised framework based on VAE [11]. CAN [16] adopts VAEs to encode both structure and attributes into the same semantic space. SIG-VAE [6] employs a hierarchical variational framework and a Bernoulli-Poisson link decoder to model the graphical structure.
– **GNN-based:** GCN [13] is a semi-supervised model to embed the nodes by employing a first-order approximation of spectral filters. GAT [25] introduce attention mechanism to aggregate neighbors with attention weights. GAE [12] uses GCNs as the encoder and reconstructs the adjacency matrix in the decoder. Gra.SAGE [5] is an inductive approach that generates embeddings by sampling and aggregating. DGI [26] learns node representations by maximizing mutual information between the patch and graph level.

Parameter Settings and Metrics. We adopt two-layer GCNs as encoder and set the number of neurons in the hidden layer and the final layer to 32 and 8, respectively. The parameter λ is set as 5. We train the model via Adam optimizer with the learning rate 0.01. For node classification, we take accuracy (ACC) as the evaluation metric. For link prediction, the Area Under Curve (AUC) and Average Precision (AP) are used to verify the performance. We repeat each experiment 10 times and report the mean and standard deviation of them.[1]

---

[1] Our reference code may be found at https://github.com/DLUTElvis/SGRLVI.

Table 3. Classification accuracy over Cora, Citeseer and Pubmed. The **bold**, ($\pm$) and (l) are the same as the definition in Table 2.

Dataset	Cora	Citeseer	Pubmed
DeepWalk	0.746 ± 0.07	0.510 ± 0.11	0.733 ± 0.07
GAE	0.761 ± 0.01	0.574 ± 0.02	0.784 ± 0.02
VGAE	0.746 ± 0.02	0.551 ± 0.02	0.759 ± 0.02
Gra.SAGE	0.783 ± 0.02	0.650 ± 0.01	0.775 ± 0.02
CAN	0.763 ± 0.03	0.652 ± 0.03	0.760 ± 0.03
SIG-VAE	0.798 ± 0.04	0.694 ± 0.04	0.776 ± 0.07
DGI	0.815 ± 0.08	$\mathbf{0.721 \pm 0.06}$	0.772 ± 0.06
SGRLVI	$\mathbf{0.818 \pm 0.01}$	0.708 ± 0.02	$\mathbf{0.797 \pm 0.01}$
GCN(l)	0.813 ± 0.01	0.703 ± 0.01	0.786 ± 0.02
GAT(l)	0.830 ± 0.01	0.725 ± 0.01	0.790 ± 0.01

4.2 Link Prediction Results

In this subsection, we conduct link prediction task to evaluate the graph structure reconstruction ability of the node representations. We process the datasets like previous works [6,12]. Specifically, we randomly select 5% of the edges for verification, 10% of the edges and the equal number of non-existent edges as the test set. Given the representations of two nodes $\mathbf{z}_i, \mathbf{z}_j$, we calculate the probability of the edge between these two nodes e_{ij} by $p(e_{ij}) = \text{sigmoid}(\mathbf{z}_i^T \mathbf{z}_j)$. Table 2 reports the AUC and AP scores of all baselines over the three datasets. Overall, GAE, VGAE, CAN, SIG-VAE, and our SGRLVI perform better than others, which is mainly benefit from the adjacency matrix reconstruction during training. Besides, SGRLVI shows superior performance than both the unsupervised and the supervised baselines, which further proves the effectiveness of our model.

4.3 Node Classification Results

To further demonstrate the performance of SGRLVI, we perform node classification task. After training the model, we randomly select 20 nodes in each category for training, all baselines are evaluated on 1000 remaining nodes, and other 500 nodes are used for verification. The results are shown in Table 3. The results show that the supervised models generally work better than the unsupervised counterparts, proving the importance of external label guidance to the improvement of model performance. Besides, SGRLVI is superior to other unsupervised methods in most cases and achieves significant promotion over Pubmed compared with all other baselines, which verifies the effectiveness of our assumption.

4.4 Evaluation on Missing Data

In this subsection, we randomly remove 10%, 30%, and 50% of node attributes and replace them with zero vectors to simulate the missing-data setting. We

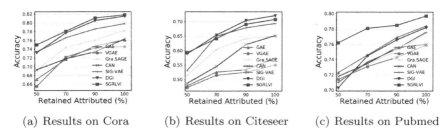

| (a) Results on Cora | (b) Results on Citeseer | (c) Results on Pubmed |

Fig. 2. Classification accuracy over Cora, Citeseer and Pubmed under the missing-data setting. The attributes of 10%, 30%, and 50% of the nodes are randomly removed.

conduct node classification experiment compared with unsupervised methods to verify the robustness under such situation. We report the node classification performance w.r.t. ACC, the experimental results are shown in Fig. 2. Easy to find that our SGRLVI shows relative high classification performance, which mainly benefit from the use of generative framework. Therefore, even if part of the data is missing, our model can learn the patterns underlies the data to obtain a robust graph representation which is useful for downstream tasks.

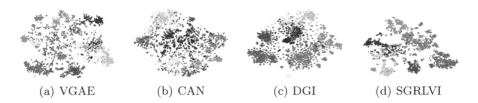

| (a) VGAE | (b) CAN | (c) DGI | (d) SGRLVI |

Fig. 3. t-SNE [15] visualization for graph representation learned by different methods on Cora dataset. Each point denotes a node, with the color represents its category. (Color figure online)

4.5 Visualization

To further verify the rationality of our proposed model, we project the learned graph representation of Cora into the two-dimensional space by t-SNE [15] visualization tool. Each point indicates a node in the graph, with its color denotes its category. We compared our model with three other unsupervised graph representation learning methods, and the visualization results are shown in Fig. 3. The results of VGAE and CAN exist a large number of points with different colors mixed together. SGRLVI obtained a better graph representation than DGI, for the node representations with the same color learned by SGRLVI are close together, with clear boundaries between different clusters.

4.6 Parameter Investigation

We investigate the parameter sensitivity with different settings of parameters. We report the results of node classification w.r.t. ACC. We tune the trade-off

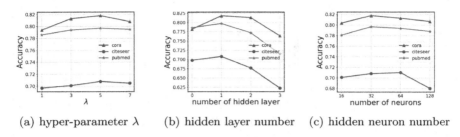

| (a) hyper-parameter λ | (b) hidden layer number | (c) hidden neuron number |

Fig. 4. Parameter analysis on hyper-parameter λ, number of hidden layer and neurons.

parameter $\lambda = \{1, 3, 5, 7\}$ and from Fig. 4 (a), it achieves the best performance when $\lambda = 5$. For the depth of the GCN encoder, we change the number of hidden layers in the range of $\{0, 1, 2, 3\}$. According to Fig. 4 (b), the performance gets the best when it equals 1, we adopt a single hidden layer GCN as the encoder. We finally evaluate the single hidden layer with different number of neurons. Easy to find in Fig. 4 (c) that the model performs well when the number is either 32 or 64, therefore, the model is insensitive to this parameter, we employ 32 hidden neurons as default for all datasets.

5   Conclusion

In this paper, we propose a novel self-supervised graph representation learning model, namely SGRLVI. By exploiting the semantic relation between node and graph level with a contrastive self-supervised method, we avoid the issue of over-dependence on data labels. By utilizing the variational inference technique to capture the general pattern underlying the data, we improve the robustness of the model under some data missing circumstances. Extensive experimental results on three widely used graph-structured datasets show the superiority of our method. Besides, the experiments in data missing scenarios further validate the robustness of SGRLVI. For future work, we plan to extend our method to solve more real applications, such as social networks and biological networks.

Acknowledgements. This work was supported by the National Science Foundation of China (Grant No. 61632019), the Fundamental Research Funds for the Central Universities (Grant No. DUT19RC(3)048, DUT20RC(3)040 and DUT20GF106).

References

1. Bhagat, S., Cormode, G., Muthukrishnan, S.: Node classification in social networks. In: SNDA, pp. 115–148 (2011)
2. Gidaris, S., Singh, P., Komodakis, N.: Unsupervised representation learning by predicting image rotations. In: ICLR (2018)
3. Grover, A., Leskovec, J.: node2vec: scalable feature learning for networks. In: SIGKDD, pp. 855–864 (2016)

4. Grover, A., Zweig, A., Ermon, S.: Graphite: iterative generative modeling of graphs. In: ICML, pp. 2434–2444 (2019)
5. Hamilton, W.L., Ying, Z., Leskovec, J.: Inductive representation learning on large graphs. In: NeurIPS, pp. 1024–1034 (2017)
6. Hasanzadeh, A., Hajiramezanali, E., Narayanan, K.R., Duffield, N., Zhou, M., Qian, X.: Semi-implicit graph variational auto-encoders. In: NeurIPS, pp. 10711–10722 (2019)
7. Hjelm, R.D., et al.: Learning deep representations by mutual information estimation and maximization. In: ICLR (2019)
8. Hou, Y., Chen, H., Li, C., Cheng, J., Yang, M.: A representation learning framework for property graphs. In: SIGKDD, pp. 65–73 (2019)
9. Jaiswal, A., Babu, A.R., Zadeh, M.Z., Banerjee, D., Makedon, F.: A survey on contrastive self-supervised learning. CoRR (2020)
10. Jang, E., Devin, C., Vanhoucke, V., Levine, S.: Grasp2vec: learning object representations from self-supervised grasping. In: CoRL, pp. 99–112 (2018)
11. Kingma, D.P., Welling, M.: Auto-encoding variational bayes. In: ICLR (2014)
12. Kipf, T., Welling, M.: Variational graph auto-encoders. arxiv (2016)
13. Kipf, T., Welling, M.: Semi-supervised classification with graph convolutional networks. In: ICLR (2017)
14. Liu, X., et al.: Self-supervised learning: generative or contrastive. CoRR (2020)
15. Maaten, L.V.D., Hinton, G.: Visualizing data using t-SNE. JMLR 9, 2579–2605 (2008)
16. Meng, Z., Liang, S., Bao, H., Zhang, X.: Co-embedding attributed networks. In: WSDM, pp. 393–401 (2019)
17. Mikolov, T., Sutskever, I., Chen, K., Corrado, G.S., Dean, J.: Distributed representations of words and phrases and their compositionality. In: NeurIPS, pp. 3111–3119 (2013)
18. Namata, G., London, B., Getoor, L., Huang, B.: Query-driven active surveying for collective classification. In: MLG (2012)
19. Perozzi, B., Al-Rfou, R., Skiena, S.: Deepwalk: online learning of social representations. In: SIGKDD, pp. 701–710 (2014)
20. Schneider, S., Baevski, A., Collobert, R., Auli, M.: wav2vec: unsupervised pretraining for speech recognition. In: ISCA, pp. 3465–3469 (2019)
21. Sen, P., Namata, G., Bilgic, M., Getoor, L., Gallagher, B., Eliassi-Rad, T.: Collective classification in network data. AI Mag. 29, 93–106 (2008)
22. Sharma, C., Chauhan, J., Kaul, M.: Learning representations using spectral-biased random walks on graphs. In: IJCNN, pp. 1–8 (2020)
23. Shijie, J., Ping, W., Peiyi, J., Siping, H.: Research on data augmentation for image classification based on convolution neural networks. In: CAC, pp. 4165–4170 (2017)
24. Timofte, R., Rothe, R., Gool, L.V.: Seven ways to improve example-based single image super resolution. In: CVPR, pp. 1865–1873 (2016)
25. Velickovic, P., Cucurull, G., Casanova, A., Romero, A., Liò, P., Bengio, Y.: Graph attention networks. In: ICLR (2018)
26. Velickovic, P., Fedus, W., Hamilton, W.L., Liò, P., Bengio, Y., Hjelm, R.D.: Deep graph infomax. In: ICLR (2019)
27. Yang, C., Liu, Z., Zhao, D., Sun, M., Chang, E.Y.: Network representation learning with rich text information. In: IJCAI, pp. 2111–2117 (2015)

Manifold Approximation and Projection by Maximizing Graph Information

Bahareh Fatemi[1], Soheila Molaei[1(✉)], Hadi Zare[1], and Shirui Pan[2]

[1] Department of Network Science and Technologies, University of Tehran,
Tehran, Iran
{bfatemi,soheila.molaei,h.zare}@ut.ac.ir
[2] Faculty of Information Technology, Monash University, Melbourne, Australia
shirui.pan@monash.edu

Abstract. Graph representation learning is an effective method to represent graph data in a low dimensional space, which facilitates graph analytic tasks. The existing graph representation learning algorithms suffer from certain constraints. Random walk based methods and graph convolutional neural networks, tend to capture graph local information and fail to preserve global structural properties of graphs. We present MAPPING (Manifold APproximation and Projection by maximizINg Graph information), an unsupervised deep efficient method for learning node representations, which is capable of synchronously capturing both local and global structural information of graphs. In line with applying graph convolutional networks to construct initial representation, the proposed approach employs an information maximization process to attain representations to capture global graph structures. Furthermore, in order to preserve graph local information, we extend a novel manifold learning technique to the field of graph learning. The output of MAPPING can be easily exploited by downstream machine learning models on graphs. We demonstrate our competitive performance on three citation benchmarks. Our approach outperforms the baseline methods significantly.

Keywords: Representation learning · Graph embedding · Manifold learning · Feature extraction

1 Introduction

Graph (network) data structure is considered to be one of the most informative but challenging data structures widely used in a variety of applications. Social networks, recommender systems, and molecular graph structures are some of the most well-known graph use-cases in which the interactions between individual units are modeled in the form of graphs [16]. Representation learning (Graph embedding) provides a framework on graph summarization by mapping each node of the graph into a lower-dimensional space. The goal is to optimize

---

B. Fatemi and S. Molaei—Equal Contribution.

© Springer Nature Switzerland AG 2021
K. Karlapalem et al. (Eds.): PAKDD 2021, LNAI 12714, pp. 128–140, 2021.
https://doi.org/10.1007/978-3-030-75768-7_11

this mapping in such a way that the geometric relationships in this space can represent the principal information of the original graph. [11].

Graph embedding approaches can be categorized into three categories: (1) Factorization based, (2) Random Walk based, and (3) Deep Learning based methods [6,8]. Early methods on graph embedding were mostly centered around matrix factorization approaches. Factorization based algorithms decompose the matrix containing node adjacency information. Laplacian eigenmaps [2], Graph Factorization algorithm [1], GraRep [3] and HOPE [17] are some of the well-known examples in this category.

The main intuition of random walk based methods is to map the nodes to a real geometrical space in a way that nodes have similar embeddings if they tend to co-occur on short random walks over the graph [11]. The dominant algorithms presented for unsupervised learning use random walk-based objectives, such as node2vec [9], DeepWalk [18] and LINE [22] or even only reconstruction of adjacency information, such as [13] and [5]. The main limitation of these methods is that they concentrate on the proximity information and neglect other structural information [20].

Several deep learning methods for learning on graph structural data such as graph neural networks [7] have been proposed. Particularly graph convolutional neural networks have demonstrated their remarkable power in graph embedding problem. Graph convolution is effective in aggregating both topological and content information of graphs. There are plenty of works focusing on the design of such networks. GCN [12] as one of the most well-known ones is limited to only two-layer neighborhoods. One main limitation of GCNs is that as the model's depth increases, its learning performance degrades greatly. From this perspective, by utilizing the strength of GCNs, Deep Graph Infomax (DGI) [24] presents an unsupervised graph embedding algorithm that is also mindful of the global structural properties of the entire graph rather than first-order proximities of the nodes.

Although DGI as a deep state-of-the-art method in this category has a considerable performance in practice, it has two limitations that we aim to cover in this work. In contrast to DGI, our approach is motivated by a solution proposed for preserving both local and global graph structures. This property results in higher performance for both classification and clustering tasks.

In this work, we address the limitation of DGI that works independently from the input graph. Our proposed method has the ability to establish a trade-off between local and global information preservation based on the inherent properties of the input graph. Besides, we focus on incorporating the novel manifold learning algorithm UMAP [14] into the field of graph embedding and introduce an integrated method for node representation learning. We take DGI [19] as our base model, to capture Global structural information, and then a manipulated version of UMAP incorporates for preserving the local properties of the graph (Fig. 1). As we will report, this combination yields an improvement to both node classification and clustering tasks' performance. This paper's main contributions are summarized as follows:

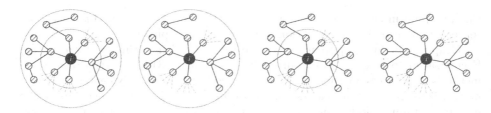

Fig. 1. A toy example of local versus global view of node *i* in graph **G** (on the right). The smaller circle shows a more local view. The large circle corresponds to a global view. Our proposed method is capable of preserving both local and global structures of a graph.

- Since graph data is unlabeled in most real-world scenarios, supervised methods seem to lose their functionality for the graph embedding problem. Therefore on the contrary to the prior state-of-the-art methods, we propose an unsupervised algorithm to learn node embeddings.
- We propose an approach for learning node embedding vectors, to be capable of providing a balance between both local and global structure preservation.
- By leveraging some conventional classification methods on the node embeddings, we are going to affirm the quality of the extracted features in terms of classification accuracy on test sets.

2   Related Works

Graph embedding algorithms can be categorized from a different perspective: topological embedding approaches that only take graph structural data into consideration and content-enhanced approaches that can also leverage available node content information. The objective of topological approaches is to preserve topological information of a graph. Factorization based methods fall into this category since their main intuition is to obtain the embedding by decomposing a form of the adjacency matrix. Laplacian eigenmaps [2] and Graph Factorization algorithm [1] reconstruct first-order proximities, GraRep [3] reconstructs *k*-order transition probabilities, HOPE [17] using different proximity measures, preserves higher-order proximities. Random walk based models DeepWalk [18], node2vec [9] and LINE [22] learn the embedding from graph topological properties carried by a collection of random walks of length *k*.

Deep learning approaches SDNE [25] and DNGR [4] that employ deep autoencoders to preserve the graph first and second-order proximities and positive point-wise mutual information (PPMI) also fall into this category. The main drawback of these methods is their inability to use the valuable content information of real-world graphs. Therefore, content enhanced embedding methods were presented.

In content-enhanced embedding approaches, the learned embeddings get augmented with node content information. Graph convolutional neural networks are the most recent approaches in this category. Graph convolution is an effective approach for combining the topological and content information of graphs and several convolution-based methods have shown promising performance. GCN [12] employs a first-order approximation of spectral graph convolution. GrapSAGE [10] presents a general inductive framework that exploits node feature information and Graph convolutional neural networks. GAT [23] proposed Graph Attention Networks for graph-based semi-supervised learning. DGI [19] uses a one-layer GCN as its encoder and obtains the embeddings in a mutual information maximization process.

Although the well-known graph convolution has been shown very effective in merging both topological and content information, it still has one main drawback. As indicated in [26] increasing the depth of the network leads to performance degradation such that a network with more than 3 layers loses its functionality in practice. This is equivalent to the loss of a Global view during the learning process. In order to attain good generative ability and mitigate the aforementioned performance degradation, we adopt a content-enhanced embedding approach motivated by DGI. MAPPING is capable of capturing both local and global structures of a graph, while the majority of the prior methods at most can preserve *k*-order proximities of graphs.

3   Problem Definition and Framework

In an input graph **G** with N nodes, we are provided with a set of node features, $\mathbf{X} = \{\vec{x}_1, \vec{x}_2, \dots, \vec{x}_N\}$ such that $\vec{x}_i \in \mathbb{R}^F$ indicates the content features associated with each node i. The interaction of individuals comes in form of the adjacency matrix $\mathbf{A} \in \mathbb{R}^{N \times N}$ ($A_{ij} = 1$ if there exists an edge between nodes i and j and otherwise $A_{ij} = 0$). **A** contains topological structure information of **G**. As we aim to map our graph from a F-dimensional space to F'-dimensional space, our objective would be to learn the encoder $\mathcal{E} : \mathbb{R}^{N \times F} \times \mathbb{R}^{N \times N} \rightarrow \mathbb{R}^{N \times F'}$ in a way that $\mathcal{E}(\mathbf{X}, \mathbf{A}) = \mathbf{E} = \{\vec{e}_1, \vec{e}_2, \dots, \vec{e}_N\}$ contains high-level embeddings ($\vec{e}_i \in \mathbb{R}^{F'}$ represents the latent features of node i).

3.1   Overall Framework

Our goal is to learn the node embedding of the input graph **G** in an unsupervised manner. In detail, our hybrid approach, to capture global structure of the graph, employs Deep Graph Infomax algorithm. For local information preservation purpose, we apply a modified version of manifold learning algorithm UMAP. Eventually, our final hybrid embedding matrix is built using Eq. 1.

$$\mathbf{E}_{Final} = \alpha(\mathbf{E}_{Local}) + \beta(\mathbf{E}_{Global}) \tag{1}$$

By adjusting the α and β parameters, a trade off can be struck between preserving global and local information, taking into account the intrinsic properties of the data as well as the down-stream application.

3.2 Proposed Method

Global Structure Preservation: In Deep Graph Infomax, the node embeddings of the input graph are learned by maximizing the mutual information between obtained representations $(\vec{e_i})$ and a graph-level summary vector $(\vec{sum})$, (i.e. we want our local representations to capture the global information of the entire graph.)

The Eq. 2 shows the standard binary cross-entropy loss objective, where $(\mathbf{X}, \mathbf{A})$, $(\tilde{\mathbf{X}}, \tilde{\mathbf{A}})$ are positive and negative examples respectively and $\mathcal{D}\left(\vec{e_i}, \vec{sum}\right)$ is a discriminator which assignes probability scores to the patch-summary pairs. We expect this score to be higher for positive examples and vice versa.

$$\mathcal{L} = \frac{1}{N+M}\left(\sum_{i=1}^{N}\mathbb{E}_{(X,A)}\left[\log\mathcal{D}\left(\vec{e_i}, \vec{sum}\right)\right] + \sum_{j=1}^{M}\mathbb{E}_{(\tilde{X},\tilde{A})}\left[\log(1-\mathcal{D}\left(\vec{\tilde{e}_j}, \vec{sum}\right)\right]\right) \tag{2}$$

First, negative examples are sampled using some corruption function, then by passing positive and negative examples through the one-layer GCN encoder it respectively obtains the patch representation matrices $\mathbf{E}$ and $\tilde{\mathbf{E}}$. The summary vector is then built by passing $\mathbf{E}$ through some readout function $(\vec{sum} = \mathcal{R}(\mathcal{E}(\mathbf{X}, \mathbf{A})))$. By applying gradient descent to maximize Eq. 2, parameters of the discriminator, readout function and the encoder get updated.

Construction of the negative examples is a simple shuffling procedure. The corruption function $\mathcal{C}$ shuffles the rows of $\mathbf{X}$. Note that the adjacency matrix remains the same in order to preserve structural similarities. The readout function simply calculates the average of positive embedding vectors as follows in which σ is a logistic sigmoid nonlinearity:

$$\mathcal{R}(\mathbf{E}) = \sigma\left(\frac{1}{N}\sum_{i=1}^{N}\vec{e_i}\right) \tag{3}$$

In order to score representation-summary pairs, a simple bi-linear scoring function is applied in which $\mathbf{W}$ is a learnable scoring matrix:

$$\mathcal{D}(\vec{e_i}, \vec{sum}) = \sigma\left(\vec{e_i}^{T}\mathbf{W}\vec{sum}\right) \tag{4}$$

The propagation rule in our one-layer encoder is defined as 5:

$$\mathcal{E}(\mathbf{X}, \mathbf{A}) = \sigma\ (\hat{\mathbf{D}}^{-\frac{1}{2}}\hat{\mathbf{A}}\ \hat{\mathbf{D}}^{-\frac{1}{2}}\mathbf{X}\ \Theta) \tag{5}$$

where σ is parametric ReLU, $\hat{\mathbf{D}}$ is the degree matrix, $\hat{\mathbf{A}} = \mathbf{A} + \mathbf{I}_N$ is the adjacency matrix containing self-loops and $\Theta \in \mathbb{R}^{F \times F'}$ is a learnable linear transformation applied to every node (Fig. 2).

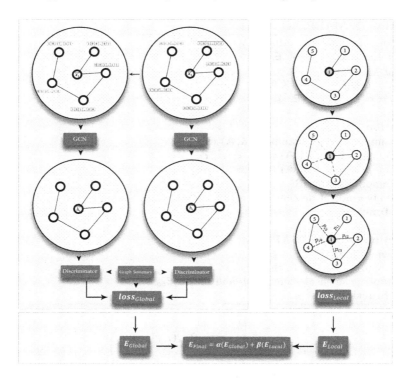

Fig. 2. A high level overview of the proposed method

Local Information Preservation: So far the Global node embedding matrix $\mathbf{E}_{Global}$ is built. In this section we aim to calculate $\mathbf{E}_{Local}$. UMAP comes within the class of K-neighbor-based graph learning algorithms such as Laplacian Eigenmaps, Isomap and T-SNE. As other K-neighbor graph-based algorithms, UMAP can be defined in two steps. At first, a particular weighted k-neighbor graph is created from the data manifold and in the second step, the low-dimensional structure of this graph is computed. Like several other dimensionality reduction techniques, UMAP can be extended to address the problem of graph representation learning. In graph data in particular, there is no obligation to estimate the data manifold using K-neighbor graphs, because graph proximity information is available in the form of adjacency matrices from the beginning. In this section, by applying a series of changes, we reconstruct the UMAP algorithm for graph domain applications.

Instead of estimating a k-neighbor graph that preserves the overall structure of the data manifold, we leverage matrix $\mathbf{A}^k, (k = 1 \dots N)$. Parameter k controls the arbitrary degree of locality we tend to preserve.

$p_{j|i}$, the node similarities in the high dimensional space which is equivalent to the edge weights in $\mathbf{A}^k$ is calculated as:

Algorithm 1. MAPPING

Input: Feature matrix $\mathbf{X}$, adjacency matrix $\mathbf{A}$
Output: Embedding graph $\mathbf{E}_{Final}$
1: $\mathbf{E}_{Global} = DGI(\mathbf{X},\mathbf{A})$
2: $\mathbf{E}_{Local} = GRAPHUMAP(\mathbf{X},\mathbf{A})$
3: $\mathbf{E}_{Final} = \alpha(\mathbf{E}_{Local}) + \beta(\mathbf{E}_{Global})$
4:
5: **function** DGI($\mathbf{X}$,$\mathbf{A}$)
6:     Sample negative examples $(\tilde{\mathbf{X}}, \tilde{\mathbf{A}})$ using $\mathcal{C}$
7:     Compute $\mathbf{E} = \mathcal{E}(\mathbf{X}, \mathbf{A}) = \{\vec{e}_1, \vec{e}_2, \dots, \vec{e}_N\}$
8:     Compute $\tilde{\mathbf{E}} = \mathcal{E}(\tilde{\mathbf{X}}, \tilde{\mathbf{A}}) = \{\vec{\tilde{e}}_1, \vec{\tilde{e}}_2, \dots, \vec{\tilde{e}}_N\}$
9:     Compute $\vec{s} = \mathcal{R}(\mathbf{E})$
10:    apply gradient descent on 2 and return $\mathbf{E}_{Global}$
11: **end function**
12:
13: **function** GRAPHUMAP($\mathbf{X}$,$\mathbf{A}$)
14:    Compute $\mathbf{A}^k$
15:    Assign weights p_{ij} to the edges of $\mathbf{A}^k$
16:    Apply gradient descent on 10 and return $\mathbf{E}_{Local}$
17: **end function**

$$p_{j|i} = \exp\left(\frac{-\max(0, d(i,j), -\rho_i)}{\sigma_i}\right) \tag{6}$$

$d(i,j)$ indicates the distance between node i and its neighbor j, under an arbitrary distance metric. ρ_i represents the distance from data point i to its first nearest neighbor and σ_i is set in a way that satisfies the following equation:

$$\sum_{j=1}^{n} \exp\left(\frac{-\max(0, d(i,j), -\rho_i)}{\sigma_i}\right) = \log_2(k) \tag{7}$$

Symmetrization of the node similarities is carried out parameters ρ_i and σ_i, make node similarities $v_{j|i} \neq v_{i|j}$. Equation 8 symmetries the node similarities.

$$p_{ij} = (p_{j|i} + p_{i|j}) - p_{j|i}p_{i|j} \tag{8}$$

p_{ij} is set as the weight of the edge between j and i. The low dimensional similarities are given by 9, where e_i and e_j are the locations of j and i in low dimensions:

$$q_{ij} = \left(1 + a\|e_i - e_j\|_2^{2b}\right)^{-1} \tag{9}$$

$$\mathcal{L} = \sum_{i \neq j} p_{ij} Log\left(\frac{p_{ij}}{q_{ij}}\right) + (1 - p_{ij})Log\left(\frac{1 - p_{ij}}{1 - q_{ij}}\right) \tag{10}$$

By minimizing cross entropy loss function Eq. 10, the node representation vectors are calculated. At the end of the training process, matrix $\mathbf{E}_{Local}$ contains the local information preserved node representation vectors. The proposed algorithm is summarized by Algorithm 1.

Table 1. Dataset statistics

Dataset	#Nodes	#Edges	#Features	#Classes
Cora	2,708	5,429	1,433	7
Citeseer	3,327	4,732	3,703	6
Pubmed	19,717	44,338	500	3

4   Experiments

4.1   Datasets

To demonstrate the efficiency of MAPPING, we conducted comprehensive experiments on three real-World citation networks Cora, Citeseer, and Pubmed [21]. Since the existing evaluation strategies for GNN models have some limitations, we used multiple data splits. The data (Table 1) is first split into a train and a test set. For the train set, q nodes were sampled and the test set contains all the remaining nodes. We used three different label sets in each experiment: A training-set of t nodes per class, a validation-set of r nodes, and a test-set. We selected t such that dataset splits to $q = 40\%$ of datasets which 20% applied for train-set, 20% for validation-set, and the remaining (60%) for test-set.

4.2   Baseline Methods

In order to evaluate MAPPING, we choose the following variety of methods for comparison:

- **DGI:** Deep Graph Infomax [19], by maximizing mutual information between patch representations of nodes and corresponding high-level summaries of graphs, obtains node representations in an unsupervised manner.
- **GCN**: GCN [12] is a semi-supervised learning variation of convolutional neural networks on graph-structured data which uses a layer-wise propagation rule that is based on a first-order approximation of spectral convolutions on graphs.
- **DeepWalk:** DeepWalk [18] as an unsupervised random walk based representation learning method inspired by SkipGram [15], uses local information obtained from truncated random walks to learn latent representations by treating walks as the equivalent of sentences.
- **node2vec:** Similar to DeepWalk, the semi-supervised method node2vec [9] preserves higher-order proximity between nodes by defining biased random walks and maximizing the probability of occurrence of subsequent nodes in fixed-length random walks.
- **UMAP:** Uniform Manifold Approximation and Projection (UMAP) [14] is a manifold learning technique for dimension reduction that at its core, is

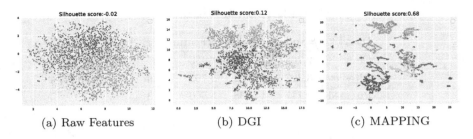

(a) Raw Features   (b) DGI   (c) MAPPING

Fig. 3. The UMAP visualization of the raw feature space, DGI and MAPPING node embedding space, for the Cora dataset and K-means clustering silhouette score.

very similar to t-SNE, but with several key theoretical footings that give the algorithm a stronger infrastructure.

- **PCA:** A statistical procedure that uses an orthogonal transformation as a dimension reduction technique.

4.3   Results

For evaluation purpose, we report the strength of embedding methods using the generated embedding as node features to classify the nodes. Table 2 shows the comparison of average accuracy results of ten runs over three citation network benchmark datasets using well-known classifiers Logistic Regression (LR), Gaussian Naive Bayes (GNB), K Nearest Neighbors (KNN), Decision Tree (DT), One-Versus-One SVM (OVO-SVM), One-Versus-Rest SVM (OVR-SVM), Multi-layer Perceptron (MLP) and Perceptron (PER).

We compare the output of MAPPING against (1) embeddings learned by merging DGI with dimensionality reduction techniques PCA and t-SNE. (2) embeddings learned by four graph neural network methods including DGI, GCN, DeepWalk, node2vec in terms of classification accuracy. To make the experimental studies more complete, we also compare our results with pure UMAP which works as a dimension reduction technique that does not involve relational information into its calculations and only works on raw features.

We investigated the outcome of simultaneously capturing local and global information generated by our proposed method. We expected that, given the nature of the data set, maintaining certain levels of local as well as global neighborhoods would lead to an increased classification accuracy. The best results are marked as bold. We can note that our unsupervised method obtains better average accuracy in comparison with other unsupervised and semi-supervised methods on all datasets. Comparing with DGI as the base model, MAPPING's contribution is exceptional with an average of 6.27% of higher performance. Particularly, on Cora dataset, it achieves 11.6% of more accuracy. We combined DGI with t-SNE and PCA in order to enhance its embeddings but our results demonstrate strong performance being achieved across all datasets. We further observe that, in most cases, MAPPING competitively exceeds in performance

Table 2. Summary of results in terms of Average classification accuracies under 10 runs for 20:60:20 (train:test:validation) data split (%).

	Input	LG	GNB	KNN	DT	OVO-SVM	OVR-SVM	MLP	PER	Average
Cora										
MAPPING	**X, A**	85	79	81	79	86	86	84	81	**83.06**
Raw Features	**X**	31	45	60	54	55	55	65	54	52.37
UMAP	**X**	68	45	34	57	65	65	63	65	57.75
DeepWalk	**A**	70	71	70	48	67	67	72	66	66.37
node2vec	**A**	76	74	73	61	74	74	75	71	72.25
DGI+t-SNE	**X, A**	78	78	80	79	52	58	81	37	67.87
DGI+PCA	**X, A**	67	56	38	66	62	61	40	59	56.12
DGI	**X, A**	36	80	82	70	75	75	85	73	72
GCN	**X, A, Y**	82	79	81	78	82	82	82	82	81
Citeseer										
MAPPING	**X, A**	70	68	70	61	69	69	71	69	**68.99**
Raw Features	**X**	39	55	53	54	67	67	67	63	58.12
UMAP	**X**	68	55	39	55	67	67	66	65	60.25
DeepWalk	**A**	50	53	55	40	50	50	57	47	50.25
node2vec	**A**	57	55	55	47	55	55	57	51	54
DGI+t-SNE	**X, A**	69	69	68	65	39	29	68	48	56.87
DGI+PCA	**X, A**	56	47	22	59	49	49	42	52	47
DGI	**X, A**	57	71	69	58	72	72	71	31	62.62
GCN	**X, A, Y**	65	61	64	66	65	65	66	66	64.75
Pubmed										
MAPPING	**X, A**	84	78	83	76	85	85	86	85	**82.75**
Raw Features	**X**	80	74	73	79	84	84	81	84	79.87
UMAP	**X**	83	73	73	76	84	84	82	82	79.62
DeepWalk	**A**	78	76	79	63	79	79	78	75	75.87
node2vec	**A**	80	77	79	67	81	81	77	75	77.12
DGI+t-SNE	**X, A**	63	65	82	76	43	51	74	60	64.25
DGI+PCA	**X, A**	83	64	77	73	84	84	80	82	78.37
DGI	**X, A**	83	78	83	76	84	84	84	79	81.37
GCN	**X, A, Y**	83	82	82	79	83	83	83	81	82

in comparison with the GCN model in the fully supervised setting. Considering the fact that GCN uses labeled data during its learning process, this is a notable achievement. We assume that this achievement is originated from the fact that, in MAPPING, every node has access to global structural properties of the graph, whereas in GCN nodes can only see their two-step local neighborhoods. In general, MAPPING performs better than recent graph networks DGI, GCN, DeepWalk, node2vec, which demonstrates its benefit on data representation and learning. Also, it generally performs better than semi-supervised methods node2vec and GCN which further indicates its capability to be conducted on real-world unlabeled graph-structured data.

We also analyzed the node clustering quality of MAPPING over Cora dataset in 2-D space. Figure 3 displays the visualization of the learned 2-D embeddings of raw features (Fig. 4(a)), features extracted from DGI (Fig. 4(b)), and features

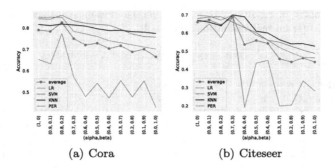

Fig. 4. The effect of (α, β) parameters on classification accuracy. (Color figure online)

extracted from our proposed method MAPPING (Fig. 3(c)) by applying UMAP as a visualization procedure respectively. Each point corresponds to a node in the input graph. The color of each node signifies its community. The clusters of the learned MAPPING model's embeddings are clearly defined and separated appropriately. In Cora, by running the K-means clustering algorithm on MAPPING's embeddings, we achieved silhouette score 68%, which remarkably exceeds DGI's.

In the estimation of node representation vectors, the alpha and beta parameters respectively, indicate the importance of preserving global and local neighborhoods. The effect of these two parameters on the accuracy of LR, SVM, KNN and PRC is illustrated in Fig. 4 (only for Cora and Citeseer due to memory limitations). The blue line in both plots indicates the average value of the other lines. The pinnacle of the average line for Cora and Citeseer locates respectively at points $(\alpha, \beta) = (0.8, 0.2)$ and $(\alpha, \beta) = (0.7, 0.3)$ which stipulates the highest gained average accuracy. Two inferences can be deduced from this figure: (1) It is not adequate to only preserve either of the two types of global and local neighborhoods alone, while by preserving a combination, a better representation can be achieved. (2) According to the observations of this study, it seems that in the application of node classification, the importance of preserving global neighborhoods is somewhat greater than local ones.

5   Conclusions

We have presented MAPPING, a novel approach for learning unsupervised representations on graph-structured data. MAPPING, captures graph's global structure using a graph convolutional architecture and encodes the nodes to a low-dimensional space by maximizing mutual information across its patch representations. Moreover, by generalizing novel dimensionality reduction technique UMAP to the field of graphs, MAPPING obtains a second set of node representation vectors so as to capture the graph's local information. Along these lines, we lessen the complicated structural information-loss which leads to an enhanced node-classification performance. The experimental evaluations demonstrate that our model achieves competitive performance across state-of-the-art methods.

References

1. Ahmed, A., Shervashidze, N., Narayanamurthy, S., Josifovski, V., Smola, A.J.: Distributed large-scale natural graph factorization. In: Proceedings of the 22nd International Conference on World Wide Web, pp. 37–48. ACM (2013)
2. Belkin, M., Niyogi, P.: Laplacian eigenmaps and spectral techniques for embedding and clustering. In: Advances in Neural Information Processing Systems, pp. 585–591 (2002)
3. Cao, S., Lu, W., Xu, Q.: Grarep: learning graph representations with global structural information. In: Proceedings of the 24th ACM International on Conference on Information and Knowledge Management, pp. 891–900. ACM (2015)
4. Cao, S., Lu, W., Xu, Q.: Deep neural networks for learning graph representations. In: Thirtieth AAAI Conference on Artificial Intelligence (2016)
5. Duran, A.G., Niepert, M.: Learning graph representations with embedding propagation. In: Advances in Neural Information Processing Systems, pp. 5119–5130 (2017)
6. Fatemi, B., Molaei, S., Zare, H., Veisi, H.: Attributed graph clustering via deep adaptive graph maximization. In: 2020 10th International Conference on Computer and Knowledge Engineering (ICCKE), pp. 376–381. IEEE (2020)
7. Gori, M., Monfardini, G., Scarselli, F.: A new model for learning in graph domains. In: Proceedings of 2005 IEEE International Joint Conference on Neural Networks, 2005, vol. 2, pp. 729–734. IEEE (2005)
8. Goyal, P., Ferrara, E.: Graph embedding techniques, applications, and performance: a survey. Knowl.-Based Syst. **151**, 78–94 (2018)
9. Grover, A., Leskovec, J.: node2vec: scalable feature learning for networks. In: Proceedings of the 22nd ACM SIGKDD International Conference on Knowledge Discovery and Data Mining, pp. 855–864. ACM (2016)
10. Hamilton, W., Ying, Z., Leskovec, J.: Inductive representation learning on large graphs. In: Advances in Neural Information Processing Systems, pp. 1024–1034 (2017)
11. Hamilton, W.L., Ying, R., Leskovec, J.: Representation learning on graphs: Methods and applications. arXiv preprint arXiv:1709.05584 (2017)
12. Kipf, T.N., Welling, M.: Semi-supervised classification with graph convolutional networks. arXiv preprint arXiv:1609.02907 (2016)
13. Kipf, T.N., Welling, M.: Variational graph auto-encoders. arXiv preprint arXiv:1611.07308 (2016)
14. McInnes, L., Healy, J., Melville, J.: Umap: uniform manifold approximation and projection for dimension reduction. arXiv preprint arXiv:1802.03426 (2018)
15. Mikolov, T., Sutskever, I., Chen, K., Corrado, G.S., Dean, J.: Distributed representations of words and phrases and their compositionality. In: Advances in Neural Information Processing Systems, pp. 3111–3119 (2013)
16. Molaei, S., Zare, H., Veisi, H.: Deep learning approach on information diffusion in heterogeneous networks. Knowl.-Based Syst. **189**, 105153 (2020)
17. Ou, M., Cui, P., Pei, J., Zhang, Z., Zhu, W.: Asymmetric transitivity preserving graph embedding. In: Proceedings of the 22nd ACM SIGKDD International Conference on Knowledge Discovery and Data Mining, pp. 1105–1114 (2016)
18. Perozzi, B., Al-Rfou, R., Skiena, S.: Deepwalk: online learning of social representations. In: Proceedings of the 20th ACM SIGKDD International Conference on Knowledge Discovery and Data Mining, pp. 701–710. ACM (2014)

19. Veličković, P., Fedus, W., Hamilton, W.L., Lio, P., Bengio, Y., Hjelm, R.D.: Deep graph infomax. In: 7th International Conference on Learning Representations (ICLR 2019) (2019)
20. Ribeiro, L.F., Saverese, P.H., Figueiredo, D.R.: struc2vec: learning node representations from structural identity. In: Proceedings of the 23rd ACM SIGKDD International Conference on Knowledge Discovery and Data Mining, pp. 385–394. ACM (2017)
21. Sen, P., Namata, G., Bilgic, M., Getoor, L., Galligher, B., Eliassi-Rad, T.: Collective classification in network data. AI Mag. **29**(3), 93–93 (2008)
22. Tang, J., Qu, M., Wang, M., Zhang, M., Yan, J., Mei, Q.: Line: large-scale information network embedding. In: Proceedings of the 24th International Conference on World Wide Web, pp. 1067–1077
23. Veličković, P., Cucurull, G., Casanova, A., Romero, A., Liò, P., Bengio, Y.: Graph attention networks. In: International Conference on Learning Representations (2018)
24. Veličković, P., Fedus, W., Hamilton, W.L., Liò, P., Bengio, Y., Hjelm, R.D.: Deep graph infomax (2018)
25. Wang, D., Cui, P., Zhu, W.: Structural deep network embedding. In: Proceedings of the 22nd ACM SIGKDD International Conference on Knowledge Discovery and Data Mining, pp. 1225–1234 (2016)
26. Zhang, J., Meng, L.: GResNet: graph residual network for reviving deep GNNs from suspended animation (2019)

Learning Attention-Based Translational Knowledge Graph Embedding via Nonlinear Dynamic Mapping

Zhihao Wang, Honggang Xu, Xin Li[(✉)], and Yuxin Deng

Shanghai Key Laboratory of Trustworthy Computing,
East China Normal University, Shanghai, China
{51184501158,51194501193}@stu.ecnu.edu.cn,
{xinli,yxdeng}@sei.ecnu.edu.cn

Abstract. Knowledge graph embedding has become a promising method for knowledge graph completion. It aims to learn low-dimensional embeddings in continuous vector space for each entity and relation. It remains challenging to learn accurate embeddings for complex multi-relational facts. In this paper, we propose a new translation-based embedding method named ATransD-NL to address the following two observations. First, most existing translational methods do not consider contextual information that have been proved useful for improving performance of link prediction. Our method learns attention-based embeddings for each triplet taking into account influence of one-hop or potentially multi-hop neighbourhood entities. Second, we apply nonlinear dynamic projection of head and tail entities to relational space, to capture nonlinear correlations among entities and relations due to complex multi-relational facts. As an extension of TransD, our model only introduces one more extra parameter, giving a good tradeoff between model complexity and the state-of-the-art predictive accuracy. Compared with state-of-the-art translation-based methods and the neural-network based methods, experiment results show that our method delivers substantial improvements over baselines on the MeanRank metric of link prediction, e.g., an improvement of 35.6% over the attention-based graph embedding method KBGAT and an improvement of 64% over the translational method TransMS on WN18 database, with comparable performance on the Hits@10 metric.

Keywords: Knowledge graph embedding · Translation-based methods · Link prediction · Attention mechanism

---

Z. Wang and H. Xu are co-first authors. This work was supported in part by the Ministry of Science and Technology of China under Grant No. 2018YFC0830400, National Natural Science Foundation of China under Grant No. 61802126, 61832015, 62072176, the Inria-CAS joint project Quasar and Shanghai Pujiang Program under Grant No. 17PJ1402200.

K. Karlapalem et al. (Eds.): PAKDD 2021, LNAI 12714, pp. 141–154, 2021.
https://doi.org/10.1007/978-3-030-75768-7_12

1   Introduction

Knowledge graphs (KGs) such as YAGO [6], WordNet [12], DBpedia [9] and Freebase [1] represent knowledge bases as directed graphs, and store relational facts in the knowledge base in terms of triplets. Entities with different types and attributes are represented as nodes, and different types of relations on entities are represented as edges. Each triplet denoted by (h, r, t) represents the relationship r directed from the head entity h to the tail entity t. Knowledge graphs become useful information resources for many AI related applications such as question answering, recommendation, information retrieval, etc. Although a real-world knowledge graph often contains millions of relational facts, it still suffers from the issue of incompleteness due to missing a lot of factual triplets. Knowledge graph completion aims to predict the most probable missing entities and relations in the knowledge base, also referred to as *link prediction*, e.g., a completion query may be like $(?, r, t)$ for head prediction, $(h, r, ?)$ for tail prediction or $(h, ?, t)$ for relation prediction. Since real-world knowledge graphs are enormous and heterogeneous containing multi-relational facts, e.g., one-to-many (or 1-N), many-to-one (or N-1), many-to-many (or N-N) relations, it remains difficult to accurately predict complex relations while scaling to large-scale knowledge graphs.

Knowledge graph embedding has recently become a promising method for knowledge graph completion. It aims to learn low-dimensional embeddings in continuous vector space for entities and relations denoted by lower case bold letters. The state-of-the-art embedding methods are generally classified as translation-based methods [3,7,8,19,23] and neural-network based methods [5,13,14,18]. Translation-based methods model relations as translation operations from the embeddings of head entities to tail entities, expecting that $\mathbf{h} + \mathbf{r} \approx \mathbf{t}$ holds in the embedding space when (h, r, t) is a valid triplet. Translational models often use simple operations and less parameters in consideration of scalability, and prove to be cost-effective approaches for knowledge graph completion. In contrast, the neural-network based methods can learn more expressive embeddings indicating sophisticated nonlinear correlations over the relational facts, yet are not always parameter efficient [13,18].

Some recent research work has shown that combining contextual information into the embedding method would effectively improve the performance of link prediction [10,13]. PTransE [10] proposes to represent relation paths via semantic composition of relation embeddings, and to model relation paths as translations between entities. Notably, an attention-based graph embedding method called KBGAT [13] is proposed recently that delivers significant improvements on the performance of link prediction over state-of-the-art baselines. The model attempts to learn contextual information of local neighbourhood entities by designing a generalized attention-based graph embedding for link prediction. Our empirical study indicates that KBGAT generally outperforms most of the state-of-the-art translational methods, although it often consumes up more time and space for training. To our knowledge, most translation-based methods treat each triplet separately and haven't combined its contextual information in the

Table 1. Complexity of translation-based models compared in the experiments.

Model	Score function	# Parameters
TransE [3]	$\|h + r - t\|_{\ell_1/\ell_2}$ $h, t \in \mathbb{R}^n, r \in \mathbb{R}^m$	$O(nN_e + mN_r)$ where $m = n$
TransH [19]	$\|h(I - w_r^\top w_r) + r - t(I - w_r^\top w_r)\|_{\ell_{1/2}}$ $h, t \in \mathbb{R}^n, M_r \in \mathbb{R}^{n \times m}$	$O(nN_e + 2mN_r)$
TransR [11]	$\|hM_r + r - tM_r\|_{\ell_{1/2}}$ $h, t \in \mathbb{R}^n, r \in \mathbb{R}^m, M_r \in \mathbb{R}^{n \times m}$	$O(nN_e + (n+1)mN_r)$
TransD [7]	$\|(r_p h_p^\top + I^{k_r \times k_e})h + r - (r_p t_p^\top + I^{m \times n})t\|_{\ell_{1/2}}$ $h, t, h_p, t_p \in \mathbb{R}^n, r, r_p \in \mathbb{R}^m$	$O(2nN_e + 2mN_r)$
TranSparse [8] (separate)	$\|M_r^t(\theta_r^1)h + r - M_r^2(\theta_r^2)t\|_{\ell_{1/2}}$ $h, t \in \mathbb{R}^n, r \in \mathbb{R}^m, M_r^t(\theta_r^1), M_r^2(\theta_r^2) \in \mathbb{R}^{*m \times n}$	$O(2nN_e + 2(1-\theta)(n+1)mN_r)$
TransMS [23]	$\| - tanh(t \otimes r) \otimes h + r - tanh(h \otimes r) \otimes t + \alpha \cdot g(h \otimes t)\|_{\ell_{1/2}}$ $h, t \in \mathbb{R}^n, r \in \mathbb{R}^m, \alpha = r_{k_{r+1}} \in \mathbb{R}^1$	$O(nN_e + (m+1)N_r)$
ATransD-NL (this paper)	$\|(h + tanh(h_p^\top h_p) \otimes \bar{t}) + r - (t + tanh(t_p^\top tr_p) \otimes h)\|_{\ell_{1/2}}$ $h, h_p, t, t_p \in \mathbb{R}^n, r, r_p \in \mathbb{R}^m$	$O((2n+1)N_e + 2mN_r)$ where $m = n$

graph into the learning process. Motivated by the observation, we aim to fill the gap by extending TransD that delivers promising results and a good balance of predictive accuracy and performance [7].

Comparing the existing approaches above, we propose a translation-based method named ATransD-NL (Attention-based Translational Knowledge Graph Embedding via Nonlinear Dynamic Semantic Projection) that extends and generalizes TransD with two major ideas. First, to capture nonlinear semantic correlations among complex relations, we further apply nonlinear functions during the dynamic projection from entity embedding spaces to relation-specific embedding spaces. Next, to learn contextual information for improving the performance on link prediction, our model further integrates attention-based embedding methods taking into account the influence of other entities when performing translation operations in the relational space. These entities may contain local neighbourhood entities or potentially linked entities via multi-hop relation paths, implicitly. As a translation-based method, we only introduce one more extra parameter than TransD indicating the cost-effectiveness of our method. We evaluate our approach on challenging benchmark datasets including WN18 and FB15k. We show that our method gives a good tradeoff between model complexity and predictive accuracy. We conduct comparison not only with translation-based methods but also with the state-of-the-art neural-network based methods. Experiment results show that our method delivers substantial improvements on the MeanRank metric of link prediction, e.g., an improvement of 35.6% over the attention-based graph embedding method KBGAT [13] and of 64% over the translational method TransMS on WN18, with comparable performance on the Hits@10 metric Table 3 and 4.

2   Related Work

The state-of-the-art knowledge graph embedding methods can be broadly classified as translation-based methods, neural network based methods, and tensor decomposition methods such as DistMult [22] and ComplEx [17]. There are some other methods earlier like structured embedding (SE) [2], semantic matching energy (SME) [4], etc. Below we mainly discuss the line of work on translation methods that this work follows and some state-of-the-art neural network based methods. We denote by N_e and N_r the number of entities and relations in the knowledge base, respectively, and denote by n and m the dimension of embedding vectors for entities and relations, respectively.

2.1 Translation-Based Methods

Originated from **TransE** [3], translational methods regard the relation $\mathbf{r}$ as a translation operation from the head entity $\mathbf{h}$ to the tail entity $\mathbf{t}$ so that $\mathbf{h}+\mathbf{r} \approx \mathbf{t}$ when the triplet (h, r, t) is valid in the knowledge base. We denote by $f_r(\mathbf{h}, \mathbf{t})$ the score function. The higher the ranking score, the higher probability that (h, r, t) corresponds to a true fact in the knowledge base. Here $f_r(h, t) = \|\mathbf{h}+\mathbf{r}-\mathbf{t}\|_{l_1/l_2}$ and the training objective is to minimize the margin-based ranking loss raised by positive and negative triplets.

TransE is effective for modelling 1-to-1 relations, but has been found less suitable for modelling more complex reflexive and multi-relational facts. Later, **TransH** [19] was proposed to model each relation r as a hyperplane to which head and tail entities are projected with translation operations over it. Let $\mathbf{w}_r$ denote the normal vector of the relation-specific hyperplane with $\|\mathbf{w}_r\|_2 = 1$. The head or tail entity embedding $\mathbf{e} \in \{\mathbf{h}, \mathbf{t}\}$ is thus defined as $\mathbf{e}_\perp = \mathbf{e} - \mathbf{w}_r^\top \mathbf{e} \mathbf{w}_r$, with $f_r(h, t) = \|\mathbf{h}_\perp + \mathbf{r} - \mathbf{t}_\perp\|_{l_1/l_2}$. **TransR** [11] models entities and relations in distinct vector spaces whereby entities are projected to the relation-specific spaces (by using a projection matrix $\mathbb{M}_r$) with $f_r(h, t) = \|\mathbf{M}_r \mathbf{h} + \mathbf{r} - \mathbf{M}_t \mathbf{t}\|_{l_1/l_2}$. Considering each relation may exhibit diverse semantic meanings in various scenarios, TransR was extended as **CTransR** by first clustering entities (h, t) according to offsets $\mathbf{h} - \mathbf{t}$, and then the datasets are trained with $f_r(h, t) = \|\mathbf{M}_r \mathbf{h} + \mathbf{r}_c - \mathbf{M}_t \mathbf{t}\|_{l_1/l_2} + \alpha \|\mathbf{r}_c - \mathbf{r}\|_{l_1/l_2}$, where the newly-added constraint ensures that the cluster-specific relation $\mathbf{r}_c$ is close to the original relation $\mathbf{r}$.

The model **TransD** further refines the model TransR/CTransR by using dynamic mapping matrix. For each triplet (h, r, t), two mapping matrices $\mathbf{M}_{rh}, \mathbf{M}_{rt} \in \mathbb{R}^{m \times n}$ are used to project entities from entity space to relational space. These matrices are defined in terms of two vectors: $\mathbf{M}_{rh} = \mathbf{r}_p \mathbf{h}_p^\top + \mathbf{I}^{m \times n}$ and $\mathbf{M}_{rt} = \mathbf{r}_p \mathbf{t}_p^\top + \mathbf{I}^{m \times n}$, where $\mathbf{r}_p$, $\mathbf{h}_p$ and $\mathbf{t}_p$ are vectors for projection, and the original vectors represent the semantic meanings of entities and relations. As aforementioned, TransD has less parameters and matrix multiplication operations compared with TransR, and meanwhile delivers promising experimental results on link prediction tasks.

Afterwards, several other refined translational models are proposed such as TransA [20], TransG [21], TranSparse [8], TransMS [23], etc. Notably, **TranSparse** deals with heterogeneous and unbalanced issues of complex relations. TranSparse replaces the projection matrix by adaptive sparse matrices $\mathbf{M}_r(\theta_r)$ for sharing the sparse transfer matrix among relations (or by $\mathbf{M}_r^e(\theta_r^e)$) where $e \in \{h, t\}$ for different triplets separately). The sparse degree θ_r is determined by the quantities of entities pairs linked by the relations. Recently, a

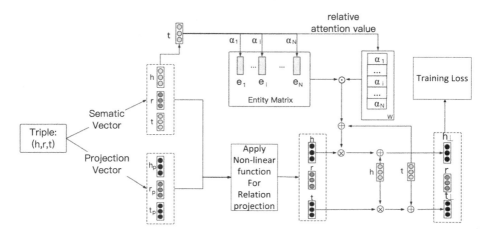

Fig. 1. Illustration of our model ATransD-NL

translation method called **TransMS** [23] aims to model complex relations by capturing multidirectional semantics from head/tail entities and relations to tail/head entities with nonlinear functions and from entities to relations with linear bias vectors. The model only introduces one more extra hyper-parameter than TransE yet gains substantial improvements against many other state-of-the-art translation-based methods.

2.2   Neural-Network Based Methods

The Single Layer Model (SLM) [4] model defines its score function with a non-liner neural network. Later the Neural Tensor Network (NTN) [4] model extends SLM by encoding a 3-way tensor and other transformations into multilayer neural network, to address the second-order correlations. We are aware of several recent work based on convolutional networks for knowledge graph embedding. The model **ConvE** [5] uses 2D convolutions over embedding for link prediction. It is formed by a single convolution layer, a projection layer and an inner product layer for computing the scores of each tail entity t with respect to input pairs (h, r). The model is highly parameter efficient while yielding competitive performance. The model **ConvKB** [14] is another knowledge graph embedding method based on convolutional neural network. It takes the entire triples as inputs which are encoded as a 3-column matrix for each and fed to 1D convolution operators. The feature maps learned by the convolution layer is concatenated into a feature vector which is used to produce the score for triples via dot product. The model also shows state-of-the-art performance on some benchmarks. ConvKB is later extended as **CapsE** [18] by further utilizing capsule network. The model **SACN** [16] is composed of an encoder WGCN and a decodor Conv-TransE and shows enhanced performance compared with ConvE. Recently, a novel attention-based embedding method for link prediction called **KBGAT** [13] delivers significant improvements on the performance of link prediction over state-of-the-art methods. The model uses an encode-decoder structure with generalized graph attention model and ConvKB as encoder and decoder, respectively. Its performance gain is mainly attributed to the attention-based model that discovers richer and more expressive feature embedding compared with other convolutional neural network based methods.

3   The Proposed Method

3.1   Model

We denote by $\mathcal{E}$ the universe of entities in the knowledge graph. Following TransD, each entity and relation is also represented by two vectors in our method. Given a triplet (h, r, t), the two vectors used are $\mathbf{h}, \mathbf{h}_p, \mathbf{t}, \mathbf{t}_p \in \mathbb{R}^n$ and $\mathbf{r}, \mathbf{r}_p \in \mathbb{R}^m$, where the subscript p denotes the projection vectors. These projection vectors are used for constructing the mapping matrices from the entity space to the relational spaces dynamically. The original vectors represent the semantic meanings of entities and relations. For ease of presentation, we set the dimensions of entities n and relations m as equal, i.e., $n = m$, following the treatment of some other translational methods.

We denote by $\mathbf{N} \subseteq \mathcal{E}$ the local and possibly multi-hop neighbourhood entities of a given triplet (h, r, t). To learn the contextual information of the given triplet, we first compute the absolute attention value of each concerned entity pair (h, t_i) for each entity $t_i \in \mathcal{E}$ in Eq. (1):

$$\beta_i = \begin{cases} a(\mathbf{h}, \mathbf{t}_i) & if \ t_i \in \mathbf{N} \\ -\infty & otherwise \end{cases} \tag{1}$$

where a is an attention function. Considering the efficiency, we choose a to be dot product of two embedding vectors. Then we compute the relative attention values over all β_i in Eq. (2):

$$\alpha_i = softmax(\beta_i) = \frac{exp(\beta_i)}{\sum_{1 \leq j \leq N} exp(\beta_j)} \tag{2}$$

Then we obtain a weight vector $\mathbf{w} = [\alpha_1, \ldots, \alpha_n]^\top \in \mathbb{R}^{N_e}$ that attempts to capture the probability for each $t_i \in \mathcal{E}$ that a one-hop or multi-hop path exists from h to t_i. Next we introduce an embedding vector $\bar{\mathbf{t}}$ that combines the contextual information of possible neighbourhood entities for the head entity h with respect to complex relations.

$$\bar{\mathbf{t}} = \mathbf{t} + \mathbf{E}\mathbf{w} \tag{3}$$

where $\mathbf{E} = [\mathbf{e}_1; \ldots; \mathbf{e}_{N_e}] \in \mathbb{R}^{n \times N_e}$ denotes the embedding matrix for all the entities in the knowledge graph, and $\mathbf{e}_i$ represents the original embedding vector for each entity $e_i \in \mathcal{E}$. Then we define

$$\mathbf{h}_\perp = \sigma(\mathbf{h}_p^\top \mathbf{h} \mathbf{r}_p) \otimes \bar{\mathbf{t}} + \mathbf{h} \tag{4}$$

$$\mathbf{t}_\perp = \sigma(\mathbf{t}_p^\top \mathbf{t} \mathbf{r}_p) \otimes \mathbf{h} + \mathbf{t} \tag{5}$$

where σ is any nonlinear function of interest. By empirical study, *tanh* turns out to be the optimal choice here among other choices like *sfotmax*, *ReLu*, etc. Here $\otimes$ is the Hadamard product. We would like to remark that, the projected vectors of entities in TransD can be rewritten as above because we set $m = n$ in our setting.

Similarly, one can also define

$$\mathbf{h}_\perp = \sigma(\mathbf{h}_p^\top \mathbf{h} \mathbf{r}_p) \otimes \mathbf{t} + \mathbf{h} \tag{6}$$

$$\mathbf{t}_\perp = \sigma(\mathbf{t}_p^\top \mathbf{t} \mathbf{r}_p) \otimes \bar{\mathbf{h}} + \mathbf{t} \tag{7}$$

where $\bar{\mathbf{h}}$ is an embedding vector that combines the context information of possible neighbourhood head entities for the tail entity t with respect to complex relations.

Figure 1 illustrates the computation process of various embeddings in Eqs. (4) and (5). Our empirical study shows that any of the two choices above, as well as a combination of the two embedding methods by using some hyper-parameter, deliver similar performance, and we consider the former choice in our experiments.

Finally, the score function of our model is given as

$$f_r(h, t) = \|\mathbf{h}_\perp + \mathbf{r} - \mathbf{t}_\perp\|_{l_1/l_2}$$

As usual, the score is expected to be lower for golden triples from the knowledge graph and higher for invalid relational facts that do not exist.

3.2   Training Objectives

Since knowledge graphs only specify golden triplets, one has to take a proper approach to sampling the set of negative triplets for model training. We denote by $\mathcal{R}$ the set of golden triplets from the knowledge graph and take $\mathcal{R}'$ as the set of negative triplets that is obtained by corrupting either the head entity or tail entity by randomly sampling entities in $\mathcal{E}$. We also use two strategies called uniform sampling (abbreviated as "unif") and Bernoulli sampling (abbreviated as "bern"), respectively [19]. Formally, given a golden triplet (h, r, t), its corresponding negative triplets are denoted by $\mathcal{R}'_{(h,r,t)}$:

$$\mathcal{R}'_{(h,r,t)} = \{(h', r, t) \notin \mathcal{R} \mid \exists h' \in \mathcal{E} : h' \neq h\}$$

$$\cup \{(h, r, t') \notin \mathcal{R} \mid \exists t' \in \mathcal{E} : t' \neq t\}$$

Then $\mathcal{R}' = \bigcup_{(h,r,t)\in\mathcal{R}} \mathcal{R}'_{(h,r,t)}$. We train our model by minimizing the margin-based ranking loss $\mathcal{L}$ as defined below. The training objective is given as follows:

$$\mathcal{L} = \sum_{(h,r,t)\in\mathcal{R}} \sum_{(h',r,t')\in\mathcal{R}'_{(h,r,t)}} \max\left(0, f_r(h,t) - f_r(h',t') + \gamma\right)$$

where γ is the margin separating golden triplets and negative triplets. The training objective is subject to the following constraints: $\|\mathbf{h}_\perp\|_2 \leq 1$, $\|\mathbf{t}_\perp\|_2 \leq 1$, $\|\mathbf{h}\|_2 \leq 1$, $\|\mathbf{t}\|_2 \leq 1$ and $\|\mathbf{r}\|_2 \leq 1$ for each embedded entity and relation. We use the Adam optimizer for minimizing the above loss.

4   Experiments and Result Analysis

4.1   Datasets and Protocols

We evaluate our model on two popular knowledge graph datasets built with WordNet [12] and FreeBase [1]. These two datasets are widely used in knowledge graph completion and representation learning. Neural network-based methods mostly use WN18RR and FB15k-237, and translation model-based methods mostly use WN18 and FB15k. The work of this paper compares two types of methods on relevant datasets. Table 2 shows the details of four datasets and the division of training set, validation set, and test set, including the number of relation and the number of entities.

Evaluation Protocols. The task of link prediction aims to predict the most probable missing head or tail entity in a test triplet. We construct negative samples by replacing head or tail entity of the golden triplets, and those corrupted triplets are not in knowledge graph. Following evaluation metrics proposed in [4], we use two popular evaluation metrics MeanRank (that is the mean rank over all the test golden triplets) and Hits@10 (the proportion of correct entities ranked in top 10), respectively.

The evaluation on MeanRank is computed as the average of MeanRank for the head and tail entity prediction over all the triplets in the testing set. Take the head prediction as an example, we first compute scores of those corrupted triplets by replacing the head entity for each triplet in the testing set, and rank their scores in an ascending order. Then we rank the scores of golden triplets, and compute the mean rank of head entity prediction. The mean rank for the tail

Table 2. Details of datasets used in the experiments.

Datasets	#Entity	#Relation	#Train	#Valid	#Test
FB15k	14,951	1,345	483,142	50,000	59,071
FB15k-237	14,541	237	272,115	17,535	20,466
WN18	40,943	18	141,442	5,000	5,000
WN18RR	40,943	11	86,835	3,034	3,134

entity is similarly computed. The lower the MeanRank, the better the model's potential generalization ability. The evaluation on Hits@10 also contains the metrics on both the head entity prediction and the tail entity prediction. For each prediction, Hits@10 is calculated as the ratio of golden triplets ranked in top 10 divided by the quantity of all the triplets in the testing set. The final evaluation on the Hits@10 metric is the average of the Hits@10 value for the head and tail entities among all the triplets in the testing set. Therefore, the higher the Hits@10, the better the model's accuracy in general. Besides, the results are also classified into two categories "Raw" and "Filter". Since the corrupted triplet may also exits in the knowledge graph, the corrupted triplets are removed from the train, valid and test sets. This evaluation setting is called "Filter", and the setting "Raw" means that the scores of all corrupted triplets are also considered.

Model Parameters. For all the experiments, we take the embedding dimension size in $[100, 200]$, the value of margin γ in $[-10, 10]$, and the learing rate of SGD in the range of $\{0,001, 0.0005, 0.0001\}$. We choose the batch size from $\{5k, 10k, 50k, 200k\}$. The best experimental results are obtained by using the following parameters: the embedding dimension $d = 100$, the learing rate is 0.0001 and the margin γ is different in four datasets, margin $\gamma = -1.2$ in WN18RR, $\gamma = 0.5$ in FB15k-237, $\gamma = -1.3$ in FB15k, $\gamma = 2.0$ in WN18. Here, we use the l_1 norm in the scoring function.

4.2   Experiment Results

We report results of our model ATransD-NL on WN18 and FB15k in Table 3, for comparison with translation-based models. **ATransD-NL** is the originally proposed model defined in Sect. 3 considering both nonlinear correlations among entities and relations and contextual information. Here for a given triplet (h, r, t), we take the neighbourhood entities $\mathbf{N}$ as the all the entities in the knowledge graph, i.e., $\mathbf{N} = \mathcal{E}$, which turns out to be empirically optimal. Experiments show that we have achieved substantial improvements of both metrics on the two benchmarks, compared with the baselines. We would like to remark that, TransMS delivers the best Hist@10 on FB15k for the Filter setting by fine-turning a hyperparameter α used in the linear bias vectors, with the need of setting different values for different relations.

We also experimented with three ablated models with details given below.

– **TransD-NL**: Only nonlinear mapping is applied without attention mechanism:

$$\mathbf{h}_\perp = \sigma(\mathbf{h}_p^\top \mathbf{h}\mathbf{r}_p) + \mathbf{h} \tag{8}$$

$$\mathbf{t}_\perp = \sigma(\mathbf{t}_p^\top \mathbf{t}\mathbf{r}_p) + \mathbf{t} \tag{9}$$

– **TransDT-NL**: Based upon TransD-NL, a simple attention-learning is used, such that the semantic vector of tail entity is embedded into the semantic representation of head entity, and vice versa:

Table 3. Experimental results of link prediction on WN18 and FB15K, compared with translational models. The best results from the two sampling strategies are taken for compared models. The best results are in bold, and the second best score is underlined. The number shown in parenthesis is the normal Hits@10 result of TransMS on FB15K by setting the hyperparameter α uniformly rather than fine-tuning α for different types of relations.

Datasets	WN18				FB15K			
Metric	MeanRank		Hits@10(%)		Mean Rank		Hits@10(%)	
	Raw	Filter	Raw	Filter	Raw	Filter	Raw	Filter
TransE [3]	263	251	75.4	89.2	243	125	34.9	47.1
TransH [19]	318	303	75.4	86.7	211	84	42.5	58.5
TransD [7]	224	212	79.6	92.5	194	67	53.4	77.3
TransR [11]	232	219	78.3	91.7	226	78	43.8	65.5
PTransE [10]	-	-	-	-	200	54	51.8	<u>84.6</u>
TranSparse [8]	223	211	80.1	93.4	190	66	53.7	79.9
TransMS [23]	427	414	82.5	94.8	171	63	**55.0**	**86.8**
							(51.8)	(79.8)
TransD-NL	274	262	72.0	81.4	311	205	40.2	51.9
TransDT-NL	195	185	**83.4**	<u>95.1</u>	147	44	54.2	80.6
ATransD	166	153	80.6	93.4	<u>145</u>	<u>36</u>	54.4	75.3
ATransD-NL(unif)	**162**	**149**	<u>82.7</u>	**95.4**	**145**	**35**	54.7	76.1
ATransD-NL(bern)	<u>165</u>	<u>152</u>	82.5	**95.4**	172	65	<u>54.8</u>	76.5

$$h_\perp = \sigma(\mathbf{h}_p^\top \mathbf{hr}_p) \otimes \mathbf{t} + \mathbf{h} \tag{10}$$

$$t_\perp = \sigma(\mathbf{t}_p^\top \mathbf{tr}_p) \otimes \mathbf{h} + \mathbf{t} \tag{11}$$

- **ATransD**: Only attention-based learning is applied:

$$h_\perp = (\mathbf{h}_p^\top \mathbf{hr}_p) \otimes \bar{\mathbf{t}} + \mathbf{h} \tag{12}$$

$$t_\perp = (\mathbf{t}_p^\top \mathbf{tr}_p) \otimes \mathbf{h} + \mathbf{t} \tag{13}$$

Besides, we have also experimented with the choice of taking $\mathbf{N} = \{t \mid \exists r \exists t.(h, r, t) \in \mathcal{R}\}$ in the attention-based embedding method. That is, we consider the set of tail entities that are linked with the head entity via 1-hop relation. We have also experimented with the model KBGAT on this dataset and obtain MeanRank = 231 and Hits@10 = 91.4% on WN18 the Filter setting, and Mean-Rank = 37 and Hits@10 = 92.1% on FB15K for the Filter setting. KBGAT delivers the best result of Hits@10 on FB15K.

In Table 4, we report results of our model ATransD-NL on WN18RR and FB15k-237, compared with neural-network based methods. Experiments show the good performance of our model on two datasets against the stat-of-the-art

Table 4. Experimental results of link prediction on WN18RR and FB15K-237, compared with models based on neural networks. The best results from the two sampling strategies are taken for compared models. The best results are in bold, and the second best score is underlined.

Datasets	WN18RR		FB15K-237	
Metric	MeanRank	Hits@10(%)	MeanRank	Hits@10(%)
DisMult [22]	7000	50.4	512	44.6
ComplEx [17]	7882	53.0	546	45.0
ConvE [5]	4464	53.1	245	49.7
ConvKB [14]	2554	52.5	257	51.7
R-GCN [15]	6700	20.7	600	30.0
KBGAT [13]	<u>1940</u>	**58.1**	210	**62.6**
TransD-NL	4230	39.9	328	43.8
TransDT-NL	**1464**	48.8	184	56.5
ATransD	2429	53.6	<u>182</u>	53.7
ATransD-NL(unif)	2104	<u>54.2</u>	**180**	56.1
ATransD-NL(bern)	2050	52.0	**203**	<u>59.7</u>

Table 5. Detailed results by the categories of relations on FB15k. The best results from the two sampling strategies are taken for compared models. The best results are in bold and the second best scores are underlined (Note that ATransD-NL would outperform TransMS on this metric if it sets the hyper-parameter α uniformly as reported in [23]).

Prediction type	Head prediction				Tail prediction			
Relation type	1-1	1-N	N-1	N-N	1-1	1-N	N-1	N-N
TransE [3]	43.7	65.7	18.2	47.2	43.7	19.7	66.7	50.0
TransH [19]	66.7	81.7	30.2	57.4	63.7	30.1	83.2	60.8
TransD [7]	86.1	95.5	39.8	78.5	85.4	50.6	94.4	81.2
TransR [11]	76.9	77.9	38.1	66.9	76.2	38.4	76.2	69.1
TranSparse [8]	83.2	85.2	51.4	80.3	82.6	60.0	85.5	82.5
TransMS [23]	**89.5**	94.4	**78.5**	**85.6**	**90.0**	**84.8**	91.7	**87.7**
ATransD-NL (unif)	85.4	**97.1**	<u>61.2</u>	<u>82.3</u>	<u>87.2</u>	<u>84.0</u>	**95.5**	<u>82.8</u>
ATransD-NL (bern)	<u>87.1</u>	<u>96.7</u>	40.4	73.9	86.7	48.3	<u>95.4</u>	77.1

neural-network based methods. We have shown advantages on both datasets, especially that we can always obtain the best results on MeanRank without degrading Hits@10 that much, even compared with the model KBGAT that contain much more parameters than our model. Our training time took around 4 h on FB15k-237 and WN18RR, and double that time on FB15k and WN18.

Table 5 classifies the results of FB15k according to several classifications of complex relations on Hits@10. In this experiments, we divide the ternary components into four categories according to the cardinality of the head entity and tail entity parameters: 1-1, 1-N, N-1, N-N. If a head entity has at most one tail entity, the given relation is 1-1. If a head entity has multiple tail entities connected to it, the given relation is 1-N. If multiple head entities have only one tail entity, The given relation is N-1. If multiple head entities have multiple tail entities, it is N-N. By calculating the average number of head entities (respectively, tail entities) that appear in the FB15k dataset for each relationship ℓ, a pair (ℓ, t) is given (respectively, a pair (h, ℓ)). If the average value is less than 1.5, the parameter is marked as 1, otherwise it is marked as N. For example, a relation with an average tail of 1.2 and an average tail of 3.2 is classified as 1-N.

According to the results of Table 5, the advantage of our method on the FB15k dataset lies in dealing with the kind of 1-N and N-1 relations. We have obtained the best results compared with other baselines on predictions in the two categories: for head prediction, our 1-N prediction is the best among all baselines, and for tail prediction, our N-1 prediction is the best among all baselines, and the other metrics are slightly worse than TransMS. Results show that the overall performance of our method on Hits@10 is comparable to TransMS on FB15k, yet still suffer from some shortcomings when it comes to dealing with 1-1 and N-N relations. Our method is designed to combine the influence of linked tail and head entities, yet entities involved in complex relations may interfere with each other during the learning process.

5   Concluding Remarks

We have presented a new model named ATransD-NL based on extending the model TransD. Our method learns the attention-based translational knowledge graph embedding via nonlinear dynamic semantic projection from entity spaces to relation-specific embedding spaces. The model captures nonlinear correlations among entities and relations. Moreover, our attention-based embedding combines contextual information into the translation-based learning process. Experimental results show that ATransD-NL achieves consistent and significant improvements on link prediction (especially on the MeanRank metric) compared with both state-of-the-art translation-based methods and neural-network based methods. As future work, we will explore more fine-grained attention-based mechanisms to better deal with N-N relations.

References

1. Bollacker, K., Evans, C., Paritosh, P., Sturge, T., Taylor, J.: Freebase: a collaboratively created graph database for structuring human knowledge. In: Proceedings of the 2008 ACM SIGMOD International Conference on Management of Data, pp. 1247–1250 (2008)
2. Bordes, A., Glorot, X., Weston, J., Bengio, Y.: A semantic matching energy function for learning with multi-relational data. Mach. Learn. **94**(2), 233–259 (2013). https://doi.org/10.1007/s10994-013-5363-6

3. Bordes, A., Usunier, N., Garcia-Duran, A., Weston, J., Yakhnenko, O.: Translating embeddings for modeling multi-relational data. In: Advances in Neural Information Processing Systems, pp. 2787–2795 (2013)
4. Bordes, A., Weston, J., Collobert, R., Bengio, Y.: Learning structured embeddings of knowledge bases. In: Proceedings of AAAI 2011, pp. 301–306 (2011)
5. Dettmers, T., Minervini, P., Stenetorp, P., Riedel, S.: Convolutional 2D knowledge graph embeddings. In: Proceedings of AAAI 2018, pp. 1811–1818 (2018)
6. Fabian, M., Gjergji, K., Gerhard, W., et al.: Yago: a core of semantic knowledge unifying Wordnet and Wikipedia. In: Proceedings of 16th International World Wide Web Conference, pp. 697–706 (2007)
7. Ji, G., He, S., Xu, L., Liu, K., Zhao, J.: Knowledge graph embedding via dynamic mapping matrix. In: Proceedings of the 53rd Annual Meeting of the Association for Computational Linguistics and the 7th International Joint Conference on Natural Language Processing (volume 1: Long Papers), pp. 687–696 (2015)
8. Ji, G., Liu, K., He, S., Zhao, J.: Knowledge graph completion with adaptive sparse transfer matrix. In: Proceedings of AAAI 2016, pp. 985–991 (2016)
9. Lehmann, J., et al.: DBpedia-a large-scale, multilingual knowledge base extracted from Wikipedia. Semant. Web 6(2), 167–195 (2015)
10. Lin, Y., Liu, Z., Luan, H., Sun, M., Rao, S., Liu, S.: Modeling relation paths for representation learning of knowledge bases. arXiv preprint arXiv:1506.00379 (2015)
11. Lin, Y., Liu, Z., Sun, M., Liu, Y., Zhu, X.: Learning entity and relation embeddings for knowledge graph completion. In: Proceedings of AAAI 2015, pp. 2181–2187 (2015)
12. Miller, G.A.: Wordnet: a lexical database for English. Commun. ACM 38(11), 39–41 (1995)
13. Nathani, D., Chauhan, J., Sharma, C., Kaul, M.: Learning attention-based embeddings for relation prediction in knowledge graphs. arXiv preprint arXiv:1906.01195 (2019)
14. Nguyen, D.Q., Nguyen, T.D., Nguyen, D.Q., Phung, D.: A novel embedding model for knowledge base completion based on convolutional neural network. arXiv preprint arXiv:1712.02121 (2017)
15. Schlichtkrull, M., Kipf, T.N., Bloem, P., van den Berg, R., Titov, I., Welling, M.: Modeling relational data with graph convolutional networks. In: Gangemi, A., et al. (eds.) ESWC 2018. LNCS, vol. 10843, pp. 593–607. Springer, Cham (2018). https://doi.org/10.1007/978-3-319-93417-4_38
16. Shang, C., Tang, Y., Huang, J., Bi, J., He, X., Zhou, B.: End-to-end structure-aware convolutional networks for knowledge base completion. In: Proceedings of the AAAI 2019, vol. 33, pp. 3060–3067 (2019)
17. Trouillon, T., Welbl, J., Riedel, S., Gaussier, E., Bouchard, G.: Complex embeddings for simple link prediction. In: Proceedings of ICML 2016, pp. 2071–2080 (2016)
18. Vu, T., Nguyen, T.D., Nguyen, D.Q., Phung, D., et al.: A capsule network-based embedding model for knowledge graph completion and search personalization. In: Proceedings of the 2019 Conference of the North American Chapter of the Association for Computational Linguistics: Human Language Technologies, pp. 2180–2189 (2019)
19. Wang, Z., Zhang, J., Feng, J., Chen, Z.: Knowledge graph embedding by translating on hyperplanes. In: Proceedings of AAAI 2014, pp. 1112–1119 (2014)
20. Xiao, H., Huang, M., Hao, Y., Zhu, X.: TransA: an adaptive approach for knowledge graph embedding. arXiv preprint arXiv:1509.05490 (2015)

21. Xiao, H., Huang, M., Hao, Y., Zhu, X.: TransG: a generative mixture model for knowledge graph embedding. arXiv preprint arXiv:1509.05488 (2015)
22. Yang, B., Yih, W.t., He, X., Gao, J., Deng, L.: Embedding entities and relations for learning and inference in knowledge bases. arXiv preprint arXiv:1412.6575 (2014)
23. Yang, S., Tian, J., Zhang, H., Yan, J., He, H., Jin, Y.: TransMS: knowledge graph embedding for complex relations by multidirectional semantics. In: Proceedings of IJCAI 2019, pp. 1935–1942 (2019)

Multi-Grained Dependency Graph Neural Network for Chinese Open Information Extraction

Zhiheng Lyu[1,2], Kaijie Shi[1,2], Xin Li[1,2], Lei Hou[1,3(✉)], Juanzi Li[1,3], and Binheng Song[1,2]

[1] Department of Computer Science and Technology, BNRist,
Tsinghua University, Beijing, China
{lvzh18,skj19,lixin18}@mails.tsinghua.edu.cn,
{houlei,lijuanzi}@tsinghua.edu.cn, songbinheng@sz.tsinghua.edu.cn
[2] Tsinghua Shenzhen International Graduate School, Tsinghua University,
Shenzhen, China
[3] KIRC, Institute for Artificial Intelligence, Tsinghua University, Beijing, China

Abstract. Recent neural Open Information Extraction (OpenIE) models have improved traditional rule-based systems significantly for Chinese OpenIE tasks. However, these neural models are mainly word-based, suffering from word segmentation errors in Chinese. They utilize dependency information in a shallow way, making multi-hop dependencies hard to capture. This paper proposes a Multi-Grained Dependency Graph Neural Network (MGD-GNN) model to address these problems. MGD-GNN constructs a multi-grained dependency (MGD) graph with dependency edges between words and soft-segment edges between words and characters. Our model makes predictions based on character features while still has word boundary knowledge through word-character soft-segment edges. MGD-GNN updates node representations using a deep graph neural network to fully exploit the topology structure of the MGD graph and capture multi-hop dependencies. Experiments on a large-scale Chinese OpenIE dataset SpanSAOKE shows that our model could alleviate the propagation of word segmentation errors and use dependency information more effectively, giving significant improvements over previous neural OpenIE models.

Keywords: Open information extraction · Dependency graph model · Graph neural network

1 Introduction

Open information extraction (OpenIE) is an important task in natural language processing (NLP), aiming to mine semi-structured fact knowledge from unstructured text. Unlike traditional relation extraction tasks with schema constraint, OpenIE imposes no limitations on relations or arguments and could extract more

© Springer Nature Switzerland AG 2021
K. Karlapalem et al. (Eds.): PAKDD 2021, LNAI 12714, pp. 155–167, 2021.
https://doi.org/10.1007/978-3-030-75768-7_13

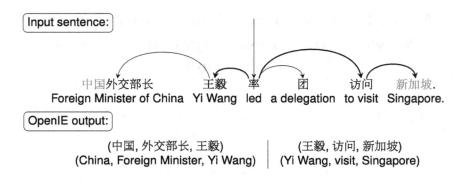

Fig. 1. An example of Chinese OpenIE. Given a sentence, two fact knowledge triples in the form of *(subject, predicate, object)* are extracted from this sentence. The segmented words and dependency edges between words are also shown.

fact knowledge from text sources. OpenIE has benefited many downstream tasks, such as knowledge base construction, question answering [13], and summarization [9].

Traditional OpenIE systems are mostly based on syntactic patterns and heuristic rules. For example, REVERB [8], ClausIE [6], OPENIE4 [19] for English and CORE [28], ZORE [21] for Chinese, leverage external NLP tools to obtain part-of-speech (POS) tags or dependency features and generate syntactic patterns to extract fact knowledge tuples[1]. The syntactic patterns in these systems are usually language-specific and cannot generalize well to other languages. Besides, these pattern-based approaches cannot handle the complexity and diversity of languages well, and the performance is usually far from satisfactory.

Recently, neural models have been employed on OpenIE tasks to conquer the limitations of syntactic pattern-based methods. Neural OpenIE methods could be divided into two categories: sequence generation and span selection. Sequence generation models, such as Neural Open IE [5], Logician [27] and IMOJIE [16], generate fact tuples directly using the encoder-decoder framework. Span selection models, such as RnnOIE [25] and SpanOIE [33], select spans of a sentence as predicates or arguments. Previous sequence generation and span selection models are mostly word-based, and thus would propagate word segmentation errors when applied to Chinese. For example, as illustrated in Fig. 1, "中国外交部长 (Foreign Minister of China)" which is composed of word "中国 (China)" and "外交部长 (Foreign Minister)" is segmented incorrectly as a whole word, which would cause word-based models to neglect the first fact knowledge triple.

It is proven that dependency knowledge benefits many information extraction tasks, such as semantic role labeling [18] and relation extraction [35]. However, existing neural OpenIE models usually integrate dependency information into neural models in a shallow way. For example, SpanOIE concatenates word embedding with corresponding dependency label embedding as the input of a

---

[1] We express *n*-ray extraction as tuple and binary extraction as triple in this paper.

sentence encoder. This ignores the topology structure of a dependency tree, making it hard to capture multi-hop dependencies. As shown in Fig. 1, the dependency between subject "王毅 (Yi Wang)" and predicate "访问 (visit)" is multi-hop, which is hard to capture for shallow integration approach.

To address these issues, we propose a multi-grained dependency graph neural network (MGD-GNN) model for Chinese OpenIE. (1) To avoid propagating of word segmentation errors, we leverage multi-granularity information and make predictions on characters rather than words. Specifically, we construct a multi-grained dependency (MGD) graph with words and characters interconnected with each other. Words are connected to their corresponding characters to provide soft hints of word boundaries rather than hard predictions. (2) To capture multi-hop dependencies, we incorporate the dependency relations between words and adopt a graph neural network (GNN) to encode the MGD graph. Deep GNN updates node representations using information from its multi-hop neighbors and thus could capture multi-hop dependencies. To the best of our knowledge, we are the first to explore character-based models and introduce GNN to Chinese OpenIE. The main contributions of this paper can be summarized as follows:

- We propose a character-based neural OpenIE model for the Chinese OpenIE task and introduce graph neural networks to neural OpenIE models.
- We propose an MGD-GNN model for Chinese OpenIE, which could alleviate word segmentation errors in Chinese and capture multi-hop dependencies.
- Experiments on SpanSAOKE, a Chinese OpenIE dataset, demonstrate that our proposed model has significant improvements over baselines.

2   Problem and Methodology

We first formulate the problem of Chinese open information extraction. As shown in Fig. 1, it is formalized as a span extraction task: given a sentence $S = \langle c_1, ..., c_N \rangle$ with N characters, the goal is to extract M fact triples $T = \{(s_1, p_1, o_1), ..., (s_M, p_M, o_M)\}$ from the sentence. s_i, p_i, o_i represent subject, predicate and object of a fact triple respectively, and are defined as spans of sentence S. Here we define span as a continuous character segment $\langle c_i, ..., c_j \rangle_{1 \leq i \leq j \leq N}$ of sentence S. Similar as [5], we only consider binary triples, which have exactly one subject and one object.

Next, we introduce our MGD-GNN model. Our model employs a two-stage pipeline extraction method, including **predicate extraction** and **argument extraction** stages following [33]. Both stages share the same neural network architecture to get character embeddings with the context encoder and our MGD-GNN. In the predicate extraction stage, our model extracts all predicate spans from the sentence. Then, in the argument extraction stage, our model predicts its corresponding subject and object for each predicate.

2.1   Context Encoder

Our model first maps each character c_i to its distributed representation $\mathbf{c}_i \in \mathbb{R}^{d_c}$ using pre-trained embeddings via word2vec [20]. We also concatenate different

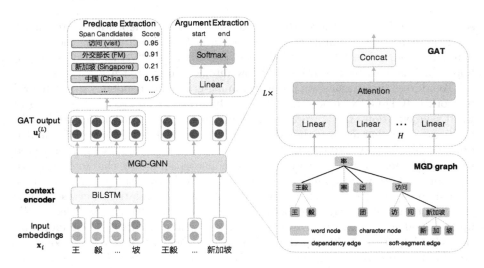

Fig. 2. The architecture of our multi-grained dependency graph neural network (MGD-GNN) model for Chinese OpenIE. The left side is the overall architecture, and the right side shows the details of MGD-GNN.

feature embeddings $\mathbf{f}_i$ for each character in predicate and argument extraction stages. After concatenation $\mathbf{x}_i = [\mathbf{c}_i; \mathbf{f}_i]$, we get an embedding sequence $\langle \mathbf{x}_1, ..., \mathbf{x}_N \rangle$ which is directly fed to a context encoder.

To get the contextualized representations for each character, we use a bidirectional LSTM (BiLSTM) [24] to model the sequence. Other character-based context encoders, such as BERT Chinese [7], could also be applied. BiLSTM processes character-based sentences from forward and backward directions.

$$\overrightarrow{\mathbf{h}}_i = \text{BiLSTM}(\mathbf{x}_i, \overrightarrow{\mathbf{h}}_{i-1}) \tag{1}$$

$$\overleftarrow{\mathbf{h}}_i = \text{BiLSTM}(\mathbf{x}_i, \overleftarrow{\mathbf{h}}_{i+1}) \tag{2}$$

where $\overrightarrow{\mathbf{h}}_i$ and $\overleftarrow{\mathbf{h}}_i$ represent the left and right hidden states of the i-th character respectively. By concatenating the hidden states from both sides, we get the final contextual representations $\mathbf{h}_i = [\overrightarrow{\mathbf{h}}_i; \overleftarrow{\mathbf{h}}_i] \in \mathbb{R}^d$.

2.2 MGD-GNN

Multi-Grained Dependency Graph Construction. To alleviate word segmentation errors in Chinese and capture multi-hop dependency knowledge, we propose to build a multi-grained dependency (MGD) graph as illustrated in Fig. 2. The word and character nodes in MGD graph are interconnected through two types of undirected edges, naming dependency edge and soft-segment edge. To build the MGD graph, we first obtain segmented words and dependency tree of a sentence using LTP [4]. We keep segmented words as word nodes in

MGD graph, and make directed dependency tree edges as undirected dependency edges between word nodes. Meanwhile, each character in the sentence is added to MGD graph as a character node. A word node has soft-segment edges connected to the character nodes that make up the word. For example, word node "新加坡 (Singapore)" in Fig. 2 has soft-segment edges with its three character nodes "新", "加"" and "坡". In this way, when we make predictions on character nodes afterward, the word segmentation result of a specific NLP tool acts as soft hints to alleviate word segmentation errors in Chinese. MGD graph incorporates dependency and word segmentation knowledge by its word-word dependency edges and word-character soft-segment edges to help the model make better decisions.

Graph Neural Network over MGD Graph. Graph neural network (GNN) has been used widely in modeling graph-structured data [12,15,31]. GNN captures multi-hop dependency features on graphs by aggregating information from neighborhoods of a node. Among these variants of GNN, we use graph attention network (GAT) [31] as our graph encoder, which is capable of controlling node weights when aggregating information from neighbors.

Formally, let $\mathcal{G} = \{\mathcal{V}, \mathcal{E}\}$ denote a graph, where $\mathcal{V}$ is the vertex set and $\mathcal{E}$ is the edge set. In our MGD graph, $\mathcal{V} = \mathcal{W} \cup \mathcal{C}$ consists of all word nodes $\mathcal{W}$ and character nodes $\mathcal{C}$, and $\mathcal{E}$ contains all dependency edges and soft-segment edges. Let $\mathbf{u}_i^l \in \mathbb{R}^d$ indicate node embedding of the i-th node at the l-th GAT layer, where d is the dimension of node embeddings. We initialize node embeddings using hidden state outputs from BiLSTM encoder,

$$\mathbf{u}_i^0 = \begin{cases} \mathbf{h}_i & \text{if } i \in \mathcal{C} \\ (\sum_{j \in \mathcal{N}_i^s} \mathbf{h}_j)/|\mathcal{N}_i^s| & \text{otherwise} \end{cases} \quad (3)$$

where $\mathcal{N}_i^s$ represents all neighbors of node i that have soft-segment edge connections with i. The initialization vector of a word node is computed by mean pooling of its character hidden states.

We use multi-head attention [30] to update node representations. Specifically, multi-head attention adopts H attention heads when aggregating information from neighbors, each transforming input to different spaces and focusing on different aspects of neighbors. For attention head h, we first calculate attention scores $e_{ij}^{(h)}$ between each pair of nodes using a feed-forward neural network,

$$e_{ij}^{(h)} = \text{LeakyReLU}(\mathbf{a}^{(h)} \cdot (\mathbf{W}^{(h)}\mathbf{u}_i^{l-1}||\mathbf{W}^{(h)}\mathbf{u}_j^{l-1})) \quad (4)$$

where $\mathbf{u}_i^{l-1} \in \mathbb{R}^d$ and $\mathbf{u}_j^{l-1} \in \mathbb{R}^d$ express the features of node i and node j at the $l-1$-th layer. $\mathbf{W}^{(h)} \in \mathbb{R}^{d_h \times d}$ is the linear projection matrix, transforming input to d_h dimensions. $||$ is the concatenation operator. $\mathbf{a}^{(h)} \in \mathbb{R}^{d_h}$ is the learnable weight vector of attention.

Then we compute normalized attention weights using a *softmax* function,

$$\alpha_{ij}^{(h)} = \text{softmax}(e_{ij}^{(h)}) = \frac{\exp\left(e_{ij}^{(h)}\right)}{\sum_{k \in \mathcal{N}_i} \exp\left(e_{ik}^{(h)}\right)} \tag{5}$$

where $\mathcal{N}_i$ denotes the set of neighbors of node i. We perform a linear combination of the features of node neighbors using normalized attention weights. The output features of node i on head h is obtained as follows,

$$\mathbf{u}_i^{(h)} = \sum_{j \in \mathcal{N}_i} \alpha_{ij}^{(h)} \mathbf{W}^{(h)} \mathbf{u}_j^{l-1} \tag{6}$$

where $\mathbf{u}_i^{(h)} \in \mathbb{R}^{d_h}$ is the d_h dimensional output features.

We concatenate outputs of H attention heads to get the updated node features $\mathbf{u}_i^l = [\mathbf{u}_i^{(1)}; \mathbf{u}_i^{(2)}; \cdots ; \mathbf{u}_i^{(H)}]$, where $\mathbf{u}_i^l$ is a $d_h \times H$ dimensional vector. For implementation convenience, we choose the dimensions of input and output representations of a GAT layer to be the same, that is, $d = d_h \times H$.

By stacking L GAT layers, each node could gather information from its L-hop neighbors. We retain character node features of the last GAT layer $\mathbf{U} = [\mathbf{u}_1^L, \mathbf{u}_2^L, \cdots, \mathbf{u}_{|\mathcal{C}|}^L]$ for triple extraction.

2.3 Triple Extraction

Predicate Extraction. Following [33], we model predicate extraction as a span classification problem. We only preserve the spans satisfying maximum length, non-overlapping and syntactic constraints [33] as candidates to classify. To incorporate syntactic features, we take POS tag embeddings $\mathbf{t}_i \in \mathbb{R}^{d_t}$ as additional feature input $\mathbf{f}_i = \mathbf{t}_i$. For a candidate span $\langle c_i, ..., c_j \rangle_{1 \le i \le j \le N}$, we select its start and end character features, and predict the probability of it being a predicate.

$$\mathbf{p} = \text{softmax}(\mathbf{W}_p[\mathbf{u}_i^L; \mathbf{u}_j^L; \mathbf{u}_i^L + \mathbf{u}_j^L; \mathbf{u}_i^L - \mathbf{u}_j^L]) \tag{7}$$

where $\mathbf{W}_p \in \mathbb{R}^{2 \times 4d}$ is the weight of linear predicate classifier.

Argument Extraction. Given each predicate span acquired in the predicate extraction phase, we extract its corresponding subject and object. We adopt relative position embeddings [32] as additional input features $\mathbf{f}_i = \mathbf{p}_i \in \mathbb{R}^{d_p}$ to indicate predicate positions. To extract the subject, we apply linear classifiers on $\mathbf{U}$ to compute the probability of each character being the start and end of a subject span.

$$\mathbf{p}_{\text{subj\_start}} = \text{softmax}(\mathbf{W}_{\text{subj\_start}}\mathbf{U}) \tag{8}$$

$$\mathbf{p}_{\text{subj\_end}} = \text{softmax}(\mathbf{W}_{\text{subj\_end}}\mathbf{U}) \tag{9}$$

where $\mathbf{W}_{\text{subj\_start}} \in \mathbb{R}^{1 \times d}$ and $\mathbf{W}_{\text{subj\_end}} \in \mathbb{R}^{1 \times d}$ are weight matrices, and $\mathbf{p}_{\text{subj\_start}} \in \mathbb{R}^{|\mathcal{C}|}$ and $\mathbf{p}_{\text{subj\_end}} \in \mathbb{R}^{|\mathcal{C}|}$ are start and end probability distributions over characters of the sentence. The object is extracted in the same way.

During training, the predicate extraction model and argument extraction model are optimized independently. The argument extraction model is trained using gold predicates of a sentence. At inference, predicates obtained by the predicate extraction model are fed to the argument extraction model, and we combine the results of two stages to get final fact triples.

3   Experiments

3.1   Experiment Settings

Dataset. We evaluate our model on the SpanSAOKE dataset, a span extraction dataset from the original SAOKE [27]. SAOKE is a large-scale sentence-level dataset for Chinese OpenIE. Each sentence in SAOKE is manually labeled with its contained facts in a unified knowledge representation format. To adapt SAOKE dataset to our span extraction setting, we filter unknown, description and concept facts because they either have missing subject, predicate and object or introduce special predicates such as ISA and DESC. Besides, we only retain binary relation facts whose triple components are spans of the sentence. The processed dataset, SpanSAOKE, contains 26,496 Chinese sentences with 53,869 facts. We randomly split SpanSAOKE into train, dev and test set. The statistics of SpanSAOKE is shown in Table 1.

Table 1. Statistics of SpanSAOKE dataset. #Sent. denotes the number of sentences and #Avg. represents average number of facts per sentence.

Split	#Sent.	#Fact	#Avg.
Train	21,196	43,216	2.04
Dev	2,649	5,311	2.01
Test	2,651	5,342	2.02
Total	26,496	53,869	2.03

Table 2. Main results on the test set of SpanSAOKE dataset.

Model	P	R	F_1
ZORE	31.5	17.7	22.7
CharLSTM	40.4	45.4	42.7
SpanOIE	41.8	44.3	43.0
WD-GNN	41.3	**47.2**	44.1
MGD-GNN	**45.0**	47.1	**46.0**

Evaluation Metrics. To evaluate the results, we need to compare the predicted triples $T_p = \{(\hat{s}_i, \hat{p}_i, \hat{o}_i) \mid 1 \le i \le P\}$ of a sentence with ground truth triples facts $T = \{(s_j, p_j, o_j) \mid 1 \le j \le M\}$. Each triple in T_p is regarded as correct if it matches one of the triples in T. Two triples $t_p = (\hat{s}_i, \hat{p}_i, \hat{o}_i)$ and $t = (s_j, p_j, o_j)$ are considered as matched if they satisfy one of the two matching conditions described in [27]: (1) $\mathbf{g}(\hat{s}_i, s_j), \mathbf{g}(\hat{p}_i, p_j), \mathbf{g}(\hat{o}_i, o_j) \ge \delta$, (2) $\mathbf{g}(\mathrm{Cat}(t_p), \mathrm{Cat}(t)) \ge \delta$, where $\mathbf{g}(\cdot, \cdot)$ is the gestalt pattern matching function [22] and $\mathrm{Cat}(\cdot)$ concatenates triple components as a whole string. We choose the threshold $\delta = 0.85$ in our experiments. The matched triples in T are excluded for later matching. We use precision, recall and F_1 score to evaluate model performance.

Parameters. The hyper-parameters in our experiments are illustrated in Table 3. All hyper-parameters are tuned on the dev set. We use the same model structure and training hyper-parameters for both predicate and argument extraction models. For our MGD-GNN model, we use character embeddings from [34] with dimension $d_c = 50$, which is pre-trained on Chinese Giga-Word using word2vec. We choose Adam optimizer [14] to minimize the cross entropy training loss for its faster convergence rate. The learning rate of Adam is decayed by 0.01 every epoch, and the training process is early stopped with a patience of 30. In the predicate extraction stage, we set the weights of positive and negative samples to 3 and 1 for handling the label imbalance problem.

Table 3. Hyper-parameter settings.

Parameter	Value	Parameter	Value
BiLSTM layers	1	BiLSTM hidden d	300
Multi-head attention H	5	Attention hidden d_h	60
GAT layers L	3	GAT dropout	0.3
Batch size	50	Learning rate	0.001

3.2   Overall Results

We compare our proposed MGD-GNN with several competitive baselines for Chinese OpenIE.

ZORE [21] is a syntactic-based system, which extracts relational tuples and their semantic patterns iteratively based on parsed dependency trees of sentences. To evaluate ZORE under our binary triple extraction setting, we identity subject and object from possibly multiple arguments of ZORE tuples. In practice, we find that taking arguments with one of *nsubj*, *SBV* dependency labels as subject and arguments with *nobj*, *VOB*, *POB* labels as object gives better F_1 score. We use this strategy to refine ZORE tuples in our experiments.

SpanOIE [33] is a span selection based neural OpenIE model. It is a word-based span classification model and achieves competitive results on English OpenIE. To adapt SpanOIE to Chinese OpenIE, we use the pre-trained Chinese word embeddings from [17] and obtain POS tags and dependency labels from LTP [4] as in our model.

CharLSTM applies a vanilla character-based BiLSTM to encode characters to extract predicates and arguments. The only difference between CharLSTM and our proposed MGD-GNN is that CharLSTM has no extra graph encoder component to integrate dependency and word segmentation knowledge.

We present the comparison results in Table 2. We observe that neural OpenIE models outperform rule-based ZORE by a large margin, showing their abilities to handle complex and diverse sentences. Without additional word and dependency information, simple CharLSTM achieves comparable performance to word-based SpanOIE, validating the effectiveness of character-based models on

Chinese OpenIE task. We hypothesis that character-based neural encoder could capture word boundary features of Chinese and avoid propagating word segmentation errors. By introducing GAT to encode multi-grained dependency graph, our proposed MGD-GNN further outperforms SpanOIE by 3 F_1, demonstrating that our MGD-GNN could utilize dependency and word boundary knowledge more effectively.

3.3   Effect of Multi-Grained Information

To demonstrate the effectiveness of multi-grained input in alleviating the propagation of word segmentation errors, we prune all character nodes from our MGD graph and keep the other components. The pruned model, word dependency graph neural network (WD-GNN), only has word-level nodes and extracts predicates and arguments based on word representations. As shown in Table 2, our multi-grained MGD-GNN outperforms single-grained WD-GNN by 1.9 F_1. This shows the effectiveness of incorporating multi-grained information and demonstrates that making predictions on characters rather than words could alleviate word segmentation errors on Chinese OpenIE.

Table 4. Results on the test set of SpanSAOKE dataset. We present precision, recall and F_1 score for both triple and predicate extraction. Subj Acc. denotes accuracy of subject in argument extraction and Obj Acc. denotes accuracy of object.

#layer	Triple			Predicate			Argument	
	P	R	F_1	P	R	F_1	Subj Acc.	Obj Acc.
0	40.4	45.4	42.7	48.4	54.4	51.2	70.5	77.6
1	43.0	45.5	44.3	50.4	53.3	51.8	70.6	77.4
2	43.9	46.4	45.1	50.8	53.7	52.2	72.0	82.6
3	**45.0**	47.1	**46.0**	**51.4**	53.8	52.5	73.0	83.2
4	43.9	**48.3**	**46.0**	50.2	**55.3**	**52.7**	**73.4**	**83.3**

3.4   Analysis of Multi-hop Dependencies

To analyze the ability to capture multi-hop dependency features, we test our MGD-GNN with different number of GAT layers and present the results in Table 4. Predicates and arguments are compared using the gestalt string matching function as the first triple matching condition in [27]. Stacking K GAT layers makes our model capable of aggregating information at most K-hop away. We find that models with GAT layers give better performance than models without GAT layers, especially on the argument extraction. This demonstrates the effectiveness of our proposed MGD-GNN model. We also observe that the 2-layer version has an improvement of 1.4 on subject accuracy and 5.2 on object accuracy compared with the 1-layer version. This observation reveals that multi-hop dependency features are crucial to Chinese OpenIE. Shallow ways of incorporating dependency information are hard to capture these multi-hop features, and thus perform poorly on Chinese OpenIE. We can also see that even with one additional GAT layer, the 4-layer version has similar performance compared with the 3-layer model on triple, predicate and argument extractions. We hypothesis that deeper GNN models such as our 4-layer model are hard to optimize and may cause overfitting on the dataset. To trade off model efficiency and performance, we choose our MGD-GNN with 3 GAT layers.

4   Related Work

During the past decades, many systems have been built to solve the OpenIE problem. TEXTRUNNER [3] is the first OpenIE system that could extract a far broader set of facts from the Web using only part-of-speech (POS) tags. After that, several systems, e.g., REVERB [8], OLLIE [23], ClausIE [6], Stanford OPE-NIE [1], OPENIE4 [19], were introduced gradually to break the limitations of previous systems. For Chinese, CORE [28] is the first Chinese OpenIE system that applies a pipeline of NLP tools, and ZORE [21] identifies Chinese relational tuples and semantic patterns simultaneously. These systems are rule-based and depend on syntactic outputs of NLP tools.

Recently, neural models have been applied to OpenIE. The sequence generation method provides an end-to-end approach to solve OpenIE tasks. Neural Open IE [5] uses encoder-decoder framework with attention [2] and copying mechanism [11] to generate fact triples. To alleviate under- and over-extraction problems and incorporate dependency information, Logician [27] leverages additional coverage mechanism [29] and gated dependency attention compare with Neural Open IE, and achieves competitive results on Chinese OpenIE. IMOJIE [16] produces one triple in each decoding phase and concatenates the sentence and previously extracted triples as encoder input. Though it increases computation cost during training and inference, this multi-turn generation strategy could adapt the number of extractions to the sentence length.

Recent works [16,33] show that span selection based models could achieve better performance at the expense of more flexible outputs. In these works, predicates and arguments are defined as spans of a sentence. RnnOIE [25] first identifies verbs and nominal predicates using heuristic rules, and then uses a bidirectional LSTM (BiLSTM) sequence tagger to extract arguments. SpanOIE [33] introduces a word-based BiLSTM span model to classify candidate predicate and argument spans. To incorporate syntactic knowledge, SpanOIE also concatenates POS tag and dependency label embeddings with word embeddings for each word. For its superiors performance, we follow the line of span selection works for Chinese OpenIE.

Graph neural network [12,15,31] has achieved state-of-the-art performance on graph-related tasks, such as node classification and graph classification. Among variants of GNN, graph attention network (GAT) stands a strong baseline compared with other models. GNN has been widely applied on various NLP tasks, such as relation extraction [35], named entity recognition [26], question answering [10]. For example, [35] uses GNN to encode pruned dependency tree to improve performance on relation extraction task. To the best of our knowledge, we are the first to introduce GNN to OpenIE task and show its effectiveness on this task.

5 Conclusion

In this paper, we explore character-based dependency graph models for the Chinese OpenIE task. Our proposed model, MGD-GNN, constructs a multi-grained dependency graph to incorporate dependency and word boundary information and employs GNN to get node representations for predicate and argument predictions. Our character-based MGD-GNN model could avoid propagating word segmentation errors and capture multi-hop dependency features. We evaluate our model on a large-scale dataset SpanSAOKE and show that our proposed model outperforms baselines significantly.

Acknowledgements. This work is supported by the NSFC Projects (U1736204, 62006136), grants from Beijing Academy of Artificial Intelligence (BAAI2019ZD0502) and the Institute for Guo Qiang, Tsinghua University (2019GQB0003).

References

1. Angeli, G., Premkumar, M.J.J., Manning, C.D.: Leveraging linguistic structure for open domain information extraction. In: Proceedings of ACL-IJCNLP, pp. 344–354 (2015)
2. Bahdanau, D., Cho, K., Bengio, Y.: Neural machine translation by jointly learning to align and translate. In: Proceedings of ICLR (2015)
3. Banko, M., Cafarella, M.J., Soderland, S., Broadhead, M., Etzioni, O.: Open information extraction from the web. Commun. ACM **51**, 68–74 (2008)
4. Che, W., Feng, Y., Qin, L., Liu, T.: N-ltp: a open-source neural chinese language technology platform with pretrained models. arXiv preprint arXiv:2009.11616 (2020)
5. Cui, L., Wei, F., Zhou, M.: Neural open information extraction. In: Proceedings of ACL, pp. 407–413 (2018)
6. Del Corro, L., Gemulla, R.: Clausie: clause-based open information extraction. In: Proceedings of WWW, pp. 355–366 (2013)
7. Devlin, J., Chang, M.W., Lee, K., Toutanova, K.: Bert: pre-training of deep bidirectional transformers for language understanding. In: Proceedings of NAACL-HLT, pp. 4171–4186 (2019)
8. Fader, A., Soderland, S., Etzioni, O.: Identifying relations for open information extraction. In: Proceedings of EMNLP, pp. 1535–1545 (2011)
9. Fan, A., Gardent, C., Braud, C., Bordes, A.: Using local knowledge graph construction to scale seq2seq models to multi-document inputs. In: Proceedings of EMNLP-IJCNLP, pp. 4186–4196 (2019)

10. Fang, Y., Sun, S., Gan, Z., Pillai, R., Wang, S., Liu, J.: Hierarchical graph network for multi-hop question answering. In: Proceedings of EMNLP, pp. 8823–8838 (2020)
11. Gu, J., Lu, Z., Li, H., Li, V.O.: Incorporating copying mechanism in sequence-to-sequence learning. In: Proceedings of ACL, pp. 1631–1640 (2016)
12. Hamilton, W., Ying, Z., Leskovec, J.: Inductive representation learning on large graphs. In: Advances in neural information processing systems (NIPS), pp. 1024–1034 (2017)
13. Khot, T., Sabharwal, A., Clark, P.: Answering complex questions using open information extraction. In: Proceedings of ACL, pp. 311–316 (2017)
14. Kingma, D.P., Ba, J.: Adam: a method for stochastic optimization. In: Proceedings of ICLR (2015)
15. Kipf, T.N., Welling, M.: Semi-supervised classification with graph convolutional networks. In: Proceedings of ICLR (2017)
16. Kolluru, K., et al.: Imojie: iterative memory-based joint open information extraction. In: Proceedings of ACL, pp. 5871–5886 (2020)
17. Li, S., Zhao, Z., Hu, R., Li, W., Liu, T., Du, X.: Analogical reasoning on Chinese morphological and semantic relations. In: Proceedings of ACL, pp. 138–143 (2018)
18. Marcheggiani, D., Titov, I.: Encoding sentences with graph convolutional networks for semantic role labeling. In: Proceedings of EMNLP, pp. 1506–1515 (2017)
19. Mausam, M.: Open information extraction systems and downstream applications. In: Proceedings of IJCAI, pp. 4074–4077 (2016)
20. Mikolov, T., Sutskever, I., Chen, K., Corrado, G.S., Dean, J.: Distributed representations of words and phrases and their compositionality. In: Advances in Neural Information Processing Systems (NIPS), pp. 3111–3119 (2013)
21. Qiu, L., Zhang, Y.: Zore: a syntax-based system for chinese open relation extraction. In: Proceedings of EMNLP, pp. 1870–1880 (2014)
22. Ratcliff, J.W., Metzener, D.E.: Pattern-matching-the gestalt approach. Dr Dobbs J. **13**, 46 (1988)
23. Schmitz, M., et al.: Open language learning for information extraction. In: Proceedings of EMNLP-CoNLL, pp. 523–534 (2012)
24. Schuster, M., Paliwal, K.: Bidirectional recurrent neural networks. IEEE Trans. Signal Process. **45**, 2673–2681 (1997)
25. Stanovsky, G., Michael, J., Zettlemoyer, L., Dagan, I.: Supervised open information extraction. In: Proceedings of NAACL-HLT, pp. 885–895 (2018)
26. Sui, D., Chen, Y., Liu, K., Zhao, J., Liu, S.: Leverage lexical knowledge for chinese named entity recognition via collaborative graph network. In: Proceedings of EMNLP-IJCNLP, pp. 3821–3831 (2019)
27. Sun, M., Li, X., Wang, X., Fan, M., Feng, Y., Li, P.: Logician: a unified end-to-end neural approach for open-domain information extraction. In: Proceedings of WSDM, pp. 556–564 (2018)
28. Tseng, Y.H., et al.: Chinese open relation extraction for knowledge acquisition. In: Proceedings of EACL, pp. 12–16 (2014)
29. Tu, Z., Lu, Z., Liu, Y., Liu, X., Li, H.: Modeling coverage for neural machine translation. In: Proceedings of ACL, pp. 76–85 (2016)
30. Vaswani, A., et al.: Attention is all you need. In: Advances in Neural Information Processing Systems (NIPS), pp. 5998–6008 (2017)
31. Veličković, P., Cucurull, G., Casanova, A., Romero, A., Lio, P., Bengio, Y.: Graph attention networks. In: Proceedings of ICLR (2018)
32. Zeng, D., Liu, K., Lai, S., Zhou, G., Zhao, J.: Relation classification via convolutional deep neural network. In: Proceedings of COLING, pp. 2335–2344 (2014)

33. Zhan, J., Zhao, H.: Span model for open information extraction on accurate corpus. In: Proceedings of AAAI, pp. 9523–9530 (2020)
34. Zhang, Y., Yang, J.: Chinese NER using lattice LSTM. In: Proceedings of ACL, pp. 1554–1564 (2018)
35. Zhang, Y., Qi, P., Manning, C.D.: Graph convolution over pruned dependency trees improves relation extraction. In: Proceedings of EMNLP, pp. 2205–2215 (2018)

Human-Understandable Decision Making
for Visual Recognition

Xiaowei Zhou[1,3(✉)], Jie Yin[2], Ivor Tsang[1], and Chen Wang[3]

[1] Australian Artificial Intelligence Institute, FEIT, University of Technology Sydney,
Sydney, Australia
`Xiaowei.Zhou@student.uts.edu.au, Ivor.Tsang@uts.edu.au`
[2] Discipline of Business Analytics, The University of Sydney, Sydney, Australia
`jie.yin@sydney.edu.au`
[3] Data61, CSIRO, Sydney, Australia
`Chen.Wang@data61.csiro.au`

Abstract. The widespread use of deep neural networks has achieved substantial success in many tasks. However, there still exists a huge gap between the operating mechanism of deep learning models and human-understandable decision making, so that humans cannot fully trust the predictions made by these models. To date, little work has been done on how to align the behaviors of deep learning models with human perception in order to train a human-understandable model. To fill this gap, we propose a new framework to train a deep neural network by incorporating the prior of human perception into the model learning process. Our proposed model mimics the process of perceiving conceptual parts from images and assessing their relative contributions towards the final recognition. The effectiveness of our proposed model is evaluated on two classical visual recognition tasks. The experimental results and analysis confirm our model is able to provide interpretable explanations for its predictions, but also maintain competitive recognition accuracy.

Keywords: Interpretability · Human-understandable decision

1 Introduction

Deep neural networks (DNNs) have made remarkable success in many areas such as computer vision, speech recognition, and natural language processing. Although DNNs have achieved human-level performance or even beaten humans on some tasks like object recognition or video games, their underlying operation mechanism still remains a "black box" for humans to fully comprehend. As a result, humans cannot fully trust the predictions made by DNNs, particularly in life-critical applications, such as auto-driving, and medical diagnosis. Therefore, not only the academia, but also the industry is regarding interpretability as one of the most important components and even a must-have one for responsible use of deep learning models [8].

© Springer Nature Switzerland AG 2021
K. Karlapalem et al. (Eds.): PAKDD 2021, LNAI 12714, pp. 168–180, 2021.
https://doi.org/10.1007/978-3-030-75768-7_14

Interpretability of machine learning refers to the ability to explain or to present the results in understandable terms to a human [6]. There have been considerate research efforts on studying the interpretability of deep learning models. A stream of research has adopted a post-hoc approach, which yields explanations by visualizing important features [1,18,19], or by creating a surrogate model with high fidelity to the original model [17,21]. However, feature visualisation based methods are found to be unreliable, as small perturbations to the input data can lead to dramatically different explanations [1,7]. Methods based on surrogate models do not reveal the essential process of the original model to enhance human understanding [14].

Recently, researchers have attempted to train an inherently interpretable model directly [5,10,23]. The aim is to learn a deep learning model from scratch that is able to give explanations. These methods often use auxiliary information as additional supervision to train the model. Such information is encoded in the loss function to regularize model training, aiming to learn certain mappings between latent features and human-understandable semantic information. Model prediction results are then augmented with auxiliary outputs, such as topic words [5], sentences [10], or object parts [5], to improve human understanding. However, due to the lack of connection between model learning and human-understandable decision making, how these models make decisions is still beyond the direct comprehension of humans. Even worse, there is no clue to assert whether or not decisions made by these models are reliable.

To fill this research gap, in this paper, we propose a new learning framework that makes predictions in a human-understandable way. Our core idea is inspired by key findings from cognitive science [2,13] that humans make decisions based on their perception. As illustrated by Fig. 1, humans parse an object into different components and then aggregate the results from these components to derive the final recognition. We call these semantic components as *conceptual parts* in this paper. A question naturally comes into our mind: Can we build a deep learning model that makes predictions following this

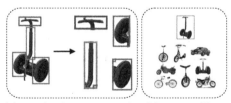

(a):parsing to components (b): recognition

Fig. 1. (a) Humans parse the object into different semantic components; (b) Results from different semantic components are aggregated for final recognition. This figure is adapted from [13].

human-understandable way? If we can link model learning with this human decision making process, it is likely that the learned deep learning model can be better understood by humans.

To achieve this goal, we design a new learning framework in analogy to the process of human-understandable decision making. Our framework comprises an automatic concept partition model and a concept-based recognition model via

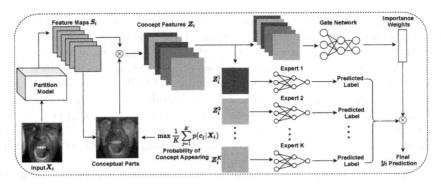

Fig. 2. The overview of the proposed method. Our model learns to partition the input image into conceptual parts, whose features are passed to the expert networks to make predictions. The gate network learns the importance weights of different conceptual parts for the final prediction.

mixture of experts [12]. The concept partition model first parses the input images into meaningful conceptual parts associated with different semantics, such as different body parts of a bird or sense organs of human faces. The features of each conceptual part are then passed to a respective expert model for recognizing the input image. A gate network is used to aggregate the predictions from different experts to obtain the final prediction results as well as the importance weights of different conceptual parts. The importance weights reflect the contributions of different conceptual parts to the final prediction. As a result, our designed model incorporates how humans perceive from images and is able to give human-understandable explanations. We show that our proposed model is able to achieve comparable or even better accuracy on two visual recognition tasks. More importantly, our model can provide human-understandable explanations by conceptual parts and their importance weights, thereby enhancing human understanding of how the model makes predictions.

The contribution of this paper is three-fold:

- We propose to inject human perception from images into model design and learning of DNN models to enable human-understandable decision making.
- Our new framework is able to give explanations through conceptual parts and their relative importance for the recognition result of each input image.
- Experiments on two visual recognition tasks verify the consistency between the learned explanations and human cognition.

2 Model Design with Human Perception

Our proposed framework trains a deep learning model by injecting the prior of how humans perceive from images. The overview of the proposed framework is shown in Fig. 2. Our model first learns to parse the input images into conceptual parts with different semantics. Features of conceptual parts are then fed to the concept-based recognition model to make the final prediction.

2.1   Concept Partition Model

We train a concept partition model to automatically segment the input images $\mathbb{X} = \{X_0, \cdots, X_N\}$ into a set of conceptual parts with different semantics using only image-level labels. Following [11], we transform the concept partition problem into the problem of estimating the probability $p(c_j|X_i)$ that each concept c_j occurrs in the input image X_i. The classical convolution neural networks, i.e., ResNet, are used as the backbone to build our partition model. We use feature maps $S_i \in R^{D \times H \times W}$ from one layer of the neural network to estimate the probability whether or not each concept appears in an input image X_i. Based on the feature maps $s_{hw} \in R^D$ at position (h, w) of S_i, the probability $p_{h,w}^j$ of the j-th concept $c_j \in R^D$ occurring at the position (h, w) is formulated as:

$$p_{h,w}^j = \frac{\exp\left(-\|(s_{hw} - c_j)/\alpha_j\|_2^2/2\right)}{\sum_j \exp\left(-\|(s_{hw} - c_j)/\alpha_j\|_2^2/2\right)}, \tag{1}$$

where $\alpha_j \in (0, 1)$ is a learnable smoothing factor for each concept c_j; c_j is a vector representing the j-th concept, which can be considered as the center of a cluster; $p_{h,w}^j > 0$ and $\sum_{j=1}^K p_{h,w}^j = 1$; K is the number of concepts.

Then, we can obtain the concept occurrence map $O = [p_{h,w}^j] \in R^{K \times H \times W}$ by assembling all the $p_{h,w}^j$'s, which indicates the probability of each concept occurring at each position. At each position, we rank the occurrence probability of each concept and assign the concept with the highest probability as the one occurring at that position, i.e., $j^* = \arg\max_j O$, where $j = 1, \cdots, K$ represents the index of concepts. Using this approach, we obtain the concept partition result for each input image. One example is given in Fig. 2, where the image is parsed into different conceptual parts represented by different colors.

As we have only image-level labels as the supervision to train the concept partition model, we obtain the concept features Z_i to predict image-level labels, based on the concept occurrence map O and feature maps S_i. Let Z_i^j be the j-th dimension of Z_i, representing concept features of the j-th concept, we have $Z_i^j = \frac{t_i^j}{\|t_i^j\|_2}$, where $t_i^j = \frac{1}{\sum_{hw} p_{hw}^j} \sum_{hw} p_{hw}^j (s_{hw} - c_j)/\sigma_j$. The concept features Z_i are passed to a classifier h constructed with several convolutional layers and fully connected layers. The cross-entropy loss is used to train the concept partition model and the classifier. The classification loss is formulated as $l_{cls} = -\frac{1}{n}\sum_{i=1}^N \hat{y}_i \log(h(Z_i))$, where N is the number of input images, and $\hat{y}_i$ is the ground-truth label of image X_i.

However, our exploration shows that using only the classification loss is insufficient to obtain meaningful concept partition results. Given the prior knowledge that all relevant concepts could occur in each image; for example, different body parts are likely to occur in most of bird images, we introduce a regularizer to incorporate such prior knowledge. Our goal is to maximize the probability of each concept appearing in the input images, which is formulated as:

$$l_r = \min \frac{1}{K * N} \sum_{i=1}^{N} \sum_{j=1}^{K} |\log(p(\boldsymbol{c}_j | \boldsymbol{X}_i) + \delta)|, \tag{2}$$

where δ is a small value (1e–5) for stable training; $p(\boldsymbol{c}_j | \boldsymbol{X}_i)$ is the probability of concept $\boldsymbol{c}_j$ occurring in the input image $\boldsymbol{X}_i$. It can be obtained through aggregating the probability of concept $\boldsymbol{c}_j$ occurring in each position, i.e., $p(\boldsymbol{c}_j | \boldsymbol{X}_i) = \max_{hw} \mathcal{G} \times \boldsymbol{O}_j$, where $\mathcal{G}$ is a 2D Gaussian kernel for smoothing, $\boldsymbol{O}_j \in R^{H \times W}$, max indicates the max pooling operation. Finally, the cross-entropy loss l_{cls} and the loss l_r in Eq. (2) are combined to train the concept partition model.

2.2 Concept-based Recognition Model

Based on concept features obtained from the concept partition model, we train a concept-based recognition model to perform the final prediction. Our concept-based recognition model involves a set of experts, each of which takes features of each concept as input to make a prediction. These concept features represent different semantic concepts, such as mouth, eyes, and cheek in the facial expression recognition task. A gate network is used to aggregate the prediction results from different experts to produce the final recognition and to estimate the importance weights of each concept. The importance weights reflect different contributions made by distinct conceptual parts to the final recognition.

For a specific recognition task, suppose the input image $\boldsymbol{X}_i$ is transformed into concept features $\boldsymbol{Z}_i$ via the concept partition model. Each concept feature $\boldsymbol{Z}_i^j$ is recognized by an expert $f_j(\boldsymbol{Z}_i^j)$. The predictions by all experts are aggregated by the importance weights to get the final recognition. The importance weights of different experts are learned by a gate network $g(\boldsymbol{Z}_i)$. Formally, we build a recognition model that can be formulated as follows:

$$f(\boldsymbol{x}_i) = \begin{bmatrix} \boldsymbol{w}_i^1, \boldsymbol{w}_i^j \cdots \boldsymbol{w}_i^K \end{bmatrix} \times \begin{bmatrix} f_1(\boldsymbol{Z}_i^1) \\ f_j(\boldsymbol{Z}_i^j) \\ \vdots \\ f_K(\boldsymbol{Z}_i^K) \end{bmatrix}. \tag{3}$$

Above, $f(\boldsymbol{x}_i)$ is the final predicted label for the input $\boldsymbol{X}_i$. The gate network $g(\boldsymbol{Z}_i)$ is parameterized by $[\boldsymbol{w}_i^1, \boldsymbol{w}_i^j, \cdots, \boldsymbol{w}_i^K]$, where $\boldsymbol{w}_i^j$ is the weight of concept feature $\boldsymbol{Z}_i^j$ for the j-th concept in $\boldsymbol{X}_i$. K is the number of experts, which is equal to the number of conceptual parts.

Each expert $f_j(\boldsymbol{Z}_i^j)$ is constructed with the same network structure but is given different concept features as input. The gate network $g(\boldsymbol{Z}_i)$ is also a neural network, which takes all concept features $\boldsymbol{Z}_i$ as input. The last layer of the gate network is a softmax layer that produces the weights summed to one. The weight $\boldsymbol{w}_i^j$ can be regarded as importance weight of each concept for the final prediction.

The learned importance weights facilitate humans to better understand how much contribution each conceptual part makes towards the final recognition.

To train the concept-based recognition model, we use the cross-entropy loss with a regularizer imposed on importance weights. Specifically, we use image-level labels as supervision to train each expert. The overall loss function for all experts is: $l_{ept} = -\frac{1}{n} \sum_{i=1}^{N} \sum_{j=1}^{K} \hat{y}_i \log(f_j(\boldsymbol{Z}_i^j))$, where N is the number of input images, $\hat{y}_i$ is the ground true label of image $\boldsymbol{X}_i$ and $f_j(\boldsymbol{Z}_i^j)$ is the predicted label for image $\boldsymbol{X}_i$ by expert j. We also use the cross-entropy loss to train the gate network. Additionally, a constraint is imposed on the value of importance weight $\boldsymbol{w}_i$, preventing one weight from dominating the prediction and generating meaningless importance weights. γ is the weighting factor for balancing the two terms. The overall loss function for training the gate network is formulated as:

$$l_g = -\frac{1}{N} \sum_{i=1}^{N} \hat{y}_i \log\left(f(\boldsymbol{Z}_i)\right) + \gamma \frac{1}{K*N} \sum_{i=1}^{N} \sum_{j=1}^{K} \left\| \boldsymbol{w}_i^j - \frac{1}{K} \right\|_2^2 . \tag{4}$$

3   Experimental Evaluation

To validate the effectiveness of our proposed model, we perform two visual recognition tasks with varying task difficulty. We use ResNet101 as the backbone to parse the images into conceptual parts. Concept features of each conceptual part are further used to train each expert and the gate network.

3.1   Facial Expression Recognition

We begin with the facial expression recognition task on the FER-2013 dataset [9]. It consists of 28,709 training images, 3,589 public test images, and 3,589 private test images. The original face images in the dataset are grey-scale images of size 48×48. All images are categorized into 7 classes. Ian [9] reported that human accuracy on FER-2013 was around 65%.

Classification Performance. We train our model using training images resized into 224×224 with three channels. We set the learning rate λ as 1e-4; use the SGD optimizer with momentum as 0.9 and weight decay as 5e-4; set weighting factor γ as 1.0; set number of experts and number of conceptual parts as 6; set the maximum iteration as 200. We compare our model with several classical baseline models (i.e. ResNet, VGG, Inception V3) on the FER2013 private test dataset. The

Table 1. Classification accuracy on FER2013 test dataset.

Model	Accuracy (%)
ResNet101	71.44
ResNet50	70.47
ResNet18	69.74
Vgg16_bn	70.35
InceptionV3	68.99
Our model	**73.67**

classification results are summarized in Table 1. All classification results are obtained by training from pre-trained weights on ImageNet. From this table, we can see that our model achieves the best classification accuracy. This validates the effectiveness of our model in facial expression recognition, beating the human accuracy of 65% on this dataset.

Explanation Results. After our model is trained, the prediction results can be explained through partitioned conceptual parts and their importance weights. Figure 3 shows three example images from FER2013 and the learned conceptual parts. The first-column images are the original images with their class labels on the top. The columns 2–7 are the identified conceptual parts with the concept name on the top. The images in the last column are the conceptual parts collated with each color indicating one concept. When we set the number of conceptual parts as 6, the partition model parses each image into 6 parts with different semantics, i.e., nose/forehead, eye, nasal bridge, mouth/eyebrow, cheek, and other part. These conceptual parts are easy for humans to recognize and understand. Furthermore, the partition results are consistent for different input images. That is to say, the concept partition model is able to identify meaningful concepts for the recognition task.

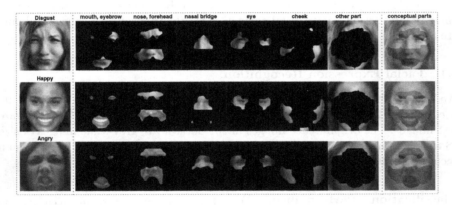

Fig. 3. The conceptual parts learned by our model on FER2013 dataset.

Table 2. Average weight of conceptual parts learned on FER2013.

Conceptual parts	Mouth/eyebrow	Nose/forehead	Nasal bridge	Eye	Cheek	Other parts
Avg. Weight	0.2127	0.1912	0.1856	0.1539	0.1444	0.1120

To show the importance of different conceptual parts for recognition, we calculate the average weight of each conceptual part for test images on FER-2013. Table 2 shows the average weight of each conceptual part. When $\gamma = 1.0$, conceptual parts are ranked as mouth/eyebrow, nose/forehead, nasal bridge, eye, cheek and other parts, according to their importance weights.

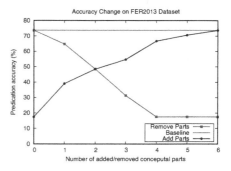

Fig. 4. Classification accuracy vs. adding/removing important conceptual parts.

We further test the effect of adding or removing the learned conceptual parts according to their associated rank in Table 2. As shown in Fig. 4, the green line indicates the baseline classification accuracy (73.67%) of our model with all concept features as input. The red line (with squares) shows the accuracy loss caused by gradually removing the most important conceptual parts from the input. As can be seen, the accuracy drops significantly to 17.47%, after the top 4 important conceptual parts are removed. The blue curve (with dots) shows the changes in accuracy when important conceptual parts are added. As we gradually add the most important conceptual parts, the prediction accuracy increases from 17.44% to 73.67%. With only 4 important conceptual parts, our method achieves the accuracy of 66.70%. This validates the effectiveness of the learned importance weights and their contribution towards the overall classification performance.

Lastly, we also conduct human evaluation to verify the consistency between our model and human decision making. We asked 30 participates to answer two questions: Q1, how important are the learned conceptual parts? Q2, what is the importance order of conceptual parts for them to make decisions? We randomly selected 60 images from FER2013 test dataset to conduct human evaluation with results summarized as follows. Firstly, we calculate the average importance score in Q1 across different images and participants. Out of 5 points, we obtain 4.21 points on average. That means that participants think the predictions made by our trained model is consistent with their perception for visual recognition. Secondly, we calculate the recall of top 4 conceptual parts considered important by participants and also selected as the top 4 important ones by our model. The average recall is 78.17%. This also exhibits a high level of consistency between our model predictions and human judgements.

3.2 Fine-Grained Bird Classification

Next, we conduct experiments on a more difficult, fine-grained bird classification task. CUB_200_2011 [20] is a fine-grained bird dataset that contains 11,788 images of 200 bird species. The training dataset has 5,994 images, and the rest of 5,774 images are for testing.

Classification Performance. We train our model on CUB_200_2011 training dataset, where images are resized into 448 × 448 as input. We set the learning rate λ as 5e-5; set number of experts and number of conceptual parts as 5; other parameters are the same as on FER-2013. We compare classification performance of our model with ResNet, VGG, InceptionV3, and MC Loss method [3] on the test dataset, as reported in Table 3. All results are obtained by training from pre-trained

Table 3. Classification accuracy on CUB_200_2011 test dataset.

Model	Accuracy (%)
ResNet101	85.02
ResNet50	84.45
ResNet18	81.67
Vgg16_bn	83.81
InceptionV3	84.05
MC (TIP2020) [3]	**87.30**
Our model	**86.66**

weights on ImageNet. We can see that our model performs better than the other three classical models except for MC method that is specially designed for fine-grained classification. Overall, our model is able to achieve better accuracy than classical models and comparable accuracy with state-of-the-art model.

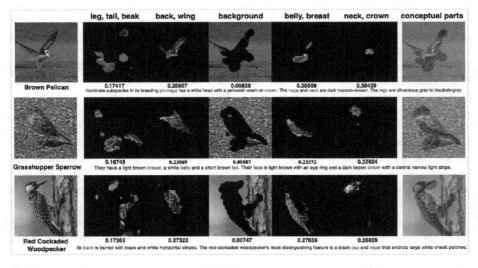

Fig. 5. The conceptual parts learned by our model on the CUB_200_2011 dataset. The text description underneath conceptual concepts are the definitions of the corresponding bird species from Wikipedia. (Color figure online)

Explanation Results. Figure 5 shows several example images from CUB_200_2011 and their important conceptual parts learned for classification. The images in the first column are the original images with their class labels on the bottom. The last column shows conceptual parts identified with each color indicating one concept. The columns 2–6 are conceptual parts with the concept definition on the top, where the values under the partition results are the importance weights of different concepts for classifying that image. The description texts are the definitions of the corresponding bird species from Wikipedia, where key features of different bird species are highlighted in red.

We notice that the important conceptual parts learned by our model have a high degree of consistency with key features of bird species given by human experts. For example, for the first image in the second row in Fig. 5, labeled as *Grasshopper Sparrow*, crown, face, breast, belly, and tail are the discriminative features that human experts use to define this bird species. These corresponding conceptual parts are also identified and attributed with high weights by our model; crown/neck and belly/breast are ranked by our method as the first and second most important concepts for classification.

Table 4 lists the average importance weights of conceptual parts learned on CUB_200_2011. We find that neck/crown is the most important concept for recognition. In contrast, background is the least important concept. We also observe that the importance weight of concept leg/tail/beak is nearly 0, when γ is 0. This is inconsistent with the definition of bird species given by human experts (see Fig. 5). This proves the necessity of adding the constraint in Eq.(4) that prevents certain weight from dominating the parameter estimation.

Based on the importance ranking of conceptual parts ($\gamma = 1$), we study the influence on recognition accuracy by adding or removing important conceptual parts. As shown in Fig. 6, the baseline accuracy (86.66%) achieved by our model is plotted as the green line. Again, this is the result using all concept features as input. As indicated by the red line (with squares), when we remove the top 3 important conceptual parts, the prediction accuracy drops markedly from 86.66% to 3.26%. The prediction accuracy shown by the blue line (with dots) increases from 0.5% to 79.84%, when the top 2 important conceptual parts are added. All results validate the

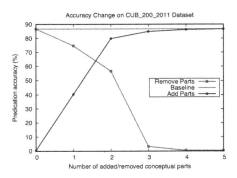

Fig. 6. Classification accuracy vs. adding/removing important conceptual parts.

effectiveness of our learned importance weights in quantifying the contributions of different conceptual parts towards the final classification.

Table 4. Average weight of conceptual parts learned on CUB_200_2011.

Conceptual parts	Leg/tail/beak	Back/wing	Background	Belly/breast	Neck/crown
$\gamma = 1$	0.1760	0.2461	0.0079	0.2404	0.3296
$\gamma = 0$	0.0044	0.2283	0.0022	0.2611	0.5039

4 Discussions on Related Work

Post-hoc explanation is the most popular method for interpreting deep learning models. This branch of methods can be divided into two categories: 1) *important feature visualization*, which yields explanations by identifying and visualizing important features for input features. [19] proposed a method called integrated gradients to attribute the important features in input images. Grad-cam++ [4] used the gradients of each class as weights to combine the features of the last layer to get important features. However, these methods are found to be unreliable; small perturbations to the input data can lead to dramatically different explanations [1,7]. 2) *surrogate models*, which create a surrogate model that can be more easily understood to mimic the performance of the original model. [17] tried to train a linear model to mimic the behaviors of the original deep learning models locally and used the learned model to explain the decisions of deep learning models. RICE [16] was proposed to give explanations for the target model by synthesizing logic program. However, methods based on surrogate models do not reveal how a decision is made by the original model [14].

Training an inherently interpretable model is another recent theme for interpreting deep learning models. [10] trained a deep classifier and an additional text generator to obtain classification results as well as text explanations for the original classifier. [23] attributed the filters in high layers to different object parts by adding a mutual information loss between object parts and filters, when training a classifier. Semantic information was used as additional supervision in [5] to train a video caption model, and the trained model could give the top important topics for explaining the results. These models augment their prediction results with auxiliary outputs, such as topic words [5], sentences [10], or object parts [5], to improve human understanding. However, due to disconnection between model learning and human decision making, it is still rather difficult for humans to fully comprehend how these models make predictions.

Our method is built upon semantic concepts for learning the partition and recognition model. We focus on training an interpretable model from a new angle of view; we inject human perception into the model design and learning process, yielding explanations that are more consistent with human cognition. Although there are many part-based object recognition methods [15,22], they solely aim at improving classification accuracy, without considering the interpretability of such models. In addition, the existing part-based methods either partition images into object-level parts rather than concept-level parts, or rely on pre-trained segmentation models with both image-level class labels and segment labels.

5 Conclusion

We proposed a new learning framework to train a deep learning model that makes predictions in a human-understandable way. Inspired by cognitive science, our framework is composed of a concept partition model, which learns conceptual parts with different semantics, and a concept-based recognition model, which makes the final prediction as well as yields relative importance weights of conceptual parts. Our method is able to provide human understandable explanations, because its design is more aligned with human-understandable decision making. Experiments on two visual recognition tasks showed that our method compares favourably to state-of-the-art methods on recognition accuracy, but also provides explanations that are highly consistent with human perception. For future work, we will investigate how to design our model in more complex classification tasks where the recognition by parts assumption may not hold.

Acknowledgments. This work is partially supported by ARC under Grant DP180100106 and DP200101328. Xiaowei Zhou is supported by a Data61 Student Scholarship from CSIRO.

References

1. Adebayo, J., Gilmer, J., Muelly, M., Goodfellow, I., Hardt, M., Kim, B.: Sanity checks for saliency maps. In: NeurIPS, pp. 9505–9515 (2018)
2. Biederman, I.: Recognition-by-components: a theory of human image understanding. Psychol. Rev. **94**(2), 115 (1987)
3. Chang, D., et al.: The devil is in the channels: mutual-channel loss for fine-grained image classification. IEEE Trans. Image Process. **29**, 4683–4695 (2020)
4. Chattopadhay, A., Sarkar, A., Howlader, P., Balasubramanian, V.N.: Gradcam++: generalized gradient-based visual explanations for deep convolutional networks. In: WACV, pp. 839–847. IEEE (2018)
5. Dong, Y., Su, H., Zhu, J., Zhang, B.: Improving interpretability of deep neural networks with semantic information. In: CVPR, pp. 4306–4314 (2017)
6. Doshi-Velez, F., Kim, B.: Towards a rigorous science of interpretable machine learning. arXiv preprint arXiv:1702.08608 (2017)
7. Ghorbani, A., Abid, A., Zou, J.: Interpretation of neural networks is fragile. In: AAAI, Vol. 33, No. 01, pp. 3681–3688 (2019)
8. Ghorbani, A., Wexler, J., Zou, J.Y., Kim, B.: Towards automatic concept-based explanations. In: NeurIPS, pp. 9273–9282 (2019)
9. Goodfellow, I.J., et al.: Challenges in representation learning: a report on three machine learning contests. In: Lee, M., Hirose, A., Hou, Z.-G., Kil, R.M. (eds.) ICONIP 2013. LNCS, vol. 8228, pp. 117–124. Springer, Heidelberg (2013). https://doi.org/10.1007/978-3-642-42051-1_16
10. Hendricks, L.A., Akata, Z., Rohrbach, M., Donahue, J., Schiele, B., Darrell, T.: Generating visual explanations. In: Leibe, B., Matas, J., Sebe, N., Welling, M. (eds.) ECCV 2016. LNCS, vol. 9908, pp. 3–19. Springer, Cham (2016). https://doi.org/10.1007/978-3-319-46493-0_1
11. Huang, Z., Li, Y.: Interpretable and accurate fine-grained recognition via region grouping. In: CVPR, pp. 8662–8672 (2020)
12. Jacobs, R.A., et al.: Adaptive mixtures of local experts. Neural Comput. **3**(1), 79–87 (1991)
13. Lake, B.M., Ullman, T.D., Tenenbaum, J.B., Gershman, S.J.: Building machines that learn and think like people. Behav. Brain Sci. **40**, (2017)
14. Laugel, T., Lesot, M.J., Marsala, C., Renard, X., Detyniecki, M.: The dangers of post-hoc interpretability: unjustified counterfactual explanations. In: IJCAI, pp. 2801–2807. AAAI Press (2019)

15. Mordan, T., Thome, N., Henaff, G., Cord, M.: End-to-end learning of latent deformable part-based representations for object detection. Int. J. Compute. Vision **127**(11–12), 1659–1679 (2019)
16. Paçacı, G., Johnson, D., McKeever, S., Hamfelt, A.: "Why did you do that?": explaining black box models with inductive synthesis. In: ICCS, Faro, Algarve, Portugal (2019)
17. Ribeiro, M.T., Singh, S., Guestrin, C.: Why should i trust you?: explaining the predictions of any classifier. In: SIGKDD, pp. 1135–1144. ACM (2016)
18. Shrikumar, A., Greenside, P., Kundaje, A.: Learning important features through propagating activation differences. In: ICML, pp. 3145–3153. JMLR. org (2017)
19. Sundararajan, M., Taly, A., Yan, Q.: Axiomatic attribution for deep networks. In: ICML, pp. 3319–3328. JMLR. org (2017)
20. Wah, C., Branson, S., Welinder, P., Perona, P., Belongie, S.: The caltech-ucsd birds-200-2011 dataset. Tech. Rep. CNS-TR-2011-001, Caltech (2011)
21. Wu, H., Wang, C., Yin, J., Lu, K., Zhu, L.: Sharing deep neural network models with interpretation. In: WWW, pp. 177–186. WWW Steering Committee (2018)
22. Zhang, H., et al.: SPDA-CNN: unifying semantic part detection and abstraction for fine-grained recognition. In: CVPR, pp. 1143–1152 (2016)
23. Zhang, Q., Nian Wu, Y., Zhu, S.C.: Interpretable convolutional neural networks. In: CVPR, pp. 8827–8836 (2018)

LightCAKE: A Lightweight Framework for Context-Aware Knowledge Graph Embedding

Zhiyuan Ning[1,2], Ziyue Qiao[1,2], Hao Dong[1,2], Yi Du[1(✉)], and Yuanchun Zhou[1]

[1] Computer Network Information Center, Chinese Academy of Sciences, Beijing, China
{ningzhiyuan,qiaoziyue,donghao,duyi,zyc}@cnic.cn
[2] University of Chinese Academy of Sciences, Beijing, China

Abstract. Knowledge graph embedding (KGE) models learn to project symbolic entities and relations into a continuous vector space based on the observed triplets. However, existing KGE models cannot make a proper trade-off between the graph context and the model complexity, which makes them still far from satisfactory. In this paper, we propose a lightweight framework named LightCAKE for context-aware KGE. LightCAKE explicitly models the graph context without introducing redundant trainable parameters, and uses an iterative aggregation strategy to integrate the context information into the entity/relation embeddings. As a generic framework, it can be used with many simple KGE models to achieve excellent results. Finally, extensive experiments on public benchmarks demonstrate the efficiency and effectiveness of our framework.

Keywords: Knowledge graph embedding · Lightweight · Graph context

1 Introduction

Recently, large-scale knowledge graphs (KGs) have been widely applied to numerous AI-related applications. 6Indeed, KGs are usually expressed as multi-relational directed graphs composed of entities as nodes and relations as edges. The real-world facts stored in KGs are modeled as triplets (head entity, relation, tail entity), which are denoted as (h, r, t).

Nevertheless, KGs are usually incomplete due to the constant emergence of new knowledge. To address this issue, a series of knowledge graph embedding (KGE) models have been proposed [14]. KGE models project symbolic entities and relations into a continuous vector space, and use scoring functions to measure the plausibility of triplets. By optimizing the scoring functions to assign higher scores to true triplets than invalid ones, KGE models learn low-dimensional representations (called embeddings) for all entities and relations,

K. Karlapalem et al. (Eds.): PAKDD 2021, LNAI 12714, pp. 181–193, 2021.
https://doi.org/10.1007/978-3-030-75768-7_15

and these embeddings are then used to predict new facts. Most of the previous KGE models use translation distance based [2,15] and semantic matching based [11,17] scoring functions which perform additive and multiplicative operations, respectively. These models have been shown to be scalable and effective.

However, the aforementioned KGE models only focus on modeling individual triplets and ignore the **graph context**, which contains plenty of valuable structural information. We argue that there are two types of important graph contexts required for successfully predicting the relation between two entities: (1) The **entity context**, i.e., for an entity, its neighboring nodes and the corresponding edges connecting the entity to its neighboring nodes. The entity context depicts the subtle differences between two entities. As an example shown in Fig. 1(a), we aim to predict whether *Joe Biden* or *Hillary Clinton* is the president of the *USA*. Both of them have the same relation *"birthplace_of"* with the *USA*, but they have distinct entity contexts. *Joe Biden's* neighboring node, *Donald Trump*, is the president of the *USA*, and *Biden* is his successor. Whereas there is no such relationship between *Hillary Clinton* and her neighboring nodes. Capturing such entity context will help predict the correct triplet (*Joe Biden, president_of, USA*). (2) The **relation context**, i.e., the two endpoints of a given relation. Relation context implicitly indicates the category of related entities. Taking Fig. fig1(b) as an example, both the *USA* and *New York* were *Donald Trump's* birthplace, but according to the context of *"president_of"*, the related tail entities {*China, Russia, . . .*} tend to be a set of countries. Since *New York* is a city and it is part of the *USA* which is a country, (*Donald Trump, president_of, USA*) is the right triplet. Moreover, entities and relations rarely appear in isolation, so considering entity context and relation context together will provide more beneficial information.

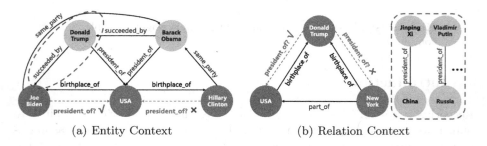

(a) Entity Context   (b) Relation Context

Fig. 1. Examples of graph context which can help the relation prediction in knowledge graph. Nodes represent entities, solid lines represent actual relations, dashed lines represent the relations to be predicted. Red dashed boxes frame the critical entity context (Figure a) and relation context (Figure b) that can provide important information for correctly predicting the relation between two entities. (Color figure online)

In order to model the graph context, some recent work has attempted to apply graph neural network (GNN) to KGE [1,8]. These GNN-based KGE models are effective to aggregate information from multi-hop neighbors to enrich the entity/relation representation. However, GNN introduces more model parameters and tensor computations, therefore making it difficult to utilize these models for large-scale real-world KGs. In addition, most GNN-based KGE models only exploit entity context or relation context individually, which may lead to information loss.

In this paper, we propose a **Light**weight Framework for **C**ontext-**A**ware **K**nowledge Graph **E**mbedding (**LightCAKE**) to address the shortcomings of existing models. LightCAKE first builds the context star graph to model the entity/relation context. It then uses non-parameterized operations like subtraction (inspired by TransE [2]) or multiplication (inspired by DistMult [17]) to encode context nodes in the context star graph. Lastly, every entity/relation node in the context star graph aggregates information from its surrounding context nodes based on the weights calculated by a scoring function. LightCAKE considers both entity context and relation context, and introduces no new parameters, making it very lightweight and capable of being used on large-scale KGs. The contributions of our work can be summarized as follows: (1) We propose a lightweight framework (LightCAKE) for KGE that explicitly model the entity context and relation context without the sacrifice in the model complexity; (2) As a general framework, we can apply many simple methods like TransE [2] and DistMult [17] to LightCAKE; (3) Through extensive experiments on relation prediction task, we demonstrate the effectiveness and efficiency of LightCAKE.

2   Related Work

Most early KGE models only exploit the triplets and can be roughly categorized into two classes [14]: translation distance based and semantic matching based. Translation distance based models are also known as **additive models**, since they project head and tail entities into the same embedding space, and treat the relations as the translations from head entities to tail entities. The objective is that the translated head entity should be close to the tail entity. TransE [2] is the first and most representative of such models. A series of work is conducted along this line such as TransR [7] and TransH [15]. On the other hand, semantic matching based models such as DistMult [17] and ComplEx [11] use multiplicative score functions for computing the plausibility of the given triplets, so they are also called **multiplicative models**. Both models are conceptually simple and it is easy to apply them to large-scale KGs. But they ignore the structured information stored in the graph context of KGs.

In contrast, **GNN-based** models attempt to use GNN for graph context modeling. These models first aggregate graph context into entity/relation embeddings through GNN, then pass the context-aware embeddings to the context-independent scoring functions for scoring. R-GCN [8] is an extension of the graph convolutional network [6] on relational data. It applies a convolution operation to the neighboring nodes of each entity and assigns them equal weights. A2N [1] uses a method similar to graph attention networks [12] to further distinguish the weights of neighboring nodes. However, this type of KGE models suffer from overparameterization since there are many parameters in GNN, which will hinder the application of such models to large-scale KGs. In addition, they don't integrate entity context and relation context, which may cause information loss.

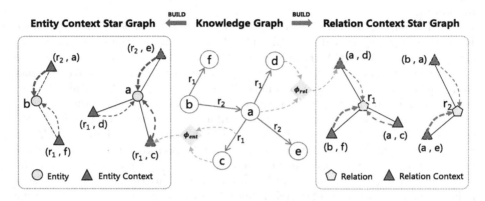

Fig. 2. Overview of LightCAKE. (1) For a KG (Middle), we build an entity context star graph (Left) for all entities and a relation context star graph (Right) for all relations. In entity/relation context star graph, each entity/relation is surrounded by its entity/relation context and they are connected to each other by solid black lines. (2) The yellow rhombus ϕ_{ent} and ϕ_{rel} denote context encoders (Details in Sect. 4.1), and the gray dashed line indicates the input and output of the encoders. (3) The blue dashed line denotes the weight α (Eq. (3)), and the green dashed line denotes the weight β (Eq. (4)). The thicker the line, the greater the weight.

3   Preliminaries

3.1   Notation and Problem Formulation

A KG can be considered as a collection of triplets $\mathcal{G} = \{(h,r,t) \mid (h,r,t) \in \mathcal{E} \times \mathcal{R} \times \mathcal{E}\}$, where $\mathcal{E}$ is the entity set and $\mathcal{R}$ is the relation set. $h, t \in \mathcal{E}$ represent the head entity and tail entity, $r \in \mathcal{R}$ denotes the relation linking from the head entity h to tail entity t. Given a triplet (h, r, t), the corresponding embeddings are e_h, e_r, e_t, where $e_h, e_r, e_t \in \mathbb{R}^d$, and d is the embedding dimension. KGE models usually define a scoring function $\psi : \mathbb{R}^d \times \mathbb{R}^d \times \mathbb{R}^d \to \mathbb{R}$. It takes the corresponding embedding (e_h, e_r, e_t) of a triplet (h, r, t) as input, and produces a score reflecting the plausibility of the triplet.

In this paper, the objective is to predict the missing links in $\mathcal{G}$, i.e., given an entity pair (h, t), we aim to predict the missing relation r between them. We refer to this task as **relation prediction**. Some related work formulates this problem as link prediction, i.e., predicting the missing tail/head entity given a head/tail entity and a relation. The two problems have proven to be actually reducible to each other [13].

3.2   Context Star Graph

Definition 1. *Entity Context*: For an entity h in $\mathcal{G}$, the entity context of h is defined as $\mathcal{C}_{ent}(h) = \{(r, t) \mid (h, r, t) \in \mathcal{G}\}$, i.e., all the (relation, tail) pairs in $\mathcal{G}$ whose head is h.

Definition 2. *Relation Context*: For a relation r in $\mathcal{G}$, the relation context of r is defined as $\mathcal{C}_{rel}(r) = \{(h, t) \mid (h, r, t) \in \mathcal{G}\}$, i.e., all the (head, tail) pairs in $\mathcal{G}$ whose relation is r.

Note that the entity context $\mathcal{C}_{ent}(h)$ only considers the neighbors of h for its outgoing edges and ignores the neighbors for its incoming edges. This is because for each triplet $(h, r, t) \in \mathcal{G}$, we create a corresponding inverse triplet (t, r^{-1}, h) and add it to $\mathcal{G}$. In this way, for entity t, $\{(r, h) \mid (h, r, t) \in \mathcal{G}\}$ can be converted to a format of $\{(r^{-1}, h) \mid (t, r^{-1}, h) \in \mathcal{G}\}$, and it is equivalent to $\mathcal{C}_{ent}(t)$. Thus, $\mathcal{C}_{ent}(\cdot)$ can contain both the outgoing and incoming neighbors for each entity.

To explicitly model entity context and relation context for a KG $\mathcal{G}$ (As shown in Fig. 2 middle), we construct an **entity context star graph** (As shown in Fig. 2 left) and a **relation context star graph** (As shown in Fig. 2 right), respectively. In the entity context star graph, all the central nodes are the entities in $\mathcal{G}$, and each entity h is surrounded by its entity context $\mathcal{C}_{ent}(h)$. Similarly, in the relation context star graph, all the central nodes are the relations in $\mathcal{G}$, and each relation r is surrounded by its relation context $\mathcal{C}_{rel}(r)$.

4   Methodology

Given the context star graph, LightCAKE can (1) encode each entity/relation context node into an embedding; (2) learn the context-aware embedding for each entity/relation by iteratively aggregating information from its context nodes.

4.1   LightCAKE Details

Denote $e_h^{(0)}$ and $e_r^{(0)}$ as the randomly initialized embedding of an entity h and a relation r respectively. The aggregation functions are formulated as:

$$e_h^{(l+1)} = e_h^{(l)} + \sum_{(r', t') \in \mathcal{C}_{ent}(h)} \alpha_{h,(r',t')}^{(l)} \phi_{ent}(e_{r'}, e_{t'}) \tag{1}$$

$$e_r^{(l+1)} = e_r^{(l)} + \sum_{(h', t') \in \mathcal{C}_{rel}(r)} \beta_{r,(h',t')}^{(l)} \phi_{rel}(e_{h'}, e_{t'}) \tag{2}$$

Here, $e_h^{(l+1)}$ and $e_r^{(l+1)}$ are the embeddings of h and r after l-iterations aggregations. $0 \leq l \leq L$ and L is the total number of iterations. $\phi_{ent}(\cdot) : \mathbb{R}^d \times \mathbb{R}^d \to \mathbb{R}^d$ is the entity context encoder, and $\phi_{rel}(\cdot) : \mathbb{R}^d \times \mathbb{R}^d \to \mathbb{R}^d$ is the relation context encoder. $\alpha_{h,(r',t')}^{(l)}$ and $\beta_{r,(h',t')}^{(l)}$ are the weights in iteration l, representing how important each context node is for h and r, respectively. We introduce the scoring function $\psi(\cdot)$ to calculate them:

$$\alpha_{h,(r',t')}^{(l)} = \frac{\exp(\psi(e_h^{(l)}, e_{r'}^{(l)}, e_{t'}^{(l)}))}{\sum_{(r'',t'') \in \mathcal{C}_{ent}(h)} \exp(\psi(e_h^{(l)}, e_{r''}^{(l)}, e_{t''}^{(l)}))} \tag{3}$$

$$\beta_{r,(h',t')}^{(l)} = \frac{\exp(\psi(e_{h'}^{(l)}, e_r^{(l)}, e_{t'}^{(l)}))}{\sum_{(h'',t'') \in \mathcal{C}_{rel}(r)} \exp(\psi(e_{h''}^{(l)}, e_r^{(l)}, e_{t''}^{(l)}))} \tag{4}$$

When Eq. (1) and Eq. (2) are iteratively executed L times, for any $h, t \in \mathcal{E}$ and $r \in \mathcal{R}$, we obtain the final context-enhanced embeddings $e_h^{(L)}, e_r^{(L)}, e_t^{(L)}$. To perform relation prediction, we compute the probability of the relation r given the head entity h and tail entity t using a softmax function:

$$p(r|h,t) = \frac{\exp(\psi(e_h^{(L)}, e_r^{(L)}, e_t^{(L)}))}{\sum_{r' \in \mathcal{R}} \exp(\psi(e_h^{(L)}, e_{r'}^{(L)}, e_t^{(L)}))} \tag{5}$$

where $\mathcal{R}$ is the set of relations, $\psi(\cdot)$ is the same scoring function used in Eq. (3) and Eq. (4). Then, we train the model by minimizing the following loss function:

$$\mathcal{L} = -\frac{1}{|\mathcal{D}|} \sum_{i=0}^{|\mathcal{D}|} \log p(r_i \mid h_i, t_i) \tag{6}$$

where $\mathcal{D}$ is the training set, and $(h_i, r_i, t_i) \in \mathcal{D}$ is one of the training triplets.

4.2 Special Cases of LightCAKE

LightCAKE is a generic framework, and we can substitute different scoring function $\psi(\cdot)$ of different KGE models into Eq. (3), Eq. (4), and Eq. (5). And we can design different $\phi_{ent}(\cdot)$ and $\phi_{rel}(\cdot)$ to encode context. In order to make the framework lightweight, we apply TransE [2] and DistMult [17], which are the simplest and most representative of the additive models and multiplicative models respectively, to LightCAKE.

LightCAKE-TransE. The scoring function of TransE [2] is:

$$\psi_{TransE}(e_h, e_r, e_t) = -\|e_h + e_r - e_t\|_2 = -\|e_t - e_r - e_h\|_2 \tag{7}$$

where $\|\cdot\|_2$ is the L2-norm. Equation (7) can be decomposed of the two following steps:6

$$e_{(h,r,t)} = \mathcal{V}_{TransE}(e_h, e_r, e_t) = e_t - e_r - e_h \tag{8}$$

$$score = \mathcal{S}_{TransE}(e_{(h,r,t)}) = -\|e_{(h,r,t)}\|_2 \tag{9}$$

where $\mathcal{V}. : \mathbb{R}^d \times \mathbb{R}^d \times \mathbb{R}^d \to \mathbb{R}^d$ and $\mathcal{S}. : \mathbb{R}^d \to \mathbb{R}$. The $e_{(h,r,t)}$ denotes the embedding of a triplet (h, r, t), $score$ denotes the score of the triplet. In Eq. (8),

TransE uses addition and subtraction to encode triplets. Moreover, the operation between e_r and e_t is subtraction, and the operation between e_h and e_t is also subtraction. So we design $\phi_{ent}(e_{r'}, e_{t'}) = e_{t'} - e_{r'}$ and $\phi_{rel}(e_{h'}, e_{t'}) = e_{t'} - e_{h'}$ to encode context, then the aggregation function of LightCAKE-TransE can be formalized as:

$$e_h^{(l+1)} = e_h^{(l)} + \sum_{(r',t') \in \mathcal{C}_{ent}(h)} \alpha_{h,(r',t')}^{(l)} (e_{t'} - e_{r'}) \tag{10}$$

$$e_r^{(l+1)} = e_r^{(l)} + \sum_{(h',t') \in \mathcal{C}_{rel}(r)} \beta_{r,(h',t')}^{(l)} (e_{t'} - e_{h'}) \tag{11}$$

Lastly, substitute $\psi_{TransE}(e_h, e_r, e_t)$ from Eq. (7) into Eq. (3), Eq. (4) and Eq. (5), we will get the complete LightCAKE-TransE.

LightCAKE-DistMult. The scoring function of DistMult [17] is:

$$\psi_{DistMult}(e_h, e_r, e_t) = \langle e_h, e_r, e_t \rangle \tag{12}$$

where $\langle \cdot \rangle$ denotes the generalized dot product. Equation (12) can be decomposed of the two following steps:

$$e_{(h,r,t)} = \mathcal{V}_{DistMult}(e_h, e_r, e_t) = e_h \odot e_r \odot e_t \tag{13}$$

$$score = \mathcal{S}_{DistMult}(e_{(h,r,t)}) = \sum_i e_{(h,r,t)}[i] \tag{14}$$

where $\odot$ denotes the element-wise product, and $e_{(h,r,t)}[i]$ denotes the i-th element in embedding $e_{(h,r,t)}$. In Eq. (13), DistMult uses multiplication to encode triplets. Moreover, the operation between e_r and e_t is multiplication, and the operation between e_h and e_t is also multiplication. So we design $\phi_{ent}(e_{r'}, e_{t'}) = e_{t'} \odot e_{r'}$ and $\phi_{rel}(e_{h'}, e_{t'}) = e_{t'} \odot e_{h'}$ to encode context, then the aggregation function of LightCAKE-DistMult can be formalized as:

$$e_h^{(l+1)} = e_h^{(l)} + \sum_{(r',t') \in \mathcal{C}_{ent}(h)} \alpha_{h,(r',t')}^{(l)} (e_{t'} \odot e_{r'}) \tag{15}$$

$$e_r^{(l+1)} = e_r^{(l)} + \sum_{(h',t') \in \mathcal{C}_{rel}(r)} \beta_{r,(h',t')}^{(l)} (e_{t'} \odot e_{h'}) \tag{16}$$

Lastly, substitute $\psi_{DistMult}(e_h, e_r, e_t)$ from Eq. (12) into Eq. (3), Eq. (4) and Eq. (5), we will get the complete LightCAKE-DistMult.

Notably, there are no extra trainable parameters introduced in LightCAKE-TransE and LightCAKE-DistMult, making them lightweight and efficient.

5   Experiments

5.1   Dataset

We evaluate LightCAKE on four popular benchmark datasets WN18RR [3], FB15K-237 [10], NELL995 [16] and DDB14 [13]. WN18RR is extracted from WordNet, containing conceptual-semantic and lexical relations among English words. FB15K-237 is extracted from Freebase, a large-scale KG with general human knowledge. NELL995 is extracted from the 995th iteration of the NELL system containing general knowledge. DDB14 is extracted from the Disease Database, a medical database containing terminologies and concepts as well as their relationships. The statistics of the datasets are summarized in Table 1.

Table 1. Statistics of four datasets. avg.$|\mathcal{C}_{ent}(h)|$ and avg.$|\mathcal{C}_{rel}(r)|$ represent the average number of entity context and relation context, respectively.

Dataset	FB15K-237	WN18RR	NELL995	DDB14
#entitiy	14,541	40,943	63,917	9,203
#relation	237	11	198	14
#train	272,115	86,835	137,465	36,561
#test	17,535	3,034	5,000	4,000
#valid	20,466	3,134	5,000	4,000
Avg.$|\mathcal{C}_{ent}(h)|$	37.4	4.2	4.3	7.9
Avg.$|\mathcal{C}_{rel}(r)|$	1148.2	7894.1	694.3	2611.5

5.2   Baselines

To prove the effectiveness of LightCAKE, we compare LightCAKE-TransE and LightCAKE-DistMult with six baselines, including (1) original TransE and Dist-Mult without aggregating entity context and relation context; (2) three state-of-the-art KGE models: ComplEx, SimplE, RotatE; (3) a classic GNN-based KGE model: R-GCN. Brief descriptions of baselines are as follows:

TransE [2]: TransE is one of the most widely-used KGE models which translates the head embedding into tail embedding by adding it to relation embedding.

DistMult [17]: DistMult is a popular tensor factorization based model which uses a bilinear score function to compute scores of knowledge triplets.

ComplEx [11]: ComplEx is an extension of DistMult which embeds entities and relations into complex vectors instead of real-valued ones.

SimplE [4]: SimplE is a simple interpretable fully-expressive tensor factorization model for knowledge graph completion.

RotatE [9]: RotatE defines each relation as a rotation from the head entity to the tail entity in the complex vector space.

R-GCN [8]: RGCN is a variation of graph neural network, it can deal with the highly multi-relational knowledge graph data and aggregate context information to entities.

To simplify, we use $\mathcal{L}$-TransE to represent LightCAKE-TransE and use $\mathcal{L}$-DistMult to represent LightCAKE-DistMult.

5.3 Experimental Settings

We use Adam [5] as the optimizer with the learning rate as 5e-3. We set the embedding dimension of entity and relation as 256, l_2 penalty coefficient as 1e-7, batch size as 512, the total number of iterations L as 4 and a maximum of 20 epochs. Moreover, we use early stopping for training, and all the training parameters are randomly initialized.

We evaluate all methods in the setting of relation prediction, i.e., for a given entity pair (h, t) in the test set, we rank the ground-truth relation type r against all other candidate relation types. We compare our models with baselines using the following metrics: (1) Mean Reciprocal Rank (MRR, the mean of all the reciprocals of predicted ranks); (2) Mean Rank (MR, the mean of all the predicted ranks); (3) Hit@3(the proportion of correctly predicted entities ranked in the top 3 predictions).

Table 2. Results of relation prediction. (Bold: best; Underline: runner-up.) The results of ComplEx, SimplE and RotatE are taken from [13]. Noted that the trainable parameters in $\mathcal{L}$-TransE and $\mathcal{L}$-DistMult are only entity embeddings and relation embeddings, for a fair comparison, we only choose those 3 traditional baselines from [13] with a small number of parameters. In addition, in order to compare context-aware KGE and context-independent KGE in the same experimental environment to prove the validity of LightCAKE, we implemented TransE and DistMult ourselves.

Method	WN18RR			FB15K-237			NELL995			DDB14		
	MRR	MR↓	Hit@3	MRR	MR↓	Hit@3	MRR	MR↓	Hit@3	MRR	MR↓	Hit@3
ComplEx	0.840	2.053	0.880	0.924	1.494	0.970	0.703	23.040	0.765	0.953	1.287	0.968
SimplE	0.730	3.259	0.755	**0.971**	1.407	<u>0.987</u>	0.716	26.120	0.748	0.924	1.540	0.948
RotatE	0.799	2.284	0.823	<u>0.970</u>	<u>1.315</u>	0.980	0.729	23.894	0.756	0.953	1.281	0.964
RGCN	0.823	2.144	0.854	0.954	1.498	0.973	0.731	22.917	0.749	0.951	1.278	0.965
TransE	0.789	1.755	0.918	0.932	1.979	0.952	0.719	16.654	0.766	0.936	1.487	0.957
$\mathcal{L}$-TransE	0.813	<u>1.648</u>	<u>0.933</u>	0.943	2.281	0.962	<u>0.793</u>	<u>9.325</u>	<u>0.831</u>	<u>0.964</u>	<u>1.184</u>	<u>0.969</u>
DistMult	<u>0.865</u>	1.743	0.922	0.935	1.920	0.979	0.712	22.340	0.744	0.937	1.334	0.958
$\mathcal{L}$-DistMult	**0.955**	**1.134**	**0.988**	0.967	**1.174**	**0.988**	**0.852**	**2.271**	**0.914**	**0.972**	**1.097**	**0.991**

5.4 Experimental Results and Analysis

The results on all datasets are reported in Table 2. We can observe that: (1) Comparing with the original TransE and DistMult, our proposed $\mathcal{L}$-TransE and

$\mathcal{L}$-DistMult consistently have superior performance on all datasets, proving that LightCAKE can greatly improve the performance of context-independent KGE models; (2) Comparing with all six KGE baselines, the proposed $\mathcal{L}$-TransE and $\mathcal{L}$-DistMult achieve substantial improvements or state-of-the-art performance on all datasets, showing the effectiveness of $\mathcal{L}$-TransE and $\mathcal{L}$-DistMult.

5.5   Ablation Study

LightCAKE utilizes both entity context and relation context. How does each context affect the performance of LightCAKE? To answer this question, we propose model variants to conduct ablation studies on $\mathcal{L}$-TransE and $\mathcal{L}$-DistMult including: (1) the original TransE and DistMult without considering entity context and relation context; (2) $\mathcal{L}_{rel}$-TransE and $\mathcal{L}_{rel}$-DistMult that just aggregate the relation context and discard the entity context; (3) $\mathcal{L}_{ent}$-TransE and $\mathcal{L}_{ent}$-DistMult that just aggregate the entity context and discard the relation context.

The experimental results of MRR on datasets WN18RR and FB15K237 are reported in Fig. 3 (a),(b),(d), and (e). $\mathcal{L}$-TransE and $\mathcal{L}$-DistMult achieve best performance compared with their corresponding model variants, demonstrating that integrating both entity context and relation context is most effective for KGE. Also, $\mathcal{L}_{rel}$-TransE and $\mathcal{L}_{ent}$-TransE are both better than TransE, $\mathcal{L}_{rel}$-DistMult and $\mathcal{L}_{ent}$-DistMult are both better than DistMult, indicating that

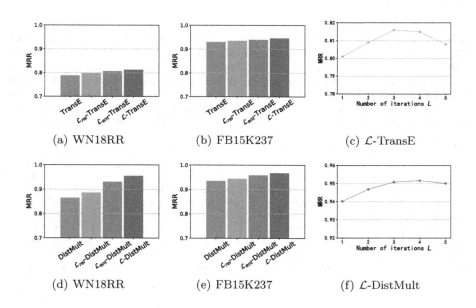

Fig. 3. The performance of model variants for (a) $\mathcal{L}$-TransE and (d) $\mathcal{L}$-DistMult on WN18RR dataset. The performance of model variants for (b) $\mathcal{L}$-TransE and (e) $\mathcal{L}$-DistMult on FB15K237 dataset. The performance of various L for (c) $\mathcal{L}$-TransE and (f) $\mathcal{L}$-DistMult on WN18RR dataset.

entity context and relation context are both helpful for KGE. $\mathcal{L}_{ent}$-TransE is better than $\mathcal{L}_{rel}$-TransE and $\mathcal{L}_{ent}$-DistMult is better than $\mathcal{L}_{rel}$-DistMult, showing that entity context contributes more to improving the model performance than relation context.

5.6   Analysis on Number of Iterations

In this section, we investigate the sensitivity of the parameter L, i.e., the number of iterations. We report the MRR on WN18RR dataset. We set that L ranges from 1 to 5. The results of $\mathcal{L}$-TransE and $\mathcal{L}$-DistMult are shown in Fig. 3 (c) and (f), we can observe that with the growth of the number of iterations, the performance raises first and then starts to decrease slightly, which may due to when further contexts are involved, more uncorrelated information are integrated into embeddings. So properly setting the number of L can help to improve the performance of our method.

5.7   Efficiency Analysis

We evaluate the efficiency of LightCAKE by comparing it with DistMult and R-GCN. We investigate the difference of DistMult, R-GCN and $\mathcal{L}$-DistMult in the views of entity context, relation context, parameter quantities (space complexity), and the MRR in WN18RR dataset. The results are shown in Table 3. We can observe that the parameter quantities of $\mathcal{L}$-DistMult are far less than R-GCN, that is because R-GCN use complicated matrix transformation to encode context information, while $\mathcal{L}$-DistMult only uses multiplication on embeddings to encode context information. Also, both DistMult and $\mathcal{L}$-DistMult achieve better prediction results than R-GCN in the relation prediction task, which may because R-GCN is overfitted due to the use of too many parameters. In summary, $\mathcal{L}$-DistMult is lighter, more efficient and more robust.

Table 3. Efficiency Analysis. Here, d is the embedding dimension, L is the number of iterations, $|\mathcal{E}|$ and $|\mathcal{R}|$ indicate the total number of entities and relations respectively.

Models	Entity context	Relation context	Space complexity	MRR
DistMult [17]	✗	✗	$\mathcal{O}(|\mathcal{E}|d + |\mathcal{R}|d)$	0.865
R-GCN [8]	✓	✗	$\mathcal{O}(L(d^2 + |\mathcal{E}|d + |\mathcal{R}|d))$	0.823
$\mathcal{L}$-DistMult	✓	✓	$\mathcal{O}(L(|\mathcal{E}|d + |\mathcal{R}|d))$	0.955

6   Conclusion

In this paper, we propose LightCAKE to learn context-aware knowledge graph embedding. LightCAKE considers both the entity context and relation context, and extensive experiments show its superior performance comparing with state-of-the-art KGE models. In addition, LightCAKE is very lightweight and efficient in aggregating context information. Future research will explore more possible context encoder, i.e. ϕ_{ent} and ϕ_{rel}, and more possible scoring functions used in Eq. (3), Eq. (4) and Eq. (5) to make LightCAKE more general and powerful.

Acknowledgments. This research was supported by the Natural Science Foundation of China under Grant No. 61836013, the Ministry of Science and Technology Innovation Methods Special work Project under grant 2019IM020100, the Beijing Natural Science Foundation(4212030), and Beijing Nova Program of Science and Technology under Grant No. Z191100001119090. Zhiyuan Ning and Ziyue Qiao contribute equally to this work. Yi Du is the corresponding author.

References

1. Bansal, T., Juan, D.C., Ravi, S., McCallum, A.: A2n: attending to neighbors for knowledge graph inference. In: Proceedings of the 57th Annual Meeting of the Association for Computational Linguistics, pp. 4387–4392 (2019)
2. Bordes, A., Usunier, N., Garcia-Duran, A., Weston, J., Yakhnenko, O.: Translating embeddings for modeling multi-relational data. In: Advances in Neural Information Processing Systems 26, vol. 26, pp. 2787–2795 (2013)
3. Dettmers, T., Pasquale, M., Pontus, S., Riedel, S.: Convolutional 2D knowledge graph embeddings. In: Proceedings of the 32th AAAI Conference on Artificial Intelligence, pp. 1811–1818 (February 2018)
4. Kazemi, S.M., Poole, D.: Simple embedding for link prediction in knowledge graphs. In: Advances in Neural Information Processing Systems, pp. 4284–4295 (2018)
5. Kingma, D.P., Ba, J.L.: Adam: a method for stochastic optimization. In: ICLR 2015: International Conference on Learning Representations (2015)
6. Kipf, T.N., Welling, M.: Semi-supervised classification with graph convolutional networks. In: International Conference on Learning Representations (ICLR) (2017)
7. Lin, Y., Liu, Z., Sun, M., Liu, Y., Zhu, X.: Learning entity and relation embeddings for knowledge graph completion. In: AAAI 2015 Proceedings of the Twenty-Ninth AAAI Conference on Artificial Intelligence, pp. 2181–2187 (2015)
8. Schlichtkrull, M., Kipf, T.N., Bloem, P., van den Berg, R., Titov, I., Welling, M.: Modeling relational data with graph convolutional networks. In: Gangemi, A., et al. (eds.) ESWC 2018. LNCS, vol. 10843, pp. 593–607. Springer, Cham (2018). https://doi.org/10.1007/978-3-319-93417-4_38
9. Sun, Z., Deng, Z.H., Nie, J.Y., Tang, J.: Rotate: knowledge graph embedding by relational rotation in complex space. In: International Conference on Learning Representations (2019)
10. Toutanova, K., Chen, D.: Observed versus latent features for knowledge base and text inference. In: Proceedings of the 3rd Workshop on Continuous Vector Space Models and their Compositionality, pp. 57–66 (2015)
11. Trouillon, T., Welbl, J., Riedel, S., Gaussier, É., Bouchard, G.: Complex embeddings for simple link prediction. In: International Conference on Machine Learning, pp. 2071–2080. PMLR (2016)
12. Veličković, P., Cucurull, G., Casanova, A., Romero, A., Liò, P., Bengio, Y.: Graph attention networks. In: International Conference on Learning Representations (2018)

13. Wang, H., Ren, H., Leskovec, J.: Entity context and relational paths for knowledge graph completion. arXiv preprint arXiv:2002.06757 (2020)
14. Wang, Q., Mao, Z., Wang, B., Guo, L.: Knowledge graph embedding: a survey of approaches and applications. IEEE Trans. Knowl. Data Eng. **29**(12), 2724–2743 (2017)
15. Wang, Z., Zhang, J., Feng, J., Chen, Z.: Knowledge graph embedding by translating on hyperplanes. In: AAAI 2014 Proceedings of the Twenty-Eighth AAAI Conference on Artificial Intelligence, pp. 1112–1119 (2014)
16. Xiong, W., Hoang, T., Wang, W.Y.: Deeppath: a reinforcement learning method for knowledge graph reasoning. In: Proceedings of the 2017 Conference on Empirical Methods in Natural Language Processing, pp. 564–573 (2017)
17. Yang, B., tau Yih, W., He, X., Gao, J., Deng, L.: Embedding entities and relations for learning and inference in knowledge bases. In: ICLR 2015: International Conference on Learning Representations (2015)

Transferring Domain Knowledge with an Adviser in Continuous Tasks

Rukshan Wijesinghe[1,2][✉], Kasun Vithanage[2], Dumindu Tissera[1,2], Alex Xavier[2], Subha Fernando[2], and Jayathu Samarawickrama[1,2]

[1] Department of Electronic and Telecommunication Engineering, University of Moratuwa, Moratuwa, Sri Lanka
jayathu@ent.mrt.ac.lk
[2] CODEGEN QBITS Lab, University of Moratuwa, Moratuwa, Sri Lanka
subhaf@uom.lk

Abstract. Recent advances in Reinforcement Learning (RL) have surpassed human-level performance in many simulated environments. However, existing reinforcement learning techniques are incapable of explicitly incorporating already known domain-specific knowledge into the learning process. Therefore, the agents have to explore and learn the domain knowledge independently through a trial and error approach, which consumes both time and resources to make valid responses. Hence, we adapt the Deep Deterministic Policy Gradient (DDPG) algorithm to incorporate an adviser, which allows integrating domain knowledge in the form of pre-learned policies or pre-defined relationships to enhance the agent's learning process. Our experiments on OpenAi Gym benchmark tasks show that integrating domain knowledge through advisers expedites the learning and improves the policy towards better optima.

Keywords: Actor-critic architecture · Deterministic policy gradient · Reinforcement learning · Transferring domain knowledge

1 Introduction

Conventional reinforcement learning approaches have been limited to domains with low dimensional discrete state and action spaces or fully observable state and action spaces, where handcrafted features are heavily used. But the emergence of deep Q-network (DQN) [15] extended its applicability to high dimensional state spaces. DQN has surpassed human-level performance in some of the challenging Atari 2600 games using only unprocessed pixels as input [15]. It was still not a generalized solution, and DQN was not suited well for the higher dimensional or continuous action spaces [12,22]. The Deep Deterministic Policy Gradient (DDPG) algorithm derived from Deterministic Policy Gradient [22] extended the Deep Q-Learning for continuous state and action space.

Although advancements in RL have reached continuous state and action spaces, they are still incapable of incorporating already known domain knowledge

© Springer Nature Switzerland AG 2021
K. Karlapalem et al. (Eds.): PAKDD 2021, LNAI 12714, pp. 194–205, 2021.
https://doi.org/10.1007/978-3-030-75768-7_16

directly into the learning process. Thus, they unnecessarily consume time and computational resources to acquire the fundamental knowledge learning from scratch, i.e., agents will follow a trial and error approach many times before successfully converging to an optimal policy. In a simulated world, this is not efficient, and it is particularly not welcomed in real-world tasks where the agents cannot make fatal mistakes during learning, such as the autonomous navigation domain. To this end, an algorithm that facilitates the incorporation of domain knowledge into the learning process enables an agent to accelerate the learning procedure by limiting the exploration space and converge to better policies.

In this paper, we propose a novel approach to train an agent efficiently in continuous and high-dimensional state-action spaces. Our approach adapts the DDPG algorithm to incorporate already available information into the training process as an adviser to accelerate it. The DDPG algorithm updates policy in each iteration with approximated policy gradients derived from the gradients of Critic-network output with respect to Actor-network parameters. However, this approach updates the policy parameters directly and does not facilitate to use of domain knowledge for the policy updating process. In contrast, we update the existing policy to a new policy based on a two-step approach. During the policy parameter update in each iteration, we first set a temporary target to the policy and then push the current policy towards it by reducing the L2 distance.

This two-fold optimization facilitates taking the adviser's suggestions into account when updating the policy. In addition, the adviser can be used to enforce the agent to explore the better regions of the state-action space. It enables the agent extract of good policies while reducing the exploration cost. We theoretically prove the convergence of the adapted DDPG algorithm and empirically show that the proposed approach itself improves over the existing DDPG algorithm with chosen benchmark tasks in the continuous domain. We further plug advisers to the adapted DDPG algorithm to show accelerated learning, validating the utility of the two-fold policy updating process.

2   Related Work

Modern foundations of RL are formed by intertwining several trial and error methods and solutions to optimal control problems with temporal methods [23]. Deep Q-networks [15] extended the applicability of RL to the continuous high-dimensional state spaces. Later, the DDPG algorithm [12,22] combined the DQN and deterministic policy gradient algorithm to handle continuous high-dimensional state-action spaces. Integrating the DDPG algorithm with actor-critic [20] architecture allows learning parameterized continuous policies.

Reinforcement learning has been an emerging trend in the autonomous navigation domain [8,9,13,30]. End-to-end trained asynchronous deep RL-based models [14] were used to do continuous control of mobile robots in mapless navigation [24]. Recent studies have shown a greater interest in reducing training time, increasing sample efficiency, and minimizing the trial and error nature of the learning process to make RL applicable to real-world applications safely and confidently. DQN and DDPG are sample-inefficient since they demand a large number of samples during the training. Nagabandi *et al.* [16] show that combining medium-sized neural networks with model predictive control and a model-free learner initialized by deep neural network dynamic models tested on MuJoCo locomotion tasks [27] achieves high sample efficiencies. Kahn *et al.* [7,29] proposed a self supervising generalized computational graph for autonomous navigation, which subsumes the advantages of both model-free and model-based methods.

To reduce the trial and error nature in training, successor-feature-based RL [31] employs knowledge transferring across similar navigational environments. Taylor *et al.* [26] introduced offline RL algorithms for transferring knowledge between agents with different internal representations. Multitask and transfer learning has been utilized in autonomous agents where they learn multiple tasks at once and apply the generalized knowledge to new domains [19]. Ross *et al.* [21] discuss the DAgger algorithm, which is similar to no-regret online learning algorithms. It uses a dataset of trajectories collected using an expert to initialize policies that can mimic the expert better. Methods of automatically mapping different tasks by analyzing agent experience have improved the training speed in RL significantly. [25].

Self-imitation learning [18] learns to reproduce the past good decisions of the agent to enhance deep exploration. Hindsight experience replay has been used with DDPG to overcome exploration bottlenecks on simulated robotics tasks in [17]. Hester *et.al* [6] proposed deep Q learning from demonstrations that use small sets of demonstration data to accelerate the learning process. The effect of function approximation errors in actor-critic settings has been addressed by employing a novel variant of Double Q-Learning [3]. Maximum entropy RL is used in off-policy actor-critic methods [10] to overcome sample inefficiency and convergence in conjunction [5]. Continuous variants of Q-learning combined with learned models have shown to be effective in addressing the sample complexity of RL [4].

Our approach uses the actor-critic architecture to deviate from the existing methods due to several reasons. First, we adapt the DDPG algorithm to incorporate domain knowledge as an adviser in continuous tasks with high dimensional state-action spaces. Secondly, we employ the adviser in data collection to enforce the agent to explore regions of state-action space with a higher return.

3   Method

The proposed adapted DDPG algorithm improves the policy in the direction of the gradient of the Q-value function. It also facilitates integrating pre-learned policies or existing relationships as advisers to transfer domain knowledge. During the training process, advisers can be deployed in; 1) data collection, as well as 2) policy updating processes. Once the adviser is involved in the data collection process, it enforces the agent to explore better regions in state and action spaces according to the adviser's perspective. When the adviser is incorporated into the policy updating process, it aids to reach better policies rapidly by selecting the best set of actions. In this section, we first introduce the adapted DDPG algorithm and show its convergence. Then we explicitly describe the proposed ways of employing an adviser for the data collection and policy updating processes to achieve an efficient training approach in continuous tasks.

3.1   Adapted Deep Deterministic Policy Gradient Algorithm

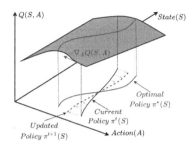

Fig. 1. Policy updating method with gradient of Q-value function.

Moving the current policy in the direction of the gradient of the Q-value function is computationally efficient than globally maximizing the Q-value function [22] in extracting the optimal policy in the continuous domain. Therefore, the proposed method can be utilized to extract better policies in tasks with high-dimensional continuous state and action spaces by improving the current policy in the direction of the gradient of the action-value (Q-value) function. The surface in Fig. 1 represents the Q-value function corresponding to a hypothetical RL problem in the continuous domain. For explanation simplicity, it only contains a single action variable (A) and a single state variable (S). Let the $\pi(S; \phi)$ be the policy function that governs the actions in given states, and it is parameterized by ϕ. Similarly, Q-value function $Q(S, A; \theta)$ is parameterized by θ.

The peak red line on the surface in Fig. 1 represents the Q-values corresponding to the state-action pairs on the optimal policy $\pi^*(S)$. Its projection on the state-action plane denotes the optimal policy. The direction of the gradient of the Q-value function with respect to actions $\nabla_A Q(S, A)$ corresponding to a point on current policy $\pi^t(S)$ at a particular time step t always leads towards either local or global optimal-policy. Therefore, the term $\nabla_A Q(S, A)$ can be used to update the current policy and obtain a better-updated policy $\pi^{t+1}(S)$ by pushing π^t in the direction of $\nabla_A Q(S, A_{\pi^t})$, as shown in Fig. 1. Thus, the corresponding policy improvement at a particular step can be represented by;

$$\pi^{t+1}(S) \leftarrow \pi^t(S) + \beta \nabla_A Q(S, A_{\pi^t}; \theta). \tag{1}$$

Here, β is the updating rate of the current policy, and it represents the degree of shift between updated and current policies. Once the updated policy $\pi^{t+1}(S)$

is set, it is used as a temporary target to optimize the current policy $\pi^t(S; \phi)$. We update ϕ by performing the gradient descent to minimize the loss L_π;

$$L_\pi = \frac{1}{n} \sum [\pi^{t+1}(S) - \pi^t(S; \phi)]^2, \tag{2}$$

which is the mean squared error between current and updated policy samples. The main advantage of this two-fold policy update is that we can plug suggestions of an adviser who has domain knowledge in between the aforementioned two steps to achieve a better-updated policy at a particular time step.

Algorithm 1. Adapted DDPG Algorithm

For each update of Actor-network $\pi(S; \phi)$ and Critic-network $Q(S, A; \theta)$ at a given time step t ;

1: Select a batch of experiences $M =< S, A, R, S' >$ randomly from memory replay buffer with the size of n
2: Set $\hat{Q}(S, A) \leftarrow R + \gamma Q^-(S', \pi^-(S'; \phi^-); \theta^-)$
3: Update θ by minimizing the loss function
  $L_Q = \frac{1}{n} \Sigma (\hat{Q}(S, A) - Q(S, A; \theta))^2$
4: Set $\pi^{t+1}(S) \leftarrow \pi^t(S; \phi) + \beta \nabla_A Q(S, A_{\pi^t}; \theta)$
5: Update ϕ by minimizing the loss function
  $L_\pi = \frac{1}{n} \Sigma (\pi^{t+1}(S) - \pi^t(S; \phi))^2$
6: Update the parameters of the target networks, θ^- and ϕ^-
  $\theta^- \leftarrow \tau\theta + (1 - \tau)\theta^-$
  $\phi^- \leftarrow \tau\phi + (1 - \tau)\phi^-$

In our approach, the Q-value function is updated with the Temporal Difference (TD) error similar to the DDPG algorithm presented in [12]. It maintains two parameterized Q-value functions known as "Q-network" $Q(s, a; \theta)$ and "target Q-network" $Q^-(s, a; \theta^-)$. Similarly, it keeps two policy functions named "policy-network" $\pi(s, a; \phi)$ and "target policy-network" $\pi^-(s, a; \phi^-)$. Algorithm 1 illustrates the steps followed in each update of the Q-network and policy-network. A soft update mechanism weighted by τ and $(1 - \tau)$ (where $0 < \tau << 1$) is used to update the target policy-network and target Q-network as shown in the last step of Algorithm 1. Maintaining two separate networks and using a soft updating mechanism enhances the stability of the learning process and supports training the Q-network without a divergence [12].

3.2 Convergence of Adapted DDPG Algorithm

Smooth Concave Functions. If $f : \mathbb{R}^n \to \mathbb{R}$ is a twice differentiable function and holds Lipschitz continuity with constant $L > 0$ then,

$$\|\nabla f(y) - \nabla f(x)\| \leq L \|x - y\| \quad \forall x, y \in \mathbb{R}^n \tag{3}$$

Here, L is a measurement for the smoothness of the function [11] and if $f(x)$ is a concave function then the following inequality is satisfied;

$$f(y) \geq f(x) + \langle \nabla f(x), (y - x) \rangle + \frac{L}{2} \|y - x\|^2 \; \forall x, y \in \mathbb{R}^n \tag{4}$$

Convergence Analysis: Let $Q : \mathbb{R}^{m+n} \to \mathbb{R}$ be the action-value function where m and n represent the number of state variables and the number of actions. S and A be the state and action vectors, such that $S \in \mathbb{R}^m$ and $A \in \mathbb{R}^n$. Let's define $(S, A) = (s_1, s_2, \ldots, s_m, a_1, a_2, \ldots, a_n)$ where $s_1, \ldots, s_m$ and $a_1, \ldots, a_n \in \mathbb{R}$. Let $A_\pi = \pi(S)$ where π is the policy function that predicts the action to be executed for a given state S. Consider the policy update at a particular time step t.

$$A_{\pi^{t+1}} = \pi^{t+1}(S) = \pi^t(S) + \beta \nabla_A Q(S, A_{\pi^t}) \tag{5}$$

If Q(S, A) is a concave and twice differentiable function with Lipschitz continuity then considering the Eq. 4;

$$Q(S, A_{\pi^{t+1}}) \geq Q(S, A_{\pi^t}) + \langle \nabla Q(S, A_{\pi^t}), ((S, A_{\pi^{t+1}}) - (S, A_{\pi^t})) \rangle$$
$$- \frac{L}{2} \|(S, A_{\pi^{t+1}}) - (S, A_{\pi^t})\|^2 \tag{6}$$

$$Q(S, A_{\pi^{t+1}}) \geq Q(S, A_{\pi^t}) + \langle \nabla_A Q(S, A_{\pi^t}), (A_{\pi^{t+1}} - A_{\pi^t}) \rangle$$
$$- \frac{L}{2} \|(A_{\pi^{t+1}} - A_{\pi^t})\|^2 \tag{7}$$

Since $A_{\pi^{t+1}} - A_{\pi^t} = \beta \nabla_A Q(S, A_{\pi^t})$ by substituting for Eq. 7

$$Q(S, A_{\pi^{t+1}}) \geq Q(S, A_{\pi^t}) + \beta \|\nabla_A Q(S, A_{\pi^t})\|^2 - \frac{L\beta^2}{2} \|\nabla_A Q(S, A_{\pi^t})\|^2 \tag{8}$$

Considering Eq. 8 and if $0 < \beta \leq \frac{2}{L}$ then,

$$Q(S, A_{\pi^{t+1}}) - Q(S, A_{\pi^t}) \geq \beta(1 - \frac{\beta L}{2}) \|\nabla_A Q(S, A_{\pi^t})\|^2 \geq 0 \tag{9}$$

Therefore, $Q((S, A_{\pi^{t+1}}) \geq Q(S, A_{\pi^t})$ at any time step. Once all the time steps from 0 to k considered in Eq. 9,

$$\sum_{t=0}^{t=k} \{Q(S, A_{\pi^{t+1}}) - Q(S, A_{\pi^t})\} \geq \sum_{t=0}^{t=k} \{\beta(1 - \frac{\beta L}{2}) \|\nabla_A Q(S, A_{\pi^t})\|^2\} \tag{10}$$

$$Q(S, A_{\pi^{k+1}}) - Q(S, A_{\pi^0}) \geq \beta(1 - \frac{\beta L}{2}) \sum_{t=0}^{t=k} \|\nabla_A Q(S, A_{\pi^t})\|^2 \tag{11}$$

If $Q(S, A_{\pi^*})$ is the Q-value at optimal policy $\pi^*(S)$ then $Q(S, A_{\pi^*}) \geq Q(S, A_{\pi^{k+1}})$. By considering Eq. 11,

$$Q(S, A_{\pi^*}) - Q(S, A_{\pi^0}) \geq Q(S, A_{\pi^{k+1}}) - Q(S, A_{\pi^0})$$

$$\geq \beta(1 - \frac{\beta L}{2}) \sum_{t=0}^{t=k} \|\nabla_A Q(S, A_{\pi^t})\|^2 \qquad (12)$$

Eq. 12 implies that as $k \to \infty$, the right hand side converges.

$$\therefore \lim_{k \to \infty} \|\nabla_A Q(S, A_{\pi^k})\|^2 = 0 \qquad (13)$$

Since $\pi^*(S) = \arg\max_A Q(S, A)$, at optimal policy $\nabla_A Q(S, A_{\pi^*}) = 0$. Therefore, the Eq. 13 implies that $\lim_{k\to\infty} A_{\pi^{k+1}} = A_{\pi^*}$.

3.3 Adapted DDPG with Actor-Critic Agent

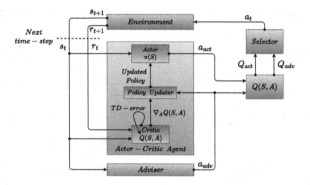

Fig. 2. Actor-Critic architecture with adviser module.

Actor-Critic architecture is generally used in model-free RL, and it consists of two main components, namely "Actor" and "Critic". The Actor decides which action to be executed by the agent and the Critic always tries to improve the Actor's performance by analyzing rewards received in each time-step. Generally, the Actor is updated with the policy gradient approach, while the Critic gets updated with the temporal difference error [20]. Our agents are based on the Actor-Critic architecture and learn a diverse set of RL benchmark tasks in the continuous domain. Figure 2 illustrates the basic block diagram of the Actor-Critic architecture that employs the adviser for data collection and policy updating processes. In our implementation, the Actor and Critic modules represent the parameterized policy and Q-value functions, respectively.

3.4 Employing an Adviser to Transfer Domain Knowledge

Although model-free value-based approaches have demonstrated state-of-the-art performance in the RL domain, the low sample efficiency is one of the main concerns that limit their applicability in real-world applications [7]. One solution to this issue is integrating domain knowledge into the learning process. Thus, the agent does not need to learn everything from scratch. Here, we propose two techniques that employ an adviser to integrate domain knowledge into the learning process. Firstly, we enforce the agent to explore better regions in the state and action spaces by enabling adviser suggestions for the data collecting process. Secondly, we improve the policy updating process by allowing the adviser to adjust the updated policy to a better policy based on the current knowledge.

Algorithm 2. Data collection with an adviser

At a given time step t;

1: Observe current state s_t
2: $a_{adv} \leftarrow f(s_t)$
3: $a_{act} \leftarrow \pi(s_t)$
4: $C \leftarrow 1 - e^{-\lambda N}$
5: $\epsilon \leftarrow \dfrac{e^{-Q(s_t, a_{adv})/T}}{e^{-Q(s_t, a_{adv})/T} + e^{-CQ(s_t, a_{act})/T}}$
6: With probability ϵ, $a_{t+1} \leftarrow a_{adv}$
7: Otherwise, $a_{t+1} \leftarrow a_{act}$
8: $a_{t+1} \leftarrow a_{t+1} + noise$

Algorithm 3. Policy updating with an adviser

For each update of Actor-network $\pi(s; \phi)$ and Critic-network $Q(s, a; \theta)$;

1: Steps 1 - 3 in Algorithm 1
2: $A_{adv} \leftarrow f(S)$
3: $A_{act} \leftarrow \pi(S; \phi)$
4: $\hat{A}(S) \leftarrow A_{act} + \beta \nabla_A Q(S, A; \theta)$
5: **for** $1 : i : n$ **do**
6:     **if** $Q(s^i, a_{adv}^i) > Q(s^i, \hat{a}^i)$ **then**
7:         $\hat{a}^i \leftarrow a_{adv}^i$
8: Steps 5 - 6 in Algorithm 1

3.5   Adviser for Data Collection Process

Here, we employ an adviser (f) as in Algorithm 2, that maps states (S) to actions (A), to make sampling more efficient by comparing the Actor's current prediction against the adviser's suggestion. In each time step, both Actor (A_{act}) and adviser (A_{adv}) suggests the action corresponding to the current state (S_t). Then, both suggestions are evaluated with respect to the current knowledge (Q-value) of the Critic module, and the adviser's action is selected for the execution with a probability of ϵ. Calculation of ϵ is adapted by the work [2], and our method deviates in several ways. We employ the softmax function to induce a higher probability corresponding to the action with a higher Q-value. By varying the softmax temperature T, it is possible to change the priority given to the adviser. The constant C $(C = 1 - e^{-\lambda N})$ is a confidence value calculated on the agent's behalf, where N is the number of episodes elapsed, and λ $(\lambda > 0)$ is the decaying constant. It enforces the agent to give higher priority to the adviser's suggestions at the beginning, and enables the agent to explore near a better policy. In the end, we add a noise signal to the selected action for exploration. The noise generation is influenced by the Ornstein-Uhlenbeck process [28], and it ensures a better exploration near the selected action.

3.6   Adviser for Policy Updating Process

Since the adapted DDPG algorithm improves the existing policy with an updated set of sampled actions, it enables integrating adviser's suggestions into the policy updating process as described in Algorithm 3. At each iteration, a batch of experiences with n samples is fetched randomly from the memory replay buffer, and the Q-value function is updated as similar to Algorithm 1. Before updating the policy, both adviser and actor suggestions for the selected batch (A_{adv} and A_{act} respectively) are calculated. Then updated set of actions ($\hat{A}(S)$) is calculated similarly to Algorithm 1. In the next step, each updated action ($\hat{a}^i$) is replaced by the adviser's action (a^i_{adv}) if the Q-value corresponding to the adviser suggested action is greater than the corresponding updated action ($\hat{a}^i$). In Algorithm 3, s^i and a^i refer to the state and action of the ith sample of the selected batch. Finally, policy parameters are updated with the modified set of updated actions corresponding to the selected batch.

Table 1. The averaged total episodes score of the trained agents for 30 runs with 500 episodes in each. The adapted DDPG surpasses the conventional DDPG in all tasks. Although the adviser performance is comparatively low, the adapted DDPG algorithm with an adviser shows the best performance.

	Pendulum	MountainCar continuous	LunarLander continuous	Bipedal walker
adviser	−508.5	12.2	−126.5	20.1
DDPG	−398.9	28.5	−65.7	100.3
Adapted DDPG	−272.7	55.1	−31.4	150.1
Adapted DDPG + adviser	**−178.3**	**93.4**	**−30.1**	**190.3**

4   Experiments

To evaluate the performance of the adapted DDPG algorithm and adviser-based agent architecture, we experiment on a diverse set of benchmark tasks in the continuous domain. It includes four OpenAI Gym [1] environments namely Pendulum-v0, MountainCarContinuous-v0, LunarLanderContinuous-v2 and Bi-pedalWalker-v2. We train three distinct agents using the DDPG, Adapted DDPG, and the Adapted DDPG with adviser algorithms in separation for each benchmark task. As the adviser of the BipedalWalker, we deploy a policy trained for a similar task, and classical control approaches (Proportional Integral and Derivative controllers) and predefined rules are used in the other three. We employed neural networks to parameterize the Q-value function and the policy function. We set the β to a lower like 0.01 to satisfy condition $\beta < \frac{2}{L}$ (see Sect. 3.2) to ensure convergence of learning as we don't have enough information about the smoothness of the Q-Value function.

In each task, we train the agents for 30 runs, where each run consists of a pre-defined number of training episodes. After each run, we test the agents for 500 episodes. We define the "total episode score" as the total reward earned by the agent in all the steps of a given episode. We take the average of such "total episode scores" gained in all 500 test episodes in a given run and average this figure again over the 30 runs. Table 1 reports this average total episode score where the adapted DDPG algorithm comfortably surpasses the conventional DDPG algorithm in all tasks. The adviser is incapable of performing the given tasks to the level of agents trained with DDPG or adapted DDPG algorithms (see the first row of Table 1). However, the combination of the adapted DDPG algorithm and the adviser attain the best performance in all the tasks. It shows that the adviser assists the adapted DDPG agents to converge towards a policy with higher scores, even though the adviser is not perfect.

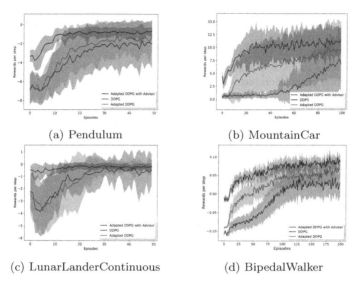

(a) Pendulum   (b) MountainCar

(c) LunarLanderContinuous   (d) BipedalWalker

Fig. 3. Reward per step of trained agents with the episode number on continuous benchmark tasks. The adapted DDPG algorithm reaches higher reward levels rapidly compared to the DDPG algorithm, and the adviser always accelerates the training speed further with a low variance in the learning curve.

We further plot the averaged training reward per step with the episode number in Fig. 3. "The rewards per step" is the total reward earned by the agent in an episode divided by the number of steps. We obtain this value for all episodes across all runs. The average of the "rewards per step" for a particular indexed episode is calculated by averaging these values belonging to the same indexed episodes across the 30 runs. It demonstrates that the adapted DDPG algorithm achieves higher reward levels rapidly than conventional DDPG. It further illustrates that incorporating an adviser during the training phase expedites the learning process significantly compared to both DDPG and adapted DDPG algorithms. It is also evident that the agent with adviser converges to better policies, achieving higher rewards compared to other methods. Additionally, the adapted DDPG with adviser shows considerably less variance than all the others.

5 Conclusion

In this paper, we adapt the DDPG algorithm to incorporate an adviser with domain knowledge to expedite the training process. The adviser in our actor-critic architecture causes the data collection and policy updating process to be more effective. We theoretically proved the convergence of the adapted DDPG algorithm and showed experimentally that the proposed adapted DDPG algorithm outperforms the standard DDPG algorithm in conventional RL benchmark tasks in the continuous domain. Additionally, we also demonstrated that the proposed two-fold policy updating mechanism of the adapted DDPG algorithm effectively incorporates domain knowledge, resulting in an accelerated convergence.

Acknowledgement. We thank Prof. Sanath Jayasena and Dr. Ranga Rodrigo for arranging insight discussions which supported this work.

References

1. Brockman, G., et al.: Openai gym. arXiv preprint arXiv:1606.01540 (2016)
2. Fernández, F., Veloso, M.: Probabilistic policy reuse in a reinforcement learning agent. In: Proceedings of the Fifth International Ioint Conference on Autonomous Agents and Ultiagent Systems, pp. 720–727 (2006)
3. Fujimoto, S., van Hoof, H., Meger, D.: Addressing function approximation error in actor-critic methods. arXiv preprint arXiv:1802.09477 (2018)
4. Gu, S., Lillicrap, T., Sutskever, I., Levine, S.: Continuous deep q-learning with model-based acceleration. In: International Conference on Machine Learning, pp. 2829–2838 (2016)
5. Haarnoja, T., Zhou, A., Abbeel, P., Levine, S.: Soft actor-critic: off-policy maximum entropy deep reinforcement learning with a stochastic actor. arXiv preprint arXiv:1801.01290 (2018)
6. Hester, T., et al.: Deep q-learning from demonstrations. In: Thirty-Second AAAI Conference on Articial Intelligence (2018)
7. Kahn, G., Villaflor, A., Ding, B., Abbeel, P., Levine, S.: Self-supervised deep reinforcement learning with generalized computation graphs for robot navigation. In: IEEE International Conference on Robotics and Automation (ICRA), pp. 1–8. IEEE (2018)
8. Kahn, G., Villaflor, A., Pong, V., Abbeel, P., Levine, S.: Uncertainty-aware reinforcement learning for collision avoidance. arXiv preprint arXiv:1702.01182 (2017)
9. Kang, K., Belkhale, S., Kahn, G., Abbeel, P., Levine, S.: Generalization through simulation: integrating simulated and real data into deep reinforcement learning for vision-based autonomous flight. In: 2019 International Conference on Robotics and Automation (ICRA), pp. 6008–6014. IEEE (2019)
10. Konda, V.R., Tsitsiklis, J.N.: Actor-critic algorithms. In: Advances in Neural Information Processing Systems, pp. 1008–1014 (2000)
11. Lee, J.D., Simchowitz, M., Jordan, M.I., Recht, B.: Gradient descent only converges to minimizers. In: Conference on Learning Theory, pp. 1246–1257 (2016)
12. Lillicrap, T.P., et al.: Continuous control with deep reinforcement learning. In: International Conference on Learning Representations (ICLR) (2016)

13. Mirowski, P., et al.: Learning to navigate in cities without a map. In: Advances in Neural Information Processing Systems, pp. 2419–2430 (2018)
14. Mnih, V., et al.: Asynchronous methods for deep reinforcement learning. In: International Conference on Machine Learning (ICML), pp. 1928–1937 (2016)
15. Mnih, V., et al.: Human-level control through deep reinforcement learning. Nature **518**(7540), 529–533 (2015)
16. Nagabandi, A., Kahn, G., Fearing, R.S., Levine, S.: Neural network dynamics for model-based deep reinforcement learning with model-free fine-tuning. In: IEEE International Conference on Robotics and Automation (ICRA), pp. 7559–7566 (2018)
17. Nair, A., McGrew, B., Andrychowicz, M., Zaremba, W., Abbeel, P.: Overcoming exploration in reinforcement learning with demonstrations. In: IEEE IEEE International Conference on Robotics and Automation (ICRA), pp. 6292–6299. IEEE (2018)
18. Oh, J., Guo, Y., Singh, S., Lee, H.: Self-imitation learning. arXiv preprint arXiv:1806.05635 (2018)
19. Parisotto, E., Ba, J.L., Salakhutdinov, R.: Actor-mimic: deep multitask and transfer reinforcement learning. arXiv preprint arXiv:1511.06342 (2015)
20. Peters, J., Schaal, S.: Natural actor-critic. Neurocomputing **71**(7–9), 1180–1190 (2008)
21. Ross, S., Gordon, G., Bagnell, D.: A reduction of imitation learning and structured prediction to no-regret online learning. In: Proceedings of the Fourteenth International Conference on Artificial Intelligence and Statistics, pp. 627–635 (2011)
22. Silver, D., Lever, G., Heess, N., Degris, T., Wierstra, D., Riedmiller, M.: Deterministic policy gradient algorithms. In: International Conference on Machine Learning (ICML) (2014)
23. Sutton, R.S., et al.: Introduction to Reinforcement Learning. vol. 2. MIT press Cambridge (1998)
24. Tai, L., Paolo, G., Liu, M.: Virtual-to-real deep reinforcement learning: continuous control of mobile robots for mapless navigation. In: International Conference on Intelligent Robots and Systems (IROS), pp. 31–36. IEEE (2017)
25. Taylor, M.E., Kuhlmann, G., Stone, P.: Autonomous transfer for reinforcement learning. In: Proceedings of the International Joint Conference Autonomous Agents and Multiagent Systems, Vol. 1, pp. 283–290 (2008)
26. Taylor, M.E., Stone, P.: Representation transfer for reinforcement learning. In: AAAI Fall Symposium: Computational Approaches to Representation Change during Learning and Development, pp. 78–85 (2007)
27. Todorov, E., Erez, T., Tassa, Y.: Mujoco: a physics engine for model-based control. In: IEEE/RSJ International Conference Intelligent Robots and Systems, pp. 5026–5033 (2012)
28. Uhlenbeck, G.E., Ornstein, L.S.: On the theory of the brownian motion. Phys. Rev. **36**(5), 823 (1930)
29. Van Hasselt, H., Guez, A., Silver, D.: Deep reinforcement learning with double q-learning. In: AAAI Conference on Artificial Intelligence (2016)
30. Wayne, G., et al.: Unsupervised predictive memory in a goal-directed agent. arXiv preprint arXiv:1803.10760 (2018)
31. Zhang, J., Springenberg, J.T., Boedecker, J., Burgard, W.: Deep reinforcement learning with successor features for navigation across similar environments. In: IEEE/RSJ International Conference on Intelligent Robots and Systems (IROS), pp. 2371–2378 (2017)

Inferring Hierarchical Mixture Structures: A Bayesian Nonparametric Approach

Weipeng Huang[1]($\boxtimes$), Nishma Laitonjam[1], Guangyuan Piao[2],
and Neil J. Hurley[1]

[1] Insight Centre for Data Analytics, University College Dublin, Dublin, Ireland
{weipeng.huang,nishma.laitonjam,neil.hurley}@insight-centre.org
[2] Department of Computer Science, Maynooth University, Maynooth, Ireland
guangyuan.piao@mu.ie

Abstract. We present a Bayesian Nonparametric model for Hierarchical Clustering (HC). Such a model has two main components. The first component is the random walk process from parent to child in the hierarchy and we apply nested Chinese Restaurant Process (nCRP). Then, the second part is the diffusion process from parent to child where we employ Hierarchical Dirichlet Process Mixture Model (HDPMM). This is different from the common choice which is Gaussian-to-Gaussian. We demonstrate the properties of the model and propose a Markov Chain Monte Carlo procedure with elegantly analytical updating steps for inferring the model variables. Experiments on the real-world datasets show that our method obtains reasonable hierarchies and remarkable empirical results according to some well known metrics.

1 Introduction

We study the problem of Hierarchical Clustering (HC) via a Bayesian Nonparametric (BNP) modelling perspective. A BNP proposes a generative model for the observed data whose dimension is not fixed, but rather is learned from the data. In the case of HC, this allows the structure of the hierarchy to be inferred along with its clusters. Considering a model for generating the data, a BNP for HC typically associates each node in the hierarchy with a particular choice of parameters. Then the model consists of two components: 1) the random process for generating a path through the hierarchy from the root node to a leaf; 2) a parent-to-child transition kernel that models how a child node's parameters are related to those of its parent. The generative model posits that each observation firstly selects a path randomly, and is then sampled through the distribution associated with the leaf node of the path.

An early example of the parametric generative probabilistic model is the Gaussian tree generative process [6], such that each non-root node is sam-

Electronic supplementary material The online version of this chapter (https://doi.org/10.1007/978-3-030-75768-7_17) contains supplementary material, which is available to authorized users.

K. Karlapalem et al. (Eds.): PAKDD 2021, LNAI 12714, pp. 206–218, 2021.
https://doi.org/10.1007/978-3-030-75768-7_17

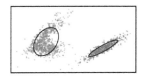
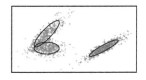

(a) Gaussian node   (b) Mixture node

Fig. 1. A simple example of the two nodes (distinguished by colour) at the first layer after the root node. Figure 1a depicts a manner of clustering the data with Gaussian-to-Gaussian node transition from root to the level 1, and Fig. 1b is the version of applying HDPMM. (Color figure online)

pled from a Gaussian distribution with the mean of the parent node and pre-defined level-wise covariances. In general, for the first component, there are some well-known random processes, e.g. the nested Chinese Restaurant Process (nCRP) [1,4], the Dirichlet Diffusion Tree (DDT) [21], the Pitman-Yor Diffusion Tree (PYDT) [15], which is a generalisation of the DDT, and the tree-structured stick-breaking construction (TSSB) [1], which generalises the (nCRP). It is proved in [15], that the nCRP and PYDT (and so also the DDT) are asymptotically equivalent—thus, this work focuses on the second component and selects the nCRP for the first component, given its simplicity. Existing BNP methods mostly utilise Gaussian-to-Gaussian (G2G) diffusion kernels for the node parameters [1,6,15,21]. Within such a setting, e.g., if a parent node's distribution (denoted by $\sim$) is Normal$(\boldsymbol{\mu}_p, \boldsymbol{\Sigma}_p)$, its child has the node parameters $\boldsymbol{\mu}_c \sim$ Normal$(\boldsymbol{\mu}_p, \boldsymbol{\Sigma}_p)$ and is associated with the data distribution Normal$(\boldsymbol{\mu}_c, \boldsymbol{\Sigma}_c)$ where $\boldsymbol{\Sigma}_p$ and $\boldsymbol{\Sigma}_c$ can be predefined hyperparameters.

G2G kernels are easy to apply but somehow lack the ability to handle more complex patterns. In our work, we investigate the Hierarchical Dirichlet Process Mixture Model (HDPMM) for the parent-to-node transition. In this setting, each node is actually a mixture distribution and the node parameters maintain the mixture weights for a global book of components. Then, the parent-to-node transition follows a Hierarchical Dirichlet Process (HDP) [28]. As a simple example, in Fig. 1, a single level of clustering is applied to the shown data points, using G2G (Fig. 1a) and HDPMM (Fig. 1b). The HDPMM fits a mixture model to the purple nodes, which can be further refined in lower levels of the tree. Note that, the HDP [29] is discussed in [1], for applications based on Latent Dirichlet Allocations. However, our setting HDPMM is more suited to general clustering.

Related Work. Our focus is on statistical methods, despite that there are many non-statistical methods for HC e.g. [5,16,19,30], to name just a few. Statistical extensions to Agglomerative clustering (AC) have been proposed in [11,12,18,26] but these are not generative models. On the other hand, Teh et al. [27] has proposed applying the Kingman's coalescent as a prior to the HC, which is similar in spirit to DDT and PYDT, however it is a backward generative process.

It is worth mentioning that the tree generation and transition kernels that we exploit have been used in a number of other contexts, beyond HC. For example, Paisley et al. [22] discuss the nested Hierarchical Dirichlet Process (nHDP) for topic modelling which, similar to the method we present here, uses the nCRP to navigate through the hierarchy, but associates nodes in the tree with topics. The nHDP first generates a global topic tree using the nCRP where each node relates to one atom (topic) drawn from the base distribution. In summary, nHDP constitutes multiple trees with one global tree and many local trees, whereas our construction has a single tree capturing the hierarchical structure of the data. Ahmed et al. [2] also proposed a model that is very similar to the nHDP, but appealed to different inference procedures.

2   Preliminary

Dirichlet Process. We briefly describe the CRP and the stick-breaking process which are two forms of the Dirichlet Process (DP).

In the CRP, we imagine a Chinese restaurant consisting of an infinite number of tables, each with sufficient capacity to seat an infinite number of customers. A customer enters the restaurant and picks one table at which to sit. The nth customer picks a table based on the previous customers' choices. That is, assuming c_n is the table assignment label for customer n and N_k is the number of customers at table k, one obtains

$$p(c_{n+1} = k \mid c_{1:n}) = \begin{cases} \frac{N_k}{n+\alpha} & \text{existing } k \\ \frac{\alpha}{n+\alpha} & \text{new } k \end{cases} \quad \theta^*_{n+1} \mid \boldsymbol{\theta}^*_{1:n} \sim \frac{\alpha}{n+\alpha}H + \sum_{k=1}^{K} \frac{N_k}{n+\alpha}\delta_{\theta_k}$$

where $\boldsymbol{c}_{1:n} = \{c_1, \ldots, c_n\}$, likewise for $\boldsymbol{\theta}^*_{1:n}$. The right hand side indicates how the parameter θ^*_{n+1} is drawn given the previous parameters, where each θ_k is sampled from a base measure H, and $\theta_1, \ldots, \theta_K$ are the unique values among $\theta^*_1, \ldots, \theta^*_n$.

Denoting a distribution by G, the stick-breaking process can be depicted by $G = \sum_{k=1}^{\infty} \beta_k \delta_{\theta_k}$, $\{\theta_k\}_{k=1}^{\infty} \sim H$ and $\boldsymbol{\beta} \sim \text{GEM}(\alpha)$. Also, GEM (named after Griffiths, Engen and McCloskey), known as a stick-breaking process, is analogous to iteratively breaking a portion from the remaining stick which has the initial length 1. In particular, we write $\boldsymbol{\beta} \sim \text{GEM}(\alpha)$ when $u_k \sim \text{Beta}(1, \alpha)$, $\beta_1 = u_1$, and $\beta_k = u_k \prod_{l=1}^{k-1}(1 - u_l)$.

Nested CRP. In the nCRP [4], customers arrive at a restaurant and choose a table according to the CRP, but at each chosen table, there is a card leading to another restaurant, which the customer visits the next day, again using the CRP. Each restaurant is associated with only a single card. After L days, the customer has visited L restaurants, by choosing a particular path in an infinitely branching hierarchy of restaurants.

Hierarchical Dirichlet Process Mixture Model. When the number of mixture components is infinite, we connect components along a path in the hierarchy using a HDP. The 1-level HDP [29] connects a set of DPs, G_j, to a common base DP, G_0. It can be simply written as $G_j \sim \mathrm{DP}(\gamma, G_0)$ and $G_0 \sim \mathrm{DP}(\gamma_0, H)$. It has several equivalent representations while we will focus on the following form: $\beta_0 \sim \mathrm{GEM}(\gamma_0)$, $\beta_j \sim \mathrm{DP}(\gamma, \beta_0)$, and $\theta_k \sim H$ to obtain $G_0 = \sum_{k=1}^{\infty} \beta_{0k} \delta_{\theta_k}$ and then $G_j = \sum_k \beta_{jk} \delta_{\theta_k}$. Hence G_j has the same components as G_0 but with different mixing proportions. It may be shown that β_j can be sampled by firstly drawing $u_{jk} \sim \mathrm{Beta}(\gamma \beta_k, \gamma(1 - \sum_{\ell=1}^{k} \beta_\ell))$ and then $\beta_{jk} = u_{jk} \prod_{\ell=1}^{k-1}(1 - u_{j\ell})$. Considering each data in x belongs to one of the mixture models and in particular x_n belongs to the group associated with G_j, the 1-level HDPMM is completed by having $x_n \sim F(\theta_{c_n})$ where $c_n \sim \mathrm{Categorical}(\beta_j)$. This process can be extended to multiple levels, by defining another level of DPs with G_j as a base distribution and so on to higher levels. In fact, we can build a hierarchy where, for any length L path in the hierarchy, the nodes in the path correspond to an L-level HDP.

3   Generative Process

We call our model BHMC, for Bayesian Hierarchical Mixture Clustering and illustrate it using Fig. 2.

Let z be the label for a certain node in the tree. We can denote the probability to choose the first child under z by w_{z1}. Also, let us denote the mixing proportion for a node z by β_z, and denote the global component assignment for the nth data item, x_n, by c_n. To draw x_n, the hierarchy is traversed to a leaf node, z, at which c_n is drawn from mixing proportions β_z. A path through the hierarchy is denoted by a vector e.g. $v = \{z_0, z_1, z_4\}$. In a finite setting, with a fixed number of branches, such a path would be generated by sampling from

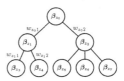

Fig. 2. One example of a BHMC hierarchy

weights $w_{z_0} \sim \mathrm{Dir}(\alpha)$, at the first level, then $w_{z_1} \sim \mathrm{Dir}(\alpha)$ at the second level and so on. The nCRP enables the path to be sampled from infinitely branched nodes. Mixing proportions, β_z, along a path are connected via a multilevel HDP. Hence, if the probability of a mixture component goes to zero at any node in the tree, it will remain zero for any descendant nodes. Moreover, the smaller γ is, the sparser the resulting distribution drawn from the HDP [20].

The generative process with an infinite configuration is shown in Algorithm 1. For a finite setting, the mixing proportions are on a finite set and Line 1 should be changed to $\beta_{z_0} \sim \mathrm{Dir}(\gamma_0/K, \ldots, \gamma_0/K)$. Correspondingly, Line 9 has to be changed to $\beta_{z'} \sim \mathrm{Dir}(\gamma \beta_z)$ according to the preliminary section.

3.1   Properties

Let us denote the tree by $\mathcal{H} = (Z_\mathcal{H}, E_\mathcal{H})$. First, the node set is $Z_\mathcal{H} = \{z_0, \ldots, z_{M-1}\}$ where M is the number of nodes. Then, $E_\mathcal{H}$ is the set of edges in the tree, where $(z, z') \in E_\mathcal{H}$ means that there exists some path v_n, such that x_n moves from z to z' in the path v_n. Hence, we infer the following variables: $\mathbf{B} = \{\beta_{z_0}, \ldots, \beta_{z_{M-1}}\}$, the mixing proportions for the components in the node z; $\theta = \{\theta_1, \ldots, \theta_K\}$, component parameters; $\mathbf{V} = \{v_1, \ldots, v_N\}$ where each $v_n = \{v_{n0}, v_{n1}, \ldots, v_{nL}\}$ is the ordered set of nodes in $Z_\mathcal{H}$ corresponding to the path of x_n; $c = \{c_1, \ldots, c_N\}$, the component label for all the observations. Denoting $\Phi = \{\gamma_0, \gamma, \alpha, H, L\}$, we focus on the marginal prior $p(\mathbf{V}, \theta, \mathbf{B} \mid \Phi)$ and obtain $p(\mathbf{V}, \theta, \mathbf{B} \mid \Phi) = p(\mathbf{B} \mid \mathbf{V}, \gamma_0, \gamma)p(\mathbf{V} \mid \alpha, L)p(\theta \mid H)$.

The first term is $p(\mathbf{B} \mid \mathbf{V}, \gamma_0, \gamma) = p(\beta_0 \mid \gamma_0) \prod_{(z,z') \in E_\mathcal{H}} p(\beta_{z'} \mid \gamma, \beta_z)$. To expand the second term, we first denote by m_z the number of children of z. Hence, the CRP probability for a set of clusters $\{z'_n\}_{n=1}^{m_z}$ under the same parent z [3,8]:

$$p(\{z'_n\}_{n=1}^{m_z} \mid \alpha) = \frac{\alpha^{m_z} \Gamma(\alpha)}{\Gamma(N_z + \alpha)} \prod_{n=1}^{m_z} \Gamma\left(N_{z'_n}\right)$$

where N_z is the number of observations in z. With this equation, one can observe the exchangeability of the order of arriving customers—the probability of obtaining such a partition is not dependent on the order. The tree $\mathcal{H}$ is constructed via $\mathbf{V}$ with the empty nodes all removed. For such a tree, the above result can be extended to

$$p(\mathbf{V} \mid \alpha) = \Gamma(\alpha)^{|\mathcal{I}_\mathcal{H}|} \prod_{z \in \mathcal{I}_\mathcal{H}} \frac{\alpha^{m_z}}{\Gamma(N_z + \alpha)} \prod_{(z,z') \in E_\mathcal{H}} \Gamma(N_{z'})$$

where $\mathcal{I}_\mathcal{H}$ is the set of internal nodes in $\mathcal{H}$. Next, we obtain $p(\theta \mid H) = \prod_{k=1}^K p(\theta_k \mid H)$. The likelihood for a single observation is

$$p(x_n \mid v_{nL}, \theta, \beta_{v_{nL}}) = \sum_{k=1}^K \beta_{v_{nL}k} f(x_n; \theta_k) + \beta^*_{v_{nL}} f^*(x_n) \tag{1}$$

where β^*_z denotes the probability of drawing a new component which is always the last element in the vector β_z. Here, $f(\cdot)$ is the corresponding density function for the distribution F and we obtain $f(x; \theta) \equiv p(x \mid \theta)$. Furthermore, $f^*(x) = \int p(x \mid \theta)dp(\theta \mid H)$. The above presentation is similar to the Polyá urn construction of DP [23], which trims the infinite setting of DP to a finite configuration. Then, one can see $p(\mathbf{X} \mid \mathbf{V}, \theta, \mathbf{B}) = \prod_n p(x_n \mid v_{nL}, \theta, \beta_{v_{nL}})$. Finally, the unnormalised posterior is $p(\mathbf{V}, \theta, \mathbf{B} \mid \mathbf{X}, \Phi) \propto p(\mathbf{X} \mid \mathbf{V}, \theta, \mathbf{B})p(\mathbf{V}, \theta, \mathbf{B} \mid \Phi)$.

Algorithm 1: Generative process (infinite)	**Algorithm 2:** MH sampler
	`// ε: stopping threshold`
1 Sample $\boldsymbol{\beta}_{z_0} \sim \text{GEM}(\gamma_0)$	**1** Sample $\boldsymbol{\beta}_{z_0}$ until $\beta_{z_0}^* < \epsilon$
2 Sample $\theta_1, \theta_2, \ldots \sim H$	**2** **for** $\boldsymbol{x}_n \in \text{SHUFFLED}(\mathbf{X})$ **do**
3 **for** $n = 1 \ldots N$ **do**	**3** $\quad$ Clean up c_n and $\boldsymbol{v}_n$
4 $\quad$ $v_{n0} \leftarrow z_0$	**4** $\quad$ Sample $\hat{\boldsymbol{v}}_n$ (and possibly
5 $\quad$ **for** $\ell = 1 \ldots L$ **do**	$\quad\quad$ new $\boldsymbol{\beta}$) through the
6 $\quad\quad$ Sample $v_{n\ell}$ using CRP(α)	$\quad\quad$ generative process
7 $\quad\quad$ $z, z' \leftarrow v_{n(\ell-1)}, v_{n\ell}$	**5** $\quad$ $s \sim \text{Unif}(0,1)$
8 $\quad\quad$ **if** z' *is new* **then**	**6** $\quad$ **if** $s \le \mathcal{A}$ **then** $\boldsymbol{v}_n \leftarrow \hat{\boldsymbol{v}}_n$
9 $\quad\quad\quad$ Sample $\boldsymbol{\beta}_{z'} \sim \text{DP}(\gamma, \boldsymbol{\beta}_z)$	
10 $\quad\quad\quad$ Attach (z, z') to the tree	**7** $\quad$ Sample c_n using a
	$\quad\quad$ Gibbs step by Eq.(2)
11 $\quad$ Sample $c_n \sim \text{Categorical}(\beta_{v_{nL}})$	**8** Update $\mathbf{B}$ by Eqs.(3) and (4)
12 $\quad$ Sample $\boldsymbol{x}_n \sim F(\theta_{c_n})$	**9** Update $\boldsymbol{\theta}$ by Eq.(5)

4   Inference

We appeal to Markov Chain Monte Carlo (MCMC) for inferring the model. One crucial property for facilitating the sampling procedure is exchangeability, which, as noted in the previous section, follows from the model's connection to the CRP.

Sampling $\mathbf{V}$ *and* $\boldsymbol{c}$. Following [4], we sample a path for a data index as a complete variable using nCRP and decide to preserve the change based on a Metropolis-Hastings (MH) step.

Recall that the component will only be drawn at the leaf. Thus,

$$p\left(c_n = k \mid \boldsymbol{x}_n, v_{nL} = z, \mathbf{B}, \boldsymbol{\theta}\right) \propto \begin{cases} \beta_{zk} f(\boldsymbol{x}_n; \theta_k) & \text{existing } k \\ \beta_z^* f^*(\boldsymbol{x}_n) & \text{new } k. \end{cases} \quad (2)$$

Sampling the set $\boldsymbol{c}$ is necessary for updating the set of mixtures at each node i.e., $\mathbf{B}$. Our MH scheme applies a partially collapsed Gibbs step. Following the principles in [7], our algorithm first samples $\mathbf{V}$ with $\boldsymbol{c}$ being collapsed out, and subsequently updates $\boldsymbol{c}$ based on $\mathbf{V}$.

Sampling $\mathbf{B}$. It is straightforward to decide the sampling for a leaf node. For non-root nodes, we would like to find out $p\left(c_n = k \mid \mathbf{B} \backslash \{\mathbf{B}_{\bar{z}}\}, v_{n(L-1)} = z\right)$ where $\mathbf{B}_{\bar{z}}$ is the set of mixing proportions in the sub-tree rooted at z excluding $\boldsymbol{\beta}_z$. We write $p\left(c_n = k \mid \mathbf{B} \backslash \{\mathbf{B}_{\bar{z}}\}, v_{n(L-1)} = z\right)$ to be

$$p\left(v_{nL} = z' \mid v_{n(L-1)} = z\right) p\left(c_n = k \mid \boldsymbol{\beta}_z, v_{nL} = z', v_{n(L-1)} = z\right).$$

Logically, $p(c_n = k \mid \mathbf{B}, \boldsymbol{v}_n) = p\left(c_n = k \mid \{\boldsymbol{\beta}_{v_{n\ell}}\}_{\ell=0}^{L}\right)$ which is then $p\left(c_n = k \mid \boldsymbol{\beta}_{v_{nL}}\right)$. We also derive $p\left(c_n = k \mid \boldsymbol{\beta}_z, v_{nL} = z', v_{n(L-1)} = z\right)$ to be

$$\int p(c_n = k \mid \boldsymbol{\beta}_{z'}) dp(\boldsymbol{\beta}_{z'} \mid \boldsymbol{\beta}_z) = \int p\left(c_n = k \mid \boldsymbol{\beta}_{z'}\right) \frac{\Gamma\left(\sum_{j=1}^{K} \gamma\beta_{zj}\right)}{\prod_{j=1}^{K} \Gamma(\gamma\beta_{zj})} \prod_{j=1}^{K} \beta_{z'k}^{\gamma\beta_{zk}-1} d\boldsymbol{\beta}_{z'}$$

$$= \frac{\Gamma\left(\sum_{j=1}^{K} \gamma\beta_{zj}\right)}{\prod_{j=1}^{K} \Gamma(\gamma\beta_{zj})} \int \prod_{j=1}^{K} \beta_{z'j}^{\mathbb{1}\{c_n=k\}+\gamma\beta_{zk}-1} d\boldsymbol{\beta}_{z'}$$

$$= \frac{\Gamma\left(\sum_{j=1}^{K} \gamma\beta_{zj}\right)}{\prod_{j=1}^{K} \Gamma(\gamma\beta_{zj})} \frac{\prod_{j=1}^{K} \Gamma(\mathbb{1}\{c_n = k\} + \gamma\beta_{zj})}{\Gamma(1 + \sum_{j=1}^{K} \gamma\beta_{zj})} = \beta_{zk}$$

given that $\Gamma(x + 1) = x\Gamma(x)$ holds when x is any complex number except the non-positive integers. Therefore, $p\left(c_n = k \mid \mathbf{B}\backslash\{\mathbf{B}_{\bar{z}}\}, v_{n(L-1)} = z\right) = \beta_{zk}$. This indicates that, marginalising out the subtree rooted at z, the component assignment is thought to be drawn from $\boldsymbol{\beta}_z$, which can be seen through induction. Therefore, it allows us to conduct the size-biased permutation,

$$\text{root } z_0 \qquad \beta_{z_0 1}, \ldots, \beta_{z_0 K}, \beta_{z_0}^* \sim \text{Dir}\left(N_{z_0 1}, \ldots, N_{z_0 K}, \gamma_0\right) \qquad (3)$$

$$\forall z' : (z, z') \in E_{\mathcal{H}} \qquad \beta_{z' 1}, \ldots, \beta_{z' K}, \beta_{z'}^* \sim \text{Dir}\left(\tilde{N}_{z' 1}, \ldots, \tilde{N}_{z' K}, \gamma\beta_z^*\right) \qquad (4)$$

where $\tilde{N}_{z'k} = N_{z'k} + \gamma\beta_{zk}$. Eq. (3) employs a Polyá urn posterior construction of the DP to preserve the exchangeability when carrying out the size-biased permutation [23]. This step in our inference enables an analytical form of node parameter update.

Sampling $\boldsymbol{\theta}$. Even though there are many options for H, we choose a Gaussian distribution in this paper. With $\theta_k := \boldsymbol{\mu}_k$, we define $f(\boldsymbol{x}; \boldsymbol{\mu}_k) = \text{Normal}(\boldsymbol{x}; \boldsymbol{\mu}_k, \boldsymbol{\Sigma})$ and $H = \text{Normal}(\boldsymbol{\mu}_0, \boldsymbol{\Sigma}_0)$, where covariance matrices $\boldsymbol{\Sigma}$ and $\boldsymbol{\Sigma}_0$ are known and fixed. Collapsing out the unused terms, we can write $p(\theta_k \mid \mathbf{X}, \boldsymbol{c}) \sim p(\theta_k \mid H) \prod_{n:c_n=k} p(\boldsymbol{x}_n \mid \theta_k)$. Given the conjugacy of a Gaussian prior with a Gaussian of known covariance, by considering $\bar{\boldsymbol{x}}_k = \frac{1}{N_k} \sum_n \mathbb{1}\{c_n = k\}\boldsymbol{x}_n$ and $N_K = \sum_n \mathbb{1}\{c_n = k\}$, we obtain $\boldsymbol{\mu}_k \mid \mathbf{X}, \boldsymbol{c} \sim \text{Normal}(\tilde{\boldsymbol{\mu}}_k, \tilde{\boldsymbol{\Sigma}}_k)$ where

$$\tilde{\boldsymbol{\mu}}_k = \tilde{\boldsymbol{\Sigma}}_k(\boldsymbol{\Sigma}_0^{-1}\boldsymbol{\mu}_0 + N_k\boldsymbol{\Sigma}^{-1}\bar{\boldsymbol{x}}_k) \qquad \tilde{\boldsymbol{\Sigma}}_k = (\boldsymbol{\Sigma}_0^{-1} + N_k\boldsymbol{\Sigma}^{-1})^{-1}. \qquad (5)$$

4.1   Algorithmic Procedure

In practice, one useful step of the inference is to truncate the infinite setting of $\boldsymbol{\beta}_{z_0}$ to a finite setting. Referring back to Eq. (1), one can have a threshold such that the sampling of $\boldsymbol{\beta}_{z_0}$ terminates when the remaining length of the stick $\beta_{z_0}^*$ is shorter than that threshold. Once a new component is initialised, each node will update its $\boldsymbol{\beta}$ by one more stick-breaking step. That is, for the root node, it samples one u from $\text{Beta}(1, \gamma_0)$, and assigns $\beta_{z_0(K+1)} = \beta_{z_0}^* u$ and the remaining stick length $1 - \sum_{k=1}^{K+1} \beta_{z_0 k}$ as a new $\beta_{z_0}^*$. For a non-root node z' inheriting from node z, we apply the results from the preliminary section such that u is sampled from $\text{Beta}(\gamma\beta_{z(K+1)}, \gamma\beta_z^*)$ and then update in the same manner as the root node.

Algorithm 2 depicts the procedure. The MH scheme samples a proposal path and the corresponding β's if new nodes are initialised. MH considers an acceptance variable $\mathcal{A}$ such that $\mathcal{A} = \min\left\{1, \frac{\mathcal{P}'\,q(\mathbf{V},\mathbf{B}|\mathbf{V}',\mathbf{B}')}{\mathcal{P}\,q(\mathbf{V}',\mathbf{B}'|\mathbf{V},\mathbf{B})}\right\}$. Here, $\mathcal{P}$ and $\mathcal{P}'$ are the posteriors at the current and proposed states, respectively. Then, q is the proposal for sampling $\mathbf{V}'$ and $\mathbf{B}'$, which in our case is the nCRP and HDP. At each iteration for a certain data index n, $\mathbf{V}$ changes to $\mathbf{V}'$ by replacing $\boldsymbol{v}_n$ with $\boldsymbol{v}'_n$. Thus, in this example, $q(\mathbf{V}' \mid \mathbf{V}) = \text{nCRP}(\boldsymbol{v}'_n; \mathbf{V}\backslash\{\boldsymbol{v}_n\})$ and vice versa. The term $q(\mathbf{B} \mid \mathbf{V})/q(\mathbf{B}' \mid \mathbf{V}')$ is cancelled out by the terms $p(\mathbf{B}' \mid \mathbf{V}')/p(\mathbf{B} \mid \mathbf{V})$ in the posterior, as $q(\mathbf{B} \mid \mathbf{V})$ and $p(\mathbf{B} \mid \mathbf{V})$ are identical. Apart from that, $\boldsymbol{\theta}$ will be updated only after all the paths are decided, and hence gain no changes. Therefore, for a specific $\boldsymbol{x}_n$, we have that

$$\mathcal{A} = \min\left\{1, \frac{p(\boldsymbol{x}_n \mid \beta_{v'_{nL}}, v'_{nL}, \boldsymbol{\theta})p(\mathbf{V}')}{p(\boldsymbol{x}_n \mid \beta_{v_{nL}}, v_{nL}, \boldsymbol{\theta})p(\mathbf{V})}\frac{\text{nCRP}(\boldsymbol{v}_n; \mathbf{V}' \setminus \{\boldsymbol{v}'_n\})}{\text{nCRP}(\boldsymbol{v}'_n; \mathbf{V} \setminus \{\boldsymbol{v}_n\})}\right\}$$

given that the likelihood for $\mathbf{X}\backslash\{\boldsymbol{x}_n\}$ remains unaltered. After the paths $\mathbf{V}$ for all the observations are sampled, the process updates $\mathbf{B}$ and $\boldsymbol{\theta}$ using the manner discussed above.

4.2 Time Complexity

Contrary to first impression, at each iteration, the algorithm complexity is only N^2 for the worst case, and log-linear for the expected case.

Assume that the maximum number of children is $M_{z_\ell}^*$ at the level ℓ, then the cost of sampling a path for a single observation with the nCRP is $O(\sum_{\ell=1}^{L} M_{z_\ell}^*)$. After that, sampling c_n is carried out with time $O(Kg(D))$ where $g(D)$ is the time for computing the Gaussian likelihood of D-dimensional data. In regard to the global variables, $\mathbf{B}$ will be updated with every node and thus achieves $O(K|Z_{\mathcal{H}}|)$. Lastly, $\boldsymbol{\theta}$ will be updated with time $O(Kg_s(D))$ where $g_s(D)$ is the complexity of sampling the Gaussian mean. In addition, we notice $\sum_{\ell=1}^{L} M_{z_\ell}^* = O(|Z_{\mathcal{H}}|)$ and $g(D) = O(g_s(D))$. Overall, for one iteration, we can summarise the complexity by $N\sum_{\ell=1}^{L} M_{z_\ell}^* + K|Z_{\mathcal{H}}| + NKg_s(D) = O((N + K)|Z_{\mathcal{H}}| + NKg_s(D))$. Since the number of components will not exceed the data size, $K \leq N$ holds. With respect to $|Z_{\mathcal{H}}|$, at each level, it is no more than N. The extreme case is that each datum is a node and extends to L levels. It follows that $|Z_{\mathcal{H}}| \leq NL$. Hence, $O((N + K)|Z_{\mathcal{H}}| + NKg_s(D)) = O\left(N^2(L + g_s(D))\right)$.

The expected number of DP components, considering γ_0, is $\gamma_0 \log N$ for sufficiently large N [20]. Likewise for the first level in the nCRP, which is then $\alpha \log N$. Let $\mathbb{E}[\cdot]$ denote the expectation over all random $\mathcal{H}$ drawn from the BHMC. If N is sufficiently large, we can have $\sum_{\ell=1}^{L} \mathbb{E}[M_{z_\ell}] = O(L\alpha \log N)$, as the expected number of nodes will be no greater than $\alpha \log N$. However, $\mathbb{E}[|Z_{\mathcal{H}}|]$ is hard to decide but is known to be $\leq \min(NL, (\alpha \log N)^L)$. If $N \gg 0$ and $\alpha > 0$, then $NL \leq (\alpha \log N)^L$. We can obtain $N(\sum_{\ell=1}^{L} \mathbb{E}[M_z] + \mathbb{E}[K]g_s(D)) =$

$O(N \log N(\alpha L + \gamma_0 g_s(D)))$ and $\mathbb{E}[K|Z_{\mathcal{H}}|] = O(\gamma_0 \log N \times NL)$. Combining the two expressions, we derive the upper bound $O(N \log N(L + g_s(D)))$ for the average case.

5 Experiments

For qualitative analysis, we use the datasets: Animals [14] and MNIST-fashion [31]. Following this, we carry out a quantitative analysis for the Amazon text data [10] since it contains the cluster labels of all the items at multiple levels.

We fix the parameters for H as $\boldsymbol{\mu}_0 = \mathbf{0}$ and $\boldsymbol{\Sigma}_0 = \mathbf{I}$. Additionally, we set the covariance matrix for F as $\boldsymbol{\Sigma} = \sigma^2 \mathbf{I}$. We set the number of levels L intuitively based on the data. For more complex cases, one can consider the theoretical results of the effective length of the nCRP [25].

5.1 Convergence Analysis

We examine the convergence of the algorithm using Animals. Figure 3 shows the unnormalised log likelihood for five individual simulations. The simulations quickly reach a certain satisfactory level. Apart from that, the fluctuation shows that the algorithm keeps searching the solution space over the iterations.

5.2 Results

Animals. This dataset contains 102 binary features, e.g. "has 6 legs", "lives in water", "bad temper", etc. Observing the heat-map of the empirical covariance of the data, there are not many influential features. Hence, we employ Principal Component Analysis (PCA) [20] to reduce it to a seven-dimensional feature space. The MCMC burns 500 runs and then reports the one with the greatest complete data likelihood $p(\mathbf{X}, c, \mathbf{V} \mid \mathbf{B}, \boldsymbol{\theta})$ amongst the following 5,000 draws (following [1]).

We apply the hyperparameters: $\alpha = 0.4, \gamma_0 = 1, \gamma = 0.5, \sigma^2 = 1, L = 4$. Figure 4 shows a rather intuitive hierarchical structure. From the left to the right, there are insects, (potentially) aggressive mammals, herbivores, water animals, and birds.

MNIST-Fashion. This data is a collection of fashion images. Each image is represented as a 28×28 vector of grayscale pixel values. For better visualisation, we sample 100 samples evenly from each class. PCA transforms the data to 22 dimensions via the asymptotic root mean square optimal threshold [9] for keeping the singular values. Using the same criterion as for Animal, we output two hierarchies with two sets of hyperparameters. We set $\alpha = 0.5, \gamma_0 = 1.5, \gamma = 2, \sigma^2 = 1, L = 5$. This follows precisely the same running settings as for Animals.

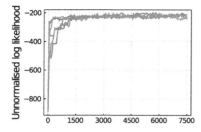

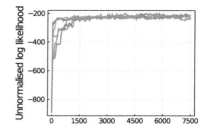

Fig. 3. Convergence analysis   **Fig. 4.** The tree of `Animals`

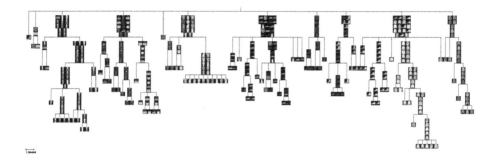

Fig. 5. The tree of `MNIST-fashion`

Figure 5 reflects a property of the CRP which is "the rich get richer". As it is grayscale data, in addition to the shape of the items, other factors affecting the clustering might be, e.g., the foreground/background colour area, the percentage of non-black colours in the image, the darkness/lightness of the item, etc. The hierarchical structure forms a hierarchy with high purity per level. Some mis-labelled items are expected in a clustering task.

Amazon. We uniformly down-sample the indices in the fashion category of `Amazon` and reserve $2,303$ entries from the data, which contain textual information about items such as their titles and descriptions[1]. The data is preprocessed via the method in [9] and only 190 features are kept, formed as the Term Frequency and Inverse Document Frequency (TF-IDF) of the items' titles and descriptions.

In this study, we adopt a Bayesian approach to dealing with the hyperparameters. Due to a compromise to the runtime efficiency, we run BHMC once with 500 runs for burn-in, during which a MH-based hyperparameter sampling procedure is performed with hyperpriors $\alpha \sim \mathrm{Ga}(3,1), \gamma_0 \sim \mathrm{Ga}(3.5,1), \gamma \sim \mathrm{Unif}(0,1.5)$ and $\sigma^2 \sim \mathrm{Unif}(0,0.1)$. We limit σ^2 as mentioned since the feature values in the data are all less than 1 and some are far less. In addition, L is fixed to be 7 (presuming that we have little information about the real number of levels).

---

[1] http://jmcauley.ucsd.edu/data/amazon: the data is available upon request.

The tree with the maximal complete likelihood in the subsequent 2,500 draws is reported.

We use the evaluation methodology in [17] to compare the clustering results against the ground-truth labels level by level. We compare 6 levels, which is the maximum branch length of the items in the ground truth. When extracting the labels from the trees (either for the algorithm outputs or the ground truth), items that are on a path of length less than 6 are extended to level 6 by assigning the same cluster label as that in their last level to the remaining levels. This is to keep the consistency of the number of items for computing metrics at each level.

For the comparison, we first consider the gold standard AC with Ward distance [30]. We adopt the existing implementations for PERCH and PYDT[2]. For PYDT which is sensitive to the hyperparameters, we applied the authors' implemented hyperparameter optimisation to gather the hyperparameters prior to running the repeated simulations. At each level, we consider four different evaluation metrics, namely, the purity, the normalised mutual information (NMI), the adjusted rand index (ARI), and the F-Measure [13]. Level 1 is the level for the root node.

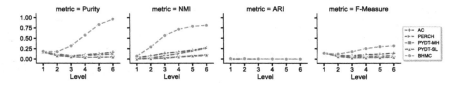

Fig. 6. Metrics on Amazon by levels

Figure 6 depicts that our method achieves clearly better scores with respect to purity and NMI. In the figure, PYDT-SL and PYDT-MH correspond to the slice and the MH sampling solutions, respectively. As the tree approaches to a lower level, our method also achieves a better performance in F-measure. For ARI, despite that PERCH performs the best, all numerical values are exceedingly close to 0. However, some theoretical work of [24] suggests that ARI is more preferred in the scenario that the data contains big and equal-sized clusters. This is opposed to our ground truth which is highly unbalanced among the clusters at each level. BHMC does show the potential to perform well according to certain traditional metrics.

6 Conclusion

This paper has discussed a new perspective for Bayesian nonparametric HC. Our model, BHMC, develops an infinitely branching hierarchy of mixture parameters,

---

[2] https://github.com/iesl/xcluster and https://github.com/davidaknowles/pydt

that are linked along paths in the hierarchy through a multilevel HDP. A nested CRP is used to select a path in the hierarchy and mixture components are drawn from the mixture distribution in the leaf node of the selected path. The evaluation shows that BHMC is able to provide good hierarchical clustering results on three real-world datasets with different types of characteristics (i.e., binary, visual and textual) and clearly performs better than other methods with respect to purity and NMI on the Amazon dataset with ground truth, which shows the promising potential of the model.

Acknowledgements. We thank the reviewers for the helpful feedback. This research has been supported by SFI under the grant SFI/12/RC/2289_P2.

References

1. Adams, R.P., Ghahramani, Z., Jordan, M.I.: Tree-structured stick breaking for hierarchical data. In: NeurIPS, pp. 19–27 (2010)
2. Ahmed, A., Hong, L., Smola, A.J.: Nested Chinese restaurant franchise processes: applications to user tracking and document modeling. In: ICML, vol. 28, pp. 2476–2484 (2013)
3. Antoniak, C.E.: Mixtures of Dirichlet processes with applications to Bayesian nonparametric problems. Ann. Stat. **2**, 1152–1174 (1974)
4. Blei, D.M., Griffiths, T.L., Jordan, M.I.: The nested Chinese restaurant process and Bayesian nonparametric inference of topic hierarchies. JACM **57**, 1–30 (2010)
5. Charikar, M., Chatziafratis, V., Niazadeh, R.: Hierarchical clustering better than average-linkage. In: SODA, pp. 2291–2304. SIAM (2019)
6. Williams, C.K.I.: A MCMC approach to hierarchical mixture modelling. In: Advances in Neural Information Processing Systems, vol. 12, pp. 680–686 (2000)
7. Dyk, D.A.V., Jiao, X.: Metropolis-hastings within partially collapsed Gibbs samplers. J. Comput. Graph. Stat. **24**(2), 301–327 (2015)
8. Ferguson, T.S.: A Bayesian analysis of some nonparametric problems. Ann. Stat. **1**, 209–230 (1973)
9. Gavish, M., Donoho, D.L.: The optimal hard threshold for singular values is $4/\sqrt{3}$. IEEE Trans. Inf. Theory **60**(8), 5040–5053 (2014)
10. He, R., McAuley, J.: Ups and downs: modeling the visual evolution of fashion trends with one-class collaborative filtering, WWW 2016, pp. 507–517 (2016)
11. Heller, K.A., Ghahramani, Z.: Bayesian hierarchical clustering. In: Proceedings of the 22nd International Conference on Machine Learning, pp. 297–304 (2005)
12. Iwayama, M., Tokunaga, T.: Hierarchical Bayesian clustering for automatic text classification. In: IJCAI, vol. 2, pp. 1322–1327 (1995)
13. Karypis, M.S.G., Kumar, V.: A comparison of document clustering techniques. In: KDD Workshop on Text Mining (2000)
14. Kemp, C., Tenenbaum, J.B.: The discovery of structural form. Proc. Natl. Acad. Sci. **105**(31), 10687–10692 (2008)
15. Knowles, D.A., Ghahramani, Z.: Pitman yor diffusion trees for Bayesian hierarchical clustering. IEEE TPAMI **37**(2), 271–289 (2015)
16. Kobren, A., Monath, N., Krishnamurthy, A., McCallum, A.: A Hierarchical algorithm for extreme clustering. In: SIGKDD, pp. 255–264. ACM (2017)
17. Kuang, D., Park, H.: Fast Rank-2 nonnegative matrix factorization for hierarchical document clustering. In: SIGKDD, pp. 739–747. ACM (2013)

18. Lee, J., Choi, S.: Bayesian hierarchical clustering with exponential family: small-variance asymptotics and reducibility. In: AISTATS, pp. 581–589 (2015)
19. Monath, N., Zaheer, M., Silva, D., McCallum, A., Ahmed, A.: Gradient-based hierarchical clustering using continuous representations of trees in hyperbolic space, pp. 714–722 (2019). https://doi.org/10.1145/3292500.3330997
20. Murphy, K.P.: Machine Learning: A Probabilistic Perspective. MIT Press, Cambridge (2012)
21. Neal, R.M.: Density modeling and clustering using Dirichlet diffusion trees. Bayesian Stat. **7**, 619–629 (2003)
22. Paisley, J., Wang, C., Blei, D.M., Jordan, M.I.: Nested hierarchical Dirichlet processes. TPAMI **37**(2), 256–270 (2015)
23. Pitman, J.: Some developments of the Blackwell-MacQueen Urn scheme. Lect. Notes-Monogr. Ser. **30**, 245–267 (1996)
24. Romano, S., Vinh, N.X., Bailey, J., Verspoor, K.: Adjusting for chance clustering comparison measures. JMLR **17**(1), 4635–4666 (2016)
25. Steinhardt, J., Ghahramani, Z.: Flexible martingale priors for deep hierarchies (2012)
26. Stolcke, A., Omohundro, S.: Hidden Markov model induction by Bayesian model merging. In: Advances in Neural Information Processing Systems, pp. 11–18 (1993)
27. Teh, Y.W., Daume III, H., Roy, D.M.: Bayesian agglomerative clustering with coalescents. In: NeurIPS, pp. 1473–1480 (2008)
28. Teh, Y.W., Jordan, M.I.: Hierarchical Bayesian nonparametric models with applications. Bayesian Nonparametrics **1**, 158–207 (2010)
29. Teh, Y.W., Jordan, M.I., Beal, M.J., Blei, D.M.: Hierarchical Dirichlet processes. J. Am. Stat. Assoc. **101**(476), 1566–1581 (2006)
30. Ward Jr., J.H.: Hierarchical grouping to optimize an objective function. J. Am. Stat. Assoc. **58**(301), 236–244 (1963)
31. Xiao, H., Rasul, K., Vollgraf, R.: Fashion-MNIST: a novel image dataset for benchmarking machine learning algorithms (2017)

Quality Control for Hierarchical Classification with Incomplete Annotations

Masafumi Enomoto[1(✉)], Kunihiro Takeoka[2], Yuyang Dong[2],
Masafumi Oyamada[2], and Takeshi Okadome[1]

[1] Graduate School of Science and Technology, Kwansei Gakuin University,
Sanda, Hyogo, Japan
tokadome@acm.org
[2] NEC Corporation, Tokyo, Japan
{k_takeoka,dongyuyang,oyamada}@nec.com

Abstract. Hierarchical classification requires annotations with hierarchical class structures. Although crowdsourcing services are inexpensive ways to collect annotations for hierarchical classification, the results are often incomplete because of the workers' limited abilities that unable to label all classes, and crowdsourcing platforms also allow suspensions during the labeling flow. Unfortunately, existing quality control approaches for refining low-quality annotations discard those incomplete annotations, and this limits the quality improvement of the results. We propose a quality control method for hierarchical classification that leverages incomplete annotations and the similarity between classes in the hierarchy for estimating the true leaf classes. Our method probabilistically models the labeling process and estimates the true leaf classes by considering the class-likelihood of samples and workers' class-dependent expertise. Our method embeds the class hierarchy into a latent space and represents samples as well as the worker's prototypical samples for classes (prototypes) as vectors in this space. The similarities between the vectors in the latent space are used to estimate the true leaf classes. The experimental results on both real-world and synthetic datasets demonstrate the effectiveness of our method and its superiority over the baseline methods.

Keywords: Crowdsourcing · Quality control · Hierarchical classification

1 Introduction

Hierarchical classification, in which the classes are organized within a given hierarchical structure, is used in many applications such as document categorization [1,6]. In hierarchical classification, there are different layers for different levels of classes. A sample can be associated with one of the classes in the leaf layer

© Springer Nature Switzerland AG 2021
K. Karlapalem et al. (Eds.): PAKDD 2021, LNAI 12714, pp. 219–230, 2021.
https://doi.org/10.1007/978-3-030-75768-7_18

through a single path and can also be associated with all classes on its path. As shown in Fig. 1a, if a sample belongs to the leaf class "Cat," it also belongs to its parent class "Animal." Hierarchical classification tasks always have a large number of classes. For example, the hierarchy of nouns in WordNet [8] contains 5,247 classes, and the average depth of the hierarchy is seven, i.e., it requires to label seven times for a sample. Because it is costly to use experts to label such a large number of classes, the use of crowdsourcing services for hierarchical classification tasks is preferable.

In reality, however, crowdsourced hierarchical classification suffers from the incompleteness of the annotations from workers, which we call an *incomplete annotation* problem. As shown in Fig. 1b, a worker may not cover all classes in an annotation of a sample for several reasons including limited ability, lack of motivation, suspensions on crowdsourcing platforms, and limited budget [13]. Since most existing hierarchical classification methods [10] require a sample to be labeled at all layers for training classifiers, the incomplete annotation problem limits the real-world applicability of applying crowdsourcing to the hierarchical classification.

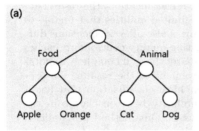
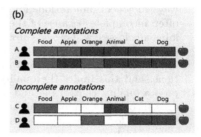

Fig. 1. (a) A class hierarchy with two intermediate classes and four leaf classes. (b) An example of complete and incomplete annotations to an image of an apple from workers. The row with white cells is an incomplete annotation. The cells in orange, blue, and white correspond to "should belong to the class", "should not belong to the class", and "missed answers". (Color figure online)

Historically, low-quality annotations by crowd workers are refined by the *quality control* process that asks a question (labeling task) to several workers. Then the high-quality labels are obtained by aggregating all annotations such as majority voting [3,14]. Unfortunately, the conventional quality control method for hierarchical classification discards those incomplete annotations, limiting the quality improvement of labels [9]. There are works on multi-label classification with incomplete annotations [7,11] but they cannot incorporate with the class hierarchy. The classes in the class hierarchy have dependencies in the parent-child relationship, and the aforementioned quality control methods cannot capture these dependencies leading to poor performance.

We may avoid collecting incomplete annotations (e.g., by forcing workers to label all classes) and obtain high-quality labels at a low-cost by aggregating complete annotations using conventional methods. However, the labeling cost of complete annotations is large because of the large number of classes. Under a situation in which a crowdsourcing platform sends the same number of queries to workers, we observed that the quality of the labels obtained by aggregating both complete annotations and incomplete annotations was higher than that obtained by aggregating only complete annotations (see Sect. 2.1). Thus, collecting and aggregating incomplete annotations is a cost-effective way to obtain high-quality labels.

In this paper, we propose a quality control method for hierarchical classification. It leverages the incomplete annotations and the similarity between classes in the hierarchy to estimate the true leaf classes. Our method is based on a generative probabilistic model for annotations that include the parameters of the latent features of the samples, the class-dependent expertise of the workers, and the true leaf classes. We make assumptions that a worker's expertise for a specific class is represented by the typical features of the samples (prototype) of this class in her experience. Her answer to this class is generated based on the similarity between the sample and the class prototype. Based on these assumptions, our method represents a sample and a prototype as embedded vectors of a latent space for the class hierarchy. The probabilities of answers are generated through the similarities in this space. This modeling allows us to estimate the parameters of true leaf classes with the similarity in the latent space of the hierarchical structure.

We conducted experiments on real and synthetic datasets for hierarchical classification. We tested the case of random injection of incomplete annotations on the open-source dataset, and our method outperformed the baseline methods when the observed ratio of answers is less than or equal to 40%. We collected a dataset of incomplete annotations by human workers and our method also worked better than the baseline methods. In addition, we tested the effectiveness of our method in estimating the class-dependent expertise by modeling prototypes on synthetic datasets. Our contributions can be summarized as follows:

- We observe that aggregating incomplete annotations is a cost-effective approach to obtaining high-quality labels for hierarchical classification.
- We propose a novel quality control method that leverages incomplete annotations based on a generative probabilistic model, in which a class hierarchy is embedded into a latent space, and the samples and the expertise of the workers are represented by latent vectors of the space.
- The results of experiments on both real and synthetic datasets show that the performance of our approach is better than those of the baseline methods.

2   Crowdsourcing with Incomplete Annotations

We first introduce a preliminary experiment to show the cost-effectiveness of the incomplete annotations used in a crowdsourced hierarchical classification. We then define the problem of quality control for hierarchical classification with incomplete annotations.

2.1   Cost-Effectiveness of Incomplete Annotations

The requesters of crowdsourcing want to obtain high-quality labels at a low-cost. Typically, complete annotations are collected and aggregated using quality control methods to achieve this goal. Although the number of classes to which a worker labeled in a complete annotation is larger than that in an incomplete annotation, the labeling cost is higher. Thus, two types (complete and incomplete) of annotations lead us to the following research question: Should we collect not only the complete annotations but also the incomplete annotations to obtain high-quality labels under the same budget?

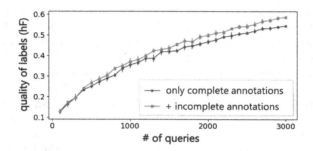

Fig. 2. The comparison result of the quality of aggregated labels by majority vote on only complete annotations and annotations including incomplete ones, varying the number of queries. The bars represent the standard deviation.

To answer the above research question, we conducted a preliminary experiment. We collected annotations from five crowd workers by using images and taxonomy in the ImageNet dataset [5]. We adopted the first two layers under the "musical instrument, instrument" category for hierarchical classification. We picked up 10 images for each leaf class and asked the workers to organize them into the taxonomy. A worker sequentially labeled the hierarchical classes from coarse to fine and was asked to skip the labeling halfway through. As a result, we collected annotations including $426/1999 \simeq 21\%$ of incomplete annotations. We created complete annotations by extracting 1573 annotations that workers did not skip labeling.

We compared the quality of the aggregated labels when using the complete annotations with that of the aggregated labels when using the annotations including incomplete ones. We estimated the true leaf classes of the samples by majority vote in a top-down manner: after choosing a class in the first layer by majority vote, we again took a majority vote with the answers for its child classes. We assumed that a crowdsourcing platform sends a single query to a worker to collect answers in each layer. We randomly selected the answers and estimated the true leaf classes 10 times for each number of queries, then reported the average of the hierarchical F1-measure (hF) [6] scores.

Figure 2 shows the comparison results of the quality of the aggregated labels by majority vote. To our surprise, as the number of queries increased, we obtained higher-quality labels when we used annotations including incomplete ones, which is statistically significant ($p < 0.05$) between the types of annotations. We therefore examined the percentage of correct answers assuming that we sent 3000 queries. We observed that the quality of the annotations including incomplete ones (46% correct) is higher than that of the complete annotations (43% correct). Incomplete annotations contain more answers for coarse-grained classes than fine-grained classes since workers labeled from coarse to fine. The answers in incomplete annotations tend to be correct since most answers are for coarse-grained classes which do not require expertise. Furthermore, incomplete annotations include annotations for a single sample from more workers. We can ask a worker regarding some classes on multiple samples instead of all classes on a single sample at the same cost, and the obtained labels are robust against incorrect answers by aggregating incomplete annotations.

2.2   Problem Definition

In a crowdsourced hierarchical classification, we ask a set of workers M to label the set of samples N using the set of classes C including both intermediate classes and leaf classes. As mentioned previously, each worker does not always cover all classes in the annotations for several reasons. Thus, we use $A = \{(i, j, k) \mid i \in N, k \in M_i \subseteq M, j \in C_{ik} \subseteq C\}$ to represent the *observed answers* where M_i is a set of the workers labeled samples i and C_{ik} is a set of the labeled classes of sample i by worker k. As observed answer (i, j, k) points to an answer from a worker, we use $\mathcal{Y}_{ijk} \in \{0, 1\}$ to represent whether worker k regards sample i should belongs to class j ($\mathcal{Y}_{ijk} = 1$) or not ($\mathcal{Y}_{ijk} = 0$). We use $\mathcal{Y}^A = \{\mathcal{Y}_{ijk}\}_{(i,j,k)\in A}$ to represent *incomplete annotations from workers*.

Given incomplete annotations from workers, $\mathcal{Y}^A$, our goal is to obtain quality-controlled labels for hierarchical classification. Since the intermediate class of a sample can be determined by its leaf class, our interest is then to estimate the true leaf classes of the samples from incomplete annotations. Provided this, the problem can be formalized as follows: Given incomplete annotations $\mathcal{Y}^A$ by workers M on samples N and class hierarchy $\mathcal{T}$, to estimate the true leaf classes of the samples, $\mathbf{Z} = \{\mathbf{z}_i\}_{i\in N}$.

We use H to denote the height of class hierarchy $\mathcal{T}$, and C_h to denote h-th layer's intermediate classes. Because we often distinguish the leaf classes and the intermediate classes, we use alias L to represent all leaf classes, that is, $L := C_H$. We use $leaf(j) \subset L$ to denote the set of the leaf classes of the subtree rooted at class j. Given leaf classes L, the true leaf class for sample i can be represented by a one-hot vector $\mathbf{z}_i \in \{0, 1\}^{|L|}$, where $z_{il} = 1$ when sample i belongs to leaf class l.

3   Modeling Labeling Process of Workers

To estimate the true leaf classes of the samples, we model the worker's labeling process that incorporates a class hierarchy. Furthermore, we infer the parameters from the observed incomplete annotations. We first introduce the assumptions of our model, then propose a probabilistic generative model in the next section.

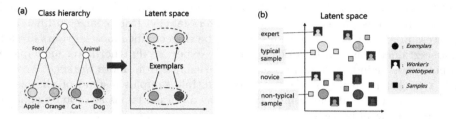

Fig. 3. Conceptual view of our method. (a) Embedding the exemplars of the leaf classes into a latent space, similar exemplars are close to each other. (b) The prototypes of the workers and the samples are distributed around the exemplars. Typical samples and prototypes of experts are close to the exemplar of the corresponding class.

In our model, for each leaf class in a hierarchy, a worker creates the *prototype* which is typical features of samples of the class by aggregating the samples based on experience. For a non-leaf intermediate class, the prototype is determined by the aggregation procedure (e.g., a weighted average) of the prototypes of the leaf classes in its descendants because we assume that a worker again aggregates samples of these leaf classes. In labeling phase, we assume that a worker determines whether a sample belongs to a class or not by considering the relevance between the sample and the prototype of the class.

Since prototypes are just personal cognitive representations of a worker, we prepare the ground-truth prototype called *exemplar* for each class, which is essentially the prototype of the oracle worker. We also assume that all samples, all prototypes, and all exemplars are from the same latent space. With this modeling, (1) the class-likelihood of a sample is represented by the *similarity* between the exemplar of the class and the sample in the latent space, and (2) the worker's *class-dependent expertise* (e.g., how she is familiar with apples) is represented by the similarity between the exemplar and the worker's prototype of the class in the latent space.

To aggregate annotations by considering the similarity between classes represented by a class hierarchy, we assume that the samples, the prototypes, and the exemplars are distributed in the latent space to reflect the similarity. To reflect the similarity between leaf classes with a common ancestor, we embed the exemplars of the leaf classes into a latent space (Fig. 3a). The similarity between parent and child classes is also reflected in the latent space. Recall that we assume that a prototype (exemplar) of an intermediate class is the weighted average of the prototypes (exemplars) of leaf classes in its descendants. In the latent space, the prototypes of the workers and the samples corresponding to a class are distributed around the exemplar of the class (Fig. 3b).

4 Proposed Method

We introduce our proposed method for solving the problem by probabilistically modeling the labeling process of workers.

4.1 Probabilistic Generative Model

We define a D dimensional Euclidean space $\mathbb{R}^D$ as the latent space to capture the similarities between exemplars, prototypes, and samples. The exemplar of leaf class l in class hierarchy $\mathcal{T}$ is represented as vector $\boldsymbol{\mu}^l \in \mathbb{R}^D$. In the same way, the prototype of class j from worker k is represented as vector $\mathbf{v}_k^j \in \mathbb{R}^D$, and the latent features of sample i is represented as vector $\mathbf{u}_i \in \mathbb{R}^D$, in the latent space. We evaluate the similarity between two points in the latent space by the Euclidean distance $\|\mathbf{p} - \mathbf{q}\|_2 = \sqrt{\sum_{d=1}^{D}(p_d - q_d)^2}$.

Given incomplete annotations $\mathcal{Y}^A$, the true leaf classes $\mathbf{Z} = \{\mathbf{z}_i\}$, the latent features of the samples $\mathbf{U} = \{\mathbf{u}_i\}$, and the prototypes on the leaf classes $\mathbf{V} = \{\mathbf{v}_k^l\}$, the joint probability distribution is defined as follows: $p(\mathcal{Y}^A, \mathbf{U}, \mathbf{V}, \mathbf{Z}) = p(\mathcal{Y}^A|\mathbf{U}, \mathbf{V})p(\mathbf{U}|\mathbf{Z})p(\mathbf{Z})p(\mathbf{V})$, $p(\mathbf{Z}) = \prod_{i \in N} p(\mathbf{z}_i)$, $p(\mathbf{U}|\mathbf{Z}) = \prod_{i \in N} p(\mathbf{u}_i|\mathbf{z}_i)$, $p(\mathbf{V}) = \prod_{k \in M} \prod_{l \in L} p(\mathbf{v}_k^l)$, and $p(\mathcal{Y}^A|\mathbf{U}, \mathbf{V}) = \prod_{(i,j,k) \in A} p(\mathcal{Y}_{ijk}|\mathbf{u}_i, \mathbf{v}_k^j)$.

The true leaf class $\mathbf{z}_i$ of sample i is generated from the following categorical distribution: $p(\mathbf{z}_i) = \text{Cat}(\mathbf{z}_i|\boldsymbol{\pi}) = \prod_{l \in L} \pi_l^{z_{il}}$, where hyper-parameter π_l denotes the prior probability that a sample belongs to leaf class l. As mentioned in Sect. 3, we assume that the latent features of the samples are distributed around the exemplars. Thus, the latent features $\mathbf{u}_i$ of sample i is generated from the normal distribution with precision parameter m centered at the exemplar of the true leaf class $\mathbf{z}_i$: $p(\mathbf{u}_i|\mathbf{z}_i) = \prod_{l \in \mathcal{L}} \mathcal{N}(\mathbf{u}_i|\boldsymbol{\mu}^l, m^{-1}\mathbf{I}_D)^{z_{il}}$ where $\mathbf{I}_D$ is an identity matrix of size D.

Similarly, we also assume that the prototypes are distributed around the exemplars. Worker k's prototype $\mathbf{v}_k^l$ of leaf class l is generated from the normal distribution with precision parameter s centered at the exemplar $\boldsymbol{\mu}^l$ of leaf class l: $p(\mathbf{v}_k^l) = \mathcal{N}(\mathbf{v}_k^l|\boldsymbol{\mu}^l, s^{-1}\mathbf{I})$. The prototype $\mathbf{v}_k^j$ of an intermediate class $j \in C$ is represented as the weighted average of the prototypes of leaf classes $leaf(j)$ in its descendant with the prior probability $\{\pi_l\}$, as follows: $\mathbf{v}_k^j = \sum_{l \in leaf(j)} (\pi_l/s(j))\mathbf{v}_k^l$, where $s(j) = \sum_{l \in leaf(j)} \pi_l$.

We assume that answers of a class from a worker are determined based on the similarities between samples and her class prototype, i.e., she labels samples to the class if they are similar. The answer $\mathcal{Y}_{ijk}$ is generated from a Bernoulli distribution of the parameter calculated using the latent features $\mathbf{u}_i$ of sample i and worker k's prototype $\mathbf{v}_k^j$ of class j: $p(\mathcal{Y}_{ijk} = 1|\mathbf{u}_i, \mathbf{v}_k^j) = f_{\alpha,\beta}^h \left(\|\mathbf{u}_i - \mathbf{v}_k^j\|_2 \right)$, where $f_{\alpha,\beta}^h : [0, \infty) \to [0, 1]$ denotes a monotonically decreasing function with hyper-parameters α and β to transform the Euclidean distance to probabilities depending on the h-th layer ($j \in C_h$). We use the scaling function $f^h_{\alpha,\beta}$ so that workers who have exemplars choose the true class in h-th layer with high probability. The setting of the scaling function and the hyper-parameters is described in Sect. 5.1.

4.2 Inference

Given annotations $\mathcal{Y}^A$, we aim at inferring the true leaf classes $\mathbf{Z}$, the latent features of the samples $\mathbf{U}$ and the prototypes $\mathbf{V}$. The EM (expectation-maximization) algorithm enables us to obtain the MAP (maximum a posteriori) estimates of parameters $\{\mathbf{U}, \mathbf{V}\}$, and the posterior probabilities of the true leaf classes $\mathbf{Z}$. The E step calculates the posterior distribution $p(\mathbf{Z}|\mathcal{Y}^A, \mathbf{U}, \mathbf{V})$ of the true leaf classes $\mathbf{Z}$ with annotations $\mathcal{Y}^A$ and the values of parameters $\{\mathbf{U}, \mathbf{V}\}$ from the last M step. Then, we take the expectation of the logarithm of the joint probability $p(\mathcal{Y}^A, \mathbf{U}, \mathbf{V}, \mathbf{Z})$ over the posterior distributions. The M step determines the values of parameters $\{\mathbf{U}, \mathbf{V}\}$ by locally maximizing the expected value using gradient descent. The cycle of the EM algorithm optimizes the value of the parameters $\{\mathbf{U}, \mathbf{V}\}$ so that the lower bound of the posterior probability $p(\mathbf{U}, \mathbf{V}|\mathcal{Y}^A)$ always increases until convergence.

4.3 Embedding Class Hierarchy

To obtain the exemplars which preserve similarity between classes, we embed class hierarchy $\mathcal{T}$ into latent space $\mathbb{R}^D$. We define the semantic distance $d(l, l')$ between two leaf classes l and l' by the length of the path connected between the corresponding nodes on class hierarchy $\mathcal{T}$. We can obtain exemplars $\boldsymbol{\mu} = \{\boldsymbol{\mu}^l; \boldsymbol{\mu}^l \in \mathbb{R}^D\}_{l \in L}$ by metric multi-dimensional scaling [2] that places each leaf class into latent space $\mathbb{R}^D$ such that the semantic distances $\{d(l, l')\}_{l \in L}$ are preserved as well as possible.

5 Experiments and Discussion

5.1 Experimental Setup

We evaluated our method on annotations on two real crowdsourced hierarchical classification datasets. We collected a dataset named "ImageNet Musical Instrument (INMI)" with the crowdsourcing workflow that allowed workers to skip labeling halfway through as described in Sect. 2.1. In the INMI dataset, five workers classified 440 samples, using the class hierarchy with a depth of two which contains six intermediate classes and 44 leaf classes. The other dataset is the TDB dataset used in [9]. In the TDB dataset, 93 workers classified 388 samples, using the class hierarchy with a depth of two which contains 21 intermediate classes and 273 leaf classes. We randomly replace the values in complete annotations to generate the incomplete annotations.

We also evaluated our method on several sets of synthetic incomplete annotations. We first created hierarchies with a depth of three, and each class has n

children ($3 \leq n \leq 6$), and the total number of leaf classes is from 27 to 216. We assumed that 10 samples belong to each leaf class. For each sample, 10 workers choose a path to a leaf class in a hierarchy. We also assumed that a worker has an expertise parameter on each leaf class $r_{kl} = p(\mathcal{Y}_{ilk} = 1|z_{il} = 1)$, which was sampled from a beta distribution. A worker chooses incorrect path uniformly, i.e., $p(Y_{ilk} \neq 1|z_{il} = 1) = (1 - r_{kl})/(|L| - 1)$. After sampling annotations from workers based on their expertise parameters, we randomly selected 20% answers from them to generate incomplete annotations.

We compared our method with majority voting (MV), DS [4], GLAD [12], CRIA$_V$ [7] and Steps-GLAD [9] (SGLAD). CRIA$_V$ leverages incomplete annotations and sample features to complete the missing answers by assuming the low-rank structure of an annotation tensor. Note that we modified the method without using the sample features since they are not available in our problem. SGLAD utilizes a class hierarchy to aggregate answers on intermediate and leaf classes based on the GLAD algorithm, which estimates the class-independent expertise of workers and the degree of difficulty of samples.

For evaluating the effectiveness of the quality control for hierarchical classification, we used the hierarchical F1-measure (hF) [6], which is widely used in hierarchical classification. It is an extension of the standard F-measure to reflect the distance between true leaf class $l_i = \arg\max_{l \in L} z_{il}$ and estimated leaf class $\tilde{l}_i = \arg\max_{l \in L} \tilde{z}_{il}$ on a class hierarchy. In our setting, hF is defined as $\sum_{i \in N} |path(l_i) \cap path(\tilde{l}_i)|/H|N|$, where $path(l)$ denotes the set of classes on the path from the root to leaf class l.

We set the scaling function $f_{\alpha,\beta}^h : [0, \infty) \to [0, 1]$ as a piecewise function that consists of three functions with the input of 0, a^h, b^h and the output of 1, α, β as follows: $f_{\alpha,\beta}^h(x) = \frac{\alpha-1}{a^h}x + 1$ $(0 \leq x \leq a^h)$, $f_{\alpha,\beta}^h(x) = \frac{\alpha-\beta}{a^h-b^h}(x - b^h) + \beta$ $(a^h < x \leq b^h)$, $f_{\alpha,\beta}^h(x) = \beta \exp(b^h - x)$ $(b^h < x)$, where a^h, b^h are assigned with a high probability value α and a low probability value β, and are defined as follows: $a^h = \frac{1}{\sum_{j \in C_h} |leaf(j)|} \sum_{j \in C_h} \sum_{l \in leaf(j)} \|\boldsymbol{\mu}^l - \boldsymbol{\mu}^j\|_2$, $b^h = \frac{1}{\sum_{j \in C_h} |leaf(pa(j))\backslash leaf(j)|} \sum_{j \in C_h} \sum_{l \in leaf(pa(j))\backslash leaf(j)} \|\boldsymbol{\mu}^l - \boldsymbol{\mu}^j\|_2$, where $pa(j)$ denotes the parent of class j. In the same way as the obtaining of prototypes, the exemplar $\boldsymbol{\mu}^j$ of an intermediate class $j \in C$ is represented as the weighted average of the exemplars of the leaf classes in its descendants.

Our probabilistic model includes the hyper-parameters D, $\boldsymbol{\pi}$, α, β, m, and s. We set D, π_l, α and β to L, $1/|L|$, 0.9 and 0.1, respectively. For each dataset, we determined the values of precision parameters m and s by grid search to maximize the approximation of model evidence $p(\mathcal{Y}^A|m, s)$. For each possible parameter pair on the searching grid, we calculated average of 10 values of the likelihood function $p(\mathcal{Y}^A|\mathbf{U}, \mathbf{V})$ using the values of parameters $\{\mathbf{U}, \mathbf{V}\}$ sampled from the prior distribution $p(\mathbf{U}, \mathbf{Z}|m)$ and $p(\mathbf{V}|s)$.

5.2 Experimental Results and Discussion

Estimation of True Leaf Classes on Human-Labeled Annotations. We first tested the performance of our method and the baseline methods on the

Table 1. Comparison results on the human-labeled incomplete annotations (hF), varying observed ratio ρ. Winners are shown in bold. In table (a), statistically significant results compared with the second-best results by paired t-test are shown with $\star$ ($p <$ 0.05).

(a) Comparison result on the TDB dataset

ρ	10%	20%	40%	60%	80%	100%
MV	0.32 ± 0.02	0.44 ± 0.01	0.51 ± 0.01	0.55 ± 0.01	0.55 ± 0.01	0.55 ± 0.00
DS	0.36 ± 0.02	0.44 ± 0.02	0.49 ± 0.01	0.51 ± 0.01	0.51 ± 0.01	0.52 ± 0.00
GLAD	0.33 ± 0.02	0.45 ± 0.01	0.52 ± 0.01	0.55 ± 0.01	0.56 ± 0.01	0.56 ± 0.01
$CRIA_V$	0.28 ± 0.03	0.43 ± 0.01	0.53 ± 0.01	0.56 ± 0.01	0.56 ± 0.01	0.56 ± 0.00
SGLAD	0.33 ± 0.02	0.45 ± 0.02	0.53 ± 0.01	0.56 ± 0.01	0.57 ± 0.01	0.57 ± 0.00
OURS	$\star\mathbf{0.40 \pm 0.02}$	$\star\mathbf{0.50 \pm 0.01}$	$\star\mathbf{0.54 \pm 0.01}$	0.56 ± 0.01	0.57 ± 0.01	$\mathbf{0.58 \pm 0.00}$

(b) Comparison result on the INMI dataset

MV	DS	GLAD	$CRIA_V$	SGLAD	OURS
0.60	0.60	0.60	0.62	0.64	**0.65**

human-labeled annotations. To generate randomly incomplete annotations using the TDB dataset, we randomly selected ρ% answers from complete annotations and varied ρ as 10% to 100%. For each ρ, we estimated the true leaf classes 10 times, then reported the average of the hF scores. We also reported the average of the 10 hF scores on the INMI dataset since our method uses gradient descent.

Table 1a shows the comparison results for true leaf class estimation on TDB dataset. Our method outperformed the baseline methods, and significant differences were observed when we could only observe less than or equal to 40% answers, i.e., $\rho \leq 40\%$. This demonstrates the effectiveness of our method can accurately estimate the true leaf classes even there are a lot of incomplete annotations since it utilizes incomplete annotations. Note that $CRIA_V$ also utilized the incomplete annotations but had less performance than our method since it cannot incorporate with a class hierarchy. There were no significant differences between our method and SGLAD when $\rho \geq 60\%$. This is because both methods considered the relation of hierarchical structure, which can help to estimate the true leaf classes accurately. Table 1b shows the comparison results for true leaf class estimation on the INMI dataset. Our method outperformed the baseline methods. This result proved that our method works on the dataset collected for a hierarchical classification task that allowed incomplete annotation.

Effectiveness of Modeling Class-Dependent Expertise. We also tested the performance of our method on the synthetic incomplete annotations assuming workers with class-independent expertise (IND) and class-dependent expertise (DEP). In both datasets, the expertise parameters $\{r_{kl}\}$ were generated from the beta distribution with $\alpha = 0.1$ and $\beta = 0.1$. The probability density of this distribution is high near 0 and 1, i.e., $p(1/3 < r_{kl}), p(2/3 < r_{kl}) = 0.47$. For the IND dataset, a single expertise parameter was sampled per worker, and annotations were generated independently of the true leaf classes. For each trial, we generated a class hierarchy and incomplete annotations and estimated the true leaf classes. Then, we reported the average of the hF scores of 10 trials.

Table 2. Comparison results on the synthetic incomplete annotations (hF). Winners are shown in bold.

	MV	SGLAD	OURS
IND	0.63 ± 0.06	0.74 ± 0.05	**0.77 ± 0.07**
DEP	0.55 ± 0.02	0.56 ± 0.02	**0.68 ± 0.02**

Table 2 shows the comparison results for true leaf class estimation on the synthetic incomplete annotations assuming workers with class-independent expertise (IND) and class-dependent expertise (DEP). Our method improved the quality of the labels by more than 10% from the MV on both datasets. On the other hand, the quality improvement by SGLAD on the DEP dataset was only 1%, which was limited compared to our method. This demonstrates the effectiveness of our method that can estimate class-dependent expertise by modeling prototypes. Recall that the worker's expertise on each leaf class is polarized into high and low levels in DEP datasets. The discovery of the worker's strong classes is, therefore, important to improve the quality of the labels. For this reason, the quality improvement by SGLAD was limited because it cannot model the worker's expertise of each class.

6   Related Work

Crowdsourcing platforms such as Amazon Mechanical Turk[1] enable us to collect many labels at a low-cost. However, the quality of collected labels is one of the important issues in research and industrial fields. Truth inference [14] and quality control for crowdsourcing [3] aimed at finding the true classes of samples accurately from given workers' annotations. There are a few quality control methods for hierarchical classification tasks [9]. The method proposed by Otani et al. [9] used answers for intermediate classes to utilize the similarities between classes and outperformed approaches that ignore a class hierarchy [4,12]. However, it assumed all annotations are complete, which is different from our problem of incomplete annotations.

Quality control methods using incomplete annotations have been proposed for multi-label classification tasks [7,11]. Li et al. [7] assumed that annotations are incomplete because workers label related classes, but they do not check all classes. Thus, this method estimated the missing answers from incomplete annotations by assuming the low-rank structure of an annotation tensor. Tu et al. [11] estimated the co-occurrence dependency between classes and workers' skills to obtain the true labels accurately by using incomplete annotations. However, the

---

[1] https://www.mturk.com/.

above researches are different from our problem as they focused on the multi-label classification without a class hierarchy.

7 Conclusion

In this paper, we studied the problem of estimating the true leaf classes of samples with incomplete annotations for hierarchical classification. We designed a novel method for quality control by utilizing incomplete annotations and modeling the labeling process that considers the class-likelihood of samples and the class-dependent expertise in the same latent space. Extensive experiments on both real and synthetic datasets demonstrated the effectiveness of our method and its superiority over the baseline methods.

References

1. Brecheisen, S., Kriegel, H.P., Kunath, P., Pryakhin, A.: Hierarchical genre classification for large music collections. In: ICME, pp. 1385–1388. IEEE (2006)
2. Cox, M.A.A., Cox, T.F.: Multidimensional scaling. In: Handbook of Data Visualization, pp. 315–347. Springer, Heidelberg (2008). https://doi.org/10.1007/978-3-540-33037-0_14
3. Daniel, F., et al.: Quality control in crowdsourcing: a survey of quality attributes, assessment techniques, and assurance actions. ACM Comput. Surv. **51**, 1–40 (2018)
4. Dawid, A.P., Skene, A.M.: Maximum likelihood estimation of observer error-rates using the EM algorithm. J. R. Stat. Soc. Ser. C Appl. Stat. **28**(1), 20–28 (1979)
5. Deng, J., et al.: ImageNet: a large-scale hierarchical image database. In: CVPR, pp. 248–255. IEEE (2009)
6. Kiritchenko, S., Matwin, S., Famili, A.F.: Functional annotation of genes using hierarchical text categorization. In: Proceedings of the ACL Workshop on Linking Biological Literature, Ontologies and Databases: Mining Biological Semantics (2005)
7. Li, S.-Y., Jiang, Y.: Multi-label crowdsourcing learning with incomplete annotations. In: Geng, X., Kang, B.-H. (eds.) PRICAI 2018. LNCS (LNAI), vol. 11012, pp. 232–245. Springer, Cham (2018). https://doi.org/10.1007/978-3-319-97304-3_18
8. Miller, G.A.: WordNet: a lexical database for English. Commun. ACM **38**(11), 39–41 (1995)
9. Otani, N., Baba, Y., Kashima, H.: Quality control for crowdsourced hierarchical classification. In: ICDM, pp. 937–942. IEEE (2015)
10. Silla, C.N., Freitas, A.A.: A survey of hierarchical classification across different application domains. Data Min. Knowl. Discov. **22**(1–2), 31–72 (2011)
11. Tu, J., et al.: Multi-label answer aggregation based on joint matrix factorization. In: ICDM, pp. 517–526. IEEE (2018)
12. Whitehill, J., et al.: Whose vote should count more: optimal integration of labels from labelers of unknown expertise. In: NeurIPS, pp. 2035–2043 (2009)
13. Yan, Y., Huang, S.: Cost-effective active learning for hierarchical multi-label classification. In: IJCAI, pp. 2962–2968 (2018)
14. Zheng, Y., et al.: Truth inference in crowdsourcing: is the problem solved? Proc. VLDB Endow. **10**(5), 541–552 (2017)

Learning from Data

Learning Discriminative Features Using Multi-label Dual Space

Ali Braytee$^{(\boxtimes)}$ and Wei Liu

School of Computer Science, University of Technology Sydney, Sydney, Australia
{ali.braytee,wei.liu}@uts.edu.au

Abstract. Multi-label learning handles instances associated with multiple class labels. The original label space is a logical matrix with entries from the Boolean domain $\in \{0, 1\}$. Logical labels are not able to show the relative importance of each semantic label to the instances. The vast majority of existing methods map the input features to the label space using linear projections with taking into consideration the label dependencies using logical label matrix. However, the discriminative features are learned using one-way projection from the feature representation of an instance into a logical label space. Given that there is no manifold in the learning space of logical labels, which limits the potential of learned models. In this work, inspired from a real-world example in image annotation to reconstruct an image from the label importance and feature weights. We propose a novel method in multi-label learning to learn the projection matrix from the feature space to semantic label space and projects it back to the original feature space using encoder-decoder deep learning architecture. The key intuition which guides our method is that the discriminative features are identified due to map the features back and forth using two linear projections. To the best of our knowledge, this is one of the first attempts to study the ability to reconstruct the original features from the label manifold in multi-label learning. We show that the learned projection matrix identifies a subset of discriminative features across multiple semantic labels. Extensive experiments on real-world datasets show the superiority of the proposed method.

Keywords: Multi-label learning · Feature selection · Label correlations

1 Introduction

Multi-label learning deals with the problem where each sample is represented by a feature vector and is associated with multiple concepts or semantic labels. For example, in image annotation, an image may be annotated with different scenes; or in text categorization, a document may be tagged to multiple topics. Formally, given a data matrix $X \in \mathbb{R}^{d \times n}$ is composed of n samples of d-dimensional feature space. The feature vector $x_i \in X$ is associated with label set $Y_i = \{y_{i1}, y_{i2}, \ldots, y_{ik}\}$ where k is the number of labels, and $y_{(i,j)} \in \{0, 1\}$ is a logical value where

© Springer Nature Switzerland AG 2021
K. Karlapalem et al. (Eds.): PAKDD 2021, LNAI 12714, pp. 233–245, 2021.
https://doi.org/10.1007/978-3-030-75768-7_19

the associated label is relevant to the instance x_i. Over the past decade, many strategies have been proposed in the literature to learn from multi-labeled data. Initially, the problem was tackled by learning binary classification models on each label independently [14]. However, this strategy ignores the existence of label correlation. Interestingly, several methods [2,10,11] show the importance of considering the label correlation during multi-label learning to improve the classification performance. However, these methods use logical labels where no manifold exist and apply traditional similarity metrics such as Euclidean distance which is mainly built for continuous data.

In this study, contrary to the majority of the methods, in addition to learning the mapping function from a feature space to multi-label space, we explore the projection function from label space to feature space to reconstruct the original feature representations. For example, in image annotation, our novel method is able to reconstruct the scene image using the projection function and the semantic labels. Initially, it is necessary to explore the natural structure of the label space in multi-labeled data. Existing datasets naturally contain logical label vectors which indicate whether the instance is relevant or not relevant to a specific label. For example, as shown in Fig. 1, both images tagged the label boat with the same weight equal to 1 (present). However, to accurately describe the labels in both images, we need to identify the importance of the labels in each image. It is clearly seen that the label boat in image (Fig. 1b) is more important than that in image (Fig. 1a). Furthermore, the label with the zero value in the logical label vector refers to different meanings, which may either be irrelevant, unrepresented or missing. Using the same example in Fig. 1, the boat and the sun labels in images (Figs. 1a and 1b) are not tagged due to their small contribution (unrepresented). Our method learns a numerical multi-label matrix in semantic embedding space during the optimization method based on label dependencies. Therefore, replacing the importance of labels using numerical values instead of the logical labels can improve the multi-label learning process.

Importantly, learning the numerical labels is essential to our novel approach that is developed based on the encoder-decoder deep learning paradigm [9]. Specifically, the input training data in the feature space is projected into the learned semantic label space (label manifold) as an encoder step. In this step, simultaneously through an optimization problem, it learns the projection function and the semantic labels in Euclidean space. Significantly, we also consider the reconstruction task of the original feature representations using the projection matrix as input to a decoder. This step imposes a constraint to ensure that the projection matrix preserves all the information in the original feature matrix. The decoder allows the original features to be recovered using the projection matrix and the learned semantic labels. In the case of image annotation, this process is similar to combining puzzle pieces to create the picture. However, in the case of logical labels where the label either exists or not, it is incapable of reconstructing the original visual feature representations. We show that the impact of the decoder in identifying the relevant features can improve multi-label classification performance. This is because the feature coefficients are estimated

(a)   (b)

Fig. 1. Two images annotated with labels: sea, sunset, and boat

based on the actual numerical labels and more importantly the weights of the relevant features results in a reduction of the reconstruction error. The proposed method is visualized in Fig. 2. We test the proposed approach on a variety of public multi-label datasets, and clearly verify that they favourably outperform the state-of-the-art methods in feature selection and data reconstruction.

We formulate the proposed approach as a constrained optimization problem to project feature representations into semantic labels with a reconstruction constraint. More precisely, the method is designed by an effective formulation of the encoder and decoder model using a linear projection to and from the learned semantic labels, respectively. This design alleviates the computational complexity of the proposed approach making it suitable for large scale datasets. To the best of our knowledge, this is the first attempt to learn the semantic label representation from the training data that can be used for data reconstruction in multi-label learning. In summary our contribution are: (1) a semantic encoder-decoder model to learn the projection matrix from original features to semantic labels that can be used for data reconstruction; (2) we extend the logical label to a numerical label which describes the relative importance of the label in a specific instance; (3) we propose a novel *Learning Discriminative Features using Multi-label Dual Space (LDFM)* which is able to identify discriminative features across multiple class labels.

2   Related Work

Label Correlation. Over the past decade until recently, it has been proven that label correlation has improved the performance of multi-label learning methods. The correlation is considered either between pairs of class labels or between all the class labels which is known as second and high order approaches, respectively [4,14,15]. However, in these models, the common learning strategy is to deal with logical labels which represents whether the label is relevant or irrelevant to an instance. The label matrix in the available multi-labeled datasets contains logical values which lack semantic information. Hence, few works reveal that transforming the labels from logical into numerical values improves the learning process.

Semantic Labels. The numerical value in the label space carries semantic information, i.e., the value may refer to the importance or the weight of an object in the image. The numerical label matrix in Euclidean space is not explicitly available in the multi-labeled data. A few works have studied the multi-label manifold by transforming the logical label space to the Euclidean label space. For example, [6] explore the label manifold in multi-label learning and reconstruct the numerical label matrix using the instance smoothness assumption. Another work [3] incorporates feature manifold learning in the multi-label feature selection method, and [7] select the meaningful features using the constraint Laplacian score in manifold learning. However, our proposed method differs from these by learning an encoder-decoder network to reconstruct the input data using the learned projection matrix along with predicting the semantic labels.

Autoencoder. Several variants use an autoencoder for multi-label learning. [5] learn the unknown labels using the entropy measure from existing labels, then the completed label matrix is used as an input layer feature set in autoencoder architecture. However, our method reconstructs the original input data in the decoder using the learned semantic labels. Further, [8] propose a stacked autoencoder for feature encoding and an extreme learning machine to improve the prediction capability. However, the authors did not take label correlation into consideration and the original logical labels are used in the learning process. In this paper, we select the discrete features that are important to detect the objects' weights during the encoding phase and simultaneously they are significant to reconstruct the original data in the decoding phase.

3   The Proposed Method

In multi-label learning, as mentioned above, the training set of multi-labeled data can be represented by $\{x_i \in X | i = 1, \cdots, n\}$, the instance $x_i \in \mathbb{R}^d$ is a d-dimensional feature vector associated with the logical label $Y_i = \{y_{i1}, y_{i2}, \dots, y_{ik}\}$, where k is the number of possible labels; and the values 0 and +1 represent the irrelevant and relevant label to the instance x_i respectively.

3.1   Label Manifold

To overcome the key challenges in logical label vectors, we first propose to learn a new numerical label matrix $\widetilde{Y} \in \mathbb{R}^{k \times n}$ which contains labels with semantic information. According to the label smoothness assumption [12] which states that if two labels are semantically similar, then their feature vectors should be similar, we initially exploit the dependencies among labels to learn $\widetilde{Y}$ by multiplying the original label matrix with the correlation matrix $C \in \mathbb{R}^{k \times k}$. Due to the existence of logical values in the original label matrix, we use the *Jaccard index* to compute the correlation matrix as follows

$$\widetilde{Y} = Y^T C \tag{1}$$

where the element $\widetilde{Y}_{i,j} = Y_{i,1}^T \times C_{1,j} + Y_{i,2}^T \times C_{2,j} + \cdots + Y_{i,n}^T \times C_{n,j}$ is initially determined as the predictive numerical value of instance x_i is related to the j-th label using the prior information of label dependencies. Following is a simple example to investigate the efficiency of using label correlation to learn the semantic numerical labels. The original logical label vectors Y of images Figs. (1a) and (1b) are shown in Fig. 2. The zero values of the original label matrix point to three different types of information. The grey and red colors in Fig. 2 refer to unrepresented and missing labels respectively. However, the white color means that images Figs. (1a) and (1b) are not labeled as "Grass". We clearly show that the predictive label space $\widetilde{Y}$ can distinguish between three types of zero values and it provides the appropriate numerical values which include semantic information. For example, due to the correlation between "Boat" and "Ocean" and "Sunset" and "Sun", the unrepresented label information for "Boat" and "Sun" in image (Fig. 1a) and "Sun" in image (Fig. 1b) is learned. Further, the missing label of the "Ocean" image (Fig. 1b) is predicted. Interestingly, we can further see in the predictive label matrix that the numerical values of the "Grass" label in both images are very small because the "Grass" label is not correlated with the other labels. Thus, this perfectly matches with the information in the original label matrix that the "Grass" object does not exist in images Figs. (1a) and (1b) as shown in Fig. 2. Therefore, based on the above example, we can learn the accurate numerical labels with semantic information. The completed numerical label matrix $\widetilde{Y}$ of the training data is learned in the optimization method of the encoder-decoder framework in the next section.

3.2   Approach Formulation

Suppose there is training data $X \in \mathbb{R}^{d \times n}$ with n samples that are associated with $Y \in \mathbb{R}^{k \times n}$ labels. The predictive numerical matrix $\widetilde{Y}$ is initialised using Eq. 1. The intuition behind our idea is that the proposed method is able to capture the relationship between the feature space and the manifold label space. Inspired by the autoencoder architecture, we develop an effective method that integrates the characteristics of both the low-rank coefficient matrix and semantic numerical label matrix. Specifically, our method is composed of encoder-decoder architecture which tries to learn the projection matrix $W \in \mathbb{R}^{k \times d}$ from the feature space X to the numerical label space $\widetilde{Y}$ in the encoder. At the same time, the decoder can project back to the feature space with $W^T \in \mathbb{R}^{d \times k}$ to reconstruct the input training data as shown in Fig. 2. The objective function is formulated as

$$\min_{W} \left\| X - W^T W X \right\|_F^2 \quad \text{s.t. } WX = \widetilde{Y} \tag{2}$$

where $\|.\|_F$ is the Frobenius norm.

Optimization Algorithm. To optimize the objective function in Eq. 2, we first substitute WX with $\widetilde{Y}$. Then, due to the existence of constraint $WX = \widetilde{Y}$,

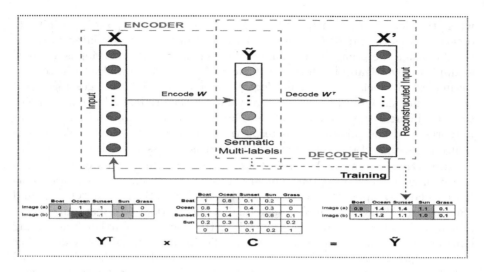

Fig. 2. Encoder-decoder architecture of our proposed method LDFM. A predicted numerical label matrix $\widetilde{Y}$ is initialized by multiplying the correlation matrix with the original logical label matrix Y. The zero values of the images in red, grey and white boxes represent missing, misrepresented, and irrelevant labels, respectively (Color figure online)

it is very difficult to solve Eq. 2. Therefore, we relax the constraint into a soft one and reformulates the objective function (2) as

$$\min_{W} \left\| X - W^T \widetilde{Y} \right\|_F^2 + \lambda \left\| WX - \widetilde{Y} \right\|_F^2 \tag{3}$$

where λ is a parameter to control the importance of the second term. Now, the objective function in Eq. 3 is non-convex and it contains two unknown variables W and $\widetilde{Y}$. It is difficult to directly solve the equation. We propose a solution to iteratively update one variable while fixing the other. Since the objective function is convex by updating one variable, we compute the partial derivative of Eq. 3 with respect to W and $\widetilde{Y}$ and set both to zero.

– **Update W:**

$$
\begin{aligned}
& -\widetilde{Y}\left(X^T - \widetilde{Y}^T W\right) + \lambda\left(WX - \widetilde{Y}\right)X^T = 0 \\
\Rightarrow\ & \widetilde{Y}\widetilde{Y}^T W + \lambda W X X^T = \widetilde{Y}X^T + \lambda\widetilde{Y}X^T \\
\Rightarrow\ & PW + WQ = R
\end{aligned}
\tag{4}
$$

where $P = \widetilde{Y}\widetilde{Y}^T$, $Q = \lambda X X^T$, and $R = (\lambda + 1)\widetilde{Y}X^T$

ALGORITHM 1: Learning Discriminative Features using Multi-label Dual Space (LDFM)

Input: Training data $X \in \mathbb{R}^{d \times n}$
Logical label matrix $Y \in \mathbb{R}^{k \times n}$
Parameters: λ and MaxIteration

Initialization:
$C_{jk} = \frac{\sum_{i=1}^{n} x_{ij} x_{jk}}{\sum_{i=1}^{n} x_{ij} + \sum_{i=1}^{n} x_{ik} - \sum_{i=1}^{n} x_{ij} x_{ik}}; j, k$ are the label vectors
$\widetilde{Y} = YC$
$t = 0$

while *MaxIteration* $> t$ **do**

$\quad$ Update W by the solving the Eq. 6 ;
$\quad$ Update $\widetilde{Y}$ by the solving the Eq. 7 ;
$\quad$ $t = t + 1$;

end

Output: Projection matrix $W \in \mathbb{R}^{k \times d}$
Rank features by $\|W_{m,:}\|_2$ in descending order and return the top ranked features

---

– **Update** $\widetilde{Y}$:

$$-WX + WW^T\widetilde{Y} + \lambda\left(-WX + \widetilde{Y}\right) = 0$$

$$\Rightarrow WW^T\widetilde{Y} + \lambda\widetilde{Y} = WX + \lambda WX \qquad (5)$$

$$\Rightarrow A\widetilde{Y} + \widetilde{Y}B = D$$

where $A = WW^T$, $B = \lambda I$, $D = (\lambda + 1)WX$ and $I \in \mathbb{R}^{k \times k}$ is an identity matrix.

Equations 4 and 5 are formulated as the well-known Sylvester equation of the form $MX + XN = O$. The sylvester equation is a matrix equation with given matrices M, N, and O and it aims to find the possible unknown matrix X. The solution of the Sylvester equation can be solved efficiently and lead to a unique solution. For more explanations and proofs, the reader can refer to [1]. Using the Kronecker products notation and the vectorization operator *vec*, Eqs. 4 and 5 can be written as a linear equation respectively

$$\left(I_d \otimes P + Q^T \otimes I_k\right) vec(W) = vec(R) \qquad (6)$$

where $I_d \in \mathbb{R}^{d \times d}$ and $I_k \in \mathbb{R}^{k \times k}$ are identity matrices and $\otimes$ is the Kronecker product.

$$\left(I_n \otimes A + B^T \otimes I_k\right) vec(\widetilde{Y}) = vec(D) \qquad (7)$$

where $I_n \in \mathbb{R}^{n \times n}$ is an identity matrix. Fortunately, in MATLAB, this equation can be solved with a single line of code *sylvester*[1]. Now, the two unknown matri-

---

[1] https://uk.mathworks.com/help/matlab/ref/sylvester.html.

ces W and $\widetilde{Y}$ can be iteratively updated using the proposed optimization rules above until convergence. The procedure is described in Algorithm 1.

Our proposed method learns the encoder projection matrix W. Thus, we can embed a new test sample x_i^s to the semantic label space by $\widetilde{y}_i = W x_i^s$. Similarly, we can reconstruct the original features using the decoder projection matrix W^T by $x_i^s = W^T \widetilde{y}_i$. Therefore, W contains the discriminative features to predict the real semantic labels. To identify these features, we rank each feature according to the value of $\|W_{m,:}\|_2 \ (m = 1, \cdots, d)$ in descending order and return the top ranked features.

Table 1. Characteristics of the evaluated datasets

Dataset	Domain	#Instance	#Training	#Test	#Features	#Labels
Scene	Image	2407	1211	1196	294	6 scenes
Emotions	Audio	593	391	202	72	6 emotions
Reference	Text (Yahoo)	5000	2000	3000	793	33 topics
Computers	Text (Yahoo)	5000	2000	3000	681	33 topics

4   Experiments

4.1   Experimental Datasets

We open source for our LDFM code for reproducibility of our experiments[2]. Experiments are conducted on four public multi-label datasets which can be downloaded from the Mulan repository[3]. The details of these datasets are summarized in Table 1. Due to space limitations, we use three evaluation metrics namely *Hamming loss*, *Average precision*, and *Micro-F1* which define in [2].

4.2   Comparing Methods and Experiment Settings

Multi-label feature selection methods attracted the interest of the researchers in the last decade. In this study, we compare our proposed method LDFM against the recent state-of-the-art multi-label feature selection methods, including GLO-CAL [15], MCLS [7], and MSSL [3]. MCLS and MSSL methods consider feature manifold learning in their studies. ML-KNN ($K = 10$) [13] is used as the multi-label classifier to evaluate the performance of the identified features. The results based on a different number of features, vary from 1 to 100 features. PCA is applied as a prepossessing step with and retain 95% of the data. To ensure a fair comparison, the parameters of the compared methods are tuned to find the optimum values. For GLOCAL, the regularization parameters λ_3 and λ_4 are tuned in

---

[2] https://github.com/alibraytee/LDFM.
[3] http://mulan.sourceforge.net/datasets-mlc.html.

$\{0.0001, 0.001, \ldots, 1\}$, the number of clusters is searched from $\{4, 8, 16, 32, 64\}$, and the latent dimensionality rank is tuned in $\{5, 10, \ldots, 30\}$. MCLS uses the defaults settings for its parameters. For MSSL, the parameters α and β are tuned by searching the grid $\{0.001, 0.01, \ldots, 1000\}$. The parameter settings for LDFM are described in the following section.

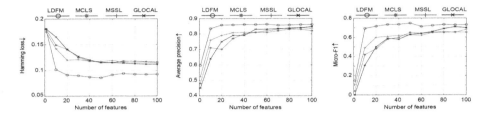

Fig. 3. Comparison of multi-label feature selection algorithms on Scene

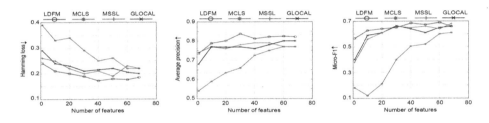

Fig. 4. Comparison of multi-label feature selection algorithms on Emotions

4.3   Results

Classification Results. Several experiments have been conducted to demonstrate the classification performance of LDFM compared to the state-of-the-art multi-label feature selection methods. Figures 3, 4, 5 and 6 show the results in terms of *Hamming loss*, *Average precision*, and *Micro-F1* evaluation metrics on the four datasets. In these figures, the classification results are generated based on top-ranked 100 features (except Emotions which only has 72 features). Based on the experiment results shown in Figs. 3, 4, 5 and 6, interestingly, it is clear that our proposed method has a significant classification improvement with an increasing number of selected features, and then remains stable. Thus, this observation indicates that it is meaningful to study dimensionality reduction in multi-label learning. Further, it highlights the stability and capability of LDFM to achieve good performance on all the datasets with fewer selected features.

The proposed method is compared with the state-of-the-art on each dataset. As shown in Figs. 3, 4, 5 and 6, LDFM achieves better results compared to

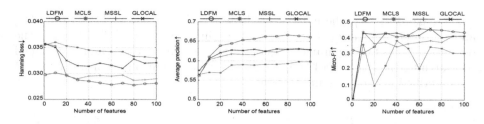

Fig. 5. Comparison of multi-label feature selection algorithms on Reference

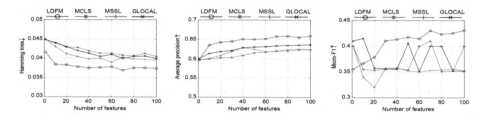

Fig. 6. Comparison of multi-label feature selection algorithms on Computers

MCLS, MSSL, and GLOCAL on almost all the evaluated datasets. Specifically, in terms of the *Hamming loss* evaluation metric, where the smaller the values, the better the performance, LDFM's features substantially improve the classification results compared to the state-of-the-art. It can be observed that MCLS has the worst results and MSSL and GLOCAL achieve comparable results as shown in Figs. 3, 4, and 6. In terms of *Average precision*, and *Micro-F1* evaluation metrics, where the larger the values, the better the performance, LDFM generally achieves better results against the compared methods in all datasets. We note that LDFM performs slightly better than MSSL and GLOCAL on the Emotions dataset using *Micro-F1* metric. We also note that the compared methods produce unstable results on the Reference and Computers datasets using the *Micro-F1* metric. In general, our proposed method demonstrates the great benefits of using label manifolds in encoder-decoder architecture to identify the discriminative features. Furthermore, using the Friedman test, we investigate whether the results that are produced by LDFM are significantly different to the state-of-the-art. In particular, we examine the Friedman test between LDFM against each compared method for each evaluation metric in the four datasets. The statistical results show that the p-value in all tests is less than 0.05 which rejects the null hypothesis that the proposed method and the compared methods have an equal performance. Finally, we explored the reconstruction capability of the decoder in LDFM using the projection matrix W to reconstruct the original data. Table 2 reports the reconstruction errors using the training logical labels (Y), training predicted labels $(\widetilde{Y})$, and the logical testing labels. It is observed that the percentage reconstruction error of the original training matrix using the logical training labels only ranges between 4% to 8% on the four dataset, and this error is dramatically decreased to 0.1% to 3% by using the predicted numerical label matrix. This observation reveals that the decoder plays an important role in selecting the important features which can be used to reconstruct the original matrix. Further, it supports our argument to reconstruct the visual images using the semantic labels and the coefficient matrix. In addition, we report the capability of reconstructing the testing data matrix using the projection matrix and the testing labels with a small error that ranges between 4% to 8%.

Table 2. Reconstruction different error values using the decoder

Dataset	Logical error	Predicted error	Testing error
Scene	0.042	0.032	0.043
Emotions	0.087	0.022	0.086
Reference	0.044	0.001	0.044
Computers	0.051	0.001	0.051

LDFM Results for Handling Missing Labels. In this experiment, to investigate the ability of the proposed method to handling missing labels, we randomly removed different proportions of labels from the samples from moderate to extreme levels: 20%, 40%, 60%, and 80%. Table 3 shows that LDFM achieved consistent improvement over the base especially on 20%, 40%, and 60% missing label levels. Further, it is superior across four datasets using four different missing label proportions.

Table 3. Results on four datasets with different missing label proportions

Dataset↓	Evaluation criteria	Base				LDFM			
	Missing label proportion →	20%	40%	60%	80%	20%	40%	60%	80%
Scene	Hamming loss	0.18	0.24	0.69	0.81	0.10	0.14	0.39	0.78
	Average precision	0.57	0.54	0.51	0.49	0.81	0.79	0.76	0.73
	Micro-F1	0.25	0.32	0.32	0.30	0.67	0.63	0.44	0.32
Computers	Hamming loss	0.04	0.05	0.18	0.91	0.03	0.04	0.17	0.91
	Average precision	0.58	0.56	0.55	0.55	0.61	0.59	0.57	0.56
	Micro-F1	0.39	0.40	0.24	0.09	0.42	0.42	0.26	0.10
Reference	Hamming loss	0.03	0.04	0.15	0.92	0.02	0.03	0.14	0.91
	Average precision	0.54	0.52	0.50	0.50	0.60	0.57	0.55	0.54
	Micro-F1	0.40	0.41	0.20	0.07	0.42	0.45	0.23	0.07
Emotions	Hamming loss	0.23	0.33	0.57	0.65	0.21	0.28	0.51	0.64
	Average precision	0.76	0.74	0.70	0.68	0.79	0.76	0.75	0.72
	Micro-F1	0.62	0.59	0.50	0.48	0.65	0.63	0.54	0.49

4.4   Parameter Sensitivity and Convergence Analysis

In this section, we study the influence of the proposed method's parameters λ and $MaxIteration$ on the classification results. First, λ controls the contribution of the decoder and encoder in the method, however the second parameter defines the number of iterations required to convergence. The parameter λ and $MaxIteration$ are tuned using a grid search from $\{0.2, 0.4, \ldots, 2\}$ and $\{1, 20, 40, \ldots, 100\}$ respectively. As shown in Fig. 7a and 7b, using the average precision metric on the two datasets, we can observe that with an increasing λ, the learning performance is improved. Further, we investigate the convergence of the LDFM optimization method. As shown in Fig. 7c and 7d using two datasets, it is clearly seen that our method converges rapidly and has around 10 iterations which demonstrates the efficacy and speed of our algorithm.

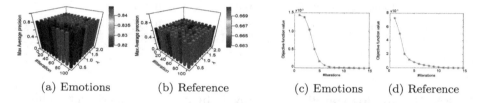

| (a) Emotions | (b) Reference | (c) Emotions | (d) Reference |

Fig. 7. LDFM Results on Emotions and Reference datasets. (a) and (b) the average precision results w.r.t different parameters. (c) and (d) convergence curves

5   Conclusion

This paper proposes a novel semantic multi-label learning model based on an autoencoder. Our proposed method learns the projection matrix to map from the feature space to semantic space back and forth. The semantic labels are predicted in the optimization method because they are not explicitly available from the training samples. We further rank the feature weights in the learned project matrix for feature selection. The proposed method is simple and computationally fast. We demonstrate through extensive experiments that our method outperforms the state-of-the-art. Furthermore, we demonstrate the efficiency of the proposed method to reconstruct the original data using the predicted labels.

References

1. Bartels, R.H., Stewart, G.W.: Solution of the matrix equation ax+ xb= c [f4]. Commun. ACM **15**(9), 820–826 (1972)
2. Braytee, A., Liu, W., Anaissi, A., Kennedy, P.J.: Correlated multi-label classification with incomplete label space and class imbalance. ACM Trans. Intell. Syst. Technol. (TIST) **10**(5), 1–26 (2019)

3. Cai, Z., Zhu, W.: Multi-label feature selection via feature manifold learning and sparsity regularization. Int. J. Mach. Learn. Cybern. **9**(8), 1321–1334 (2017). https://doi.org/10.1007/s13042-017-0647-y
4. Che, X., Chen, D., Mi, J.: A novel approach for learning label correlation with application to feature selection of multi-label data. Inf. Sci. **512**, 795–812 (2020)
5. Cheng, Y., Zhao, D., Wang, Y., Pei, G.: Multi-label learning with kernel extreme learning machine autoencoder. Knowl.-Based Syst. **178**, 1–10 (2019)
6. Hou, P., Geng, X., Zhang, M.L.: Multi-label manifold learning. In: AAAI, pp. 1680–1686. Citeseer (2016)
7. Huang, R., Jiang, W., Sun, G.: Manifold-based constraint Laplacian score for multi-label feature selection. Pattern Recognit. Lett. **112**, 346–352 (2018)
8. Law, A., Ghosh, A.: Multi-label classification using a cascade of stacked autoencoder and extreme learning machines. Neurocomputing **358**, 222–234 (2019)
9. Ranzato, M., Boureau, Y.L., Chopra, S., LeCun, Y.: A unified energy-based framework for unsupervised learning. In: Artificial Intelligence and Statistics, pp. 371–379 (2007)
10. Wang, H., Liu, W., Zhao, Y., Zhang, C., Hu, T., Chen, G.: Discriminative and correlative partial multi-label learning. In: IJCAI, pp. 3691–3697 (2019)
11. Wang, L., Liu, Y., Qin, C., Sun, G., Fu, Y.: Dual relation semi-supervised multi-label learning. In: Proceedings of the AAAI Conference on Artificial Intelligence, vol. 34, pp. 6227–6234 (2020)
12. Xu, L., Wang, Z., Shen, Z., Wang, Y., Chen, E.: Learning low-rank label correlations for multi-label classification with missing labels. In: 2014 IEEE International Conference on Data Mining, pp. 1067–1072. IEEE (2014)
13. Zhang, M.L., Zhou, Z.H.: ML-KNN: a lazy learning approach to multi-label learning. Pattern Recognit. **40**(7), 2038–2048 (2007)
14. Zhang, M.L., Zhou, Z.H.: A review on multi-label learning algorithms. IEEE Trans. Knowl. Data Eng. **26**(8), 1819–1837 (2013)
15. Zhu, Y., Kwok, J.T., Zhou, Z.H.: Multi-label learning with global and local label correlation. IEEE Trans. Knowl. Data Eng. **30**(6), 1081–1094 (2018)

AutoCluster: Meta-learning Based Ensemble Method for Automated Unsupervised Clustering

Yue Liu[1,2,3(✉)] [iD], Shuang Li[1], and Wenjie Tian[1]

[1] School of Computer Engineering and Science, Shanghai University, Shanghai, China
{yueliu,aimeeli,twenjie}@shu.edu.cn
[2] Shanghai Institute for Advanced Communication and Data Science,
Shanghai, China
[3] Shanghai Engineering Research Center of Intelligent Computing System,
Shanghai 200444, China

Abstract. Automated clustering automatically builds appropriate clustering models. The existing automated clustering methods are widely based on meta-learning. However, it still faces specific challenges: lacking comprehensive meta-features for meta-learning and general clustering validation index (CVI) as objective function. Therefore, we propose a novel automated clustering method named AutoCluster to address these problems, which is mainly composed of Clustering-oriented Meta-feature Extraction (CME) and Multi-CVIs Clustering Ensemble Construction (MC^{2}EC). CME captures the meta-features from spatial randomness and different learning properties of clustering algorithms to enhance meta-learning. MC^{2}EC develops a collaborative mechanism based on clustering ensemble to balance the measuring criterion of different CVIs and construct more appropriate clustering model for given datasets. Extensive experiments are conducted on 150 datasets from OpenML to create meta-data and 33 test datasets from three clustering benchmarks to validate the superiority of AutoCluster. The results show the superiority of AutoCluster for building an appropriate clustering model compared with classical clustering algorithms and CASH method.

Keywords: Clustering · Automated machine learning · Meta-learning · Model selection · Clustering ensemble

1 Introduction

Clustering, one of the most popular unsupervised learning methods, divides instances into clusters where instances in same cluster are similar while in different clusters are dissimilar [7]. However, algorithm selection and hyperparameter optimization are still two of the most challenging tasks for clustering problem.

In order to build high-quality clustering models, automated clustering as the subtask of Automated Machine Learning (AutoML) [20] has been proposed to address the above challenges. Existing automated clustering methods are widely

© Springer Nature Switzerland AG 2021
K. Karlapalem et al. (Eds.): PAKDD 2021, LNAI 12714, pp. 246–258, 2021.
https://doi.org/10.1007/978-3-030-75768-7_20

based on meta-learning [1,3,4,6,9,13]. They learned from prior experience how different clustering models perform across datasets to speed up model design for given datasets [17,20]. Despite the recent progress of meta-learning used in automated clustering, it still faces two specific problems: lacking comprehensive meta-features for meta-learning and general clustering validation index (CVI) as objective function in optimization process.

Meta-features play an important role in selecting promising algorithms or configurations in meta-learning based automated clustering. Most of the existing meta-features are extracted from labeled data, while applicable meta-features proposed for automated clustering are still incomprehensive. Reference [3] first studied meta-learning in clustering algorithm selection but they only used statistical meta-features. Later, meta-features from instances distance, link constraints, internal measures, and correlation for clustering datasets are proposed respectively [1,4,13,19]. However, data distribution and the learning schema of clustering model are also important to characterize clustering datasets in meta-learning, which are related to the intrinsic features of clustering datasets.

Moreover, clustering validation indexes (CVIs) are used to measure the quality of clustering results. CVIs with different measuring criteria are suitable for specific clustering datasets and algorithms. Therefore, no general CVI is consistently superior to others in clustering validation [2,11], which is one of the biggest challenges for model optimization in automated clustering. Reference [4] and [13] combined multiple CVIs and ranked algorithms based on their performance to choose the appropriate one. However, they are not robust since the error selection under any CVI can affect the overall algorithm ranking. In addition, the use of internal CVIs remains uncertain in hyperparameter optimization process for clustering evaluation. Hence, these methods still can not alleviate the dilemma of lacking general CVI.

In this paper, we propose a novel meta-learning based automated clustering method named AutoCluster to address the above problems, which is composed of Clustering-oriented Meta-feature Extraction (CME) and Multi-CVIs Clustering Ensemble Construction (MC^2EC). The contributions of our work are highlighted as follows.

- In order to provide a more comprehensive characterization of clustering datasets, we propose CME for meta-learning. It extracts clustering-oriented meta-features from spatial randomness of data distribution and landmarker, i.e. running simple landmark clustering algorithms to fleetly capture the learning scheme, to enhance meta-learning.
- In order to alleviate the dilemma of lacking general CVI, we propose MC^2EC for clustering model construction. It optimizes hyperparameters of promising algorithms suggesting by meta-learning under different CVIs and combines them to construct an ensemble model. Therefore, it provides a collaborative mechanism to balance the measuring criteria of different CVIs for discovering better clusters.
- In order to effectively build an appropriate clustering model for given datasets, we incorporate CME with MC^2EC, and propose AutoCluster. It determines

promising clustering algorithms through CME-enhanced meta-learning under multiple CVIs, and performs automated ensemble construction based on these algorithms through MC²EC to provide appropriate clustering model.
– Finally, extensive experiments are conducted with a wide range of datasets from OpenML [18] and clustering benchmarks [5,16], as well as various clustering algorithms from scikit-learn [12]. The results show the superiority of AutoCluster for building appropriate clustering model compared with classical clustering algorithms and CASH method.

The remainder of this paper is as follows: Sect. 2 presents the proposed automated clustering: AutoCluster. The extensive experiments for AutoCluster are analyzed in Sect. 3. Finally, we conclude this work in Sect. 4.

2 AutoCluster: Toward Automated Unsupervised Clustering

2.1 The Goal and Process of AutoCluster

Automated clustering automatically builds appropriate clustering model for given datasets. In the process, AutoCluster has two specific problems: i) What meta-features can characterize unlabeled clustering datasets? ii) How to measure clustering result impartially? Thus, the goal of AutoCluster can be defined as follows.

Definition 1 (AutoCluster). For $i = 1, 2, ..., d$, let x_i denote a feature vector of instance i without target value from clustering dataset D. Given a set of clustering algorithms $A = \{A^1, A^2, ..., A^m\}$, and let the hyperparameters of each clustering algorithm A^i have domain θ^i. The goal of AutoCluster is to discover more reasonable clusters π^* as Eq. 1.

$$\pi^* = \underset{\pi \in Comb(A^*, \lambda^*)}{\arg\min} \ C\left(\left\{\underset{A^i \in A, \lambda^i \in \theta^i}{\arg\min} \ CVI_j\left(A^i, \lambda^i, D\right) | j = 1, ..., c\right\}\right) \quad (1)$$

where $CVI_j\left(A^i, \lambda^i, D\right)$ denotes CVI measured by clustering algorithm A^i with hyperparameter λ^i on dataset D, and c is the number of CVI in AutoCluster. Since AutoCluster handles the dilemma of lacking general CVI for model optimization by clustering ensemble, $C(\cdot)$ represents consensus function to combine clustering model A^* with hyperparameter λ^* optimized with individual CVI.

As shown in Fig. 1, it is mainly composed of Clustering-oriented Metafeature Extraction (CME) enhanced meta-learning and Multi-CVIs Clustering Ensemble Construction (MC²EC). For CME, traditional and clustering-oriented meta-features are extracted from data distribution and landmarker. The performance data with multiple CVIs is collected for meta-decision data and meta-auxiliary data. For MC²EC, it optimizes hyperparameters of promising algorithms suggesting from CME-enhanced-meta-learning under CVI metrics respectively through grid search and combines these clustering results to construct

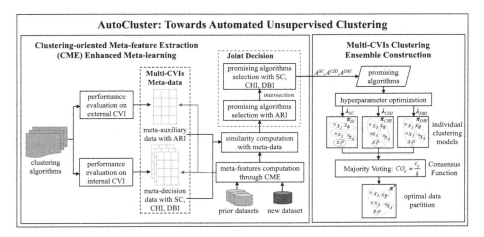

Fig. 1. The process of AutoCluster

ensemble model through Majority Voting. Multiple CVIs and hyperparameter optimization process ensure AutoCluster can obtain diverse individual models with high quality for clustering ensemble to discover better data partition.

2.2 Enhanced Meta-learning for Finding Promising Algorithms

As the fundament of AutoCluster, meta-data is composed of meta-feature matrix and performance data. Here, we introduce the formulation of them respectively.

Clustering-Oriented Meta-feature Extraction (CME) for Similarity Computation. CME extracts five new clustering-oriented meta-features from data distribution and landmarker to provide a more comprehensive characterization for given datasets. It also extracts 19 traditional meta-features from [10]. The summary of meta-features is depicted in the supplementary material[1].

The first meta-feature is from data distribution. Different clustering algorithms are suitable for the data with a specific distribution. Thus, data distribution is an important meta-feature in promising algorithm selection. Hopkins Statistic tests the spatial randomness of data distribution and also for cluster tendency which is defined as Eq. 2.

$$H = \frac{\sum_{i=1}^{d} u_i}{\sum_{i=1}^{d} u_i + \sum_{i=1}^{d} w_i} \tag{2}$$

where u_i represents the distance from d' sampling instances placed at random in the subspace of the entire h-dimensional sample space where $d' \gg d$ to its nearest neighbor in dataset and w_i represents the distance from a randomly selected

---

[1] The supplementary material of this paper is available at https://github.com/wj-tian/AutoCluster.

instance to its nearest neighbor. For example, the hopkins statistic of regularly spaced and clustered datasets are always around 0.01 to 0.3 and 0.7 to 0.99 respectively. Thus, it can be an important meta-feature to describe datasets. Moreover, the learning properties of landmark clustering algorithms reflect the relative performance on given datasets, which are captured by learning scheme of landmark clustering algorithms. Specially, three different clustering algorithms are applied to extract landmarker meta-features: 1) The distance of the instances to their closest cluster center through KMeans (partition-based), which measures the compactness of cluster. 2) The number of leaves in the hierarchical tree through Agglomerative Clustering (hierarchy-based). 3) The reachability distances of instances and distance at which each instance becomes a core point through OPTICS (density-based), which measures density around an instance.

We denote $F_i = \{f_1, f_2, ... f_{24}\}$ to be a feature vector of the enhanced meta-features of dataset D_i, where $\{f_1, ..., f_{19}\}$ are the traditional meta-features from [10] and $\{f_{20}, ..., f_{24}\}$ are clustering-oriented meta-features proposed by us. The distance metric between datasets determines how to find promising algorithms or configurations from the nearest dataset. In AutoCluster, we define p-norm distance in meta-features space to measure the similarity of datasets, which is computed as Eq. 3.

$$d_F = ||F_i - F_j||_p \tag{3}$$

Multi-performance Data for Promising Algorithms Selection. Since AutoCluster develops a collaborative mechanism based on clustering ensemble to address the problem of lacking general CVI, every entry in performance data records the performance measured by different CVIs. We employ three internal CVIs, including two center-based representatives, Calinski-Harabasz Index (CHI) and Davies-Bouldin Index (DBI), and one non-center-based representative, Silhouette Coefficient (SC). They measure intra-cluster compactness and inter-cluster separation of a cluster from different criteria. Meanwhile, the prior datasets in meta-data have ground true labels. Thus, auxiliary information is extracted from external CVI. We apply Adjusted Rand index (ARI) to create meta-auxiliary data. The promising algorithm selection by ARI provides auxiliary information for internal CVIs as performing intersection to conduct joint decision and provide more promising algorithms.

2.3   Multi-CVIs Clustering Ensemble for Model Construction

No general CVI as objective function to measure clustering model impartially is one of the biggest challenges for model optimization in automated clustering. The important improvement of AutoCluster is to employ clustering ensemble to address this problem. In order to obtain better ensemble model, it requires diverse (making uncorrelated errors) and high-quality individuals for combination [15]. The application of multiple CVIs ensures the diverse generation of individuals. For high-quality individuals, MC^{2}EC adopts grid search for a better configuration of promising algorithms. Suppose that a collection of optimized models through grid search are obtained. The ensemble method apply in MC^2EC is Majority Voting [8] to combine them, which is not conditioned by any particular clustering algorithm. It assumpts that neighboring instances in ground true cluster are still likely to partition into same cluster, and then formulates consensus function based on co-association matrix to render pairs of instances voting for association in each partition produced by different clustering model. Each (i,j)th entry of instance x_i and x_j in co-association matrix is calculated as Eq. 4.

$$CO_{ij} = \frac{c_{ij}}{n_{SC} + n_{CHI} + n_{DBI}} \qquad (4)$$

where n represents the number of optimized models with each CVI, and c_{ij} counts instance pair (x_i, x_j) co-occurring in same cluster. For final ensemble model, Majority Voting compares CO_{ij} in co-association matrix with a defined threshold θ. Here, the final partition is formed with multiple CVIs.

3 Experiments

3.1 Datasets and Clustering Algorithms

The evaluation of AutoCluster used 150 datasets in OpenML [18] sorted by most runs and selected by filtering with no more than 5000 samples and 50 features to create meta-data. Besides, 33 datasets are collected to test AutoCluster from Clustering basic benchmark [5][2], Fundamental clustering problem suite (FCPS) [16][3] and Tomas Barton's clustering benchmark[4]. More dataset information is described in Table 1. Six clustering algorithms are involved: KMeans, Affinity Propagation, Mean Shift, Agglomerative Clustering, DBSCAN, and Birch, which are implemented in scikit-learn [12] and corresponding hyperparameter spaces are depicted in the supplementary material.

Table 1. The summary of test datasets

No.	AutoCluster	Classes	Data points	Dimensions	No.	AutoCluster	Classes	Data points	Dimensions
1	a1	20	3000	2	18	Lsun	3	400	2
2	Aggregation	7	788	2	19	Lsun3D	4	404	3
3	aml28	5	804	2	20	Pathbased	3	300	2
4	Atom	2	800	3	21	R15	15	600	2
5	balance-scale	3	625	4	22	s1	15	5000	2
6	Compound	6	399	2	23	s2	15	5000	2
7	curves1	2	1000	2	24	s3	15	5000	2
8	curves2	2	1000	2	25	smile1	4	1000	2
9	D31	13	1232	2	26	Target	6	770	2
10	dietary_survey_IBS	2	400	42	27	Tetra	4	400	3
11	dim32	16	1024	32	28	unbalanced	8	6500	2
12	Flame	2	240	2	29	WingNut	2	1016	2
13	fourty	40	1000	2	30	zelnik1	3	299	2
14	gaussians1	2	100	2	31	zelnik5	4	512	2
15	Hepta	7	212	3	32	zelnik6	3	238	2
16	hypercube	8	800	3	33	zoo	7	101	16
17	Jain	2	373	2					

---

[2] http://cs.uef.fi/sipu/datasets/.
[3] https://www.uni-marburg.de/fb12/arbeitsgruppen/datenbionik/data.
[4] https://github.com/deric/clustering-benchmark.

3.2 Experimental Setup

All datasets are preprocessed by removing missing values, one-hot encoding for categorical features, and z-score standardization for all features. Our experiments selected three most similar datasets to perform majority selection of promising algorithms. 2-norm is used to compute the distance between datasets. The clustering ensemble is based on OpenEnsemble [14] and the threshold is set as 0.5. Moreover, our experiment is repeated 10 times to take the mean, and ARI is computed to measure the performance of AutoCluster to build clustering model. Then, the distance between hyperparameter configurations (HCD) is used to measure the diversity of individuals as defined in Eq. 5.

$$HCD = \frac{2}{N(N-1)} \sum_{i=1}^{N} \sum_{j=i+1}^{N} d^\lambda \left(\lambda^i, \lambda^j \right) \tag{5}$$

where $N = n_{SC} + n_{CHI} + n_{DBI}$, and $d^\lambda \left(\lambda^i, \lambda^j \right)$ is equal to 1 when the algorithms of hyperparameter configuration λ^i and λ^j are different while the ratio of different hyperparameter values when λ^i and λ^j are for the same algorithm.

All procedures are run on Linux operating system with Intel(R) Xeon(R) Gold 6130 CPU @ 2.10 GHz processor. The process of running on a specific dataset is limited to a single CPU core.

3.3 Experimental Results and Analysis

The Performance Evaluation of AutoCluster. When tackling a specific problem by unsupervised clustering, many users lack enough experience to choose right algorithm or hyperparameter. It leads them to tend to choose algorithms with high reputations such as KMeans, and leave hyperparameter as default value or tuning the number of clusters with trial and error. In this basic experiment, we compare with six clustering algorithms with default hyperparameter values (KM-d, AP-d, MS-d, AC-d, DB-d, BI-d), and three KMeans algorithms with the number of clusters from 2 to 20 under SC, CHI, DBI respectively (SC-K, CHI-K, DBI-K). The result shown in Table 2 can be observed that Auto-Cluster obtains the highest ARI on 15/33 test datasets (the other test datasets are also close to the best methods), and the average ARI (0.776) dramatically surpasses the compared methods. Moreover, AutoCluster has a more stable prediction on these datasets since other compared methods only can perform well on few datasets, and these compared methods have more test datasets performing significantly worse. Thus, AutoCluster on different categories of datasets to automatically discover appropriate clusters is effective.

The Performance Comparison of AutoCluster with CASH. In this experiment, we compare with the most classic method named CASH to verify the superior of AutoCluster, in which the clustering algorithm selection is viewed as a super-hyperparameter and executed with hyperparameter optimization simultaneously. The objective function respectively sets as internal CVI

Table 2. The comparison with default hyperparameter values and simple optimization. For these compared methods, if the performance is highest, the corresponding entries are bolded, and if the performance is significantly worse than the highest performance on this dataset (lower than 0.3), the corresponding entries are underlined.

No.	AutoCluster	KM-d	AP-d	MS-d	AC-d	DB-d	BI-d	SC-K	CHI-K	DBI-K
1	0.904	<u>0.424</u>	<u>0.558</u>	<u>0.091</u>	<u>0.092</u>	<u>0.000</u>	<u>0.174</u>	**0.936**	0.930	0.835
2	**0.991**	<u>0.664</u>	<u>0.391</u>	<u>0.628</u>	<u>0.377</u>	<u>0.000</u>	<u>0.396</u>	0.762	<u>0.377</u>	0.762
3	0.996	<u>0.333</u>	<u>0.043</u>	0.975	0.859	0.999	0.959	0.975	<u>0.396</u>	**1.000**
4	0.528	**0.576**	0.519	0.548	<u>0.067</u>	0.568	<u>0.149</u>	0.547	0.537	0.535
5	0.121	**0.136**	0.034	0.000	0.121	0.000	0.111	0.127	0.095	0.104
6	**0.745**	0.456	<u>0.319</u>	0.722	0.484	0.740	0.734	0.721	0.721	0.721
7	1.000	<u>0.249</u>	<u>0.010</u>	1.000	1.000	1.000	0.777	1.000	<u>0.099</u>	1.000
8	0.523	0.249	<u>0.061</u>	<u>0.179</u>	<u>0.019</u>	<u>0.000</u>	<u>0.130</u>	<u>0.199</u>	<u>0.098</u>	<u>0.098</u>
9	0.704	<u>0.670</u>	0.736	<u>0.254</u>	<u>0.126</u>	<u>0.319</u>	<u>0.274</u>	0.885	**0.976**	0.885
10	1.000	<u>0.360</u>	<u>0.137</u>	<u>0.689</u>	1.000	<u>0.000</u>	0.784	1.000	1.000	1.000
11	1.000	<u>0.514</u>	<u>0.363</u>	0.000	<u>0.123</u>	0.883	<u>0.175</u>	1.000	1.000	1.000
12	**0.635**	<u>0.205</u>	<u>0.128</u>	0.524	<u>0.289</u>	<u>0.013</u>	<u>0.278</u>	0.425	<u>0.204</u>	0.426
13	**0.771**	<u>0.277</u>	0.614	<u>0.000</u>	<u>0.045</u>	<u>0.000</u>	<u>0.078</u>	0.636	0.628	0.625
14	1.000	<u>0.262</u>	<u>0.498</u>	1.000	1.000	1.000	1.000	1.000	1.000	1.000
15	1.000	0.958	0.720	<u>0.000</u>	<u>0.269</u>	1.000	<u>0.354</u>	1.000	1.000	1.000
16	1.000	1.000	1.000	<u>0.000</u>	<u>0.222</u>	1.000	<u>0.426</u>	1.000	1.000	1.000
17	**0.893**	<u>0.167</u>	<u>0.105</u>	<u>0.000</u>	0.569	<u>0.000</u>	0.531	<u>0.553</u>	<u>0.075</u>	<u>0.300</u>
18	0.528	0.390	<u>0.239</u>	0.420	<u>0.282</u>	<u>0.000</u>	0.559	**0.583**	0.326	**0.583**
19	0.864	<u>0.449</u>	<u>0.293</u>	**0.881**	0.602	<u>0.532</u>	0.788	0.599	0.599	0.599
20	**0.614**	0.404	<u>0.235</u>	<u>0.063</u>	0.414	<u>0.000</u>	0.468	0.480	<u>0.200</u>	<u>0.205</u>
21	0.989	<u>0.264</u>	<u>0.693</u>	<u>0.000</u>	<u>0.041</u>	<u>0.264</u>	<u>0.099</u>	**0.993**	**0.993**	**0.993**
22	<u>0.330</u>	0.500	<u>0.388</u>	<u>0.182</u>	<u>0.133</u>	<u>0.000</u>	<u>0.189</u>	0.551	**0.739**	0.552
23	0.916	<u>0.585</u>	<u>0.203</u>	<u>0.113</u>	<u>0.121</u>	<u>0.000</u>	<u>0.232</u>	0.938	0.938	**0.938**
24	0.685	0.481	<u>0.172</u>	<u>0.098</u>	<u>0.110</u>	<u>0.000</u>	<u>0.217</u>	**0.727**	0.727	0.727
25	0.749	<u>0.659</u>	<u>0.538</u>	<u>0.331</u>	<u>0.326</u>	1.000	<u>0.290</u>	0.707	<u>0.674</u>	<u>0.681</u>
26	<u>0.691</u>	<u>0.612</u>	<u>0.462</u>	<u>0.628</u>	<u>0.252</u>	1.000	<u>0.259</u>	<u>0.563</u>	<u>0.178</u>	<u>0.299</u>
27	1.000	<u>0.607</u>	<u>0.295</u>	<u>0.000</u>	<u>0.332</u>	<u>0.687</u>	0.713	1.000	1.000	1.000
28	0.998	1.000	<u>0.193</u>	<u>0.612</u>	<u>0.124</u>	<u>0.610</u>	<u>0.127</u>	0.998	<u>0.450</u>	0.998
29	1.000	<u>0.226</u>	<u>0.073</u>	<u>0.000</u>	1.000	<u>0.000</u>	<u>0.476</u>	<u>0.670</u>	<u>0.264</u>	<u>0.156</u>
30	<u>0.308</u>	<u>0.280</u>	<u>0.266</u>	<u>0.000</u>	<u>0.024</u>	**0.630**	<u>0.112</u>	<u>0.288</u>	<u>0.260</u>	<u>0.269</u>
31	0.773	0.721	<u>0.360</u>	<u>0.000</u>	<u>0.310</u>	1.000	<u>0.226</u>	<u>0.603</u>	<u>0.301</u>	<u>0.437</u>
32	0.768	0.822	0.615	**0.829**	0.580	0.813	0.675	0.779	<u>0.323</u>	0.779
33	0.586	0.691	<u>0.480</u>	<u>0.034</u>	<u>0.448</u>	<u>-0.052</u>	0.677	<u>0.355</u>	**0.818**	<u>0.356</u>
Ave	**0.776**	0.491	0.356	0.327	0.356	0.424	0.407	0.715	0.574	0.663

in AutoCluster (CASH-SC, CASH-CHI, and CASH-DBI). The result shown in Table 3 can be observed that AutoCluster performs better than CASH methods, which obtains the highest average ARI (0.776) and highest performance on 21/33 datasets. Moreover, AutoCluster is more stable over this experiment with one worse performance. CASH-SC, CASH-CHI, and CASH-DBI are uncomparable where the number of bolded/underlined entries is 12/5, 12/13, and 4/15 respectively. Thus, it can conclude that it is infeasible to directly introduce AutoML methods in supervised learning to unsupervised clustering problems to discover optimal partition.

Why Does AutoCluster Work Well? AutoCluster performs the effectiveness and superiority of discovering appropriate clusters for users automatically in comparison with classical clustering algorithms and CASH method. In this section, ablation studies of AutoCluster are conducted to interpret why Auto-Cluster works well and whether its components are reasonable.

The Importance of CME. In order to discover which meta-features are more important to algorithm selection, F-test is employed to evaluate the significant influence of meta-features on each algorithm. For each CVI and each algorithm, two groups of samples in meta-data are divided according to whether this algorithm is selected as a promising one under this CVI. Then, we calculate F-statistic between these two groups, which reflects whether the meta-features of any algorithm selected differ significantly from those of the algorithm not selected. We show the result of CHI as Fig. 2. From this figure, CME has a significant influence on algorithm selection, such as Agglomerative Clustering (No. 20 (*Hopkins Statistic*)), DBSCAN (No. 22 (*Agglomerative Clustering*)) and Mean Shift (No. 21 (*KMeans*)), which illustrates the importance of CME in AutoCluster. The remaining experimental results are described in the supplementary material.

Table 3. The comparison with three CASH methods

No.	AutoCluster	CASH-SC	CASH-CHI	CASH-DBI	No.	AutoCluster	CASH-SC	CASH-CHI	CASH-DBI
1	0.904	**0.936**	0.916	0.424	18	0.528	**0.584**	0.319	0.582
2	**0.991**	0.776	0.259	0.771	19	**0.864**	0.597	0.599	0.591
3	**0.996**	0.975	0.104	0.988	20	**0.614**	0.476	0.143	0.539
4	0.528	**0.540**	0.244	0.518	21	0.989	**0.993**	0.993	0.851
5	0.121	**0.166**	0.082	0.022	22	0.330	0.548	**0.757**	0.595
6	0.745	0.721	0.721	**0.768**	23	0.916	0.938	**0.938**	0.224
7	**1.000**	1.000	0.068	0.000	24	0.685	0.711	**0.726**	0.110
8	**0.523**	0.199	0.068	0.000	25	**0.749**	0.733	0.649	0.411
9	0.704	0.855	**0.976**	0.792	26	0.691	0.242	0.105	0.049
10	1.000	1.000	1.000	0.501	27	1.000	**1.000**	1.000	0.667
11	1.000	1.000	1.000	0.464	28	**0.998**	0.998	0.287	0.353
12	**0.635**	0.425	0.203	0.009	29	1.000	0.693	0.081	-0.001
13	0.771	0.806	**0.845**	0.748	30	0.308	0.225	0.201	0.238
14	1.000	1.000	1.000	1.000	31	**0.773**	0.617	0.217	0.233
15	1.000	1.000	1.000	0.901	32	0.768	0.779	0.222	**0.858**
16	1.000	1.000	1.000	1.000	33	0.586	0.177	0.129	0.266
17	**0.893**	0.553	0.053	0.877	Ave	**0.776**	0.705	0.512	0.495

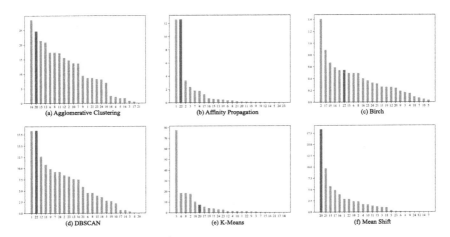

Fig. 2. The F-statics of meta-features grouped by the selected algorithm under CHI metric. No. 1–19 are for traditional meta-features from [10]. No. 20 is for the meta-feature of hopkins statistic and No. 21–24 are for landmarker meta-features

The Effectiveness of Clustering Ensemble. Clustering ensemble requires diverse and high-quality individual clustering models to construct ensemble model. Multiple CVIs and meta-learning with grid search ensure these two conditions in AutoCluster. HCD evaluates the diversity of hyperparameter configuration of individuals. The result shown in Fig. 3 illustrates the scores of HCD in clustering ensemble of all test datasets are higher than 0.75 and 31/33 datasets are higher than 0.9. Therefore, the high diversity of individuals in clustering ensemble makes AutoCluster effective to perform better.

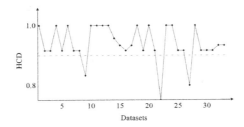

Fig. 3. The HCD of clustering ensemble on test datasets

Fig. 4. The comparison of disabling meta-learning and multiple CVIs

The Comparison of Disabling Meta-learning and Multiple CVIs. The key components of AutoCluster are CME-enhanced meta-learning and MC^{2}EC. Thus, we compare with the following methods: 1) No-D, No-K, No-A, No-B: They disable meta-learning. No-D is heterogeneous ensemble with six clustering algorithms with default hyperparameters involved in AutoCluster. No-K, No-A, and No-B are isomorphic clustering ensemble of KMeans, Agglomerative Clustering, and Birch with different numbers of clusters. 2) SC-E, CHI-E, DBI-E: These methods disable multiple CVIs. They execute meta-learning and clustering ensemble under a single CVI. As shown in Fig. 4, AutoCluster achieves drastic improvement with meta-learning and multiple CVIs. It has the highest performance on 18/33 test datasets, and the best average ARI of these two categories of methods are 0.532 and 0.675 respectively, which are well below 0.776 of AutoCluster. Since the methods disabling meta-learning lead to the individuals with low quality and the methods disabling multiple CVIs lead to the individuals with low diversity, they both fail to discover promising partition. Therefore, the result can show the necessity of meta-learning and multiple CVIs in AutoCluster.

The Comparison of Selection Strategy of Individuals in Clustering Ensemble. The evaluation of the individual selection based on ARI in AutoCluster is compared with three methods: 1) No-ARI: It directly executes clustering ensemble from meta-learning under internal CVIs. 2) SIM: It selects individuals based on similarity of clustering result measuring by Normalized Mutual Information. When it is greater than 0.8, one of these two models is removed from ensemble. 3) MAJ: When the algorithms are selected by majority (2/3) CVIs, their corresponding models are selected for final ensemble.

Fig. 5. The comparison of selection strategy of individual models

The result is shown in Fig. 5. Since the measure criteria of internal CVIs are difficult to fit the ground true label, it is important for meta-learning to provide external auxiliary information for clustering ensemble. No-ARI directs to select individuals without auxiliary information, where it only performs best on 11 test datasets, worse than AutoCluster of 23, and the average ARI is 0.131 lower than AutoCluster. SIM inevitably removes promising individuals when the high similarity between them. Thus, it fails to surpass AutoCluster where the number of best datasets is 7 and the average ARI is 0.526. MAJ method also can remove promising individuals since different criteria of CVIs. It leads the bad performance compared with AutoCluster with 12 best datasets and the average ARI of 0.567. Therefore, it shows the feasibility of individual selection to provide external auxiliary information in AutoCluster. In addition, the runtime of AutoCluster on 33 test datasets is also considered to verify the efficiency of this method, which is depicted in the supplementary material.

4 Conclusions

Automated clustering based on meta-learning faces its specific problems: lacking comprehensive meta-features and general CVI. This paper proposes a novel automated clustering method named AutoCluster, mainly composed of CME and MC^{2}EC. CME extracts five clustering-oriented meta-features to extend traditional meta-features from spatial randomness and learning properties of clustering algorithms. MC^{2}EC develops a collaborative mechanism to balance the measuring criterion of different CVIs based on clustering ensemble. Extensive experiments are conducted with a wide range of datasets from OpenML and three- clustering benchmarks. The results show that AutoCluster has strong ability to construct appropriate clustering model than compared methods.

Meta-learning and clustering ensemble are important to promote the performance of AutoCluster. Hence, applying meta-features into manifold space or importance-weighted space is promising in future works. For clustering ensemble, optimized clustering ensemble methods like evidence accumulation also can be applied to improve the performance of AutoCluster.

Acknowledgment. This work is supported by the National Natural Science Foundation of China (No. 52073169) and the State Key Program of National Nature Science Foundation of China (Grant No. 61936001).

References

1. Adam, A., Blockeel, H.: Dealing with overlapping clustering: a constraint-based approach to algorithm selection. In: Meta-Learning and Algorithm Selection workshop-ECMLPKDD2015, vol. 1, pp. 43–54 (2015)
2. Arbelaitz, O., Gurrutxaga, I., Muguerza, J., PéRez, J.M., Perona, I.: An extensive comparative study of cluster validity indices. Pattern Recogn. **46**(1), 243–256 (2013)
3. De Souto, M.C., et al.: Ranking and selecting clustering algorithms using a meta-learning approach. In: 2008 IEEE International Joint Conference on Neural Networks, pp. 3729–3735 (2008)
4. Ferrari, D.G., De Castro, L.N.: Clustering algorithm selection by meta-learning systems: a new distance-based problem characterization and ranking combination methods. Inf. Sci. **301**, 181–194 (2015)
5. Fränti, P., Sieranoja, S.: K-means properties on six clustering benchmark datasets. Appl. Intell. **48**(12), 4743–4759 (2018)
6. Garg, V., Kalai, A.T.: Supervising unsupervised learning. Adv. Neural Inf. Process. Syst. **31**, 4991–5001 (2018)
7. Jain, A.K.: Data clustering: 50 years beyond k-means. Pattern Recogn. Lett. **31**(8), 651–666 (2010)
8. Jamali, N., Sammut, C.: Majority voting: material classification by tactile sensing using surface texture. IEEE Trans. Robot. **27**(3), 508–521 (2011)

9. José-García, A., Gómez-Flores, W.: Automatic clustering using nature-inspired metaheuristics: a survey. Appl. Soft Comput. **41**, 192–213 (2016)
10. Li, Y.F., Wang, H., Wei, T., Tu, W.W.: Towards automated semi-supervised learning. In: AAAI, vol. 33, pp. 4237–4244 (2019)
11. Liu, Y., Li, Z., Xiong, H., Gao, X., Wu, J.: Understanding of internal clustering validation measures. In: ICDM, pp. 911–916 (2010)
12. Pedregosa, F., et al.: Scikit-learn: machine learning in python. J. Mach. Learn. Res. **12**, 2825–2830 (2011)
13. Pimentel, B.A., de Carvalho, A.C.: A new data characterization for selecting clustering algorithms using meta-learning. Inf. Sci. **477**, 203–219 (2019)
14. Ronan, T., Anastasio, S., Qi, Z., Sloutsky, R., Naegle, K.M., Tavares, P.H.S.V.: Openensembles: a python resource for ensemble clustering. J. Mach. Learn. Res. **19**(1), 956–961 (2018)
15. Topchy, A., Jain, A.K., Punch, W.: Combining multiple weak clusterings. In: Proceedings of the Third IEEE International Conference on Data Mining, pp. 331–338 (2003)
16. Ultsch, A.: Clustering with som: U* c. In: Proceedings of the Workshop on Self-Organizing Maps, 2005 (2005)
17. Vanschoren, J.: Meta-learning: a survey. CoRR abs/1810.03548 (2018)
18. Vanschoren, J., Van Rijn, J.N., Bischl, B., Torgo, L.: OpenML: networked science in machine learning. ACM SIGKDD Explor. Newsl. **15**(2), 49–60 (2014)
19. Vukicevic, M., Radovanovic, S., Delibašić, B., Suknovic, M.: Extending meta-learning framework for clustering gene expression data with component based algorithm design and internal evaluation measures. Int. J. Data Min. Bioinform. **14**, 101–119 (2016)
20. Zöller, M., Huber, M.F.: Benchmark and survey of automated machine learning frameworks. J. Artif. Intell. Res. **70**, 409–472 (2021)

BanditRank: Learning to Rank Using Contextual Bandits

Phanideep Gampa[1]($\boxtimes$) and Sumio Fujita[2]

[1] Indian Institute of Technology (BHU) Varanasi, Varanasi, India
gampa.phanideep.mat15@iitbhu.ac.in
[2] Yahoo Japan Corporation, Tokyo, Japan
sufujita@yahoo-corp.jp

Abstract. We propose an extensible deep learning method that uses reinforcement learning to train neural networks for offline ranking in information retrieval (IR). We call our method BanditRank as it treats ranking as a contextual bandit problem. In the domain of learning to rank for IR, current deep learning models are trained on objective functions different from the measures they are evaluated on. Since most evaluation measures are discrete quantities, they cannot be used by gradient descent algorithms without approximation. BanditRank bridges this gap by directly optimizing a task specific measure, such as mean average precision (MAP). Specifically, a contextual bandit whose action is to rank input documents is trained using a policy gradient algorithm to directly maximize a reward. The reward can be a single measure, such as MAP, or a combination of several measures. The notion of ranking is also inherent in BanditRank, similar to the current *listwise* approaches. To evaluate the effectiveness of BanditRank by answering five research questions, we conducted a series of experiments on datasets related to three different tasks, i.e., non-factoid, and factoid question answering and web search. We found that BanditRank performed better than strong baseline methods in respective tasks.

Keywords: Contextual bandits · Policy gradient · REINFORCE · Information retrieval · Learning to rank · Question answering · Web search

1 Introduction

Learning to rank is an important sub-field of information retrieval and involves designing models that rank documents corresponding to a query in order of their

---

P. Gampa—Work conducted while the first author was in research internship at Yahoo! JAPAN Research.

Electronic supplementary material The online version of this chapter (https://doi.org/10.1007/978-3-030-75768-7_21) contains supplementary material, which is available to authorized users.

K. Karlapalem et al. (Eds.): PAKDD 2021, LNAI 12714, pp. 259–271, 2021.
https://doi.org/10.1007/978-3-030-75768-7_21

relevance. Considering the type of learning approach used, all ranking models can be classified into three categories, i.e., pointwise, pairwise, and listwise. The ranking models are either trained on indirect objective functions, such as classification related functions, or direct objective functions related to the evaluation measures. Direct optimization of IR measures has been a long standing challenge in the learning-to-rank domain. If we only consider bounded IR measures such as MAP, a theoretical justification is provided regarding the superiority of direct optimization techniques according to [27], which states that if an algorithm can directly optimize an IR measure on training data, the ranking function learned with the algorithm will be one of the best ranking functions one can obtain in terms of expected test performance with respect to the same IR measure. Several algorithms have been developed that use direct optimization, and they can be grouped into three categories. The algorithms in the first category try to optimize surrogate objective functions, which are either upper bounds of IR measures [7,43,47] or smooth approximations of IR measures [11,34]. The algorithms in the second category smoothly approximate the true gradient of the evaluation measures, similar to LambdaRank [4,5,9,46]. The algorithms in the third category directly optimize evaluation measures in the form of rewards without any approximation using reinforcement learning such as MDPRank [41,50]. However, the sequential decision process of MDPRank causes a problem so serious in exploration that learning with only 46 weight parameters requires more than 10,000 epochs for convergence when training neural networks.

Deep learning [18] models have been proven to be effective with state-of-the-art results in many machine learning applications such as speech recognition, computer vision, and natural language processing, which leads to the introduction of neural networks in IR. Neural networks have been used for functions such as automatic feature extraction and comparison and aggregation (Compare-Aggregate) of local relevance [12,14,15,23,35]. However, they are generally trained on objective functions such as cross entropy, which are not directly related to the IR evaluation measures. They do not have information about the measures that they are going to be evaluated on, i.e., the objective functions indirectly optimize the evaluation measures. Since most evaluation measures such as MAP, mean reciprocal rank (MRR), and normalized discounted cumulative gain (nDCG) are not differentiable, they cannot be used as the objective functions for training neural networks.

For leveraging the efficacy of neural networks and the superiority of direct optimization, we propose an extensible deep learning method called BanditRank. BanditRank formulates ranking as a contextual bandit problem and trains neural networks using the policy gradient algorithm [31] for directly maximizing target measures. Contextual bandit is a type of multi-armed bandit, optimization problem formalization used in decision-making scenarios in which an action has to be taken by an agent depending on the provided context. The exact details of the formulation are provided in Sect. 3. BanditRank follows the listwise approach by treating a query and the corresponding candidate documents as a single instance for training.

The major contributions of this paper are summarized as follows.

- For training neural networks by directly optimizing evaluation measures using gradient descent algorithms, we formulate the ranking problem as a contextual bandit and introduce a new deep learning method called *BanditRank* for offline ranking.
- BanditRank is the listwise deep learning method that uses reinforcement learning to train neural networks for offline ranking purposes. We enabled this by introducing a hybrid training objective in order to solve the exploration problem when the number of possible actions is large.

2   Related Work

BanditRank is similar to BanditSum [8], which was proposed previously for extractive summarization tasks in NLP. BanditSum introduces a theoretically grounded method based on contextual bandit formalism for training neural-network-based summarizers with reinforcement learning. We have adapted the formulation of ranking as a contextual bandit from that of BanditSum. Adaptation of the contextual bandit framework to the ranking problem is not straightforward at all. For example, a naive application of BanditSum suffers from inadequate exploration when the number of actions is very large, which is prevalent in ranking tasks. Thus, we propose the use of hybrid loss for leveraging the feedback from a supervised loss function as explained in Sect. 3.4. Reinforcement learning was used for directly optimizing measures such as BLEU [24] and ROUGE [21] in different tasks of natural language processing such as summarization and sequence prediction [1, 19].

In the domain of learning-to-rank for IR, MDPRank [41] uses reinforcement learning for ranking by formulating ranking as the sequential decision process. Since such sequential processes are affected by the order of the decisions, they may be biased towards selecting documents with a low relevance level at the beginning. Consequently, MDPRank is not suitable for training neural networks because the training with only 46 weight parameters requires more than 10,000 epochs for convergence. In contrast, BanditRank is suitable for deep architectures, and all the best results of BanditRank were achieved in less than 30 epochs of training. For a query q with n_q candidate documents, the search space of BanditRank is $n_q P_M$ while that of MDPRank is $n_q!$, and $n_q P_M \ll n_q!$ for a small M.

The policy gradient algorithm was also used to train the generator of IRGAN [37], but the rewards for the generator depend on a scoring function learned by the discriminator. Both Bandits [16, 17, 20] and MDPs [48] were used to model the process of interaction between a search engine and user with the user providing implicit relevance feedback. As stated earlier in this paper, we focus on offline learning issues in this paper and refrain from further mentioning these lines of studies.

An overview of approaches that use reinforcement learning for different IR tasks such as query reformulation, recommendation, and session search can be found in a previous paper [50].

3   BanditRank Formulation

We formulate ranking as a contextual bandit trained using policy gradient reinforcement learning. A bandit is a decision-making problem in which an agent repeatedly chooses one out of several actions and receives a reward based on this choice. The goal of the agent is to maximize the cumulative reward it achieves by learning the actions that yield good rewards. The term *agent* is generally used to refer to an entity that interacts with an environment. Contextual Bandit is a variant of the bandit problem that forms a subclass of Markov decision processes with the length of each episode being one.

Now, we can formulate the ranking problem as a contextual bandit with the environment being the dataset of queries and documents. A set of query-document pairs corresponding to a single query is treated as a context, and each permutation of the candidate documents is treated as a different action. Formally, given a query q and its candidate documents $d = \{d_1, d_2, \ldots, d_{n_q}\}$, each context is the set c given by $c = \{(q, d_1), (q, d_2), \ldots, (q, d_{n_q})\}$, where n_q is the number of candidate documents of q, and the cardinality of c is given by $n_c = n_q$. Given c, the action of the agent is given by the permutation $a_c = (d_{k_1}, d_{k_2}, \ldots, d_{k_{n_q}})$ of the candidate documents, where $k_t \in \{1, 2, \ldots, n_q\}$ and $k_t \neq k_{t'}$ for $t \neq t'$. The reward is given by a scalar function $R(a_c, g_c)$ like MAP, that takes action a_c and the ground-truth permutation g_c corresponding to c as the input.

The action taken by the agent is determined by its *policy*. In the current formulation, a policy is a neural network $p_\theta(.|c)$ parameterized by θ. For each input c, $p_\theta(.|c)$ encodes a probability distribution over permutations of the candidate documents. The goal is to find θ that cause the network to assign high probability to the permutations that can yield good rewards induced by R. This can be achieved by maximizing the following objective function with respect to θ:

$$J(\theta) = E[R(a_c, g_c)], \tag{1}$$

where the expectation is taken over c paired with g_c and a_c generated according to $p_\theta(.|c)$.

3.1   Structure of Policy $p_\theta(.|c)$

Similar to the approach used for extractive summarisation [8], $p_\theta(.|c)$ is decomposed into a deterministic function π_θ, which contains all the network's parameters, and μ, a probability distribution induced by the output of π_θ defined as follows.

$$p_\theta(.|c) = \mu(.|\pi_\theta(c)) \tag{2}$$

Provided a c corresponding to a q, the network π_θ outputs a real valued vector of document affinities within the range $[0, 1]$. The affinity score of a document d_i given by $\pi_\theta(c)_i$ represents the network's propensity to keep the document at the top position in the output permutation.

Provided the above document affinities $\pi_\theta(c)$, μ implements a process of repeated sampling without replacement by repeatedly normalizing the set of affinities of documents not yet selected. In total, M unique documents are sampled following an ϵ-greedy strategy, yielding an ordered subset of the candidate documents. According to the prescribed definition of μ, the probability $p_\theta(a_c|c)$ of producing a permutation a_c corresponding to c according to (2) is given by

$$p_\theta(a_c|c) = \prod_{i=1}^{M} \left(\frac{\epsilon}{n_c - i + 1} + \frac{(1-\epsilon)\pi_\theta(c)_{k_i}}{z(c) - \sum_{l=1}^{i-1} \pi_\theta(c)_{k_l}} \right), \tag{3}$$

where k_t is the index to the t-th document in a_c, d_{k_t} and $z(c) = \sum_{m=1}^{n_c} \pi_\theta(c)_m$. We define $M = min(n_c, M')$, where M' is an integer hyperparameter that depends on the environment or dataset. At test time, we output all the candidate documents sorted in descending order according to their affinity scores.

3.2 Policy Gradient Reinforcement Learning

The gradient of objective function (1) cannot be calculated directly as a_c is discretely sampled while calculating $R(a_c, g_c)$. After a reformulation of the expectation term according to the REINFORCE algorithm [42], the gradient of that function can be calculated using the following equation.

$$\nabla_\theta J(\theta) = E[\nabla_\theta \log p_\theta(a_c|c) R(a_c, g_c)], \tag{4}$$

where the expectation is over the same variables as (1).

The expectation in (4) is empirically calculated by first sampling a context-true permutation pair (c, g_c) from the dataset or environment $D(c, g_c)$, sampling B permutations $a_c^1, a_c^2, \ldots, a_c^B$ from $p_\theta(.|c)$ using the sampling method mentioned in Sect. 3.1 and finally taking the average. Empirically, the inner expectation of (4) is given by the following.

$$\nabla_\theta J_c(\theta) \approx \frac{\sum_{i=1}^{B} \nabla_\theta \log p_\theta(a_c^i|c) R(a_c^i, g_c)}{B} \tag{5}$$

Given the expression for $p_\theta(a_c|c)$ (3), gradient (5) can be calculated by any automatic differentiation package. As mentioned in Sect. 3.1, we sample $M = min(n_c, M')$ number of documents from the candidate documents during training time. Therefore, we take reward feedback from an M-length ordered subset. Since we cannot efficiently explore the whole action space for a large M as the number of possible actions or permutations would then become $_{n_c}P_M$[1], we choose M on the basis of the average number of relevant documents per query in the dataset.

The gradient estimate in (5) is prone to have high variance [31]. We use a baseline function, which is subtracted from all rewards. This can significantly reduce the variance of the estimate [31] by acting as an advantage function,

---

[1] Permutation $_nP_r$ is an increasing function of r.

and it ensures that the permutations with low rewards receive negative rewards. Using a baseline r_{base}, the sample-based estimate (5) becomes the following.

$$\nabla_\theta J_c(\theta) \approx \frac{\sum_{i=1}^{B} \nabla_\theta \log p_\theta(a_c^i|c)[R(a_c^i, g_c) - r_{\text{base}}]}{B} \tag{6}$$

For choosing the baseline function, we follow the terminology of self-critical reinforcement learning, in which the test time performance of the current model is used as the baseline [8,25]. Therefore, while calculating the gradient estimate (6) after sampling the context-true permutation pair (c, g_c), we greedily generate a permutation using the current model similar to the test time action.

$$a_c^{\text{greedy}} = \arg\max_{a_c} p_\theta(a_c|c) \tag{7}$$

The baseline for a c is then calculated by setting $r_{\text{base}} = R(a_c^{\text{greedy}}, g_c)$.

3.3   Reward Function R

As mentioned earlier, the reward function can be a single target measure or a combination of several measures. For the question answering datasets, the following reward function was used.

$$R(a_c, g_c) = \frac{AP(a_c, g_c) + RR(a_c, g_c)}{2} \tag{8}$$

For the web search dataset, the following reward function was used.

$$R'(a_c, g_c) = \frac{AP(a_c, g_c) + nDCG@10(a_c, g_c)}{2}, \tag{9}$$

where the measures average precision (AP), reciprocal rank (RR), and nDCG@10 are traditional IR measures.

3.4   Hybrid Training Objective

As mentioned in Sect. 3.2, the problem of exploring the action space when M is large can be tackled using a hybrid loss, which is a combination of reinforcement learning loss and a standard supervised learning loss such as binary cross entropy, that can compensate the cost incurred due to the inefficient exploration. The hybrid loss function is given as follows.

$$L_{\text{hybrid}} = \gamma L_{\text{rl}} + (1 - \gamma)L_{\text{sl}}, \tag{10}$$

where L_{rl} is the loss given by the reinforcement-learning algorithm, which is the negative of (1), and L_{sl} is a supervised loss such as binary cross entropy. The notation γ is a scaling factor accounting for the difference in magnitude between L_{rl} and L_{sl}. It is a hyperparameter lying between 0 and 1.

4   Experiments

We conducted experiments on four different datasets in the domains of question answering and web search. For the question answering task, we tested BanditRank on InsuranceQA [10], which is a community question answering dataset (closed domain, non-factoid), and on WikiQA [44], which is a well studied factoid, open-domain question answering dataset. For the web search task, we conducted our experiments on the benchmark MQ2007 [26] and the Yahoo! Learning-to-Rank [6] datasets.

4.1   Model Architecture

For question answering datasets, we used Multi Cast Attention Networks (MCAN) [33] for feature extraction followed by two feed forward highway layers with a sigmoid unit in the output layer. For web-search datasets, we used feed forward network with three highway layers for encoding the policy of the agent. We tune the γ parameter from set of values in $[0, 0.25, 0.5, 0.75, 1.0]$. The hyper parameters B and M' are chosen from the set of values $[5, 10, 15, 20, 25, 30]$. We optimized the model using the Adam optimizer with the beta parameters set to $(0, 0.999)$, and a weight decay of 10^{-6} was used for regularization. The exact details of the datasets and the hyper parameters of the models are given in the supplementary material[2].

4.2   Results

InsuranceQA and WikiQA. The results given in Table 1 indicate the superiority of BanditRank over the previous compare-aggregate based deep learning methods on both InsuranceQA and WikiQA datasets.

MQ2007 and Yahoo! Learning to Rank. The results given in Table 2 and Table 3 show that BanditRank clearly outperformed strong baselines like ListNet, AdaRank, Coordinate Ascent, RankSVM, PGRank and MDPRank on MQ2007 and Yahoo! LTR [6] datasets.

---

[2] We also plan to release the code used for experiments post the publication of paper.

Table 1. Precision@1 for InsuranceQA dataset, and MAP, MRR for WikiQA dataset. Best results are in bold, and second best are underlined.

InsuranceQA	test-1	test-2
IR model [2]	0.5510	0.5080
QA-CNN [29]	0.6133	0.5689
LambdaCNN [29,49]	0.6294	0.6006
IRGAN [37]	0.6444	0.6111
CNN with GESD [10]	0.6530	0.6100
Attentive LSTM [32]	0.6900	0.6480
IARNN-Occam [36]	0.6890	0.6510
IARNN-Gate [36]	0.7010	0.6280
Comp-Agg (MULT) [38]	0.7520	0.7340
Comp-Agg (SUBMULT+NN) [38]	0.7560	0.7230
BanditRank ($\gamma = 1$)	<u>0.8494</u>	<u>0.8283</u>
BanditRank ($\gamma = 0.75$)	**0.8572**	**0.8522**
WikiQA	MAP	MRR
CNN-Cnt [44]	0.6520	0.6650
QA-CNN [29]	0.6890	0.6960
NASM [22]	0.6890	0.7070
Wang et al. [40]	0.7060	0.7230
He and Lin [13]	0.7090	0.7230
NCE-CNN [28]	0.7010	0.7180
BIMPM [39]	0.7180	0.7310
Comp-Agg [38]	0.7430	0.7550
Comp-Clip [3]	0.7540	0.7640
Comp-Clip (LM) [45]	0.7480	0.7680
Comp-Clip (LM+LC) [45]	<u>0.7590</u>	<u>0.7720</u>
BanditRank ($\gamma = 0.75$)	0.6663	0.6730
BanditRank ($\gamma = 1$)	0.7043	0.7160
(Pretrained features from BERT)		
BanditRank-BERT-base ($\gamma = 1$)	0.7437	0.7589
BanditRank-BERT-large ($\gamma = 1$)	**0.7649**	**0.7807**

Table 2. Comparing BanditRank to the baseline LTR methods from [30] on the Yahoo dataset. Best results are in bold, and second best are underlined.

Yahoo! LTR	NDCG@10	ERR
BM25F-SD	0.73214	0.42853
RankSVM	0.75924	0.43680
PGRank [30]	<u>0.77082</u>	<u>0.45440</u>
BanditRank ($\gamma = 1$)	**0.78210**	**0.46011**

Table 3. Results of MQ2007 dataset. Best results are in bold. Statistically significant differences compared with the best model according to paired t-test are denoted as *, and Wilcoxon signed rank test are denoted as [+]. The metric NDCG is denoted by N. ($p - value < 0.05$)

MQ2007	P@1	P@3	P@10	MAP	N@1	N@3	N@10	MRR
ListNet	0.446*[+]	0.409*[+]	0.366*[+]	0.452*[+]	0.391*[+]	0.392*[+]	0.435	0.556*[+]
AdaRank	0.474*[+]	0.434*[+]	0.379*[+]	0.471*[+]	0.432*[+]	0.426*[+]	0.457	0.577*[+]
Coordinate Ascent	0.474*[+]	0.435*[+]	0.382*[+]	0.474*[+]	0.418*[+]	0.420*[+]	0.449	0.574*[+]
LambdaMART	0.477*[+]	0.444*	0.390*[+]	0.477*[+]	0.431*[+]	0.434*[+]	0.470*	0.582*[+]
MDPRank	0.415*[+]	0.400*[+]	0.360*[+]	0.431	0.376*[+]	0.386*[+]	0.419	0.534*[+]
BanditRank ($\gamma = 1$)	0.460	0.432	0.382	0.468	0.412	0.413	0.454	0.572
BanditRank ($\gamma = 0.5$)	**0.498**	**0.457**	**0.393**	**0.483**	**0.447**	**0.437**	**0.473**	**0.597**

5 Conclusion

We proposed an extensible listwise deep learning method *BanditRank* for search ranking. It can directly optimize IR evaluation measures using the policy gradient algorithm. BanditRank is successfully applied to the question answering tasks like InsuranceQA and WikiQA, outperforming the previous best compare-aggregate methods. In the web search task, we reported statistically significant improvements over the five strong listwise baselines on MQ2007 dataset along with improvements over good ranking baselines on Yahoo! LTR dataset. We showed how BanditRank is extensible and applicable to diverse tasks. Thus, we answered the four RQs as follows:

RQ1 Does BanditRank perform better than strong baselines when applied to the non-factoid question answering task?: Yes.

RQ2 Does BanditRank perform better than strong baselines when applied to the factoid question answering task?: Yes using pretrained features from BERT.

RQ3 Does BanditRank perform better than strong learning-to-rank baselines, when applied to the web search task?: Yes on the MQ2007 dataset with statistical significance, and Yahoo! LTR dataset.

RQ4 Is the hybrid training objective effective in each task?: Yes in many contexts, but not always. When the number of relevant answers is limited, it is not effective.

Future work can involve modifying the structure of the policy network discussed in Sect. 3.1 for efficiently addressing the issue of exploration when the number of actions is large. For example, we could use adaptive exploration strategies instead of a simple ϵ-greedy strategy for exploring the action space. There is also the possibility of defining reward functions as the weighted average of different measures with trainable weights for better feedback. Regarding the theoretical aspects, we can compare the directness of BanditRank to other algorithms such as LambdaRank. Developing scenarios for applying BanditRank to online learning to rank tasks is also a natural expansion of the current study on offline learning.

References

1. Bahdanau, D., et al.: An actor-critic algorithm for sequence prediction. arXiv preprint arXiv:1607.07086 (2016)
2. Bendersky, M., Metzler, D., Croft, W.B.: Learning concept importance using a weighted dependence model. In: Proceedings of the third ACM International Conference on Web Search and Data Mining, pp. 31–40. ACM (2010)
3. Bian, W., Li, S., Yang, Z., Chen, G., Lin, Z.: A compare-aggregate model with dynamic-clip attention for answer selection. In: Proceedings of the 2017 ACM on Conference on Information and Knowledge Management, pp. 1987–1990. ACM (2017)
4. Burges, C.J., Ragno, R., Le, Q.V.: Learning to rank with nonsmooth cost functions. Adv. Neural Inf. Process. Syst. **19**, 193–200 (2007)
5. Burges, C.J.: From ranknet to lambdarank to lambdamart: an overview. Learning **11**(23–581), 81 (2010)
6. Chapelle, O., Chang, Y.: Yahoo! learning to rank challenge overview. In: Proceedings of the Learning to Rank Challenge, pp. 1–24 (2011)
7. Chapelle, O., Le, Q., Smola, A.: Large margin optimization of ranking measures. In: NIPS Workshop: Machine Learning for Web Search (2007)
8. Dong, Y., Shen, Y., Crawford, E., van Hoof, H., Cheung, J.C.K.: BanditSum: extractive summarization as a contextual bandit. arXiv preprint arXiv:1809.09672 (2018)
9. Donmez, P., Svore, K.M., Burges, C.J.: On the local optimality of lambdarank. In: Proceedings of the 32nd International ACM SIGIR Conference on Research and Development in Information Retrieval, pp. 460–467. ACM (2009)
10. Feng, M., Xiang, B., Glass, M.R., Wang, L., Zhou, B.: Applying deep learning to answer selection: a study and an open task. In: 2015 IEEE Workshop on Automatic Speech Recognition and Understanding (ASRU), pp. 813–820. IEEE (2015)
11. Guiver, J., Snelson, E.: Learning to rank with softrank and gaussian processes. In: Proceedings of the 31st Annual International ACM SIGIR Conference on Research and Development in Information Retrieval, pp. 259–266. ACM (2008)
12. Guo, J., Fan, Y., Ai, Q., Croft, W.B.: A deep relevance matching model for ad-hoc retrieval. In: Proceedings of the 25th ACM International on Conference on Information and Knowledge Management, pp. 55–64. ACM (2016)
13. He, H., Lin, J.: Pairwise word interaction modeling with deep neural networks for semantic similarity measurement. In: Proceedings of the 2016 Conference of the North American Chapter of the Association for Computational Linguistics: Human Language Technologies, pp. 937–948 (2016)

14. Hu, B., Lu, Z., Li, H., Chen, Q.: Convolutional neural network architectures for matching natural language sentences. In: Advances in Neural Information Processing Systems, pp. 2042–2050 (2014)
15. Huang, P.S., He, X., Gao, J., Deng, L., Acero, A., Heck, L.: Learning deep structured semantic models for web search using clickthrough data. In: Proceedings of the 22nd ACM International Conference on Information & Knowledge Management, pp. 2333–2338. ACM (2013)
16. Katariya, S., Kveton, B., Szepesvari, C., Wen, Z.: DCM bandits: learning to rank with multiple clicks. In: International Conference on Machine Learning, pp. 1215–1224 (2016)
17. Kveton, B., Szepesvari, C., Wen, Z., Ashkan, A.: Cascading bandits: learning to rank in the cascade model. In: International Conference on Machine Learning, pp. 767–776 (2015)
18. LeCun, Y., Bengio, Y., Hinton, G.: Deep learning. Nature **521**(7553), 436 (2015)
19. Lee, G.H., Lee, K.J.: Automatic text summarization using reinforcement learning with embedding features. In: Proceedings of the Eighth International Joint Conference on Natural Language Processing (Volume 2: Short Papers), pp. 193–197 (2017)
20. Li, C., Grotov, A., Markov, I., Eikema, B., de Rijke, M.: Online learning to rank with list-level feedback for image filtering. CoRR abs/1812.04910 (2018)
21. Lin, C.Y.: Rouge: a package for automatic evaluation of summaries. In: Text Summarization Branches Out (2004)
22. Miao, Y., Yu, L., Blunsom, P.: Neural variational inference for text processing. In: International Conference on Machine Learning, pp. 1727–1736 (2016)
23. Pang, L., Lan, Y., Guo, J., Xu, J., Xu, J., Cheng, X.: Deeprank: a new deep architecture for relevance ranking in information retrieval. In: Proceedings of the 2017 ACM on Conference on Information and Knowledge Management, pp. 257–266. ACM (2017)
24. Papineni, K., Roukos, S., Ward, T., Zhu, W.J.: Bleu: a method for automatic evaluation of machine translation. In: Proceedings of the 40th Annual Meeting on Association for Computational Linguistics, pp. 311–318. Association for Computational Linguistics (2002)
25. Paulus, R., Xiong, C., Socher, R.: A deep reinforced model for abstractive summarization. arXiv preprint arXiv:1705.04304 (2017)
26. Qin, T., Liu, T.Y.: Introducing letor 4.0 datasets. arXiv preprint arXiv:1306.2597 (2013)
27. Qin, T., Liu, T.Y., Li, H.: A general approximation framework for direct optimization of information retrieval measures. Inf. Retr. **13**(4), 375–397 (2010)
28. Rao, J., He, H., Lin, J.: Noise-contrastive estimation for answer selection with deep neural networks. In: Proceedings of the 25th ACM International on Conference on Information and Knowledge Management, pp. 1913–1916. ACM (2016)
29. Santos, C.D., Tan, M., Xiang, B., Zhou, B.: Attentive pooling networks. arXiv preprint arXiv:1602.03609 (2016)
30. Singh, A., Joachims, T.: Policy learning for fairness in ranking. In: Advances in Neural Information Processing Systems, pp. 5426–5436 (2019)
31. Sutton, R.S., McAllester, D.A., Singh, S.P., Mansour, Y.: Policy gradient methods for reinforcement learning with function approximation. In: Advances in Neural Information Processing Systems, pp. 1057–1063 (2000)

32. Tan, M., Dos Santos, C., Xiang, B., Zhou, B.: Improved representation learning for question answer matching. In: Proceedings of the 54th Annual Meeting of the Association for Computational Linguistics (Volume 1: Long Papers), vol. 1, pp. 464–473 (2016)

33. Tay, Y., Tuan, L.A., Hui, S.C.: Multi-cast attention networks for retrieval-based question answering and response prediction. arXiv preprint arXiv:1806.00778 (2018)

34. Taylor, M., Guiver, J., Robertson, S., Minka, T.: Softrank: optimizing non-smooth rank metrics. In: Proceedings of the 2008 International Conference on Web Search and Data Mining, pp. 77–86. ACM (2008)

35. Wan, S., Lan, Y., Xu, J., Guo, J., Pang, L., Cheng, X.: Match-SRNN: modeling the recursive matching structure with spatial RNN. arXiv preprint arXiv:1604.04378 (2016)

36. Wang, B., Liu, K., Zhao, J.: Inner attention based recurrent neural networks for answer selection. In: Proceedings of the 54th Annual Meeting of the Association for Computational Linguistics (Volume 1: Long Papers), vol. 1, pp. 1288–1297 (2016)

37. Wang, J., et al.: IRGAN: a minimax game for unifying generative and discriminative information retrieval models. In: Proceedings of the 40th International ACM SIGIR Conference on Research and Development in Information Retrieval, pp. 515–524. ACM (2017)

38. Wang, S., Jiang, J.: A compare-aggregate model for matching text sequences. arXiv preprint arXiv:1611.01747 (2016)

39. Wang, Z., Hamza, W., Florian, R.: Bilateral multi-perspective matching for natural language sentences. arXiv preprint arXiv:1702.03814 (2017)

40. Wang, Z., Mi, H., Ittycheriah, A.: Sentence similarity learning by lexical decomposition and composition. arXiv preprint arXiv:1602.07019 (2016)

41. Wei, Z., Xu, J., Lan, Y., Guo, J., Cheng, X.: Reinforcement learning to rank with Markov decision process. In: Proceedings of the 40th International ACM SIGIR Conference on Research and Development in Information Retrieval, pp. 945–948. ACM (2017)

42. Williams, R.J.: Simple statistical gradient-following algorithms for connectionist reinforcement learning. Mach. Learn. **8**(3–4), 229–256 (1992)

43. Xu, J., Li, H.: AdaRank: a boosting algorithm for information retrieval. In: Proceedings of the 30th Annual International ACM SIGIR Conference on Research and Development in Information Retrieval, pp. 391–398. ACM (2007)

44. Yang, Y., Yih, W.T., Meek, C.: WikiQA: a challenge dataset for open-domain question answering. In: Proceedings of the 2015 Conference on Empirical Methods in Natural Language Processing, pp. 2013–2018 (2015)

45. Yoon, S., Dernoncourt, F., Kim, D.S., Bui, T., Jung, K.: A compare-aggregate model with latent clustering for answer selection. arXiv preprint arXiv:1905.12897 (2019)

46. Yuan, F., Guo, G., Jose, J.M., Chen, L., Yu, H., Zhang, W.: Lambdafm: learning optimal ranking with factorization machines using lambda surrogates. In: Proceedings of the 25th ACM International on Conference on Information and Knowledge Management, pp. 227–236. ACM (2016)

47. Yue, Y., Finley, T., Radlinski, F., Joachims, T.: A support vector method for optimizing average precision. In: Proceedings of the 30th Annual International ACM SIGIR Conference on Research and Development in Information Retrieval, pp. 271–278. ACM (2007)

48. Zeng, W., Xu, J., Lan, Y., Guo, J., Cheng, X.: Multi page search with reinforcement learning to rank. In: Proceedings of the 2018 ACM SIGIR International Conference on Theory of Information Retrieval, pp. 175–178. ACM (2018)
49. Zhang, W., Chen, T., Wang, J., Yu, Y.: Optimizing top-n collaborative filtering via dynamic negative item sampling. In: Proceedings of the 36th International ACM SIGIR Conference on Research and Development in Information Retrieval, pp. 785–788. ACM (2013)
50. Zhao, X., Xia, L., Tang, J., Yin, D.: Reinforcement learning for online information seeking. arXiv preprint arXiv:1812.07127 (2018)

A Compressed and Accelerated SegNet for Plant Leaf Disease Segmentation: A Differential Evolution Based Approach

Mohit Agarwal$^{(\boxtimes)}$, Suneet Kr. Gupta, and K. K. Biswas

Bennett University, Greater Noida, India
{ma8573,suneet.gupta}@bennett.edu.in

Abstract. SegNet is a Convolution Neural Network (CNN) architecture consisting of encoder and decoder for pixel-wise classification of input images. It was found to give better results than state of the art pixel-wise segmentation of images. In proposed work, a compressed version of SegNet has been developed using Differential Evolution for segmenting the diseased regions in leaf images. The compressed model has been evaluated on publicly available street scene images and potato late blight leaf images from PlantVillage dataset. Using the proposed method a compression of 25x times is achieved on original SegNet and inference time is reduced by 1.675x times without loss in mean IOU accuracy.

Keywords: SegNet · Differential Evolution · Model compression · Diseased leaf dataset

1 Introduction

While a large number of researchers have been interested in identifying plant diseases from leaf images [16,21,23], there has been a growing interest in applying segmentation techniques to determine the extent of the disease on the leaves. Several researchers have used Machine Learning techniques such as Support Vector Machine (SVM) for segmenting the diseased region and background from leaf images [15,17,28]. Researchers have also used deep learning methods like Convolution Neural Networks (CNN) for disease region segmentation [9,19]. Very recently, Badrinarayanan et al. [5] have proposed SegNet, a highly efficient encoder- decoder CNN architecture for pixel wise segmentation of images. In this paper we have attempted to use SegNet for segmenting diseased regions of leaf images. Keeping in mind the fact that large computational power would not be readily available in agricultural fields, we have developed a compressed SegNet model using differential evolution [8] whose memory requirement are a fraction of the original SegNet without any compromise on pixel wise accuracy.

The motivation for carrying out compression [7,10,11] is driven by the idea that the farmer should be enabled to run the deep learning models on edge devices like tiny mobiles instead of seeking access to large computing devices.

© Springer Nature Switzerland AG 2021
K. Karlapalem et al. (Eds.): PAKDD 2021, LNAI 12714, pp. 272–284, 2021.
https://doi.org/10.1007/978-3-030-75768-7_22

This is essential if someone wants to deploy the solution on a farm rover robot for continuous pixel based segmentation of videos captured by robot to figure out extent of disease visible on the leaves, and to spray fungicide at affected areas.

The main contributions of this paper can be summarized as below:

- Development of a SegNet based pixel wise diseased region segmentation method of plant leaves.
- Compression and acceleration of SegNet using Differential Evolution.
- Development of novel fitness function which helps to compress the SegNet.

The rest of the paper is organized as follows: A brief summary of previous research on compression of various CNN models is presented in Sect. 2. Section 3 introduces the SegNet model and development of its compressed version. The experimental setup and results of disease segmentation on uncompressed and compressed version are presented in Sect. 4. Section 5 concludes the paper.

2   Literature Review

To employ popular CNN models on edge devices, a number of researchers have developed compressed versions of these architectures. Most researchers have used techniques such as matrix factorization, flattened convolution, network pruning, huffman coding etc. for this purpose.

Anwar et al. [4] described a model compression technique using pruning which also accelerates the model. Pruning has been done by two methods: evolutionary particle filtering and activation sum voting. Authors mention that 70% pruning was obtained with less than 1% loss in accuracy on CIFAR-10 network. He et al. [13] proposed pruning of pre-trained CNN models like GoogleNet and ResNet by single layer pruning, whole network pruning and multi-branch pruning.

Gong et al. [10] have developed compression models based on Matrix factorization methods and Vector quantization methods for reducing the number of nodes in the dense layer based on the fact that 90% of contribution of CNN weights are from dense layers. The authors showed that the model could be compressed 16–24% while compromising the accuracy by 1%. Cheng et al. [7] have proposed pruning by discarding the redundant weights and came up with compression ratio of 4.94 on AlexNet. Han et al. [11] described a 3-stage mechanism to compress a CNN model employing K-means clustering and Huffman coding.

Han et al. [12] also described a CNN pruning process in which all weights of convolution and Dense layers which are less than a threshold value are made zero. Li et al. [18] proposed compression by discarding low ranked filters from each convolution layer based on their L1-norm. Liu et al. [20] have described network slimming technique during training which prunes unimportant channels using sparsity-based regularization. Zhang et al. [27] have explored various methods for CNN compression and acceleration from the structure, algorithmic and implementation perspectives.

SegNet is a Encoder Decoder based deep learning model introduced by Badri-narayanan et al. [5] in 2017. They have used VGG16 like architecture for encoder portion and then added same number of CNN layers for upscaling the input in the decoder part of network to finally predict a class for each pixel in the input images. Authors have demonstrated that segmentation carried out by SegNet outperforms state of art segmentation networks using CamVid road scene segmentation and SUN RGB-D indoor scene segmentation datasets [6,25].

A number of researchers have demonstrated the use of SegNet for other segmentation applications. Alqazzaz et al. [3] demonstrated the usage of SegNet in detecting brain tumor from MRI slices. Nguyen et al. [24] employed SegNet for hand gestures recognition using RGB-depth kinect camera. Manickam et al. [22] showed that SegNet segmentation for detecting persons in aerial imagery gave superior results when compared with VGG16, ResNet and GoogleNet usage.

In this paper a meta heuristics based approach has been used to compress the SegNet model to enable it to be employed on edge devices. An application has been shown for diseased region identification on potato leaves. The next section presents the SegNet structure and the compression technique.

3   Proposed Model

SegNet is basically an encoder-decoder type deep learning network (Fig. 1). The encoder part has 16 layers. The 1st and 2nd convolution layer consist of 64 filters each. The 3rd and 4th layer has 128 filters each, 5th, 6th and 7th layer have 256 filters each and 8th layer onward till 13th layer have 512 filters each. Each convolution block is followed by a max pooling layer in the encoder part. This is followed by the decoder part which starts with an upsampling layer followed by convolution blocks in the reverse order, and finally ending with a softmax layer.

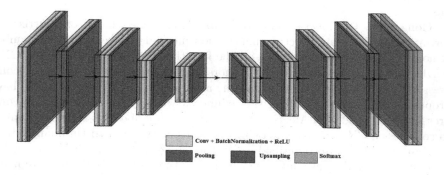

Fig. 1. Figure depicting architecture of SegNet.

3.1   Differential Evolution Based Compressed SegNet

Although SegNet is a derived version of the famous VGG architecture, the number of parameters is still 14.7 Million, and is not employable on tiny edge devices. For many small applications, using Differential Evolution (DE) it is possible to reduce the size still further by retaining only those nodes and filters whose contribution to meaningful segmentation is significant.

DE is a stochastic process for generating solution to a nonlinear problem, introduced by Storn and Price in 1996 [26]. A basic flow diagram of the process is illustrated in Fig. 2.

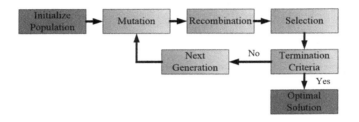

Fig. 2. Differential Evolution flow chart.

The various steps involved in DE process applied to SegNet compression are explained below.

Initial Population. To start with a random vector is created to represent the various filters and nodes of the hidden layers of SegNet. The elements of the vector (neurons) were randomly assigned values 1 and 0, with 1 denoting that the particular node is being retained and 0 denoting that the node is being discarded. Initially a population pool of 100 such vectors is created. A sample initial vector is shown in Fig. 3 with different colors representing the various layers of CNN.

Fig. 3. A sample vector randomly initialized with 0's and 1's representing various filters in SegNet. (Color figure online)

Mutation in DE. For each target vector, three random vectors are chosen from the population. The difference of two of the random vectors is multiplied by a mutation factor (a value between 0 and 1) and added to the third random vector in population. If the values of the new vector (named donor vector) do not lie between lower bound and upper bound, these are re-scaled to the desired range. For SegNet compression, the donor vector is computed as shown in equation (1).

$$v_i^{g+1} = w_{ri1}^g + F \times (w_{ri2}^g - w_{ri3}^g) \tag{1}$$

Here v_i^{g+1} is the donor vector of $(g+1)^{th}$ generation and w_{ri1}^g, w_{ri2}^g, and w_{ri3}^g are three random vectors for a target vector at index i of the population. Here i takes on values 1, 2, 3, ..., N. F is the mutation factor and chosen as 0.5 and N is chosen as 100. The values taken on by the donor vector range between -0.5 to 1.5. These are normalized to 1 for values more than 0.5, and normalized to 0 otherwise. A representative mutation vector is shown in Fig. 4.

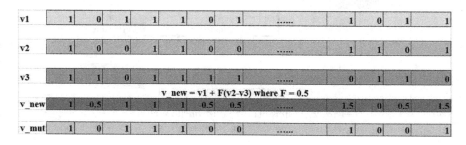

Fig. 4. Process showing generation of mutation vector using 3 random vectors.

Recombination in DE. In this step, a trial vector is created using the target vector and the donor vector based on a recombination factor. A random number is generated between 0 and 1 for each index position of vectors, and if it is found to be greater than recombination factor, then the trial vector element at that index position is retained in the target vector, else it is taken from donor vector. The recombination factor has been chosen as 0.7 for SegNet compression. The recombination step follows Eq. (2).

$$f_{i,k}^{g+1} = \begin{cases} v_{i,k}^{g+1} & rand() \leq R_p \ or \ i = I_{rand} \\ w_{i,k}^g & rand() > R_p \ and \ i \neq I_{rand} \end{cases} \tag{2}$$

Here $v_{i,k}^{g+1}$ is the ith element of kth donor vector of $(g+1)^{th}$ generation. Similarly $w_{i,k}^{g+1}$ is the ith element of kth target vector of $(g)^{th}$ generation and $f_{i,k}^{g+1}$ is the ith element of kth trial vector of $(g+1)^{th}$ generation. R_p is the recombination factor and I_{rand} is random integer between 0 and L. I_{rand} ensures that all elements are not picked from target vector. Also $i = 1, 2, 3, \ldots, L$ (size of the vector), and $k = 1, 2, 3, \ldots, N$.

Selection in DE. The selection between the original vector and the new vector created after mutation and recombination is done based on a fitness function. Some terms are being introduced here for further usage.

- M = number of hidden layers in original model.
- $\alpha = \{\alpha_1, \alpha_2, \alpha_3, \ldots \alpha_M\}$ represent number of hidden units at hidden layers 1, 2, 3 $\ldots M$ in original model.
- $\beta = \{\beta_1, \beta_2, \beta_3 \ldots \beta_M\}$ represent number of hidden units at hidden layers 1, 2, 3 $\ldots M$ in compressed model.
- $IoU(k)$ denotes the mean intersection over union value of predicted images with ground truth after k compression steps.

Then the Linear Programming Formulation (LPP) can be formed as:

$$Maximize\ X = w \times \frac{\alpha_i}{\beta_i} + (1 - w) \times IoU(k) \tag{3}$$

Subject to:

$$\beta_i \leq \alpha_i,\ \forall\ 1 \leq i \leq M \tag{4}$$

The constraint (4) ensures that nodes in compressed layer are lesser than nodes in original model at layer i for all $i = 1, 2, 3 \ldots M$. Term X in Eq. (3) denotes value for minimization at layer i during compression process. This process is iterated layer wise for $i = 1, 2, 3 \ldots M$ over entire model. Here w is a relative weighting factor to provide balance between compression and model performance, it's value is taken as 0.5 for experimental purpose. The flow of SegNet compression process is shown in Fig. 5.

The DE process was implemented on this pool of vectors using pixel wise classification accuracy of the model on a chosen test set as fitness criteria. Depending on final optimum vector, a new compressed SegNet is created and weights copied from original SegNet where vector element has a value 1. This model is further trained for 10–50 epochs and tested on the test data set. If the evaluation metrics such as sensitivity, specificity, F1-score or accuracy shows deterioration by more than a preset threshold say 2%, then the last model is taken as the final model, else the DE process is repeated with this compressed model as original model.

4   Experimental Setup and Results

For developing the compressed version of SegNet, the dataset used was the same as used by proposers of SegNet [5]. This is the CamVid road scenes dataset [1], and comprises of 367 training and 233 testing RGB images (day and dusk scenes) at 360×480 resolution. The challenge is to segment 11 classes such as pedestrians, roads, building, cars, etc. The performance analysis is done using mean of the predictive accuracy over the eleven classes, and mean intersection over union (mIoU) over all classes.

4.1 Testing the Compressed SegNet Model

The experiments were performed using NVIDIA DGX v100 machine using python. This machine is equipped with 40600 CUDA cores, 5120 tensor cores, 128 GB RAM and 1000 TFLOPS speed on UBUNTU 20.0.1 operating system.

The various parameters used in Differential Evolution compression process are given in Table 1. The SegNet was trained for 500 epochs on this dataset [1] and mean IoU and predicted images were saved. Since predicted images were lacking in clarity, training was carried out on a reduced version of SegNet with its consecutive convolution layers in a convolution block merged into a single layer [2]. This model does not support index transferring proposed in SegNet and has encoder and decoder part with total 8 convolution layers and filters numbering 64, 128, 256, 512, 512, 256, 128 and 64 in these layers from input

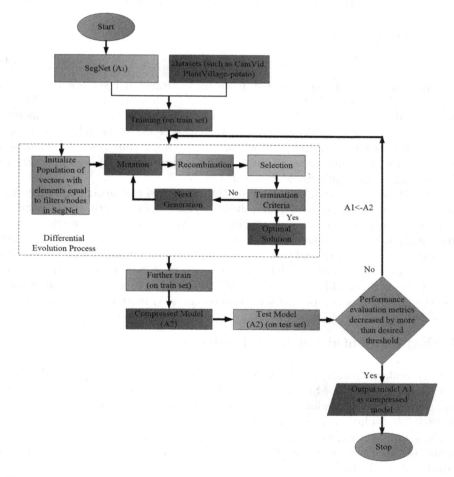

Fig. 5. Process flow diagram of iteratively compressing SegNet along with training.

Table 1. Parameters used for Differential Evolution Algorithm.

S.No.	Parameter	Remark
1	Gene value	Binary (0–1)
2	Poplulation Size	100
3	Maximum Iterations	50
4	Mutation Factor	0.5
5	Recombination Factor	0.7
6	Termination Criteria	Change in fitness < 0.000001

side. Training for 500 epochs reported better accuracy and predicted images were closer to ground truth segmentation results. This could be attributed to the reason that model being simpler it can learn better in same number of epochs. On SegNet similar accuracy was obtained on training for 1000 epochs. This mini

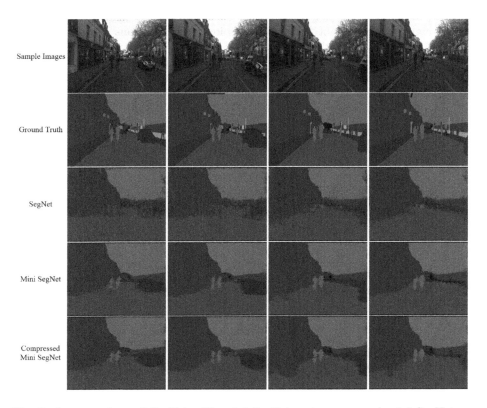

Fig. 6. A comparison of SegNet with mini SegNet and compressed mini SegNet on street scene images with ground truth pixel wise segmentation.

version of SegNet was used to obtain the compressed model using DE approach. This also gave good accuracy and predicted images were very close to ground truth. A visual comparison of three models has been shown in Fig. 6.

Comparison of performance evaluation metrics before and after compression are given in Tables 2 and 3. It is observed that the compressed SegNet model can be accommodated in 4% of space required for the original SegNet without any perceptible change in mean IoU measure, while the inference time per sample reduces to 40% of original inference time, making it feasible for deployment on mobiles and edge devices.

4.2 Segmentation Using Other CNN Models

The process of pixel wise segmentation was also tried using UNet and Fully Convolution Network (FCN) with VGG16 like model but the results were not good. UNet model gave only 3% mIoU after 100 epochs and weight file size was 121,307 KB. FCN model gave mIoU of 50.39% after 100 epochs and weight file size was even more 524,726 KB. Thus these two solutions could not predict the output segmented images correctly and were far from ground truth. Even their weight files were not suitable to deploy these solutions on mini edge or mobile devices. These models were also experimented to be compressed using DE based approach and they could be satisfactorily compressed but mIoU was still not good, so could not generate output segmented images for their compressed versions.

Table 2. Comparison of performance and size before and after compression of SegNet.

Model	mIoU	Size	Percent decrease in size
SegNet	79.16%	115,283 KB	–
Mini SegNet	84.21%	21,434 KB	81.40%
Mini SegNet Compressed	83.89%	4,501 KB	96.09%

Table 3. Comparison of inference time for 1 sample before and after compression of SegNet.

Model	Prediction time	Percent decrease
SegNet	10.72 s	–
Mini SegNet	7.26 s	32.27%
Mini SegNet Compressed	6.40 s	40.3%

4.3   Identification of Diseased Region of Leaves Using SegNet

The compressed SegNet was also used for disease identification in potato crop from leaf images. The data consisted of 256×256 sized images of potato leaves affected by late blight disease from PlantVillage dataset [14]. The diseased regions were manually annotated and marked with brown color with RGB value (128, 128, 0), the greener part of leaf was marked green with RGB value (0, 255, 0). The background was assigned black color with RGB value (0, 0, 0). A set of sample images and their corresponding annotations are shown in Fig. 7[1]. The data was transformed in HSV space, and Hue value was checked for green, brown and yellow parts of the image. This color assignment agrees with human color perception and also helps in manual annotation of diseased portions in raw images of leaves affected by disease.

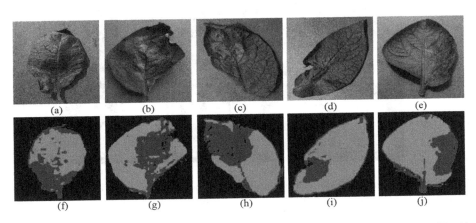

Fig. 7. (a), (b), (c), (d), (e) sample images of potato late blight. (f), (g), (h), (i), (j) the corresponding annotations in 3 colors (Color figure online)

Results on Leaf Dataset. In this experiment, the Compressed SegNet version was created by merging the consecutive convolution layers of SegNet in a convolution block. This model was trained and tested on the dataset by dividing images in ratio of 81:6:13 for training, validation and testing. This model was also compressed using the proposed DE method and results are presented in Table 4. The results on leaves dataset were better than CamVid dataset as it had only 3 classes whereas previous dataset had 11 classes and thus classification could be more accurate due to better distinguishing ability of model in lesser classes. It can be observed that even with a very highly compressed SegNet model, the mean IoU measure is hardly compromised. Mini Compressed SegNet mIoU in Table 4 was better by 0.01%, this was due to fitness criteria based on choosing

<hr>

[1] Dataset can be downloaded from: https://github.com/mohit-aren/Leaf_colormap

best neurons giving higher mIoU and discard unwanted nodes. Figure 8 presents visual comparison of ground truth and predicted outputs. It is clear that the diseased portions have been correctly identified by the compressed SegNet.

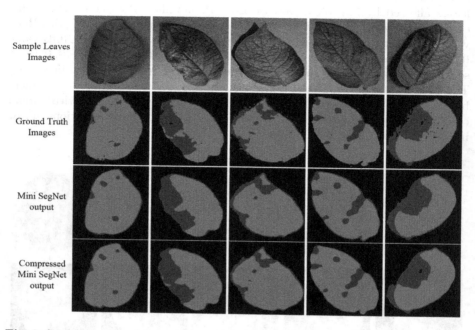

Fig. 8. Sample images of dataset, ground truth annotations of same images, predicted outputs of mini SegNet, predicted output of compressed mini SegNet.

Table 4. Comparison of performance and size before and after compression of mini SegNet on leaf dataset.

Model	mIoU	Size	Percent decrease
Mini SegNet	94.64%	21,433 KB	–
Mini SegNet Compressed	94.74%	4,842 KB	77.41%

5   Conclusion

This paper has proposed a scheme for developing a compressed SegNet architecture using differential evolution, so that it can be deployed on tiny mobile/edge devices. To illustrate this, an application of compressed SegNet for identification of diseased regions in potato leaves has been presented. It has been shown that without compromising on mean IoU, tremendous saving on storage can be achieved, as well as having a large reduction in inference time when a test sample is presented to the system. This has tremendous possibilities in the agriculture arena, and can help the farmer in proper treatment of diseased crops. Possibly a farm rover robot can move around in the field collecting images which can be processed by edge devices and help the farmer to selectively spray fungicide on the affected leaf portions. The leaves used were on a plain background but in actual scenarios they can have complex backgrounds like stems, other leaves, soil, etc. This could be experimented as a future version of this research work.

References

1. Street scene images dataset (2007). http://mi.eng.cam.ac.uk/research/projects/VideoRec/CamSeq01/
2. Keras segnet: simplified segnet model (2018). https://github.com/imlab-uiip/keras-segnet
3. Alqazzaz, S., Sun, X., Yang, X., Nokes, L.: Automated brain tumor segmentation on multi-modal MR image using SegNet. Comput. Visual Media **5**(2), 209–219 (2019)
4. Anwar, S., Hwang, K., Sung, W.: Structured pruning of deep convolutional neural networks. ACM J. Emerg. Technol. Comput. Syst. (JETC) **13**(3), 1–18 (2017)
5. Badrinarayanan, V., Kendall, A., Cipolla, R.: Segnet: a deep convolutional encoder-decoder architecture for image segmentation. IEEE Trans. Pattern Anal. Mach. Intell. **39**(12), 2481–2495 (2017)
6. Brostow, G.J., Fauqueur, J., Cipolla, R.: Semantic object classes in video: a high-definition ground truth database. Pattern Recogn. Lett. **30**(2), 88–97 (2009)
7. Cheng, Y., Wang, D., Zhou, P., Zhang, T.: A survey of model compression and acceleration for deep neural networks. arXiv preprint arXiv:1710.09282 (2017)
8. Feoktistov, V.: Differential Evolution. Springer, Boston (2006). https://doi.org/10.1007/978-0-387-36896-2
9. Ganesh, P., Volle, K., Burks, T., Mehta, S.: Deep orange: mask R-CNN based orange detection and segmentation. IFAC-PapersOnLine **52**(30), 70–75 (2019)
10. Gong, Y., Liu, L., Yang, M., Bourdev, L.: Compressing deep convolutional networks using vector quantization. arXiv preprint arXiv:1412.6115 (2014)
11. Han, S., Mao, H., Dally, W.J.: Deep compression: compressing deep neural networks with pruning, trained quantization and huffman coding. arXiv preprint arXiv:1510.00149 (2015)
12. Han, S., Pool, J., Tran, J., Dally, W.: Learning both weights and connections for efficient neural network. In: Advances in Neural Information Processing Systems, pp. 1135–1143 (2015)
13. He, Y., Zhang, X., Sun, J.: Channel pruning for accelerating very deep neural networks. In: Proceedings of the IEEE International Conference on Computer Vision, pp. 1389–1397 (2017)
14. Hughes, D., Salathé, M., et al.: An open access repository of images on plant health to enable the development of mobile disease diagnostics. arXiv preprint arXiv:1511.08060 (2015)
15. Islam, M., Dinh, A., Wahid, K., Bhowmik, P.: Detection of potato diseases using image segmentation and multiclass support vector machine. In: 2017 IEEE 30th Canadian Conference on Electrical and Computer Engineering (CCECE), pp. 1–4. IEEE (2017)

16. Johannes, A., Picon, A., Alvarez-Gila, A., Echazarra, J., Rodriguez-Vaamonde, S., Navajas, A.D., Ortiz-Barredo, A.: Automatic plant disease diagnosis using mobile capture devices, applied on a wheat use case. Comput. Electron. Agric. **138**, 200–209 (2017)

17. Lee, U., Chang, S., Putra, G.A., Kim, H., Kim, D.H.: An automated, high-throughput plant phenotyping system using machine learning-based plant segmentation and image analysis. PLoS ONE **13**(4), (2018)

18. Li, H., Kadav, A., Durdanovic, I., Samet, H., Graf, H.P.: Pruning filters for efficient convnets. arXiv preprint arXiv:1608.08710 (2016)

19. Lin, K., Gong, L., Huang, Y., Liu, C., Pan, J.: Deep learning-based segmentation and quantification of cucumber powdery mildew using convolutional neural network. Front. Plant Sci. **10**, 155 (2019)

20. Liu, Z., Li, J., Shen, Z., Huang, G., Yan, S., Zhang, C.: Learning efficient convolutional networks through network slimming. In: Proceedings of the IEEE International Conference on Computer Vision, pp. 2736–2744 (2017)

21. Ma, J., Du, K., Zheng, F., Zhang, L., Gong, Z., Sun, Z.: A recognition method for cucumber diseases using leaf symptom images based on deep convolutional neural network. Comput. Electron. Agric. **154**, 18–24 (2018)

22. Manickam, R., Rajan, S.K., Subramanian, C., Xavi, A., Eanoch, G.J., Yesudhas, H.R.: Person identification with aerial imaginary using SegNet based semantic segmentation. Earth Sci. Inform. 1–12 (2020)

23. Mohanty, S.P., Hughes, D.P., Salathé, M.: Using deep learning for image-based plant disease detection. Front. Plant Sci. **7**, 1419 (2016)

24. Nguyen, H.D., Na, I.S., Kim, S.H.: Hand segmentation and fingertip tracking from depth camera images using deep convolutional neural network and multi-task Seg-Net. arXiv preprint arXiv:1901.03465 (2019)

25. Song, S., Lichtenberg, S.P., Xiao, J.: Sun RGB-D: a RGB-D scene understanding benchmark suite. In: Proceedings of the IEEE Conference on Computer Vision and Pattern Recognition, pp. 567–576 (2015)

26. Storn, R., Price, K.: Differential evolution-a simple and efficient heuristic for global optimization over continuous spaces. J. Global Optim. **11**(4), 341–359 (1997)

27. Zhang, Q., Zhang, M., Chen, T., Sun, Z., Ma, Y., Yu, B.: Recent advances in convolutional neural network acceleration. Neurocomputing **323**, 37–51 (2019)

28. Zhou, J., Fu, X., Zhou, S., Zhou, J., Ye, H., Nguyen, H.T.: Automated segmentation of soybean plants from 3d point cloud using machine learning. Comput. Electron. Agric. **162**, 143–153 (2019)

Meta-context Transformers for Domain-Specific Response Generation

Debanjana Kar[1], Suranjana Samanta[2(✉)], and Amar Prakash Azad[2]

[1] Department of Computer Science and Engineering, IIT Kharagpur, India
debanjana.kar@iitkgp.ac.in
[2] IBM Research India, Bangalore, India
{suransam,amarazad}@in.ibm.com

Abstract. Transformer-based models, such as GPT-2, have revolutionized the landscape of dialogue generation by capturing the long-range structures through language modeling. Though these models have exhibited excellent language coherence, they often lack relevance and terms when used for domain-specific response generation. In this paper, we present DSRNet (Domain Specific Response Network), a transformer-based model for dialogue response generation by reinforcing domain-specific attributes. In particular, we extract meta attributes from context and joinly model with the dialogue context utterances for better attention over domain-specific keyterms and relevance. We study the use of DSRNet in a multi-turn multi-interlocutor environment for domain-specific response generation. In our experiments, we evaluate DSRNet on Ubuntu dialogue datasets, which are mainly composed of various technical domain related dialogues for IT domain issue resolutions and also on CamRest676 dataset, which contains restaurant domain conversations. We observe that the responses produced by our model carry higher relevance due to the presence of domain-specific key attributes that exhibit better overlap with the attributes of the context. Our analysis shows that the performance improvement is mostly due to the infusion of key terms along with dialogues which result in better attention over domain-relevant terms.

Keywords: Response generation · Meta context · Transformers

1 Introduction

Transformer-based pertained language models, such as BERT [3], GPT-2 [2,14], XLNet [20], have revolutionized the landscape of natural language processing lately. These models have achieved state-of-the-art performance on many tasks, such as natural language understanding (NLU), sentence classification, named entity recognition and question answering. The ability to capture the long-range temporal dependencies in the input sequences is one of the key reason behind

---

D. Kar—Work done during internship at IBM Research India.

© Springer Nature Switzerland AG 2021
K. Karlapalem et al. (Eds.): PAKDD 2021, LNAI 12714, pp. 285–297, 2021.
https://doi.org/10.1007/978-3-030-75768-7_23

the success of these models. However, generated responses tend to be either generic, out-of-context or disproportionately short. Some of the previous works attributed such behavior to various causes e.g. prevalence of generic utterances in training data, inadequate sized model architecture to capture the long term temporal dependence, absence of low frequency words in vocabulary, exposure bias in training models. In a domain specific multi-turn and multi interlocutor dialogue environment, where multiple users converse over a common channel simultaneously, often regarding a common subject, the above stated problems exacerbate in the generated response.

In this paper, we propose DSRNet (Domain Specific Response Network), a transformer based model, where in we alleviate some of the highlighted issues stated above by explicitly inserting meta-contextual attributes to capture the context better. In particular, we extract various meta attributes such as keywords, queries etc. The input of DSRNet includes the context which constitutes of a predefined number of previous utterances (before the response) in the conversation, and the meta-attributes. The meta attributes are composed of conversation topic, query, entities which are extracted from the conversation at hand using traditional NLP approaches. We have evaluated DSRNet on the Ubuntu-IRC corpus (multi-interlocutor conversation) [6] to generate response utterances which clearly indicate improved response text in terms of alignment with context utterances of the conversation topic.

We have experimented on technical domain specific datasets namely Ubuntu 1.0 dataset (direct conversation) [8], Ubuntu-IRC [6] (mainly pertains to the IT domain) and CamRest676 [18] which contains restaurant related conversations. For domain specific environment, it is of great importance to have the response aligned with the context instead of being generic. We also extended the Ubuntu-IRC, Ubuntu 1.0 and CamRest676 dataset with meta-contexts and intend to release both the source code and extended datasets for future research. To the best of our knowledge, our approach is the first to consider explicitly meta-context attributes and leverage it in a transformer based model to generate dialogue responses in a multi-turn dialogue environment. The key contributions of our work are as follows:

- We propose a novel approach, DSRNet, a GPT-2 based model which jointly models the meta-contextual attributes along with context, for domain specific multi-turn and multi-interlocutor dialogue response generation.
- We extend Ubuntu 1.0, Ubuntu-IRC and CamRest676 datasets with meta-context attributes for better context capturing.

2   Related Literature

Pre-trained transformer based language models have shown tremendous advances in the state-of-the-art across a variety of natural language processing (NLP) tasks ([3,5,13,20]). GPT-2 [2,14] is one of the most known auto-regressive language models and closest to our work which learns language granularity from large amounts of open web text data. Other variants to ground language generation on prescribed control codes are CTRL [5] and Grover [21] or latent variables such as Optimus [7]. GPT-2 first investigated massive Transformer-based auto-regressive language models with large-scale text data for pre-training. After fine-tuning, GPT-2 achieves drastic improvements on several generation tasks. One drawback of GPT-2 is the lack of high-level semantic controlling ability in language generation. To alleviate this issue CTRL [5] was introduced to train the model based on pre-defined codes such as text style, content description, and task-specific behaviour, mean while Grover [21] generates news article conditioned on authors, dates etc. Unlike these, SC-GPT [12] models the text generation more explicitly which is applied to task-oriented dialogue NLG in a few shot setting.

In Dialogue domain, DialoGPT [22] and CGRG [19] extended GPT-2 for chit chat dialogue system. SC-GPT [12] and SOLOIST [11] are pre-trained models for NLG module that converts a dialogue act into response in natural language. DLGNet [10] is a large transformer model trained on dialogue dataset and achieves good performance on multi-turn dialogue response generation. Our work is very close to SC-GPT and DLGNet as we also build our model on GPT2 to generate response for multi-turn dialogues. Moreover, our model reinforces better relevancy by explicit inclusion of context related meta-attributes which helps to capture the context better and its content in the generated responses.

3   Proposed Approach

We model the task of response generation by incorporating fine-grained meta-contextual attributes to capture domain specific goals in the generated utterances more effectively. We infuse various contextual attributes, called as meta-attributes, in our response generation model to capture the content of the conversation context better in the input. Given M training samples $S = (C_m, x_m)_{m=1}^{M}$, where C is the input context and x refers to the corresponding response text; our aim is to build a neural model which maximises the likelihood of the generated response conditioned on the context (C) and meta-attributes ($f(C)$), i.e., $p_\theta(x|C, f(C))$ parameterized by θ (model parameters). The decoder being auto-regressive allows us to express the likelihood as

$$p_\theta(x|C) = \prod_{w=1}^{W} p_\theta(x_w|x_{<w}, C, f(C)) \tag{1}$$

where W is the number of tokens in the ground-truth response. To capture domain-related terms in a better way, we extract the domain-specific keywords (entities or topic words) and the query vector information (question detected or not) of the utterances and perform a massive language fine tuning, allowing the model to learn domain-related word representations along with their contexts more effectively.

Prior to fine tuning the model, we perform pre-training on a massive Ubuntu-dialogue (technical IT support) domain related corpus with the aim to further enrich the model with domain-related knowledge. To achieve that, we have adopted the pre-trained GPT-2 architecture [14] and have trained it further on a language modeling task using the Ubuntu 2.0 Corpora [8].

3.1 Meta-attribute Infused Fine Tuning

The next utterance in a conversation is usually constructed using key-terms from the context. In DSRNet, we propose infusing meta-attributes, namely topics/entities and queries, from conversations in the training procedure. We process the mined meta-attributes and generate input instances as shown in Table 1. Thus the meta-attribute $f(C)$ in Eq. 1 can be represented as: $f(C) = [q_m; t_m]_{m=1}^{M}$ where, M is the number of instances in the dataset, q_m represents the query detection result on the mth instance and t_m represents the topics/entities extracted from the context (c_m) of the mth instance. In the sections to follow, we provide a detailed description of our meta-attribute mining algorithms. The proposed architecture of DSRNet is illustrated in Fig. 1, where the explicit dependency of the generation of one word in the response, on the context, meta-attributes and the previous words generated in the response are highlighted.

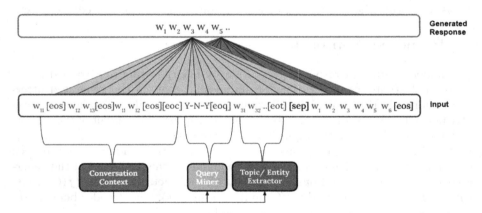

Fig. 1. The proposed Meta-Context Transformer. Special tokens [eoc], [eoq], [eot] separate the dialogue history, contextual queries and topics in the context respectively. The token [sep], separates the context from the ground truth that needs to be generated and the token [eos] marks the end of an utterance in our work.

Topic Infusion - While we are aware of the domain of the corpora, more than often it has been observed that each conversation has a topical premise of it's own. We have used Latent Dirichlet Allocation [1] method to determine the topics of the conversations in the corpus. We have treated each conversation in the corpus as a document in our approach. The number of topics rendering the optimal coherence score was found to be 80 for the Ubuntu-IRC data-set. Using the learnt topic model, we determine the dominant topic of each utterance and have appended the 10 most contributing topic keywords from the corpus to each utterance. Using the statistical approach of topic modeling has not only saved us annotation labour, it has also supported the model with consistent performance as reported in Tables 2 and 3.

Query Mining and Infusion - We observe that query utterances, if present, directly influences the response utterance in the conversation. We infuse the identified queries as a part of meta-attributes in the form of query vector. To identify both implicit and explicit queries [4,16], we have adopted a semi-supervised approach where we augment some lexical rules along with an SVM model. The lexical rules are apt to capture queries containing question words (mainly constituting 5W1h question words, for e.g.'?', when, how, where), along with a set of other curated verb and adverb based question words (e.g.,could, did, kindly, please). On the other hand, the SVM model is trained on the NPS Chat dataset[1] to detect the implicit queries which captures the informal query utterances. Our algorithm identifies an utterance as negative query only when both the lexical and SVM model yields a negative label. We report a precision score of 86.90%, a recall of 92.03% and an F-Score of 89.39% over randomly sampled 100 instances from the Ubuntu datasets. We append a sequence of 'Y-N's to the context - 'Y's confirming the presence of a query at that index in the context and 'N's confirming otherwise as shown in Table 1.

Entity Extraction - Although we capture the topical preference of the conversation by infusing topic keywords with each utterance in the context, we realise that the domain-keywords in the context may still not be captured in the generation. This is mainly because i) the topic keywords do not necessarily pertain to that particular conversational context, ii) it is often tricky to capture the optimal number of topics for topic modeling and may result in topic overlaps - hence, losing out on important topic words, iii) dialogues are known to be privy to noisy text - with spelling mismatches and word distortions, some important domain words in the context may not be captured accurately. To tackle these issues, we experiment with entity phrases extracted from the context, instead of topic words and report the performance in Tables 2 and 3. To extract entity words from conversations we adopt the weakly supervised method as outlined in [9].

4   Experimental Studies

In this section, we evaluate our proposed model DSRNet on three different dialog corpus. We address the following questions during the evaluation process, which are (i) How efficient is DSRNet when compared with other state-of-the-art (SOTA) methods of generating response in dialog settings and (ii) How efficiently DSRNet is able to improve the task of response generation using the meta-contexual information.

---

[1] http://faculty.nps.edu/cmartell/NPSChat.htm.

Dataset - We have used three popular, publicly available datasets to train and evaluate our proposed model DSRNet. For the domain language pre-training step, our data of choice were the training samples from the Ubuntu 2.0 dataset[2] due to it's massive collection of annotated examples. For fine-tuning the domain pre-trained language model for the response generation task, we have used the original Ubuntu 1.0 dataset[8] and the Ubuntu-IRC dataset[6]. For the generic restaurant enquiry response generation, we skip the domain pretraining step and fine-tune the vanilla GPT-2 on the CamRest676 dataset[18]. It is important to note, that natural language generation for domains like restaurant, hotels and travel are much simpler than the same in a technical domain, which contains plenty of non-English keywords and commands (e.g. sudo apt-get, html, css, Windows). We tackle most of these challenges through our domain language pre-training and meta-attribute learning steps. Moreover, the dialogues in the Ubuntu-IRC dataset occur in a multi-turn, multi-locutor setting. We adhere to the following pre-processing rules to tackle some of the challenges of the dataset: i) We extract the conversations from the dataset using the disentanglement annotations provided (for the Ubuntu-IRC dataset), and we remove, ii) slang from the dialogues, iii) usernames, timestamps, iv) non-English utterances from the corpus (we encountered a number of Spanish utterances in the conversations), v) instances of utterance repetitions, and vi) one word utterances which are neither questions nor commands. Ubuntu 1.0 corpus has about 3 million training instances, when the context has 3 utterances. Camrest676 has 2515 training instances and Ubuntu-IRC dataset has 22,582 training instances. For Ubuntu 1.0, we randomly sample 10,000 conversations for the development and test sets and have used the rest of the dataset for fine-tuning purposes. We use 1000 samples from each of the other two dataset for testing purposes. Ubuntu-IRC corpus [6] has 153 conversation files, each of which have several parallel conversations. We use the annotation provided by Kummerfeld et al. and extract 4621, 392 and 298 conversations for training, evaluation and test split respectively. The entire dataset with extracted meta-contextual attributes along with the source code will be made available online for research purposes[3].

4.1   Experimental Settings

We fine-tune the pre-tuned GPT-2 model on our dataset using 2 V100 cores using 100 GPU memory in each of the them, for most of the experimental settings. We fine-tune the proposed DSRNet on CamRest676 dataset for 20 epochs and on Ubuntu 1.0, Ubuntu-IRC datasets for 5 epochs. We did not observe a considerable improvement in performance on training DSRNet beyond the reported number of epochs.

Dataset Format for Fine-Tuning - We create the training instances by considering contexts in a sliding window fashion, containing three consecutive utterances. The context is followed by the query feature vector (Y represents pres-

---

[2] https://github.com/rkadlec/ubuntu-ranking-dataset-creator.
[3] https://github.com/DebanjanaKar/DSRNet.

ence of query, N represent absence of query) obtained from query mining. This is followed by (i) the list of entities extracted from the context or (ii) list of 10 dominant topic words that represent the context. Typical examples of training instances, using both query information and entities are shown in Table 1.

Table 1. Examples of training samples from Ubuntu-IRC and CamRest676 dataset used in DSRNet.

Dataset	Training example
Ubuntu IRC (with queries + topics)	however, in addition, I've observed something odd with this machine [eos] Sorry, I've had a quick look over a few Ubuntu mirrors, but can't find a Live PPC CD of Warty... are they available? [eos] any reboot short of turning the mains power off causes the bios to fail to recognize the hard drive [eos][eoc] N-Y-N [eoq] connect, fail, reinstal, final, enter, normal, command, address, account, mode [eot] [sep] there's no ppc live cd yet [eos]
CamRest676 (with queries + entities)	I need to find an expensive restauant that's in the south section of the city. [eos] There are several restaurants in the south part of town that serve expensive food. Do you have a cuisine preference? [eos] No I don't care about the type of cuisine. [eos] [eoc] N-Y-N [eoq] city, town, type, south section, expensive restauant, expensive food, south part, several restaurant, cuisine preference [eot] [sep] Chiquito Restaurant Bar is a Mexican restaurant located in the south part of town. [eos]

Topic Modeling - We apply a standard LDA based topic-modeling to extract the dominant topics in the corpus. Figure 2 shows the topic distribution (number of topics 40 in this case, for better visual aid) at utterance level and conversation level respectively. It is clear from the figure that better discriminating topics are obtained at the conversation level. Before feeding the data to the topic modeling algorithm, we heavily preprocessed our data by tokenizing, stemming and removing common words from the corpus. To remove common words, we considered a list of English stopwords, words with either very low or very high tf-idf scores and words with noun, adjective, verb and adverb part-of-speech tags. We set the number of topics to 80 in all the three corpus, after observing the coherence score.

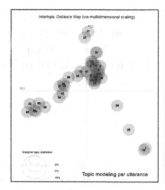 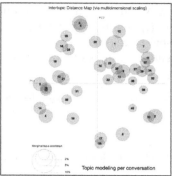

Fig. 2. Figure shows the topic distribution when topic modeling is done at (i) utterance level (left) & (ii) conversation level (right) for 40 topics.

Evaluation Metrics - We use the nlg-eval toolkit[4] and MultiTurnDialogZoo toolkit[5] for obtaining the values for the following metrics during our experimentation:

1. BLEU and ROUGE - BLEU has been used for measuring the coherency of the generated output. It mainly looks at the precision of the common n-grams in the ground truth and the generated output. Similarly, ROUGE measures the recall of the common n-grams occurring in ground truth and generated output.
2. DISTINCT - It measures the diversity of the words in the generated output. We consider the average of DISTINCT-1 and DISTINCT-2 scores.
3. BERT-Score (BERTSc) - BERT score represents the cosine similarity of the pair-wise similar words in the embedded space.

However, it is difficult to compare the generated response with the ground truth, as an utterance can be represented correctly with many different sentences. Many a times, human evaluation is the best way to determine the correctness of a generated output.

Baseline - The following methods have been adapted as baselines which gives state-of-the-art (SOTA) results on response generation in dialog settings: (i) Vanilla GPT-2 - We use the GPT-2, finetuned on the training set, as one of the baselines. As the proposed DSRNet is an improved version of GPT-2, it gives an idea about how strong the modification of the proposed method is with respect to the base model. (ii) HRED, VHRED - We train [17] and [15] on each of the datasets for 20 epochs. Both of the methods model the task using recurrent encoders and decoders in a hierarchical fashion. (iii) DLGNet - We consider DLGNet for comparison purposes for the Ubuntu 1.0 dataset. The evaluation metric values are directly taken from the paper.

---

[4] https://github.com/Maluuba/nlg-eval.
[5] https://github.com/gmftbyGMFTBY/MultiTurnDialogZoo.

4.2   Performance and Evaluation

We choose the training parameters based on the performance on the evaluation data split and report the metric results on the test split. We show the results of different variations of DSRNet - (i) DSRNet (qstn) - having question detection only as the meta-attribute (ii) DSRNet (qstn+top) - having question detection and dominant topic words as meta attribute, and (iii) DSRNet (qstn+ent) - having question detection and entities extracted from context as meta attribute. We keep topic words and entities as complimentary information in forming the meta attribute of an instance. The results for three dataset are explained below.

Table 2 shows the performance of different models on generating response for CamRest676 dataset. We compare the performance of DSRNet with that of vanilla GPT-2, HRED and VHRED.

Table 2. Performance evaluation on CamRest676 dataset (best in bold).

Method	BLEU	ROUGE	BERTSc	Distinct
HRED	0.0332	0.0582	−0.0007	0.0655
VHRED	0.0298	0.0722	0.0049	0.0667
GPT-2	0.1382	0.1496	0.242	0.4164
DSRNet (qstn)	0.211	0.2581	0.3296	0.4356
DSRNet (qstn+top)	0.1896	0.1679	0.2254	0.4002
DSRNet (qstn+ent)	**0.260**	**0.2587**	**0.6603**	**0.4533**

Table 3 shows the performance of different models on generating response for Ubuntu 1.0 & Ubuntu IRC dataset. We compare the performance of DSRNet with that of vanilla GPT-2 and DLGNet. We can observe that DLGNet is having better BLEU score than DSRNet for Ubuntu 1.0, which is mostly because of the fact that DLGNet considered a much larger sequence length (1024) while training their model. This is particularly helpful for Ubuntu 1.0 dataset, where the length of utterance can be very long.

Table 3. Performance evaluation on Ubuntu 1.0 and Ubuntu-IRC dataset.

Method	Ubuntu 1.0				Ubuntu IRC			
	BLEU	ROUGE	BERTSc	Distinct	BLEU	ROUGE	BERTSc	Distinct
HRED	0.017	0.048	NA	0.046	0.118	0.11	0.070	0.0003
VHRED	0.017	0.086	NA	0.089	0.111	0.21	0.210	0.002
DLGNet	**0.028**	**0.219**	–	0.495	–	–	–	–
GPT-2	0.012	0.024	0.401	0.202	0.085	0.103	0.235	0.219
DSRNet (qs)	0.015	0.023	0.034	0.480	0.106	0.145	0.290	0.468
DSRNet (qs+top)	0.016	0.024	0.034	0.5051	0.1287	**0.1890**	0.298	0.443
DSRNet (qs+ent)	0.020	0.032	**0.033**	**0.556**	**0.148**	0.157	**0.347**	**0.470**

It is evident from the tables that adding of meta contextual attributes does help in generating better responses. Addition of entities is performing better than that of topic words, as entities are directly picked up from the corresponding context, whereas, a dominant topic word may come from outside of the given context. It is also to be noted, that the repetition of generated response for multiple similar context is least for DSRNet, which is evident from the DISTINCT score. Also, as per our observation, DSRNet produced more semantically and syntactically meaningful responses.

4.3   Result Analysis

Importance of Query mining and Entity Extraction - The idea of including the query mining as the meta-attribute is that the questions in the context should get more importance while generating the response. We observed the response generated without using the question mining in the context and also when the question detection module gives erroneous results. In both these cases, the quality of the generated response degrades, as seen in Fig. 3. Similarly, as shown in Fig. 3, DSRNet exhibits improved performance for all the three datasets when entities are being used as the meta attribute.

question in meta context	"Context": "chrome is chromium with flash built in and google branding [eos] chromium is the open source version of chrome yah? [eos] other google added stuff too i am sure [eos][eoc] N-Y-N [eoq]" "Generated": "yes but chromium is open source and google brands it not chromium"	entity in meta context	"Context": "unfortunately for now it requires a bit of work [eos] you will only be able to read the ntfs files [eos] b) you wont be able to write to an ntfs partition [eos][eoc] N-N-N [eoq] able, eos, work, bit, ntfs partition, ntfs file [eot] ", "Generated": "then you might want to do what i suggested, mount it via ntfs"
no question in meta context	"Context": "chrome is chromium with flash built in and google branding [eos] chromium is the open source version of chrome yah? [eos] other google added stuff too i am sure [eos] [eoc] N-N-N [eoq]" "Generated": "i've got two choices i just wanted to add it in /"	no entity in meta context	"Context": "anyone know a decent dc++ client other than linux dc [eos] I have a folder that windows reads as hidden, is there any way i can change this using ubuntu? [eos] I've just read about amue crashing all the time when updating. What's the eta on a solution, anyone know? [eos] ", "Generated": "thank you, i'll check that out so yes it does"
faulty question in meta context	"Context": "chrome is chromium with flash built in and google branding [eos] chromium is the open source version of chrome yah? [eos] other google added stuff too i am sure [eos] [eoc] N-N-N [eoq]" "Generated": "and firefox 3.8 is out? what are you trying"		

Fig. 3. Examples of generated response from DSRNet with and without using questions (left) and entities (right) as a meta attribute.

Ablation Study - Most of the time the generated response of DSRNet is meaningful to the context, with proper grammar. However, at times, the generated output suffers from repetition of words and sentences. We believe that this problem is mostly due to the small training size of the dataset. Also, there has been some responses, which end abruptly, such as *"there are two moderately priced mid-range restaurants, both in the centre part of town. do you"*. This is mainly because of the sequence length that we are considering during training of DSR-Net, due to computational effectiveness of the model training. With a large value of sequence length, this problem can be resolved.

There are also some of the generated responses, which are highly relevant with the context, but is different from the ground truth utterance, such as:

Context: "There are several good restaurants in the south part of town. Do you have a preference for the type of food or price range? [eos] Yes, I'd like a restaurant that serves portuguese food. [eos] nandos is the only restaurant i can find in the south that serves portuguese. [eos] [eoc] Y-N-N [eoq] town, type, preference, nandos, south part, price range, portuguese food, restaurant [eot]"
Generated: "what is the address?"
True: "Nandos sounds great. Thank you."

These generated outputs will reduce the metric scores used for evaluation, but a human evaluation may consider the generated output as a relevant one.

5   Conclusion

In this paper, we have proposed DSRNet, a transformer-based model, for dialogue response generation by explicitly infusing domain-specific attributes. To infuse meta-contextual information in DSRNet, we have extracted meta attributes, namely conversation topics, key entities and queries from the conversation context which enables better relevance of the generated response. We conducted thorough investigation over Ubuntu-IRC, Ubuntu 1.0, and CamRest676 dataset and reported multiple performance metrics, such as BLEU, ROGUE, and semantic similarity scores. Our evaluation results indicate that DSRNet shows improvement with other existing models in terms of generating responses which are more relevant to the conversation context. The generated responses have better presence of domain-specific key attributes that exhibit better overlap with the attributes of the context.

References

1. Blei, D.M., Ng, A.Y., Jordan, M.I.: Latent Dirichlet allocation. J. Mach. Learn. Res. **3**, 993–1022 (2003)
2. Budzianowski, P., Vulic, I.: Hello, it's gpt-2-how can i help you? towards the use of pretrained language models for task-oriented dialogue systems. EMNLP-IJCNLP **2019**, 15 (2019)
3. Devlin, J., Chang, M.W., Lee, K., Toutanova, K.: BERT: Pre-training of deep bidirectional transformers for language understanding. In: Proceedings of the 2019 Conference of the North American Chapter of the Association for Computational Linguistics: Human Language Technologies, Volume 1 (Long and Short Papers), pp. 4171–4186. Association for Computational Linguistics, Minneapolis, Minnesota, June 2019. 10.18653/v1/N19-1423, https://www.aclweb.org/anthology/N19-1423
4. Forsythand, E.N., Martell, C.H.: Lexical and discourse analysis of online chat dialog. In: International Conference on Semantic Computing (ICSC 2007), pp. 19–26 (2007)

5. Keskar, N.S., McCann, B., Varshney, L.R., Xiong, C., Socher, R.: Ctrl: a conditional transformer language model for controllable generation (2019)
6. Kummerfeld, J.K., et al.: A large-scale corpus for conversation disentanglement. In: Proceedings of the 57th Annual Meeting of the Association for Computational Linguistics, pp. 3846–3856 (2019)
7. Li, C., Gao, X., Li, Y., Peng, B., Li, X., Zhang, Y., Gao, J.: Optimus: Organizing sentences via pre-trained modeling of a latent space. In: Proceedings of the 2020 Conference on Empirical Methods in Natural Language Processing (EMNLP), pp. 4678–4699. Association for Computational Linguistics, Online, November 2020
8. Lowe, R., Pow, N., Serban, I.V., Pineau, J.: The Ubuntu dialogue corpus: a large dataset for research in unstructured multi-turn dialogue systems. In: 16th Annual Meeting of the Special Interest Group on Discourse and Dialogue, p. 285 (2015)
9. Mohapatra, P., et al.: Domain knowledge driven key term extraction for IT services. In: Pahl, C., Vukovic, M., Yin, J., Yu, Q. (eds.) ICSOC 2018. LNCS, vol. 11236, pp. 489–504. Springer, Cham (2018). https://doi.org/10.1007/978-3-030-03596-9_35
10. Olabiyi, O., Vienna, V., Mueller, E.T.: Dlgnet: a transformer-based model for dialogue response generation. ACL **2020**, 54 (2020)
11. Peng, B., Li, C., Li, J., Shayandeh, S., Liden, L., Gao, J.: Soloist: Few-shot task-oriented dialog with a single pre-trained auto-regressive model (2020)
12. Peng, B., et al.: Few-shot natural language generation for task-oriented dialog. In: Proceedings of the 2020 Conference on Empirical Methods in Natural Language Processing: Findings, pp. 172–182 (2020)
13. Peters, M., et al.: Deep contextualized word representations. In: Proceedings of the 2018 Conference of the North American Chapter of the Association for Computational Linguistics: Human Language Technologies, Volume 1 (Long Papers) pp. 2227–2237. Association for Computational Linguistics, New Orleans, Louisiana, June 2018
14. Radford, A., Narasimhan, K., Salimans, T., Sutskever, I.: Improving language understanding by generative pre-training (2018)
15. Serban, I., Sordoni, A., Bengio, Y., Courville, A.C., Pineau, J.: Building end-to-end dialogue systems using generative hierarchical neural network models. In: AAAI (2016)
16. Shrestha, L., McKeown, K.: Detection of question-answer pairs in email conversations. In: Proceedings of the 20th International Conference on Computational Linguistics. COLING 2004, p. 889-es. Association for Computational Linguistics, USA (2004)
17. Sordoni, A., Bengio, Y., Vahabi, H., Lioma, C., Grue Simonsen, J., Nie, J.Y.: A hierarchical recurrent encoder-decoder for generative context-aware query suggestion. In: Proceedings of the 24th ACM International on Conference on Information and Knowledge Management. CIKM 2015, pp. 553–562, Association for Computing Machinery, New York, NY, USA (2015)
18. Wen, T.H., et al.: A network-based end-to-end trainable task-oriented dialogue system. In: EACL, Valencia, Spain, pp. 438–449. Association for Computational Linguistics, April 2017
19. Wu, Z., et al.: A controllable model of grounded response generation (2020)

20. Yang, Z., Dai, Z., Yang, Y., Carbonell, J., Salakhutdinov, R.R., Le, Q.V.: Xlnet: Generalized autoregressive pretraining for language understanding. In: Wallach, H., Larochelle, H., Beygelzimer, A., d'Alché-Buc, F., Fox, E., Garnett, R. (eds.) Advances in Neural Information Processing Systems, vol. 32. Curran Associates, Inc. (2019)
21. Zellers, R., et al.: Defending against neural fake news. Neurips (2020)
22. Zhang, Y., et al.: Dialogpt: large-scale generative pre-training for conversational response generation. In: Proceedings of the 58th Annual Meeting of the Association for Computational Linguistics: System Demonstrations, pp. 270–278 (2020)

A Multi-task Kernel Learning Algorithm for Survival Analysis

Zizhuo Meng[1,2]([✉]), Jie Xu[1], Zhidong Li[1], Yang Wang[1], Fang Chen[1], and Zhiyong Wang[2]

[1] University of Technology Sydney, Ultimo, NSW 2007, Australia
{zizhuo.meng,jie.xu,zhidong.li,yang.wang,fang.chen,
zhiyong.wang}@uts.edu.au
[2] The University of Sydney, Camperdown, NSW 2006, Australia
{zizhuo.meng,zhiyong.wang}@sydney.edu.au

Abstract. Survival analysis aims to predict the occurring times of certain events of interest. Most existing methods for survival analysis either assume specific forms for the underlying stochastic processes or linear hypotheses. To cope with non-linearity in data, we propose a unified framework that combines multi-task and kernel learning for survival analysis. We also develop optimization methods based on the Pegasos (Primal estimated sub-gradient solver for SVM) algorithm for learning. Experiment results demonstrate the effectiveness of the proposed method for survival analysis, on synthetic and real-world data sets.

Keywords: Survival analysis · Multi-task learning · SVM · Kernel method

1 Introduction

Survival analysis, also known as time-to-event analysis, is crucial in areas such as finance, engineering, medicine, etc. Its goal is to predict the occurring times of certain events of interest, especially the first occurrences. However, those events might not always be observed during the course of study due to time limitations or missing data, which is known as censoring. Censoring distinguishes survival analysis from standard regression models.

A fundamental problem of survival analysis is to understand the relationship between the underlying distribution of event occurrence, and sample characteristics. Most of the previous work has tackled this problem by assuming specific forms for the underlying stochastic process [2], but such assumption may not hold in practise. To overcome such issues, multi-task learning models [12,13] are introduced, whose tenet is to learn a shared representation across related tasks to reduce the prediction error of individual tasks. Existing multi-task learning models have been built with a linear hypothesis, which may not cope with non-linear structures in data.

© Springer Nature Switzerland AG 2021
K. Karlapalem et al. (Eds.): PAKDD 2021, LNAI 12714, pp. 298–311, 2021.
https://doi.org/10.1007/978-3-030-75768-7_24

In this paper, we propose a method that combines multi-task and kernel learning, which is a structured way for non-linear transformation. Our contributions are two-fold: (i) a unified framework that consolidates multi-task and kernel learning, which is capable of utilizing complex non-linearity exhibited in data; (ii) Pegasos (Primal Estimated sub-Gradient SOlver for SVM) based optimization algorithms for learning parameters. By exploiting the non-linearity in data, our approach achieves competitive performance on both synthetic and real-world data sets. The rest of the paper is organized as follows: Sect. 2 reviews existing survival analysis methods, Sect. 3 presents our method, Sect. 4 demonstrates our experimental results and Sect. 5 concludes our work.

2 Related Work

Previous works on survival analysis can be briefly categorized as follows: statistical models, multi-task learning models, and deep learning models.

Statistical models can be categorized into parametric and non-parametric models. Both assume specific forms of distribution for the underlying process of event occurrence. For example, the Cox proportional hazard model (CPH) [2] and its variants [17,18], assume the log of the hazard rate to be a linear function of covariates, while other models make different assumptions about data, e.g., the Weibull [11] distribution. These parametric models suffer from rigid assumptions that may not hold in practice. Non-parametric models relax those assumptions. Examples include Kaplan–Meier estimator [6], which is generalised as Random Survival Forest [5], Bayesian non-parametric (BNP) like beta/gamma process [14,15], Lomax delegate racing (LDR) [19], Gaussian process [1,3] etc. As their numbers of parameters grow with data size, the inference can have issues in both complexity and efficiency.

Multi-task learning models for survival analysis overcome the weakness of the statistical models by the multi-task formulation. Specifically, it translates the original problem into a series of related tasks. Zhou *et al.* propose a multi-task learning formulation for predicting the progression of the Alzheimer's disease measured via the clinical scores at multiple time points [20]. Li *et al.* propose an MTLSA (Multi-Task Learning model for Survival Analysis) model that transforms the original cancer survival analysis problem into a multi-task learning problem by decomposing the regression component into related classification tasks [13]. Similarly, Li *et al.* propose a survival analysis approach to model the turnover and career progression behaviors in talent management based on multi-task learning [12]. Despite the effectiveness of these models, they are built with linear hypotheses, which may not cope with data with non-linear structures.

Recently medical practitioners have employed nonlinear survival analysis, such as deep learning models to understand the relationship between patients' covariates and the effectiveness of different treatment options. Katzman presented a treatment recommender system based on Cox proportional hazards deep neural network to provide personalized treatment recommendations [7]. DeepHit is another attempt to tackle survival analysis using the deep neural network [10].

It learns the distribution of survival time directly, without assumptions about the underlying stochastic process. These methods consider the survival analysis problem as a point-wise prediction problem, which ignores the relationship between adjacent time slices.

3 Methods

In this section, we present our unified framework that consists of objective functions that incorporate the hinge loss for SVM and ranking constraints for survival analysis, and also Pegasos (Primal estimated sub-gradient solver for SVM) [16] based algorithms for optimization.

3.1 Problem Formulation

This paper only considers the non-recurring case of survival analysis, i.e., the event of interest occurs only once. Each sample i has either a survival time (O_i) or a censored time (C_i) but not both. Specifically a sample can be represented as a triplet $(\vec{X}_i, T_i, \delta_i)$, where $\vec{X}_i \in \mathbb{R}^{1 \times M}$ is a feature vector; δ_i is a censoring indicator that is set to 1 for censored samples and 0 otherwise; T_i denotes the *observed time* and is equal to the survival time O_i for uncensored samples and C_i otherwise. A sample is considered right-censored if only if $T_i = \min(O_i, C_i)$.

We use the following notations for any matrix $\boldsymbol{X}$: the entry at i-th row and j-th column is denoted as $x_{i,j}$, while the i-th row in $\boldsymbol{X}$ is expressed as $\boldsymbol{x}_{i,\cdot}$ and the j-th column of $\boldsymbol{X}$ is written as $\boldsymbol{x}_{\cdot,j}$. We follow the practice of [13] to transform the original survival data for multi-task learning. Let N be the number of samples, and T be $\max(T_i)$, the transformation defines the following:

- $\boldsymbol{Y} \in \mathbb{R}^{N \times T}$: the target (ground truth) matrix that indicates the "survival state" (alive or not) of N samples for T intervals where: $y_{i,j} = 1$ if sample i is alive at time interval j and -1 otherwise. As the status of some samples may be unknown, e.g., censored samples ($\delta_i = 0$). For those samples, we set the values of corresponding cells to "-1" respectively, according to [13];
- $\boldsymbol{X} \in \mathbb{R}^{N \times M}$: the feature matrix that contains N samples. Each row $\boldsymbol{x}_{i,\cdot}$ is a feature vector for one sample;
- $\boldsymbol{W} \in \mathbb{R}^{M \times T}$: the coefficient matrix for predicting samples' survival behavior. Each column $\boldsymbol{w}_{\cdot,j}$ defines the weights for a time interval;
- $\hat{\boldsymbol{Y}} \in \mathbb{R}^{N \times T}$: the predicted survival matrix.

3.2 Learning Framework

We learn an SVM model for each task, which associates with the predictions for the corresponding time interval. In other words, an SVM is trained for each time interval based on the same $\boldsymbol{X}$. The coefficients of the SVM at the j-th interval are specified by the column $\boldsymbol{w}_{\cdot,j}$ of $\boldsymbol{W}$. The goal is to estimate $\boldsymbol{W}$ by optimizing the following objective function:

$$\min_{\boldsymbol{W}} \frac{\lambda}{2}||\boldsymbol{W}||^2 + \mathcal{L} \tag{1}$$

where $\mathcal{L}$ defines the constraints in the formulation. In our problem there are two constraints: $\mathcal{L}_1$ that enforces the minimization of the discrepancy between $\boldsymbol{Y}$ and $\hat{\boldsymbol{Y}}$, and $\mathcal{L}_2$ that relates to non-recurring setting in which once the survival state of a sample becomes "1" it will not change back to "-1". It is termed as the "Non-increasing" constraint, which can be expressed as follows: $y_{i,j} \geqslant \max(\{y_{i,k}|k > j\}), \forall i \in \{1, 2, ..., N\}, j, k \in \{1, 2, ..., M\}$.

Objective Functions. We design two objective functions for basic and kernel learning respectively. For the **basic learning** the SVM at the j-th interval classifies each input vector $\boldsymbol{x}_{i,\cdot}$ by learning the following linear model:

$$\hat{y}_{i,j} = \langle \boldsymbol{x}_{i,\cdot}, \boldsymbol{w}_{\cdot,j} \rangle, \tag{2}$$

where $\langle \boldsymbol{x}_{i,\cdot}, \boldsymbol{w}_{\cdot,j} \rangle$ denotes the inner product of $\boldsymbol{x}_{i,\cdot}$ and $\boldsymbol{w}_{\cdot,j}$.

$\mathcal{L}_1$ computes the error of each SVM individually:

$$\mathcal{L}_1 = \sum_{i=1}^{N} \sum_{j=1}^{M} max\{0, 1 - y_{i,j}\hat{y}_{i,j}\} \tag{3}$$

$\mathcal{L}_2$ ensures that a "returning-back-to-life" action incurs a loss:

$$\mathcal{L}_2 = \sum_{i=1}^{N} \sum_{j=1}^{M} \sum_{k>j}^{M} \mathbb{1}[\hat{y}_{i,k} > 0, \hat{y}_{i,j} < 0](\hat{y}_{i,k} - \hat{y}_{i,j}) \tag{4}$$

For the **kernel learning**, SVMs are trained based on the inner products of training samples specified by a kernel operator, as per the Representer Theorem [8]. Instead of learning a classifier from training samples $\boldsymbol{x}_{i,\cdot}$, we can do so by learning it from some implicit mapping $\phi(\boldsymbol{x}_{i,\cdot})$ of the samples. By applying the mapping $\phi(\cdot)$, we rewrite (2) as:

$$\hat{y}_{i,j} = \langle \phi(\boldsymbol{x}_{i,\cdot}), \boldsymbol{w}_{\cdot,j} \rangle, \tag{5}$$

After plugging (5) into (3), (4), the $\mathcal{L}_1$ and $\mathcal{L}_2$ for kernel algorithm are calculated in a similar way as the basic version. Note that the mapping $\phi(\cdot)$ is never specified explicitly but rather through a kernel operator $K(x, x') = \langle \phi(x), \phi(x') \rangle$ that yields the inner products after the application of mapping $\phi(\cdot)$.

Optimization. We develop two multi-task sub-gradient descent algorithms based on the Pegasos algorithm [16]. As shown in Algorithm 1, the **Basic algorithm** optimizes (1) iteratively. Let $\boldsymbol{W}^{(t)}$ be to the weight matrix at the t-th

---

Algorithm 1. Multi-task Basic Learning Algorithm

Require: λ, Maximum iteration: T_m
Input: Feature Matrix: $\boldsymbol{X} \in \mathbb{R}^{N \times M}$, Target Matrix: $\boldsymbol{Y} \in \mathbb{R}^{N \times T}$

1: Set $\boldsymbol{W}^{(1)} \in \mathbb{R}^{M \times D} \leftarrow 0$
2: **for** t=1,2...T_m **do**
3:     Set the step size $\eta^{(t)} \leftarrow \frac{1}{\lambda t}$
4:     Select a sample $\boldsymbol{x}_{i,\cdot}^{(t)} \in \mathbb{R}^M$ randomly, i.e. the i-th row of $\boldsymbol{X}$
5:     Make prediction : $\hat{\boldsymbol{y}}_i^{(t)} \leftarrow (\boldsymbol{W}^{(t)})^T \boldsymbol{x}_{i,\cdot}$
6:     Compute the sub-gradient of $\mathcal{L}_1, \mathcal{L}_2$ w.r.t $\boldsymbol{W}^{(t)}$ according to $\hat{\boldsymbol{y}}_i^{(t)}$

$$\frac{\partial \mathcal{L}_1}{\partial \boldsymbol{W}^{(t)}} + \frac{\partial \mathcal{L}_2}{\partial \boldsymbol{W}^{(t)}} = \sum_{j \in P}^m \frac{\partial (1 - y_{i,j} \hat{y}_{i,j}^{(t)})}{\partial \boldsymbol{W}^{(t)}} + \sum_{j=1}^m \sum_{k \in Q} \frac{\partial (\hat{y}_{i,k}^{(t)} - \hat{y}_{i,j}^{(t)})}{\partial \boldsymbol{W}^{(t)}}$$

$$= - \sum_{j \in P}^m (y_{i,j} \frac{\partial \hat{y}_{i,j}^{(t)}}{\partial \boldsymbol{W}^{(t)}}) + \sum_{j=1}^m \sum_{k \in Q} \frac{\partial (\hat{y}_{i,k}^{(t)} - \hat{y}_{i,j}^{(t)})}{\partial \boldsymbol{W}^{(t)}}.$$

where

-   $P = \{j \| y_{i,j}^{(t)} \hat{y}_{i,j}^{(t)} < 1\}$
-   $Q = \{k \| \hat{y}_{i,j}^{(t)} < 0, \hat{y}_{i,k}^{(t)} > 0 \wedge j < k\}$
-   $\frac{\partial \hat{y}_{i,j}^{(t)}}{\partial \boldsymbol{W}^{(t)}}$ is a $\boldsymbol{0}$ matrix $\in \mathbb{R}^{M \times T}$, except the j-th column which is set to $\boldsymbol{x}_{i,\cdot}^{(t)}$

7:     $\boldsymbol{W}^{(t+1)} \leftarrow \boldsymbol{W}^{(t)} - \eta^{(t)} \nabla_{W^{(t)}} (\frac{\lambda}{2} \|\boldsymbol{W}^{(t)}\|^2 + \mathcal{L}_1 + \mathcal{L}_2)$
8: **end for**
9: Output: $\boldsymbol{W}^{(T_m+1)}$

---

iteration. At each iteration, the sub gradients of $\mathcal{L}_1$ and $\mathcal{L}_2$ w.r.t $\boldsymbol{W}^{(t)}$ are computed, which are used to update $\boldsymbol{W}^{(t)}$.

The **Kernel algorithm** also follows the iterative sub gradient optimization strategy optimize (1). Let $\boldsymbol{W}^{(1)} = 0$ and $\mathcal{L} = \mathcal{L}_1 + \mathcal{L}_2$, $\boldsymbol{W}^{(t)}$ is updated as follows:

$$\boldsymbol{W}^{(t+1)} = \boldsymbol{W}^{(t)} - \nabla_{W^{(t)}} \frac{1}{\lambda t} (\frac{\lambda}{2} \|\boldsymbol{W}^{(t)}\|^2 + \mathcal{L}) = (1 - \frac{1}{t}) \boldsymbol{W}^{(t)} - \frac{1}{\lambda t} \frac{\partial \mathcal{L}}{\partial \boldsymbol{W}^{(t)}}$$

$$= (1 - \frac{1}{t})[(1 - \frac{1}{t-1}) \boldsymbol{W}^{(t-1)} - \frac{1}{\lambda(t-1)} \frac{\partial \mathcal{L}}{\partial \boldsymbol{W}^{(t-1)}}] - \frac{1}{\lambda t} \frac{\partial \mathcal{L}}{\partial \boldsymbol{W}^{(t)}}$$

$$= \frac{t-2}{t} \boldsymbol{W}^{(t-1)} - \frac{1}{\lambda t} (\frac{\partial \mathcal{L}}{\partial \boldsymbol{W}^{(t)}} + \frac{\partial \mathcal{L}}{\partial \boldsymbol{W}^{(t-1)}}) = \cdots = - \frac{1}{\lambda t} \sum_{j=1}^t \frac{\partial \mathcal{L}}{\partial \boldsymbol{W}^{(j)}}$$

(6)

The value of $\frac{\partial \mathcal{L}}{\partial \boldsymbol{W}^{(j)}}$ requires $\frac{\partial \mathcal{L}_1}{\partial \boldsymbol{W}^{(t)}}, \frac{\partial \mathcal{L}_2}{\partial \boldsymbol{W}^{(t)}}$. Note that $\mathcal{L}_1$ reflects the classification performance at each iteration. To aid the calculation, we define an auxiliary matrix $\boldsymbol{\alpha}^{(t)} \in \mathbb{R}^{N \times T}$, which is initialized to 0 at $t = 1$. At iteration t, $\alpha_{i,j}^{(t)}$ records the times that, until iteration t, the i-th sample has been selected and is mis-classified at the j-th time interval. For the i-th sample, the corresponding vector $\boldsymbol{\alpha}_{i,\cdot}^{(t)}$ is updated as: $\alpha_{i,j}^{(t+1)} = \alpha_{i,j}^{(t)} + \mathbb{1}[y_{i,j}\hat{y}_{i,j}^{(t)} < 1], \forall j \in \mathbb{N}^M$. As such the derivative for $\mathcal{L}_1$ w.r.t $\boldsymbol{W}^{(t)}$ is:

$$\frac{\partial \mathcal{L}_1}{\partial \boldsymbol{W}^{(t)}} = -\sum_{i=1}^{N} \boldsymbol{x}_{i,\cdot} \otimes (\boldsymbol{\alpha}_{i,\cdot}^{(t)} \odot \boldsymbol{Y}_{i,\cdot}) \tag{7}$$

where $\odot$ is element-wise multiplication, and $\otimes$ is the outer product of two vectors. Likewise, we define a tensor $\boldsymbol{\beta} \in \mathbb{R}^{N \times T \times T}$ to keep track of the 'returning-back-to-life' predictions, so as to aid the computation of $\frac{\partial \mathcal{L}_2}{\partial \boldsymbol{W}^{(t)}}$:

$$\beta_{i,j,k} = \begin{cases} 0 & j > k \\ |\{k \,|\, \hat{y}_{i,j} < 0 \wedge \hat{y}_{i,k} > 0\}| & j < k \\ (-1) * \sum_{p=j+1}^{M} \beta_{i,j,p} & j = k \end{cases} \tag{8}$$

where there is one matrix $\boldsymbol{\beta}_i \in \mathbb{R}^{T \times T}$ associated for each sample $\boldsymbol{x}_{i,\cdot}$. The value at the j-th row and k-th column of $\boldsymbol{\beta}_i$ keeps the number of violated predictions for the "Non-increasing" constraint, i.e., $|\{(j,k) \,|\, \hat{y}_{i,j} < 0 \wedge \hat{y}_{i,k} > 0\}|$ for $j < k$. Thus we have

$$\frac{\partial \mathcal{L}_2}{\partial \boldsymbol{W}^{(t)}} = \sum_{i=1}^{N} \sum_{j=1}^{M} \boldsymbol{x}_{i,\cdot} \otimes \boldsymbol{\beta}_{i,j,\cdot}^{(t)} \tag{9}$$

Combing (7) and (9) yields:

$$\frac{\partial \mathcal{L}}{\partial \boldsymbol{W}^{(t)}} = \frac{1}{\lambda t} \sum_{i=1}^{N} \sum_{j=1}^{M} -\boldsymbol{x}_{i,\cdot} \otimes (\boldsymbol{\alpha}_{i,\cdot}^{(t)} \odot \boldsymbol{Y}_{i,\cdot}) + \boldsymbol{x}_{i,\cdot} \otimes \boldsymbol{\beta}_{i,j,\cdot}^{(t)} \tag{10}$$

Applying the kernel operator $\phi\{\cdot\}$ to x in (6) and (10) yields:

$$\boldsymbol{W}_{kernel}^{(t)} = \frac{1}{\lambda t} \sum_{i=1}^{N} \sum_{j=1}^{M} \phi\{\boldsymbol{x}_{i,\cdot}\} \otimes (\boldsymbol{\alpha}_{i,\cdot}^{(t)} \odot \boldsymbol{Y}_{i,\cdot}) - \phi\{\boldsymbol{x}_{i,\cdot}\} \otimes \boldsymbol{\beta}_{i,j,\cdot}^{(t)} \tag{11}$$

The prediction for a specific sample c's survival status : $\hat{\boldsymbol{y}}_c$ can be written as:

$$\hat{\boldsymbol{y}}_c^{(t)} = \phi\{\boldsymbol{x}_{c,\cdot}\} \boldsymbol{W}_{kernel}^{(t)} = \frac{1}{\lambda t} \sum_{i=1}^{N} \sum_{j=1}^{M} K_{c,i}(\boldsymbol{\alpha}_{i,\cdot}^{(t)} \odot \boldsymbol{Y}_{i,\cdot}) - K_{c,i}\boldsymbol{\beta}_{i,j,\cdot}^{(t)} \tag{12}$$

where $K_{c,i}$ is a scalar from the kernel matrix $\boldsymbol{K}$. The kernel optimization algorithm is outlined in the Algorithm 2. Unlike the Algorithm 1, Algorithm 2 does not learn the weight matrix explicitly. Instead it relies on $\boldsymbol{\alpha}$ and $\boldsymbol{\beta}$ for making predictions as per (12).

4   Experiments

In this section, we describe the experiment setup, including data sets, baseline methods, and evaluation metrics, followed by the experiment results that demonstrate the effectiveness of the proposed approach.

Algorithm 2. Multi-task Kernel Learning Algorithm

Require: λ, T_m
Input: $\boldsymbol{X} \in \mathbb{R}^{N \times M}, \boldsymbol{Y} \in \mathbb{R}^{N \times T}$

1: Set $\boldsymbol{\alpha}^{(1)} \in \mathbb{R}^{N \times T} \leftarrow 0, \boldsymbol{\beta}^{(1)} \in \mathbb{R}^{N \times T \times T} \leftarrow 0$
2: Compute the kernel matrix $\boldsymbol{K}$
3: **for** t=1,2...,T_m **do**
4:    Set the step size $\eta^{(t)} \leftarrow \frac{1}{\lambda t}$
5:    Select a sample $\boldsymbol{x}_{c,\cdot}^{(t)} \in \mathbb{R}^M$ randomly, i.e., a random row of $\boldsymbol{X}$
6:    Make prediction:
        $\hat{\boldsymbol{y}}_c = \frac{1}{\lambda t} \sum_{i=1}^{N} \sum_{j=1}^{M} K_{c,i}(\boldsymbol{\alpha}_{i,\cdot}^{(t)} \odot \boldsymbol{y}_{i,\cdot}) - K_{c,i}\boldsymbol{\beta}_{i,j}^{(t)}$
7:    Update $\boldsymbol{\alpha}^{(t)}$: according to $\hat{\boldsymbol{y}}_c$:
8:    **for** j=1,2,...,M **do**
9:        $\alpha_{c,j}^{(t+1)} = \alpha_{c,j}^{(t)} + \mathbb{1}[y_{c,j}\hat{y}_{c,j} < 1]$
10:   **end for**
11:   Update $\boldsymbol{\beta}^{(t)}$: according to $\hat{\boldsymbol{y}}_c$:
12:   **for** $(j,k) \in \{(j,k)\|\hat{y}_{i,j}^{(t)} < 0, \hat{y}_{i,k}^{(t)} > 0 \wedge j < k\}$ **do**
13:       $\boldsymbol{\beta}_{c,j,k} = \boldsymbol{\beta}_{c,j,k} + \mathbb{1}[\hat{y}_{c,j} < 0, \hat{y}_{c,k} > 0]$
14:   **end for**
15:   Update the diagonal entries of $\boldsymbol{\beta}_{c,\cdot,\cdot}$.
16:   **for** j=1,2,...,M **do**
17:       $\beta_{c,j,j}^{(t+1)} = (-1) * \sum_{p=j+1}^{M} \beta_{c,j,p}^{(t+1)}$
18:   **end for**
19: **end for**
20: Output: $\boldsymbol{\alpha}^{(T_m+1)}, \boldsymbol{\beta}^{(T_m+1)}$

4.1   Experiment Setup

Dataset Description. We use synthetic and public real-world sets for evaluation. The descriptive statistics for all public sets are summarized in Table 1.

– **SYNTHETIC** To demonstrate the capability of our method in coping with highly heterogeneous patient cohorts expected in medical data, we construct the following synthetic survival model that exhibits the relationship between patient survival times and their covariates, based on [1]:

$$x_i \sim \mathcal{N}(\mathbf{0}, \mathbf{I}),$$
$$T_i \sim \exp((\gamma_1^T x_i)^2 + \gamma_2^T x_i). \tag{13}$$

We assume that the survival times T_i are exponentially distributed with a mean parameter that is a sum of a quadratic function $((\gamma_1^T x_i)^2)$ and a linear function $(\gamma_2^T x_i)$ of the covariates $x_i \in \mathbb{R}^M$ for patients. The hyperparameters γ_1 and γ_2 are set to 10 respectively. Negative mean parameters are rejected and each T_i is rounded to the closest integer. To simulate censoring, a subset (50%) of the generated samples have their survival time T_i set to a random time between $[1, T_i]$, which is denoted as C_i. Different synthetic data are generated by varying the sample numbers N, and the feature size F.

- **METABRIC** The Molecular Taxonomy of Breast Cancer International Consortium (METABRIC) data set consists of gene expression profiles and clinical features for determining breast cancer subgroups. Pre-processing work includes replacing missing values with mean values for real-valued features, and mode for categorical features.
- **CLINIC** This data set tracks the patient clinic status [9]. The goal of survival analysis is to estimates survival over a 180-day period for seriously ill hospitalized adults.
- **SEER** The Surveillance, Epidemiology, and End Results Program (SEER) data set provides information on breast cancer patients from the San Jose-Monterey, Los Angeles, Rural Georgia, and Alaska Natives SEER registries for 1992–2016.

Table 1. Descriptive statistics of public data sets

Data set	No. features		No. samples		Event time		Censor time	
	Categorical	Numerical	Uncensored	Censored	Max	Min	Max	Min
METABRIC	15	6	888 (44.8%)	1,093 (55.2%)	745	0	768	2
CLINIC	5	9	4,238 (87.8%)	590 (12.2%)	47	1	50	1
SEER	8	6	17,126 (21.4%)	63,055 (78.6%)	297	0	299	0

Baseline Methods. The following popular baseline methods are chosen:

- **The Cox model**: The Cox proportional hazards (CPH) model regards the log-hazard of an individual as a linear function of its static covariates and a population-level baseline hazard that changes over time.
- **The Weibull method**: Weibull Analysis is to fit a Weibull distribution for the sample set so that the parameterized distribution can be used to estimate the life characteristics of samples.
- **MTLSA**: the MTLSA (Multi-Task Learning for Survival Analysis) model [13] formulates the survival analysis problem as a series binary classification problems in a multi-task learning setting. We use its Matlab implementation[1] for evaluation. Note that this is the closest baseline to our approach.
- **DeepHit**: DeepHit [10] is a deep learning model that learns the distribution of the first hitting time of an underlying stochastic process. We use its Python implementation[2] for evaluation.
- **DeepSurv**: DeepSurv [7] combines CPH and neural network for modeling interactions between a patient's covariates and treatment effectiveness to provide personalized treatment recommendations. We use its Python implementation [3] for evaluation.

---

[1] https://github.com/MLSurvival/MTLSA.
[2] https://github.com/chl8856/DeepHit.
[3] https://github.com/jaredleekatzman/DeepSurv.

These methods output the following values:

- **Hazard rate**: the risk of failure (i.e., the probability of experiencing the event) given that the participant has survived up to a specific time. Cox, Weibull, and DeepSurv output hazard rates, where the event is the death of samples.
- **Occurrence probability**: the probability of an event occurring at a given time, which is a value between 0 and 1. DeepHit predicts the probabilities of an event, which is the death of samples.
- **Survival status**: the survival status at a given time, which is often a binary value indicating if the sample is alive or dead. MTLSA and our proposed approach produces survival status for samples

Evaluation Metric. The concordance index (C-index) [4] is chosen to compare the performance of all methods, which reflects how well a model predicts the ordering of samples according to their survival times. The C-index that compares a pair of bi-variate observations $(T_1, \hat{T}_1)$, and $(T_2, \hat{T}_2)$, where T_i is the ground true survival time while $\hat{T}_i$ is the prediction, is defined as:

$$C = P_E(\hat{T}_1 \geq \hat{T}_2 | T_1 \geq T_2), \tag{14}$$

where P_E is empirical probability calculated from prediction and ground truth data. As there are three types of output values, it is necessary to transform them for fair comparisons. The following strategies were applied to baseline outputs:

- **Direct comparison** The predicted survival time $\hat{T}_i$ can be calculated as per the predicted survival status. As such C-index values can be computed for each pair by using the prediction and ground truth values as per (14).
- **Hazard rate transformation** The hazard rate $h_i(t)$ reveals the estimated death"occurrence" for sample i at time unit t. We need to convert it into survival probabilities to compute C-index values. Let $S_i(t)$ be the predicted probability that sample i is alive until time t:

$$S_i(t) = P(\hat{y}_{i,1} > 0 \wedge \cdots \wedge \hat{y}_{i,t} > 0 | h_i(1), \cdots, h_i(t)) \tag{15}$$

At each time t, we use a Poisson distribution with parameter $h_i(t)$ to express the probability of the death event occurring. Its probability mass function of is given by: $f(k|\lambda) = \frac{\lambda^k e^{-\lambda}}{k!}$, where λ is set to $h_i(t)$, and k is 0. Thus at each time t, the death probability is $e^{-h_i(t)}$. Combining them yields:

$$S_i(t) = P(\hat{y}_{i,1\cdots t} > 0 | h_i(1), \cdots h_i(t)) = exp\{-\sum_{\tau=1\ldots t} h_i(\tau)\}. \tag{16}$$

Once $S_i(t)$ is obtained for all samples, we search for a survival time T_ρ that maximizes the C-index for the test set.

- **Occurrence probability conversion** Let $p_i(t)$ be the death probability at time t for the i-th sample, we can calculate $S_i(t)$, the predicted probability that sample i is alive until time t, as: $S_i(t) = 1 - \sum_{\tau=1\ldots t} p_i(\tau)$. After $S_i(t)$ is obtained for all samples, we follow similar strategy in Hazard rate transformation to compute the average C-index value for the test set.

4.2 Results

Synthetic Data. We create 15 synthetic sets by setting the feature size F to 15, 20, and 25, and the sample number N to 500, 1000, 2000, 5000, 10000 respectively. Each data set is split into training and test data randomly with a ratio of 9:1. We compare both the basic and kernel algorithms (Gaussian kernel is chosen due to its popularity) with baseline methods. Hyper-parameters are set empirically. Table 2 summarizes the C-index values obtained by all methods where higher C-index values indicate better performance. It can be seen from Table 2 our kernel algorithm achieves the best performance in most cases. Specifically, it beats the basic algorithm and the MTLSA method. The proposed kernel algorithm appears to have utilized non-linear patterns inside the data for survival analysis. Though DeepHit is also a nonlinear method for survival analysis, its performance, in general, suggests the proposed kernel method can better cope with the data distribution governed by the synthetic model.

Table 2. The C-index values of all methods on synthetic sets

	Methods	N = 500	N = 1000	N = 2000	N = 5000	N = 10000
F = 15	Cox	0.4059	0.5248	0.5376	0.4893	0.5418
	WeibullAFT	0.4152	0.0.5419	0.5540	0.5055	0.5611
	MTLSA	0.4587	0.5887	0.5784	0.5196	0.5830
	DeepSurv	0.5602	0.5156	0.5804	0.5042	0.5130
	DeepHit	**0.6314**	0.5727	0.5489	0.5581	0.5154
	Ours (basic)	0.5975	0.6473	0.5830	0.5813	0.5795
	Ours (kernel)	0.6193	**0.6565**	**0.6610**	**0.5938**	**0.6120**
F = 20	Cox	0.5332	0.4903	0.4693	0.5504	0.5089
	WeibullAFT	0.5332	0.4982	0.4711	0.5712	0.5283
	MTLSA	0.6276	0.5168	0.5381	0.5812	0.5459
	DeepSurv	0.4888	0.4819	0.5116	0.5012	0.4954
	DeepHit	0.5689	0.5354	0.5455	0.5539	0.4714
	Ours (basic)	0.6146	0.5269	0.5653	0.5640	0.5545
	Ours (kernel)	**0.7340**	**0.6234**	**0.6200**	**0.5968**	**0.5819**
F = 25	Cox	0.3996	0.4847	0.6121	0.4721	0.5323
	WeibullAFT	0.3703	0.4969	0.6197	0.4906	0.5485
	MLTSA	0.5088	0.5467	0.6255	0.5077	0.5618
	DeepSurv	0.3430	0.3623	0.5228	0.5105	0.5264
	DeepHit	0.5732	0.5496	0.5871	0.4807	0.4819
	Ours (basic)	0.5855	0.5727	0.5916	0.5387	0.5521
	Ours (kernel)	**0.6447**	**0.6655**	**0.6563**	**0.5733**	**0.5875**

To learn how the kernel algorithm behaves during testing, we plot the loss values and the standard variation of C-index on some test sets in Fig. 1. Specifically we record the loss values and compute the standard deviation values of the C-index whenever 10% training data is sampled. As depicted in Fig. 1, the loss and the C-index standard deviation values exhibit a decreasing trend as more data is sampled. This shows that the proposed algorithm optimizes the C-index while minimizing the loss.

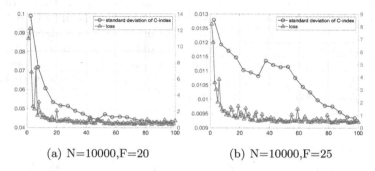

(a) N=10000,F=20   (b) N=10000,F=25

Fig. 1. The loss and standard deviation values of C-index observed on test sets.

Public Data. For each data set we use 10 fold cross-validation and report the mean C-index values for comparison. Furthermore, we also compare all the methods in the perspective of resource allocation, e.g., hospital triage.

C-index Evaluation. We compare with baseline methods the performance of different kernels, namely the Linear, Polynomial, Sigmoid and Gaussian kernels. Table 3 lists the C-index values obtained by all the methods on public data sets. As shown in the table, our algorithm with Gaussian kernel achieves the best performance on all public sets, amongst all kernels. Furthermore, it also outperforms all baseline methods on all data sets. Out of all baseline methods, MTLSA's performance comes top on the METABRIC set, whilst DeepHit is the best baseline on CLINIC and SEER sets. Our algorithm beats MTLSA (the closet baseline) on all sets. As a non-linear method, it also achieves slightly better performance than DeepHit on CLINIC and SEER sets. Note that CLINIC and SEER set contain more samples than the METABRIC set. This shows that our algorithm's performance is stable on data sets of different sizes.

Wastage-Survival Evaluation. All samples can be ranked by their predicted survival status, based on the imminence of respective death events. Resource wastage is caused by the early treatment of samples of lower priority. Let $t_r \in \{1 \cdots T\}$ be the time to treat samples. We assume that at each time only the top ($\lceil \frac{N}{T} \rceil$) living samples can be treated. A sample i is considered still alive at time t_r if $T_i \geq t_r$. The accumulated wastage $\psi_m(t_r)$ at time t_r for method m can be formulated as

$$\psi_m(t_r) = \sum_{j=1}^{t_r} \sum_{i \in S_j^m} \log(T_i - j), \tag{17}$$

where S_j^m represents the ranked sample list produced at time unit j by method m, and the $\log(\cdot)$ function is used for numerical control. We then use the aggregated Wastage-Survival(W-S) score to compare all methods, which favours methods that generate low wastage and high survival numbers. We choose to compute the W-S score at $T/2$. For method m, the wastage at $T/2$ is defined as $w_m = c/\psi_m(T/2)$ where c is a scaling constant, and the survival number s_m at $T/2$ is defined as $1 - D_m/N$ where D_m is the number of death occurrence at $T/2$. The aggregated score for m is $F_m = 2 * w_m * s_m/(w_m + s_m)$. We use the ratio $WS_m = F_m/F_{\text{Ours}}$ as the W-S measurement for method m. As a result $WS_m < 1$ indicates our method performs better. The W-S scores for all methods are shown in Fig. 2, which suggests our method achieves the best performance in general, as most of the bars are under the horizontal line of 1.

Table 3. C-index values of all methods on public sets

	Metabric	Clinic	Seer
Cox	0.6774	0.5729	0.6988
Weibull	0.6779	0.5640	0.6993
MTLSA	0.6918	0.5631	0.7695
DeepSurv	0.5658	0.5098	0.4972
DeepHit	0.6810	0.5923	0.7833
Ours (Linear)	0.6469	0.5823	0.6460
Ours (Polynomial)	0.6854	0.5492	0.7590
Ours (Sigmoid)	0.7104	0.5666	0.7196
Ours (Gaussian)	**0.7233**	**0.5955**	**0.7879**

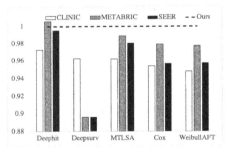

Fig. 2. W-S measurement

5   Conclusions

We have presented a multi-task kernel learning framework for survival analysis. We have also developed Pegasos based optimization algorithms for learning parameters. Experiment results demonstrate the effectiveness of our method.

References

1. Alaa, A.M., van der Schaar, M.: Deep multi-task Gaussian processes for survival analysis with competing risks. In: Proceedings of the 31st International Conference on Neural Information Processing Systems, pp. 2326–2334. Curran Associates Inc. (2017)
2. Cox, D.R.: Regression models and life-tables. J. Roy. Stat. Soc. Ser. B (Methodological) **34**(2), 187–202 (1972)
3. Fernández, T., Rivera, N., Teh, Y.W.: Gaussian processes for survival analysis. In: Advances in Neural Information Processing Systems, pp. 5021–5029 (2016)
4. Harrell, F.E., Califf, R.M., Pryor, D.B., Lee, K.L., Rosati, R.A.: Evaluating the yield of medical tests. JAMA **247**(18), 2543–2546 (1982)
5. Ishwaran, H., Kogalur, U.B., Blackstone, E.H., Lauer, M.S., et al.: Random survival forests. Ann. Appl. Stat. **2**(3), 841–860 (2008)
6. Kaplan, E.L., Meier, P.: Nonparametric estimation from incomplete observations. J. Am. Stat. Assoc. **53**(282), 457–481 (1958)
7. Katzman, J.L., Shaham, U., Cloninger, A., Bates, J., Jiang, T., Kluger, Y.: Deepsurv: personalized treatment recommender system using a cox proportional hazards deep neural network. BMC Med. Res. Methodol. **18**(1), 24 (2018)
8. Kimeldorf, G., Wahba, G.: Some results on tchebycheffian spline functions. J. Math. Anal. Appl. **33**(1), 82–95 (1971)
9. Knaus, W.A.: The support prognostic model: objective estimates of survival for seriously ill hospitalized adults. Ann. Internal Med. **122**(3), 191–203 (1995)
10. Lee, C., Zame, W.R., Yoon, J., van der Schaar, M.: Deephit: a deep learning approach to survival analysis with competing risks. In: Thirty-Second AAAI Conference on Artificial Intelligence (2018)
11. Lee, E.T., Wang, J.: Statistical Methods for Survival Data Analysis, vol. 476. Wiley, New York (2003)
12. Li, H., Ge, Y., Zhu, H., Xiong, H., Zhao, H.: Prospecting the career development of talents: a survival analysis perspective. In: Proceedings of the 23rd ACM SIGKDD International Conference on Knowledge Discovery and Data Mining, pp. 917–925. ACM (2017)
13. Li, Y., Wang, J., Ye, J., Reddy, C.K.: A multi-task learning formulation for survival analysis. In: Proceedings of the 22nd ACM SIGKDD International Conference on Knowledge Discovery and Data Mining, pp. 1715–1724. ACM (2016)
14. Li, Z., et al.: Water pipe condition assessment: a hierarchical beta process approach for sparse incident data. Mach. Learn. **95**(1), 11–26 (2013). https://doi.org/10.1007/s10994-013-5386-z
15. Lin, P., et al.: Data driven water pipe failure prediction: a bayesian nonparametric approach. In: Proceedings of the 24th ACM International on Conference on Information and Knowledge Management, pp. 193–202. ACM (2015)
16. Shalev-Shwartz, S., Singer, Y., Srebro, N., Cotter, A.: Pegasos: primal estimated sub-gradient solver for SVM. Math. Program. **127**(1), 3–30 (2011)
17. Simon, N., Friedman, J., Hastie, T., Tibshirani, R.: Regularization paths for cox's proportional hazards model via coordinate descent. J. Stat. Software **39**(5), 1 (2011)
18. Tibshirani, R.: The lasso method for variable selection in the cox model. Stat. Med. **16**(4), 385–395 (1997)

19. Zhang, Q., Zhou, M.: Nonparametric Bayesian Lomax delegate racing for survival analysis with competing risks. In: Advances in Neural Information Processing Systems, pp. 5002–5013 (2018)
20. Zhou, J., Yuan, L., Liu, J., Ye, J.: A multi-task learning formulation for predicting disease progression. In: Proceedings of the 17th ACM SIGKDD International Conference on Knowledge Discovery and Data Mining, pp. 814–822. ACM (2011)

Meta-data Augmentation Based Search Strategy Through Generative Adversarial Network for AutoML Model Selection

Yue Liu[1,2,3](✉) 🆔, Wenjie Tian[1], and Shuang Li[1]

[1] School of Computer Engineering and Science, Shanghai University, Shanghai, China
{yueliu,twenjie,aimeeli}@shu.edu.cn
[2] Shanghai Institute for Advanced Communication and Data Science,
Shanghai, China
[3] Shanghai Engineering Research Center of Intelligent Computing System,
Shanghai 200444, China

Abstract. Automated machine learning (AutoML) attempts to automatically build appropriate learning model for given dataset. Despite the recent progress of meta-learning to find good instantiations for AutoML framework, it is still difficult and time-consuming to collect sufficient meta-data with high quality. Therefore, we propose a novel method named Meta-data Augmentation based Search Strategy (MDASS) for AutoML model selection, which is mainly composed of Meta-GAN Surrogate model (MetaGAN) and Self-Adaptive Meta-model (SAM). Meta-GAN employs Generative Adversarial Network as surrogate model to collect effective meta-data based on the limited meta-data, which can alleviate the dilemma of meta-overfitting in meta-learning. Based on augmented meta-data, SAM self-adaptively builds multi-objective meta-model, which can select the algorithms with proper trade-off between learning performance and computational budget. Furthermore, for new datasets, MDASS combines promising algorithms and hyperparameter optimization to perform automated model selection under time constraint. Finally, the experiments on various classification datasets from OpenML and algorithms from scikit-learn are conducted. The results show that GAN is promising to incorporate with AutoML and MDASS can perform better than the competing approaches with time budget.

Keywords: AutoML · Meta-learning · Data augmentation · Generative Adversarial Network · Automated model selection

1 Introduction

Automated machine learning (AutoML) attempts to automate tedious but core tasks efficiently [22] to enable the wide-spread use of machine learning by non-experts [21]. It has shown promising results in various tasks, such as feature engineering [14], hyperparameter optimization [12], model selection [5,7,18].

K. Karlapalem et al. (Eds.): PAKDD 2021, LNAI 12714, pp. 312–324, 2021.
https://doi.org/10.1007/978-3-030-75768-7_25

Despite recent progress, model selection in high-dimension configuration space is still a major challenge in AutoML. To improve the efficiency, meta-learning is widely adopted [5,6,21] to learn how different configuration performs across datasets (meta-knowledge) from previous learning experience of prior datasets (meta-data) [19]. The existing researches pay more attention to how to better extract meta-knowledge. However, meta-data also plays an important role in meta-learning. It is difficult to collect sufficient meta-data with high quality for the following problems: i) Since some configurations may spend several hours or days to evaluate performance, collecting sufficient meta-data endures more computational budget. ii) The datasets provided by repositories such as UCI [2] and OpenML [20] are still limited and their distribution cannot cover the entire distribution of prior datasets. Meta-learning cannot work well when the distribution of new datasets is significantly different from prior datasets.

To efficiently collect meta-data, the existing promising methods limit maximum runtime or memory consumption in model evaluation. The models exceeding runtime or memory remain missing entries in meta-data. Afterward, the techniques such as collaborative filtering [16,21] or matrix factorization [7] are to complete meta-data. However, these techniques require at least one model on each dataset trained successfully, and become poor when facing numerous missing entries. Reference [11] showed that some datasets are not reliable for being used in hyperparameter optimization, and presented an interesting approach of using probabilistic encoder to generate inexpensive and realistic data for optimization benchmarking. Therefore, it is significant to directly generate meta-data through generative model for meta-learning. Generative Adversarial Network (GAN), a framework to estimate generative models through adversarial training [1,8,15], has been prevailing for data augmentation since its origin [9]. One of the key features of GAN is its generic nature in sampling from an unspecified distribution underlying the given data. This feature fits well when augmenting data from datasets collected without knowing a prior distribution, such in the case of meta-learning. Therefore, meta-data augmentation based on GAN is a promising approach to improve the generalization of meta-learning.

In this paper, we investigate the approach of incorporating GAN with meta-learning for meta-data augmentation, named as Meta-GAN Surrogate Model (MetaGAN). We then further propose Meta-data Augmentation based Search Strategy (MDASS) for AutoML model selection, composed of MetaGAN and Self-Adaptive Meta-model (SAM). Our contributions are summarized as follows:[1]

– In order to efficiently collect sufficient meta-data with high quality, we propose MetaGAN for meta-data augmentation. It employs GAN as surrogate model to generate meta-data described by meta-features and specific class that indicates the algorithm performance. Therefore, it drastically reduces the computational budget to collect meta-data and renders meta-learning more comprehensive to exploit the underlying distribution of meta-data.

---

[1] The supplementary material of MDASS is available at https://github.com/wj-tian/MDASS.

- In order to effectively extract meta-knowledge, we propose SAM for algorithm selection. It self-adaptively builds meta-model based on augmented meta-data through appropriate classification and regression learners with two meta-tasks: selecting high-performance algorithms and predicting runtime order respectively. This model can determine which algorithms have the right trade-off between computational budget and performance.
- In order to build appropriate models for new datasets under time constraint, MDASS combines promising algorithms selection and hyperparameter optimization, and employs Bayesian optimization to search the best model with the runtime order in the expected budget.
- Finally, we conduct experiments on a wide range of datasets from OpenML [20] and various machine learning algorithms from scikit-learn [17]. The results show that MetaGAN can significantly improve the performance of meta-learning and MDASS can obtain an appropriate model with the proper trade-off between runtime and accuracy within certain time constraints.

The remainder of this paper is organized as follows. Section 2 describes the main idea of MDASS. Section 3 analyzes the experiments and results. Finally, we present the conclusion and future work in Sect. 4.

2   Meta-data Augmentation Based Search Strategy for AutoML Model Selection (MDASS)

The framework of MDASS as shown in Fig. 1 is to perform meta-data augmentation through MetaGAN and determine which algorithms have proper trade-off between runtime and accuracy through SAM. We construct accurcacy-oriented meta-data and runtime-oriented meta-data from hand-picked datasets and specific algorithms. Then, accuracy-oriented meta-data is to execute augmentation through MetaGAN (see Sect. 2.1). Based on augmented meta-data, SAM is built for promising algorithm selection and runtime order prediction (see Sect. 2.2). Given a new dataset, MDASS computes its meta-feature for SAM, and the promising algorithms are evaluated in this order until exceeding the time budget to build appropriate model. Here, we employ the commonly used Bayesian optimization [5,12,18] to evaluate algorithm performance. Notably, when all promising algorithms have been evaluated sequentially but time budget is not exceeding, MDASS executes the process of joint optimization, i.e. combining promising algorithm selection and corresponding hyperparameter optimization.

2.1   MetaGAN for Meta-data Augmentation

Problem Formulation. Meta-data, the foundation of meta-learning, in MDASS can be defined as $M = \langle F, A, R \rangle$, where $F \in \mathbb{R}^{m \times l}$, $A \in \mathbb{R}^{m \times n}$ and $R \in \mathbb{R}^{m \times n}$ are the matrixes of meta-features, accuracy and runtime. l, m and n represent the number of meta-features, hand-picked datasets and specific algorithms.

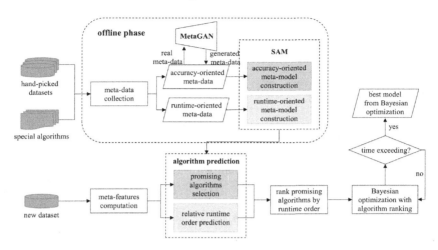

Fig. 1. The framework of Meta-data Augmentation-based Search Strategy (MDASS).

Let $d_1, ..., d_m$ be a set of hand-picked datasets. They are the instantiations sampled from an unknown datasets distribution $d_i \sim p(d)$. For each dataset d_i, it has an associated objective function $f_d : F_i \in \mathbb{R}^l \to \mathbb{R}^n$ mapping the meta-features to algorithm performance. Every entry A_{ij} in accuracy matrix records the accuracy of algorithm a_j on dataset d_i. The runtime of corresponding learning model is recorded as R_{ij}. Here, we name the combination of F and A as accuracy-oriented meta-data while F and R as runtime-oriented meta-data. For new datasets, most meta-learning methods are to compute the similarity with prior datasets in meta-feature space, or build machine learning model to learn f_d on meta-data and predict the algorithm performance. However, the distribution $p(d)$ is unknown and the hand-picked datasets are hard to represent this distribution. It limits the capacity of meta-learning learning from prior experience. Therefore, the main intention of MetaGAN is to approximate $p(d)$ with GAN and sample new meta-data that belongs to datasets $d^* \sim p(d|F, A)$.

MetaGAN: Conditional WGAN-Based Meta-data Augmentation. Hand-picked datasets can be viewed as instantiations sampling from unspecified distribution. GAN fits well when augmenting data from datasets sampled without knowing prior distribution. Since unconditional GAN has no control for the generated data mode, MetaGAN is based on conditional setting. Figure 2 shows the process of MetaGAN. To perform conditional meta-data augmentation, accuracy matrix is first labeled through the following steps: i) Each entry A_{ij} is normalized by $A_{ij}^* = \max \{A_{ij}\}_{j=1}^n - A_{ij}$. ii) The normalized accuracy is labeled by predefined threshold ξ. When A_{ij}^* is greater than ξ, it is labeled as $True$, representing algorithm a_j on dataset d_i has promising performance and vice versa.

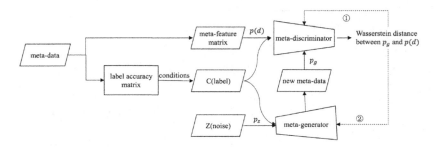

Fig. 2. The process of MetaGAN. It is built with conditional WGAN through gradient descent method. ① represents $\nabla_{\theta_d} \left[\frac{1}{m} \sum_{i=1}^{m} D\left(x^i, c^i; \theta_d\right) - \frac{1}{m} \sum_{i=1}^{m} D\left(G\left(z^i\right), c^i; \theta_d\right) \right]$, and ② represents $-\nabla_{\theta_g} \frac{1}{m} \sum_{i=1}^{m} D\left(G\left(z^i\right), c^i; \theta_g\right)$.

Afterwards, MetaGAN is conditioned on class to direct accuracy-oriented augmentation under WGAN framework. It is formulated as two adversarial model: meta-generator G and meta-discriminator D. G implicitly learns the distribution of dataset from input meta-data and transforms a prior noise distribution $p_z(z)$ into meta-features space of datasets drawn from the learned distribution. D estimates the Wasserstein distance explicating the minimum cost of transporting mass for transforming the dataset distribution $p(d)$ into the learned distribution of generator p_g. Thus, the objective function of MetaGAN can be defined as Eq. 1, which also reflects the quality of meta-data augmentation.

$$V(G, D) = \mathrm{E}_{x \sim p(d)} D(x, c) - \mathrm{E}_{G(z) \sim p_g} D(G(z), c) \qquad (1)$$

where c is specific class of condition. G and D are trained simultaneously, alternating between the training phases of them as optimizing their parameters shown in Fig. 2. When MetaGAN has been trained, meta-data can be collected with an insignificant cost, bounded only by the computational overhead of MetaGAN.

2.2 Self-Adaptive Meta-Model for Algorithm Selection

The intention of MDASS experienced by users is to efficiently obtain high-performance learning model. Therefore, it requires the algorithms with higher performance but lower runtime in optimization process. We propose SAM to solve these two meta-tasks: high-accuracy algorithm selection and runtime order prediction. It can carry out the trade-off of learning performance and runtime to provide more time for promising algorithms in evaluation process. The key problem for SAM is to achieve global optimum under different algorithm prediction. Since there exists no such universal learner outperforming other learners consistently, different algorithm prediction should be considered separately.

For accuracy-oriented meta-model, it is built on augmented accuracy-oriented meta-data to predict whether an algorithm has promising performance. Therefore, it can be viewed as multi-output classification problem. We consider the classification learners with best performance in accuracy matrix, and base learners of each output is from these learners. The objective function is defined as Eq. 2, which can be achieved by minimizing the loss of each output greedily.

$$f_j^{acc} = \operatorname*{arg\,min}_{f_j^{acc} \in \Gamma^{cla}} \frac{1}{m} \sum_{i=1}^{m} \left| f_j^{acc}(F_i) - A_{ij} \right| \tag{2}$$

where f_j^{acc} represents the base learner of each algorithm selection and Γ^{cla} represents the collection of classification learners. For runtime-oriented meta-model, it is built on runtime-oriented meta-data to predict algorithm runtime ranking. Similar to accuracy-oriented meta-model, this meta-model can be viewed as multi-output regression problem, and we consider regression learners from three different categories including ridge regression, support machine vector regression and random forest for base learners. Then, this meta-model can be achieved by minimizing the objective function as Eq. 3.

$$f_j^{time} = \operatorname*{arg\,min}_{f_j^{time} \in \Gamma^{reg}} \sum_{i=1}^{m} \left(f_j^{time}(F_i) - R_{ij} \right)^2 \tag{3}$$

where f_j^{time} represents the base learner of each runtime prediction and Γ^{reg} represents the collection of regression learners.

Therefore, SAM applys appropriate learners to adapt each algorithm prediction. When the SAM trained, MDASS can evaluate high-performance algorithms in the order of runtime ranking for new datasets. Notably, SAM is built on augmented meta-data through MetaGAN and this process is performed offline only. Runtime on it does not increase the runtime of MDASS employed by users, bounded only with the automated model selection stage for new given datasets.

3   Experiments

3.1   Generation of Meta-Data

We ran all experiments on 405 classification datasets sorted by most runs in OpenML [20]. They were selected by filtering with no more than 10000 samples and 300 features. The same preprocessings were applied to all datasets: removing missing value, one-hot encoding for categorical features and standardizing all features. All involved algorithms were implemented by scikit-learn [17]: Kernel SVM, Linear SVM, Gaussian naive Bayes, Bernoulli naive Bayes, k-Nearest Neighbor (KNN), Decision Tree, Random Forest, Adaboost, Stochastic Gradient Descent (SGD), Latent Discriminant Analysis (LDA), Quadratic Discriminant Analysis (QDA), Passive Aggressive, and Extra Tree. We generated accuracy matrix through running Bayesian optimization implemented in Hyperopt [3] to search best performance for a specific algorithm on each dataset and we set the labeled threshold as 0.025. Moreover, the meta-features used in experiments are divided into five groups: simple, statistical, information-theoretic, complexity

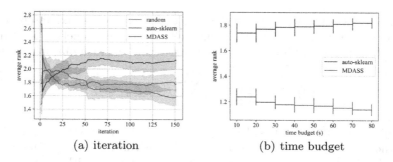

(a) iteration   (b) time budget

Fig. 3. Comparison of random method and auto-sklearn in a time-constrained setting. (a) the average rank with the increasing of iteration, 1 is the best and 3 is the worst. (b) the average rank with different time budget, 1 is the best and 2 is the worst.

and landmarkers. More details of meta-feature and algorithm hyperparameter space are listed in supplementary material. We only used six meta-features which are more relevant to runtime prediction for runtime-oriented meta-model: number of instances, number of features, dataset ratio and their log values.

3.2 Experimental Setup

In our experiments, hold-out-validation is used to evaluate MDASS, and repeated 10 times. Therefore, out of the 405 total datasets, 10% ($\approx$40) were identified to comprise the held-out test set and the remaining datasets were to build MDASS in each hold-out-validation. Six base learners were to formulate accuracy-oriented meta-model: Adaboost, Extra Tree, Random Forest, Decision Tree, Kernel SVM with the highest performance in accuracy matrix and the commonly used learner in literature [5,13], KNN. For generator and discriminator of MetaGAN, they both had same neural architecture: Multi-layer Perceptron with two hidden layers, and ReLU activation function was applied for all layers except the output layer which was linear combination. For hidden layers, we used dropout strategy with the probability of 0.2. Besides, MetaGAN used RMSProp optimizer with global learning rate of 1e$-$4 and decay rate of 0.9. We set the weight clipping parameter as 0.01. We augmented the number of meta-data to approximate 600 though MetaGAN. All procedures are ran on Windows server with Intel(R) Xeon(R) CPU E5-2609 v4 @1.79 GHZ processor.

3.3 Experimental Evaluation and Analysis

The Performance Validation of MDASS. When tackling a specific domain problem through machine learning, many users tend to quickly construct learning model to mining the knowledge in domain data. MDASS provides a novel AutoML system to efficiently obtain high-quality model with an automatic manner for given datasets. Therefore, it can alleviate the challenge of model design.

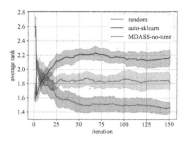

Fig. 4. Comparison of random method and auto-sklearn without time-constrain setting

In order to evaluate the effectiveness of MDASS, we compare it with the following baselines: i) random: randomly searching algorithm and hyperparameters to evaluate for each dataset in test set. ii) auto-sklearn [5]: Since ensemble and preprocessing are not involved in MDASS, we disabled automated ensemble construction and preprocessing in auto-sklearn. Figure 3 shows the average rank of these methods from optimization iteration and time budget within a time-constrained setting. From Fig. 3(a), the average rank of MDASS is decreasing consistently, while auto-sklearn is increasing. Besides, auto-sklearn is the best at the beginning 10 iteration, but MDASS can surpass auto-sklearn quickly and become the optimal about 52 iteration. From Fig. 3(b), MDASS is better than auto-sklearn consistently and the performance of MDASS turns better with the increasing of time budget. We can conclude that MDASS performs surprisingly well than other methods with time budget, since random performs optimization in all configuration space and auto-sklearn selects multiple hyperparameter configurations to warm-start optimization according to the distance between datasets, while MDASS only optimizes in promising model space, i.e., higher accuracy and lower runtime algorithms. In addition, Fig. 3 also illustrates that MDASS can search better individual model than auto-sklearn and random method since we both disable ensemble construction in these methods.

In addition, we disable the time-constrain setting and joint optimization (CASH) is applied to the promising algorithms with corresponding hyperparameters. The result is shown in Fig. 4. We can observe that MDASS turns lowest in earlier phase and the average rank of MDASS drops faster than that with time-constrain setting, surpassing random method from the beginning of optimization consistently. When considering time budget, the initial evaluated algorithm may not be the best model, which consume a few iterations.

Why Dose MDASS Work Well? MDASS performs well in comparison with random and auto-sklearn method. In this section, we execute ablation studies of MDASS to interpret why MDASS works well and whether its components, MetaGAN and SAM, are reasonable.

The Improvement of MetaGAN and SAM. MDASS contains two main components: MetaGAN and SAM. In order to evaluate the improvement of MDASS, we compare four variants of our method: original version without MetaGAN and SAM (Original), MetaGAN only, SAM only and combining MetaGAN and SAM on the following five metrics: 1) Single Best Rate (SB): overlap of one best algorithms, i.e. the best algorithm in predicted promising algorithms of SAM. 2) Two Best Rate (TB): overlap of two best algorithms. 3) Single Worst Rate (SW): overlap of one worst algorithm. 4) Global Accuracy (GA): average accuracy of each algorithm prediction. 5) Global BER (GB): the balanced error rate (the average of false negative and false positive rates) of each algorithm prediction. For these metrics, the greater values of SB, TB, GA and lower values of SW, GB indicate better performance of each component. Here, we evaluate the accuracy-oriented meta-model (promising algorithm selection), and the performance of runtime-oriented meta-model is evaluated in the following experiment. We set base learner of Original and MetaGAN variants as Random Forest.

Table 1 shows the comparison of these methods. We observe that Meta-GAN+SAM achieves superior performance over all variants. Compared with Original, the SB, TB, GA of MetaGAN+SAM have increased by 0.074, 0.057, 0.035 and SW, GB have reduced by 0.031, 0.018 respectively, which reveals our approach can find out more promising algorithms. Besides, both two components can yield improvement over Original. MetaGAN can improve meta-learning to search more promising algorithms. Compared with Original, SB, TB and GA are increased from 0.795, 0.879 and 0.788 to 0.828, 0.918 and 0.796 while SW and GB are reduced from 0.113 and 0.105 to 0.085 and 0.098 respectively, which demonstrates the promise of GAN to meta-learning. SAM adapts meta-data using multi-learners for each algorithm selection instead of single learner for all outputs, which also improves the performance of MDASS. Compared with Original, SB, TB and GA have increased by 0.026, 0.021 and 0.014 while SW has reduced by 0.013. Fig. 5 shows algorithm selection performance of MDASS on 39 datasets of meta-test set in one of 10 hold-out-validation. We can observe that the lighter algorithms (top) have larger probability to be selected (bottom).

Moreover, in order to demonstrate the superiority of GAN, we compare Meta-GAN with the following synthetic data methods: 1) Resample: random sampling from origin meta-data repeatedly. 2) SMOTE [4]: random resampling for one class and SMOTE for another, and vice versa. 3) ADASYN [10]: random resampling for one class and ADASYN for another, and vice versa. The result is shown in Table 2. Resample method only generates same meta-data, which is easy to introduce more variance for algorithm selection. Thus, it is uncompetitive with MetaGAN, even with Original. Since SMOTE and ADASYN are both based on the distance of instances in meta-data, they perform poorer than MetaGAN on all metrics, even for comparison to Resample method with higher SW and GB which are both more than 0.12. MetaGAN generates meta-data through learning the distribution of origin meta-data, maintaining the highest performance and lower variance (more robust for meta-learning) in four competing approach.

The Runtime Prediction Performance of SAM. In order to evaluate the runtime prediction performance in MADSS, we compare the runtime-oriented meta-model in SAM and the polynomial regression models with the factor of 2 to 4

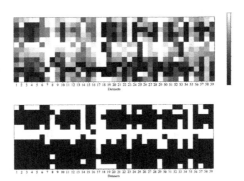

Fig. 5. Algorithm performance heatmap. Top: Each block is colored by the performance of each algorithm on 39 datasets. The lighter block represents the performance of corresponding algorithm is better, and vice versa. Bottom: the selection result of SAM on these 39 datasets. The white block represents the algorithm selected by SAM while the black block represents the algorithm is not selected as promising one. The algorithms represented by y-axis from bottom to top in these two figures are Kernel SVM, Linear SVM, Gaussian naive Bayes, Bernoulli naive Bayes, KNN, Decision Tree, Random Forest, Adaboost, SGD, LDA, QDA, Passive Aggressive, and Extra Tree.

Table 1. Four variants compasion on held-out test set.

Variants	Metrics				
	SB(↑)	SW(↓)	TB(↑)	GA(↑)	GB(↓)
Original	0.795 ± 0.044	0.113 ± 0.042	0.879 ± 0.057	0.788 ± 0.021	0.105 ± 0.017
MetaGAN	0.828 ± 0.030	0.085 ± 0.036	0.918 ± 0.050	0.796 ± 0.020	0.098 ± 0.017
SAM	0.821 ± 0.055	0.100 ± 0.047	0.900 ± 0.053	0.802 ± 0.019	0.101 ± 0.016
MetaGAN+SAM	**0.869 ± 0.051**	**0.082 ± 0.047**	**0.936 ± 0.052**	**0.823 ± 0.021**	**0.087 ± 0.018**

Table 2. The comparison of different synthetic data.

Methods	Metrics				
	SB(↑)	SW(↓)	TB(↑)	GA(↑)	GB(↓)
Resample	0.785 ± 0.066	0.090 ± 0.015	0.879 ± 0.051	0.794 ± 0.051	0.102 ± 0.015
SMOTE	0.797 ± 0.053	0.121 ± 0.067	0.882 ± 0.040	0.792 ± 0.022	0.099 ± 0.020
ADASYN	0.800 ± 0.045	0.128 ± 0.018	0.885 ± 0.048	0.794 ± 0.022	0.100 ± 0.018
MetaGAN	**0.828 ± 0.030**	**0.085 ± 0.036**	**0.918 ± 0.050**	**0.796 ± 0.020**	**0.098 ± 0.017**

using in OBOE [21] on algorithm runtime prediction. In our experiment, the polynomial regression with the factor of 3 and 4 performs much worse than the factor of 2. Therefore, we only present the performance of the model with factor of 2. Table 3 show the runtime prediction accuracy (the distance of actual time and predicted time less than 1 s) of these method. We can observe that SAM performs better than OBOE in 9 algorithms runtime prediction and the accuracy of other algorithm prediction is similar with OBOE. Although runtime

Table 3. Runtime prediction accuracy.

Algorithm type	Methods	
	With factor of 2	SAM
Kernel SVM	0.846 ± 0.055	**0.872 ± 0.053**
Linear SVM	0.635 ± 0.034	**0.685 ± 0.050**
Gaussian naive Bayes	0.972 ± 0.027	**0.977 ± 0.021**
Bernoulli naive Bayes	0.982 ± 0.020	**0.990 ± 0.012**
KNN	**0.967 ± 0.023**	0.967 ± 0.033
Decision Tree	0.751 ± 0.051	**0.851 ± 0.043**
Random Forest	0.849 ± 0.068	**0.892 ± 0.043**
Adaboost	0.454 ± 0.040	**0.523 ± 0.083**
SGD	0.821 ± 0.070	**0.846 ± 0.068**
LDA	**0.941 ± 0.046**	0.923 ± 0.049
QDA	**0.890 ± 0.043**	0.885 ± 0.043
Passive Aggressive	0.895 ± 0.045	**0.928 ± 0.039**
Extra Tree	**0.951 ± 0.035**	0.946 ± 0.045

prediction for machine learning algorithm is difficult to predict, the performance of SAM on multiple algorithms is a roughly good prediction and the predicted runtime ranking for MDASS is acceptable.

4   Conclusion

Meta-learning is widely used in AutoML. However, collecting sufficient meta-data with high quality is difficult. In this work, we incorporate GAN with meta-learning for meta-data augmentation, and propose Meta-data Augmentation-based Search Strategy (MDASS) for AutoML model selection, which is composed of Meta-GAN Surrogate Model (MetaGAN) and Self-adaptive Meta-model (SAM). MetaGAN generates effective meta-data through WGAN as surrogate model. SAM applies hybrid learners to self-adaptively build meta-models to search the algorithms with right trade-off of runtime and accuracy. For a new dataset, MDASS evaluate promising algorithms with Bayesian optimization under time constraint. Our experiments, evaluated on 405 classification datasets from OpenML and 13 classifiers from scikit-learn, show that our proposed method both can yield improvement than comparative methods.

This work demonstrates the promise of GAN for AutoML research. However, there still many lefts to future research. The well-known dilemma is the stable training process of GAN, and the neural architecture of GAN using in MDASS should be elaborate to design. Therefore, one obvious direction is to adapt neural architecture search for GAN. Moreover, MetaGAN can incorporate with semi-supervised learning and be extended to semi-AutoML through the adversarial training of meta-model, generator and discriminator.

Acknowledgment. This work is supported by the National Natural Science Foundation of China (No. 52073169) and the State Key Program of National Nature Science Foundation of China (Grant No. 61936001).

References

1. Arjovsky, M., Chintala, S., Bottou, L.: Wasserstein gan. arXiv preprint arXiv:1701.07875 (2017)
2. Asuncion, A., Newman, D.: UCI machine learning repository (2007)
3. Bergstra, J., Yamins, D., Cox, D.D.: Hyperopt: A python library for optimizing the hyperparameters of machine learning algorithms. In: Proceedings of the 12th Python in Science Conference, vol. 13, p. 20. Citeseer (2013)
4. Chawla, N.V., Bowyer, K.W., Hall, L.O., Kegelmeyer, W.P.: Smote: synthetic minority over-sampling technique. J. Artif. Intell. Res. **16**, 321–357 (2002)
5. Feurer, M., Klein, A., Eggensperger, K., Springenberg, J., Blum, M., Hutter, F.: Efficient and robust automated machine learning. In: Advances in Neural Information Processing Systems, pp. 2962–2970 (2015)
6. Feurer, M., Springenberg, J.T., Hutter, F.: Initializing Bayesian hyperparameter optimization via meta-learning. In: Twenty-Ninth AAAI Conference on Artificial Intelligence (2015)
7. Fusi, N., Sheth, R., Elibol, M.: Probabilistic matrix factorization for automated machine learning. In: Advances in Neural Information Processing Systems, pp. 3348–3357 (2018)
8. Goodfellow, I., et al.: Generative adversarial nets. In: Advances in Neural Information Processing Systems, pp. 2672–2680 (2014)
9. Gui, J., Sun, Z., Wen, Y., Tao, D., Ye, J.: A review on generative adversarial networks: algorithms, theory, and applications (2020)
10. He, H., Bai, Y., Garcia, E.A., Li, S.: Adasyn: adaptive synthetic sampling approach for imbalanced learning. In: 2008 IEEE International Joint Conference on Neural Networks, pp. 1322–1328 (2008)
11. Klein, A., Dai, Z., Hutter, F., Lawrence, N., Gonzalez, J.: Meta-surrogate benchmarking for hyperparameter optimization. In: Advances in Neural Information Processing Systems, pp. 6270–6280 (2019)
12. Klein, A., Falkner, S., Bartels, S., Hennig, P., Hutter, F.: Fast Bayesian optimization of machine learning hyperparameters on large datasets. In: Artificial Intelligence and Statistics, pp. 528–536. PMLR (2017)
13. Li, Y.F., Wang, H., Wei, T., Tu, W.W.: Towards automated semi-supervised learning. Proceedings of the AAAI Conference on Artificial Intelligence, vol. 33, pp. 4237–4244 (2019)
14. Luo, Y., et al.: Autocross: automatic feature crossing for tabular data in real-world applications. In: Proceedings of the 25th ACM SIGKDD International Conference on Knowledge Discovery & Data Mining, pp. 1936–1945 (2019)
15. Mirza, M., Osindero, S.: Conditional generative adversarial nets. arXiv preprint arXiv:1411.1784 (2014)
16. Mısır, M., Sebag, M.: Alors: an algorithm recommender system. Artif. Intell **244**, 291–314 (2017)
17. Pedregosa, F., et al.: Scikit-learn: machine learning in python. J. Mach. Learn. Res. **12**, 2825–2830 (2011)

18. Thornton, C., Hutter, F., Hoos, H.H., Leyton-Brown, K.: Auto-weka: combined selection and hyperparameter optimization of classification algorithms. In: Proceedings of the 19th ACM SIGKDD International Conference on Knowledge Discovery and Data Mining, pp. 847–855 (2013)
19. Vanschoren, J.: Meta-learning: a survey. arXiv preprint arXiv:1810.03548 (2018)
20. Vanschoren, J., Van Rijn, J.N., Bischl, B., Torgo, L.: Openml: networked science in machine learning. ACM SIGKDD Expl. Newslett 15(2), 49–60 (2014)
21. Yang, C., Akimoto, Y., Kim, D.W., Udell, M.: Oboe: Collaborative filtering for automl model selection. In: Proceedings of the 25th ACM SIGKDD International Conference on Knowledge Discovery & Data Mining, pp. 1173–1183 (2019)
22. Zöller, M.A., Huber, M.F.: Benchmark and survey of automated machine learning frameworks. arXiv preprint arXiv:1904.12054 (2019)

Tree-Capsule: Tree-Structured Capsule Network for Improving Relation Extraction

Tianchi Yang[1], Linmei Hu[1], Luhao Zhang[2], Chuan Shi[1(✉)], Cheng Yang[1] , Nan Duan[3], and Ming Zhou[3]

[1] Beijing University of Posts and Telecommunications, Beijing, China
{yangtianchi,hulinmei,shichuan}@bupt.edu.cn
[2] Meituan, Beijing, China
zhangluhao@meituan.com
[3] Microsoft Research, Beijing, China
{nanduan,mingzhou}@microsoft.com

Abstract. Relation extraction benefits a variety of applications requiring relational understanding of unstructured texts, such as question answering. Recently, capsule network-based models have been proposed for improving relation extraction with better capability of modeling complex entity relations. However, they fail to capture the syntactic structure information of a sentence which has proven to be useful for relation extraction. In this paper, we propose a **T**ree-structured **C**apsule network based model for improving sentence-level **R**elation **E**xtraction (TCRE), which seamlessly incorporates the syntax tree (Generally, syntax trees include constituent trees and dependency trees.) information (constituent tree is used in this work). Particularly, we design a novel tree-structured capsule network (Tree-Capsule network) to encode the constituent tree. Additionally, an entity-aware routing algorithm for Tree-Capsule network is proposed to pay attention to the critical relevant information, further improving the relation extraction of the target entities. Experimental results on standard datasets demonstrate that our TCRE significantly improves the performance of relation extraction by incorporating the syntactic structure information.

1 Introduction

Relation extraction aims to identify the relation between two entities in a sentence. It plays an important role in many natural language processing (NLP) tasks, such as question answering [15] and knowledge base population [21]. Compared with traditional approaches focusing on human-designed feature engineering, neural models have achieved significant improvement for relation extraction, such as Convolutional Neural Networks (CNN) [7,16,17] and Recurrent Neural Networks (RNN) [8,21,22]. Recently, some researchers begin to explore capsule networks [10] for many NLP tasks [1,4,14], including relation extraction [3,18,19], which could benefit from the stronger representing capability of

© Springer Nature Switzerland AG 2021
K. Karlapalem et al. (Eds.): PAKDD 2021, LNAI 12714, pp. 325–337, 2021.
https://doi.org/10.1007/978-3-030-75768-7_26

capsules. For example, [18] and [19] developed capsule network with attention mechanism for multi-instance multi-label relation extraction. [3] proposed an attribute-driven capsule network for relation extraction. However, they fail to take into account the syntax tree information (e.g., dependency tree and constituent tree) which captures the long-range semantic relations among words and has proven to be very useful in relation extraction [8,13,20].

Generally, syntax trees include dependency trees and constituent trees. Since capsule network is good at modeling part-whole relationships [10] and constituent trees encompass part-whole semantic relationships between words and phrases in the sentences, we explore to extend the capsule network to encode the constituent tree. However, it is nontrivial due to the following three challenges. (1) Traditional capsule networks are based on word sequence, which cannot be directly applied to the constituent tree. How can we develop a new structured capsule network that aggregates information from child to parent constituents based on the constituent tree? (2) Different types of constituents contribute in different patterns to the parent constituent. For example, a Noun (NN) contributes more to a Noun Phrase (NP), while a Verb (VB) provides more information to a Verb Phrase (VP). In addition, some constituents may contain irrelevant and noisy information. How can we capture the difference of constituent types and pay attention to critical information in encoding the constituent tree? (3) Existing capsule networks are usually shallow (no more than 2 layers) due to the gradient vanishing problem, while the constituent tree is quite deep (10 on average). How can we avoid the gradient vanishing problem?

To address the above challenges, in this paper, we propose a **Tree-C**apsule network based model for **R**elation **E**xtraction (TCRE), which improves the performance by seamlessly incorporating the constituent tree information. Specifically, we design a novel tree-structured capsule network (called Tree-Capsule network) to encode the constituent tree information. Tree-Capsule network has a carefully designed part-whole aggregation mechanism tailored to the constituent tree. In this way, it can aggregate information from child to parent constituents along the constituent tree. During the aggregation, it considers that different types of constituents (e.g., NN, VB, NP and VP) contribute in different patterns to their parent constituents. In addition, an entity-aware routing algorithm for the Tree-Capsule network is proposed to pay attention to the critical relevant information for the relation extraction of the target entities. At last, to address the gradient vanishing problem caused by the deep structure of the constituent tree, a new activation function (hardtanh with the gradient as constant 1 around 0) is also introduced. In summary, the main contributions of this paper are as follows:

1) To the best of our knowledge, we are the first to propose tree-structured capsule (Tree-Capsule) network, which is tailored to encode the syntactic structure information (constituent tree) for improving sentence-level relation extraction.

2) In the Tree-Capsule network, we design a new part-whole aggregation mechanism for encoding the constituent tree. It considers different types of constituents during information aggregation along the tree. In addition, an entity-aware routing algorithm is proposed to pay attention to the target entity relevant information. A new activation function is also introduced for alleviating gradient vanishing problem.

3) Extensive experimental results have demonstrated that our proposed model significantly improves the performance of relation extraction by incorporating the syntactic structure information.

2   Our Proposed Model

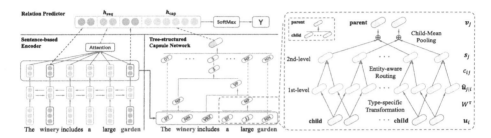

Fig. 1. Illustration of our model TCRE (left) and the new part-whole aggregation mechanism (right) of Tree-Capsule network (taking $N_c = 2$ as an example).

In this section, we will detail our proposed model TCRE, which improves the sentence-level relation extraction by taking full advantage of the syntax tree information (i.e., constituent tree in this paper). As shown in the left of Fig. 1, TCRE contains three parts: (1) Sequence-based encoder first encodes the sequence of words in a sentence. (2) Tree-Capsule network then encodes the constituent tree structure based on above. (3) Finally, relation prediction module combines the sequential and structural information for improving relation extraction. In the following, we will describe them in detail.

2.1   Sequence-Based Encoder

In this paper, we apply BiLSTM as sequence encoder to deeply learn the semantic meaning of a sentence. Note that we can also try state-of-the-art sentence encoder such as BERT [5] to obtain better performance.

The input of the BiLSTM is constructed by concatenating the word embeddings $\boldsymbol{w} \in \mathbb{R}^{D_w}$ and position embeddings $\boldsymbol{p}_1$ and $\boldsymbol{p}_2 \in \mathbb{R}^{D_p}$ which are respectively adopted for incorporating relative distances to the two target entities [16]. Formally, for each word w_i of the sentence $S = \{w_1, \cdots, w_n\}$, the representation is initialized as $\hat{\boldsymbol{w}}_i = \boldsymbol{w}_i \circ \boldsymbol{p}_{i1} \circ \boldsymbol{p}_{i2} \in \mathbb{R}^{D_w + 2D_p}$, where $\circ$ represents concatenation operator.

Then, we adopt BiLSTM to read the input sentences in two directions and then sum the states for each time step i. The corresponding state vector $\boldsymbol{h}_i$ of BiLSTM are taken as the i-th word embedding with contextual information.

To pay attention to critical words in the sentence, we apply an attention mechanism [22]. Let $H = [\boldsymbol{h}_1, \cdots, \boldsymbol{h}_n]$ be the matrix consisting of the output state vectors of BiLSTM, the sentence embedding $\boldsymbol{h}_{\text{sent}}$ is the weighted sum of these output vectors:

$$\boldsymbol{\alpha} = \text{softmax}(\boldsymbol{a}^T \cdot tanh(H)) , \tag{1}$$

$$\boldsymbol{h}_{\text{sent}} = H \cdot \boldsymbol{\alpha}^T . \tag{2}$$

where $\boldsymbol{a}$ is a trained parameter vector.

Finally, we concatenate the sentence embedding with the embeddings of the two target entities. The embedding (output by BiLSTM) of the first word in an entity mention is taken as the entity embedding (e.g., *Barack* for *Barack Hussein Obama*). The final sequence-based representation for relation extraction is computed as, $\boldsymbol{h}_{\text{seq}} = \boldsymbol{h}_{\text{head}} \circ \boldsymbol{h}_{\text{sent}} \circ \boldsymbol{h}_{\text{tail}}$. $\boldsymbol{h}_{\text{head}}$ and $\boldsymbol{h}_{\text{tail}}$ represent the embeddings of the target head entity and tail entity, respectively.

2.2   Tree-Capsule Network

In this subsection, we detail our Tree-Capsule network which encodes the constituent tree structure to improve the relation extraction. Specifically, in the Tree-Capsule network, we first construct primary capsules as the input. Then we design a part-whole aggregation mechanism which considers different constituent types with a type-specific transformation during information aggregation along the tree. Additionally, an entity-aware routing algorithm is proposed to pay attention to the target entity relevant information. Finally, we add an attentive fully connected capsule layer also with the entity-aware routing algorithm, to combine the information of all the constituents in the tree for relation extraction.

Primary Capsule Layer. After running BiLSTM, we obtain the scalar-output hidden representations $H \in \mathbb{R}^{n \times D_h}$ of words with contextual information. To transform the scalar-output features into vector-output capsules, while preserving the local order of words and semantic representations of words [14], a linear layer with multi-channels is applied as,

$$U_i = H \cdot W_i + \boldsymbol{b}_i, \quad i = 1, \cdots, N_c , \tag{3}$$

where the transformation matrix $W_i \in \mathbb{R}^{D_h \times D_c}$ projects the output word embedding (states of BiLSTM) $H \in \mathbb{R}^{n \times D_h}$ into a capsule matrix $U_i \in \mathbb{R}^{n \times D_c}$, where n is the number of words in the sentence. Then we stack the capsule matrices

$\{U_i|i = 1, \cdots, N_c\}$ as a capsule tensor $U \in \mathbb{R}^{n \times N_c \times D_c}$, where N_c is the capsule number (i.e., number of channels) and D_c denotes the capsule size. Consequently, each word is represented by a group of N_c primary capsules.

Part-Whole Aggregation. Given the primary capsules as input, we now aim to obtain groups of higher-level capsules representing the higher-level constituents in the constituent tree using a part-whole aggregation mechanism from bottom to up. The right of Fig. 1 shows one single part-whole aggregation process from child constituents to parent constituent. It includes three steps: type-specific transformation, entity-aware routing and child-mean pooling. We will detail them as follows.

In the first step, i.e., *type-specific transformation*, each capsule in the groups of capsules representing the child constituents is transformed to predict several 1st-level potential capsules that encode higher-level information, considering its distinct constituent type (e.g., NN and VB). Formally, a child capsule $\boldsymbol{u}_i \in \mathbb{R}^{D_c}$ produces a group of 1st-level potential capsules $\{\hat{\boldsymbol{u}}_{j|i}|j = 1, \cdots, N_c\}$ by the transformation matrices $\{W_j^\tau|j = 1, \cdots, N_c\}$ w.r.t. the constituent type τ of u_i:

$$\hat{\boldsymbol{u}}_{j|i} = \boldsymbol{u}_i \cdot W_j^\tau + \boldsymbol{b}_{j|i}^\tau , \qquad (4)$$

where $\boldsymbol{b}_{j|i}$ is the capsule bias term, $W_j^\tau \in \mathbb{R}^{D_c \times D_c}$ is the type-specific transformation matrix of the j_{th} channel. Note that capsules with transformation matrices allow networks to learn potential part-whole relationships automatically [19].

The second step, i.e., *entity-aware routing algorithm*, aims to filter out the noisy and irrelevant 1st-level potential capsules, and get the 2nd-level potential capsules $\boldsymbol{s}_j$ representing relation features. The traditional routing algorithm [10] fails to focus on the entities, which are critical for relation extraction. Therefore, in this work, we present an entity-aware routing algorithm, which pays attention to the relevant and critical information for the entity relation extraction. Formally, we first compute the averaged capsule $\hat{\boldsymbol{e}}$ of the two groups of capsules respectively representing the two target entities. Then we can guide the routing process with $\hat{\boldsymbol{e}}$ as follows:

$$\boldsymbol{s}_j = \text{hardtanh}(\sum_{i=1}^{N_c} \sigma(\hat{\boldsymbol{e}}^T \cdot \hat{\boldsymbol{u}}_{j|i}) \cdot c_{ij} \cdot \hat{\boldsymbol{u}}_{j|i}) , \qquad (5)$$

where $\sigma(\cdot)$ denotes the sigmoid function, and the coupling coefficient c_{ij} is iteratively determined as shown in Algorithm 1. Note that the *hardtanh*[1] activation function, whose gradient around 0 is constant 1, is introduced to replace the traditional Squash function for alleviating the gradient vanishing problem caused by the deep structure of the constituent tree.

The final step, *child-mean pooling*, is to get the parent capsule $\boldsymbol{v}_j$ that represent the j-th channel of the parent constituent by averaging the 2nd-level potential capsules $C(v_j)$ corresponding to the j-th channel:

$$\boldsymbol{v}_j = \frac{1}{|C(v_j)|} \sum_{k \in C(v_j)} \boldsymbol{s}_k . \qquad (6)$$

---

[1] $\text{hardtanh}(x) = \min(\max(x, -1), 1)$.

The group of capsules from all channels $\{v_j | j = 1, \cdots, N_c\}$ together represent a parent constituent.

The above shows a single part-whole aggregation process. After recursively repeating the process along the constituent tree from bottom to up, we obtain the groups of capsules representing all the constituents in the tree.

Algorithm 1. Entity-aware Routing Algorithm

Require: The 1st-level potential capsules, $\hat{u}_{j|i}$;
Ensure: The 2nd-level potential capsules, s_j;
 1: Initialize the logits of coefficients: $b_{ij} \leftarrow 0$
 2: **for** r iteations **do**
 3:     $\forall$ 1st-level indicator i: $c_i \leftarrow \text{softmax}(b_i)$
 4:     Calculate the averaged entity capsule: $\hat{e} = \frac{1}{2N_c} \sum_{j=1}^{N_c} (\hat{u}_{j|\text{head}} + \hat{u}_{j|\text{tail}})$
 5:     $\forall$ 2nd-level indicator j: $s_j \leftarrow \text{hardtanh}(\sum_{i=1}^{N_c} \sigma(\hat{e}^T \cdot \hat{u}_{j|i}) c_{ij} \hat{u}_{j|i})$
 6:     $\forall$ 1st-level and 2nd-level indicator i, j: $b_{ij} \leftarrow b_{ij} + \hat{u}_{j|i} \cdot s_j$
 7: **end for**
 8: **return** s_j

Attentive Fully Connected Capsule Layer. With the groups of capsules representing the constituents in the constituent tree, we now leverage all the capsules to generate the structural sentence embedding. All the capsules in the tree are flattened into a list of capsules and fed into the attentive fully connected capsule layer. Specifically, each capsule is first transformed with a shared transformation matrix $W \in \mathbb{R}^{1 \times D_c \times 3D_h}$ and fed into the entity-aware routing algorithm (as shown in Algorithm 1) which pays attention to the critical relevant information related to the relation extraction of the entities. Finally, we get one single capsule $h_{\text{cap}} \in \mathbb{R}^{3D_h}$.

Through the Tree-Capsule network, we obtain the structural sentence embedding represented by the single capsule h_{cap} that encodes the constituent tree structure information.

2.3   Relation Prediction

In this subsection, we concatenate the sequence-based sentence embedding and the structural sentence embedding for relation prediction through a softmax layer:

$$P(y|S) = \text{softmax}(W_s \cdot [h_{\text{seq}} \circ h_{\text{cap}}] + b_s) , \tag{7}$$

where W_s and b_s are the parameters. During model training, we exploit the cross-entropy loss over training data with the L_2-norm of model parameters. For model optimization, we adopt the gradient descent algorithm.

3 Experiments

3.1 Datasets and Evaluation Metrics

We conduct extensive experiments on 2 benchmark datasets for sentence-level relation extraction. **SemEval** 2010 Task 8 [6] contains 10.7k instances of 19 relation classes over entity pairs: 9 directed relations and a special "Other" class. For evaluation metrics, we follow the convention and report the official macro-average F1 score, which is based on the 9 actual relations (excluding the "Other" relation) and takes their directions into consideration. **TACRED** [21] is a large scale dataset with 106k instances, containing 41 relation classes and a special "no relation" class. The average length of sentences (36.4) and depth of the corresponding constituent trees (14.2) are larger than those of SemEval, indicating more complexity of contexts. For convenient and fair comparison, following [21], we use the same "entity mask" strategy where each subject entity is replaced with a special "SUBJ-<NER>" token (similarly to object entity). The micro-average Precision, Recall and F1 score are reported.

3.2 Baselines

We compare our model TCRE with several state-of-the-art models, which can be divided into four categories: **Semantic-based models** include: SVM [9] and LR [21], achieving the top performance in SemEval and TACRED with a variety of handcrafted features, respectively; CNN [17] based on word embeddings and position embeddings. Att-BiLSTM [22] which employs a word-level attention over BiLSTM outputs; PA-LSTM [21] proposing a position-aware attention mechanism over LSTM outputs. **Syntactic structure-based models** consist of: SDP-LSTM [13], which applies a neural sequence model on the shortest path between the subject and object entities in the dependency tree. Tree-LSTM [11], which generalizes the LSTM to arbitrary syntax tree structures. **Capsule-based models** include: Att-CapNet (CNN) [19], which integrates an attentive capsule network based on CNN for relation extraction. Att-CapNet (RNN) [19] integrating an attentive capsule network based on RNN for relation extraction. **Pretrained models** include: BERT [5], which uses a multi-layer bidirectional Transformer encoder and is trained with a masked language model. We fine-tuned BERT-base (768 hidden dimensions) until convergence. Note that we also try to apply BERT to replace Att-BiLSTM as our sentence encoder in our model TCRE, named TCRE-BERT.

3.3 Experimental Settings

In our experiments, we use the 50-dimension word vectors pre-trained in Glove setting with randomly initialized 5-dimensional position embeddings for SemEval. For TACRED, following [20], we use the 300-dimension word vectors with 30-dimensional position, POS and NER embeddings. For all sentences of the two datasets, we use Stanford Parser tool[2] to obtain their constituent trees. For fair comparison with capsule-based methods, following [10], we set the hidden dimension $D_h = 256$, capsule number and size $N_c = D_c = 16$ for the two datasets. L$_2$ regularization ($5e-6$) and dropout (0.7 for SemEval and 0.5 for TACRED) are adopted to avoid overfitting. To train our model efficiently, we apply a two-step training strategy, i.e., we first optimize the parameters of the sequence-based encoder for 10 epochs with learning rate 1.0 by AdaDelta optimizer and then fine-tune all the parameters with learning rate 0.1 for SemEval and 0.01 for TACRED by Adam optimizer until convergence.

3.4 Overall Performance

Table 1. Performance on SemEval and TACRED. The notation * means our model significantly outperforms the baselines based on t-test (p < 0.05).

Method	SemEval	TACRED		
	F1(%)	P(%)	R(%)	F1(%)
SVM [9]/LR [21]	82.2	**73.5**	49.9	59.4
CNN [17]	79.5	70.3	54.2	61.2
LSTM [21]/Att-BiLSTM [22]	81.7	65.7	59.9	62.7
PA-LSTM [21]	82.7	65.7	64.5	65.1
SDP-LSTM [13]	82.4	66.3	52.7	58.7
Tree-LSTM [11]	83.8	66.0	59.2	62.4
SPTree [8]	84.4	–	–	–
Tree-LSTM + Att-BiLSTM	82.6	–	–	–
Att-CapNet (CNN) [19]	79.9	68.1	57.3	62.2
Att-CapNet (RNN) [19]	83.3	66.0	61.2	63.7
TCRE	**84.6***	68.9	**64.6***	**66.7***
BERT-base [5]	87.3	69.9	64.5	67.1
TCRE-BERT	**87.6***	69.8	**65.7***	**67.6***

We present the comparison results in Table 1. We can find that our TCRE achieves the best performance against all the baselines except BERT-base. TCRE-BERT with BERT-base as sentence encoder also achieves better performance than BERT-base. These demonstrate that our proposed model significantly improves relation extraction by seamlessly incorporating the constituent tree information. Besides, we have the following detailed observations: (1) Compared to semantic-based models (e.g., CNN), the syntactic structure-based models (e.g., SDP-LSTM) generally perform better on dataset SemEval due to the

---

[2] https://nlp.stanford.edu/software/lex-parser.html.

incorporated syntactic structure information, but fail to perform well on dataset TACRED. The reason could be that the sentences in TACRED are much longer than those in SemEval, containing sufficient semantic information for relation extraction. They may fail to consider noise when introducing the syntax tree information. Nevertheless, our TCRE can successfully reduce the noise in the constituent tree by entity-aware routing algorithm. (2) Capsule-based models (i.e., Att-CapNet(CNN) and (RNN)) outperform those semantic-based models (i.e., Att-CapNet(CNN) vs CNN and Att-CapNet(RNN) vs Att-BiLSTM), due to the stronger capability of capsules for representation. However, generally they achieve inferior performance to the syntactic based model Tree-LSTM. (3) Our TCRE significantly outperforms the capsule-based baselines, indicating that our proposed Tree-Capsule network successfully encodes the constituent tree information for improving relation extraction. (4) Note that Tree-LSTM + Att-BiLSTM achieves lower performance (82.6%) on SemEval than our model TCRE based on Att-BiLSTM, which demonstrates that our Tree-Capsule Network can better encode the syntactic tree information. (5) Finally, BERT-base obtains good performance since it was pretrained on a large corpus. TCRE-BERT performs better than BERT-base, which verifies that our proposed Tree-Capsule network can consistently improve relation extraction.

Different from the traditional capsule networks which can only model a sequence structure [3,10,18,19], we integrate syntactic structure information into the capsule network to form a tree structure. Since the traditional capsule network uses Att-BiLSTM as the basic encoder [18,19], for fair comparison, we also choose it as the sequence encoder in our TCRE model, which limits the performance of our model, failing to outperform some SOTA BERT-based methods[2,12]. Note that we also tried an advanced sequence encoder, i.e., BERT-base as the basic encoder in our TCRE-BERT, which outperforms most SOTA methods. This further verifies the effectiveness of our TCRE.

3.5 Comparison of TCRE Variants

Table 2. Performance of TCRE variants on SemEval.

Variants	F1(%)
TCRE	84.6
TCRE w/o Att-BiLSTM	76.2
TCRE w/o Tree-Capsule	81.7
TCRE w/o Type	84.0
TCRE w/o Entity-aware Routing	84.2
TCRE w/ squash	82.9

In this subsection, we evaluate several variants of TCRE on the test set of SemEval to verify the effectiveness of its delicate designs. The performance comparisons of TCRE variants are shown in Table 2. As we can see, the performance of *TCRE w/o Tree-Capsule* which does not consider the constituent tree information drops largely by around 3.0% on macro-F1, while the performance of *TCRE w/o Att-BiLSTM* (removing the sequence-based sentence embedding) is further lower. We believe that though the part-whole aggregation mechanism learns syntactic structure information, it loses sequential information, which may be the key information for relation extraction. However, we find that syntactic structure information is complementary to the sequential information, since TCRE (combining the syntactic structure and sequential information) achieves the best performance, outperforming the models using individual syntactic structure information or sequential information. Next, *TCRE w/o Type* ignoring the different types of constituents during part-whole aggregation and *TCRE w/o Entity-aware Routing* using traditional routing algorithm both achieve inferior performance, compared to TCRE. It validates the necessity of considering the constituent types and entity-aware routing in Tree-Capsule network. At last, the performance of *TCRE w/squash* which replaces our hardtanh with the traditional activation function Squash[3] decreases substantially by 2%, due to the gradient vanishing problem caused by the deep structure of the constituent tree.

3.6   Case Study

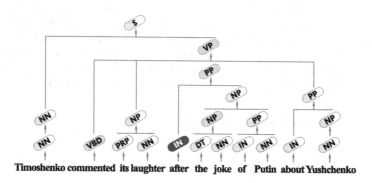

Fig. 2. An example of relation extraction for the entity "laughter" and "joke".

In this subsection, we present a practical case in the SemEval test set to illustrate how the proposed Tree-Capsule network improves relation extraction. As shown in Fig. 2, the relation of the input sentence is mistakenly predicted as "Other" by the sequence-based encoder Att-BiLSTM, while it is correctly labeled as "Cause-Effect" by our model TCRE which incorporates the constituent tree information. Specifically, TCRE pays attention to the constituent "IN" (Preposition)

---

[3] $Squash(x) = \frac{\|x\| \cdot x}{1 + \|x\|^2}$, x is a capsule.

corresponding to word "after" with the entity-aware routing algorithm, and successfully predicts the relation "Cause-Effect". It conforms to our intuition that "after" may indicate a cause-effect relationship.

3.7 Parameter Analysis

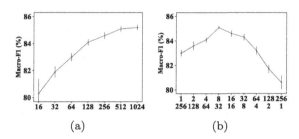

(a)   (b)

Fig. 3. Influence of (a) hidden dimension, (b) capsule number and capsule size.

In this subsection, we study the influence of hyper-parameters on the test set of SemEval dataset. Figure 3 (a) shows the average F1 score and its variance (error bars) of our TCRE over running 10 times with different hidden dimensions from 16 to 1,024. We can see that the relation extraction performance of TCRE grows (the variance decreases) with the increase of the hidden dimension, because higher dimension brings stronger representation capability. To study the influence of capsule number N_c and size D_c, we fix the product of them as 256, i.e., $N_c \cdot D_c = 256$. As shown in Fig. 3 (b) where the two lines of the x-axis represent N_c and D_c, respectively, we can see that the model performance first increases, achieves the best performance when $N_c = 8, D_c = 32$, and then begins to fall rapidly to the worst performance 80.6% on Macro-F1 when $N_c = 256, D_c = 1$. This is because the representation capability of the capsules is much lower as the size of capsules degrades to 1.

4   Conclusion

In this paper, we propose a novel Tree-Capsule network based method TCRE for improving sentence-level relation extraction by seamlessly incorporating the constituent tree information. Specifically, we design a new Tree-Capsule network with a part-whole aggregation mechanism, tailored to encode the constituent tree. It considers the different constituent types. Additionally, an entity-aware routing algorithm for Tree-Capsule network is proposed to pay attention to target entity relevant information for the relation extraction. We also introduce a new activation function for the Tree-Capsule network to address the gradient vanishing problem caused by the deep structure of the constituent tree. Extensive experiments demonstrate that our proposed model significantly improves the performance of relation extraction.

Acknowledgments. This work is supported in part by the National Natural Science Foundation of China (No.U20B2045, 61806020, 61772082) and the National Key Research and Development Program of China (2018YFB1402600).

References

1. Aly, R., Remus, S., Biemann, C.: Hierarchical multi-label classification of text with capsule networks. In: ACL 2019, pp. 323–330 (2019)
2. Baldini Soares, L., FitzGerald, N., Ling, J., Kwiatkowski, T.: Matching the blanks: distributional similarity for relation learning. In: ACL 2019, pp. 2895–2905 (2019)
3. Chen, J., Gong, X., Chen, X., Ma, Z.: Attribute-driven capsule network for entity relation prediction. In: Lauw, H.W., Wong, R.C.-W., Ntoulas, A., Lim, E.-P., Ng, S.-K., Pan, S.J. (eds.) PAKDD 2020. LNCS (LNAI), vol. 12084, pp. 675–686. Springer, Cham (2020). https://doi.org/10.1007/978-3-030-47426-3_52
4. Chen, Z., Qian, T.: Transfer capsule network for aspect level sentiment classification. In: ACL 2019, pp.547–556 (2019)
5. Devlin, J., Chang, M., Lee, K., Toutanova, K.: BERT: pre-training of deep bidirectional transformers for language understanding. In: NAACL-HLT 2019, pp. 4171–4186 (2019)
6. Hendrickx, I., et al.: SemEval-2010 task 8: multi-way classification of semantic relations between pairs of nominals. In: SemEval@ACL 2010, pp. 33–38 (2010)
7. Hu, L., Zhang, L., Shi, C., Nie, L., Guan, W., Yang, C.: Improving distantly-supervised relation extraction with joint label embedding. In: EMNLP-IJCNLP 2019, pp. 3821–3829 (2019)
8. Miwa, M., Bansal, M.: End-to-end relation extraction using LSTMs on sequences and tree structures. In: ACL 2016, pp. 1105–1116 (2016)
9. Rink, B., Harabagiu, S.: UTD: classifying semantic relations by combining lexical and semantic resources. In: SemEval@ACL 2010, pp. 256–259 (2010)
10. Sabour, S., Frosst, N., Hinton, G.E.: Dynamic routing between capsules. NIPS 2017, pp. 3859–3869 (2017)
11. Tai, K., Socher, R., Manning, C.: Improved semantic representations from tree-structured long short-term memory networks. In: ACL 2015, pp. 1556–1566 (2015)
12. Wu, S., He, Y.: Enriching pre-trained language model with entity information for relation classification. In: CIKM 2019, pp. 2361–2364 (2019)
13. Xu, Y., Mou, L., Li, G., Chen, Y., Peng, H., Jin, Z.: Classifying relations via long short term memory networks along shortest dependency paths. In: EMNLP 2015, pp. 1785–1794 (2015)
14. Yang, M., Zhao, W., Ye, J., Lei, Z., Zhao, Z., Zhang, S.: Investigating capsule networks with dynamic routing for text classification. In: EMNLP 2018, pp. 3110–3119 (2018)
15. Yu, M., Yin, W., Hasan, K.S., dos Santos, C., Xiang, B., Zhou, B.: Improved neural relation detection for knowledge base question answering. In: ACL 2017, pp. 571–581 (2017)
16. Zeng, D., Liu, K., Chen, Y., Zhao, J.: Distant supervision for relation extraction via piecewise convolutional neural networks. In: EMNLP 2015, pp. 1753–1762 (2015)
17. Zeng, D., Liu, K., Lai, S., Zhou, G., Zhao, J.: Relation classification via convolutional deep neural network. In: COLING 2014, pp. 2335–2344 (2014)
18. Zhang, N., Deng, S., Sun, Z., Chen, X., Zhang, W., Chen, H.: Attention-based capsule networks with dynamic routing for relation extraction. In: EMNLP 2018, pp. 986–992 (2018)

19. Zhang, X., Li, P., Jia, W., Zhao, H.: Multi-labeled relation extraction with attentive capsule network. In: AAAI 2019, pp. 7484–7491 (2019)
20. Zhang, Y., Qi, P., Manning, C.D.: Graph convolution over pruned dependency trees improves relation extraction. EMNLP **2018**, 2205–2215 (2018)
21. Zhang, Y., Zhong, V., Chen, D., Angeli, G., Manning, C.D.: Position-aware attention and supervised data improve slot filling. In: EMNLP 2017, pp. 35–45 (2017)
22. Zhou, P., et al.: Attention-based bidirectional long short-term memory networks for relation classification. In: ACL 2016, pp. 207–212 (2016)

Rule Injection-Based Generative Adversarial Imitation Learning for Knowledge Graph Reasoning

Sheng Wang[1], Xiaoying Chen[1,2(✉)], and Shengwu Xiong[1]

[1] Wuhan University of Technology, Wuhan, China
{wsheng,xiongsw}@whut.edu.cn
[2] Hubei Credit Information Center, Wuhan, China

Abstract. Knowledge graph reasoning is a crucial part of knowledge discovery and knowledge graph completion tasks. The solution based on generative adversarial imitation learning (GAIL) has made great progress in recent researches and solves the problem of relying heavily on the design of the reward function in reinforcement learning-based reasoning methods. However, only the semantic feature is considered in existing GAIL-based methods, which is not enough to assess the quality of reasoning paths. While logical rules contain rich factual logic that can be used for reasoning. Thus, we introduce the first-order predicate logic rule in our model called Rule Injection-based Generative Adversarial Path Reasoning. The key idea is to train the generator to learn reasoning strategies by imitating the demonstration from both semantic and rule levels. Particularly, we design a path discriminator and a logic rule discriminator to distinguish paths respectively from these two levels. Furthermore, both discriminator feedback to the generator a self-adaptively reward by assessing the quality of the generated reasoning path. Extensively experiments on two benchmarks show that our method improves the performance than the state-of-the-art baseline and some cases study also confirmed the explainability of our model.

Keywords: Query answering · Rule Injection · Reinforcement learning · Knowledge graph reasoning

1 Introduction

The knowledge graphs (KGs) store massive reality facts in the form of triples (h, r, t), supporting many tasks such as query answering [1], recommendation system [2], and natural language understanding [3]. However, due to the incompleteness of the KGs, the performance of many downstream tasks will be limited. To solve the problem, knowledge graph reasoning methods, which discover new or complement missing facts from existing ones, have attracted greater attention. In this paper, we focus on the query answering task and hope to complete the query triples automatically by multi-hop path reasoning.

Multi-hop path reasoning obtains the reasoning result by finding the path from the KGs and uses the reasoning path as evidence of the reasoning process. Among the solutions, the reinforcement learning (RL) based methods have achieved excellent results.

The original version of this chapter was revised: A spelling mistake in the name of the author Xiaoying Chen has been corrected. The correction to this chapter is available at
https://doi.org/10.1007/978-3-030-75768-7_34

But the most common problem of RL-based methods is how to design a reasonable reward to fit the specific knowledge graph. Because they are sensitive to reward, and a small difference may have a great impact on performance. Thus, Li et al. proposed a reasoning method based on generative adversarial imitation learning called DIVINE [4], which gives the agent a self-adaptively reward through imitating the demonstrations extracted from KG and without any extra reward engineering.

However, the generator in the DIVINE will only imitate learning the relation semantic feature extracted by convolutional neural networks from the demonstrations. It doesn't make full use of the effective information in the demonstrations. While logical rules contain rich factual logic that can be used for reasoning [5]. Combining logic rules and imitation learning allows the generator to additionally imitate learning from demonstrations at the perspective of rules. Thus, we present a Rule Injection-based Generative Adversarial Path Reasoning model called RIGAPR. Our goal is to assess the quality of the path at semantic and logical rule levels and train the generator to learn reasoning strategies by imitating the demonstrations at both levels.

More specifically, we designed a representation method of path logical rule, and it can be used to extract the logical rule contained in the paths. For example, given a query triple (*MichaelJordan, AthletePlaysSport, ?*), a possible two-hop reasoning path: Michael Jordan $\xrightarrow{\textit{Plays for Team}}$ Chicago Bulls $\xrightarrow{\textit{Team Plays Sport}}$ Basketball . We combine the atomic triples and logical connectives (e.g. conjunction $\wedge$ and implication $\Rightarrow$) to form the logical rule as $\forall a, b, c : (a, PlaysforTeam, b) \wedge (b, TeamPlaysSport, c) \Rightarrow (a, AthletePlaysSport, c)$. Furthermore, we model the path logical rule by t-norm fuzzy logic [6] and its truth value is used to measure the reasonableness of the rule. In this way, RIGAPR trains the generator to learn some reasoning strategies by imitating the demonstration from semantic and rule levels respectively, improved the reasoning performance and explainability (explain in sematic and logical rule levels).

In general, the main contributions of this paper can be summarized as follows:

- We introduce the logical rules into path reasoning and design an extraction method of the rule contained in the paths. To the best of our knowledge, we are the first to combine logic rules and GAIL for path reasoning.
- We propose a Rule Injection-based Generative Adversarial Path Reasoning model and design a logical rule discriminator to calculate the truth values of the logical rule contained in the paths.
- We design a reward enhancement method to enhance the reward at the semantic and logical rule levels without any extra reward engineering.
- Extensively experiments on two benchmarks show that our method improves the performance than the state-of-the-art baselines and some reasoning cases also confirmed the explainability of our model.

2   Related Work

Knowledge graph reasoning that discovers new or complements missing facts from existing ones has attracted greater attention. The most commonly embedding based reasoning methods maps entities and relations to a continuous vector space and designs a score function to evaluate the truth value of triples. They mainly including translation-based methods [7–9], tensor factorization based methods [10], and semantic matching based methods [11, 12]. Although they achieved good performance, most of them only returned the entity with the highest score among all entities as the reasoning result, which lacks explainability.

Consequently, a series of path-based reasoning methods are proposed, which use the reasoning path as an explanation of the reasoning process. The Path Ranking Algorithm (PRA) [13] and its extension [14] adopt random walk and restart strategies to get some relation paths between entity pairs. However, the path obtained through random walks often contains noisy information. Thus, some researchers combine reinforcement learning and path reasoning, hoping to guide the agent to generate a path by learning a reasoning strategy. DeepPath [15] is the first to introduce RL into path reasoning and regards the path finding as a Markov decision process. AttPath [16] incorporates LSTM and graph attention mechanism as memory components on the basis of DeepPath, getting rid of the pretraining process. But the state of MDP requires the target entity to be known in advance in DeepPath. MINERVA [1] is an RL-based end-to-end model for a more challenging query answering task. And MARLPaR [17] is proposed to solve the entity selection problem when a 1-N relation occurs, which uses two agents in path reasoning, one agent for relations reasoning and the other for entities reasoning.

However, most of the RL-based reasoning models mentioned above use the terminal reward function, which may cause the sparse reward problem. To address this, Lin et al. [18] used a pre-training well-embedding model as the reward shaping function and Qiu et al. [19] proposed a potential-based reward shaping function for natural language question answering task. Different from them, Li et al. [4] first proposed a generative adversarial imitation learning based path reasoning model DIVINE, which gives the agent a self-adaptively reward through the demonstrations extracted from KG and without any extra reward engineering.

3 Methodology

In this section, we first formalize the query answering task we focus on in this paper. Then, we elaborate our proposed RIGAPR model, which consists of three parts: path generator for reasoning, demonstration extraction module, and rule injection adversarial imitation learning. Finally, we introduce the reward enhancement method and the training procedure.

3.1 Problem Formulation

A knowledge graph $\mathcal{G}$ with entity set $\mathcal{E}$ and relation set $\mathcal{R}$ can be defined as $\mathcal{G} = \{(h, r, t) | h \in \mathcal{E}, t \in \mathcal{E}, r \in \mathcal{R}\}$, where the triple (h, r, t) represent a fact of relation r from head entity h to tail entity t.

For query answering task, given a query triple $(e_s, r_q, ?)$ with practical question meaning, e.g., the query triple $(MichaelJordan, AthletePlaysSport, ?)$, it expresses the question of "what sports does Jordan plays?", our goal is to automatically reason a path to find the answer entity set $\{e_a\}$. Such as the reasoning path:

Michael Jordan $\xrightarrow{Plays\ for\ Team}$ Chicago Bulls $\xrightarrow{Team\ Plays\ Sport}$ Basketball and we can find the answer of the example query triple is Basketball. More especially, the tripe $(e_s, r_q, e_a) \notin \mathcal{G}$.

In the following sections, we elaborate our proposed RIGAPR model. It solves the problem that existing GAIL-based methods are insufficient to discriminate the path. As illustrated in Fig. 1, the main idea is to extract semantic feature and logical rule contained in the expert demonstrations, training the generator (agent) to learn reasoning strategies by imitating the demonstration from both semantic (sematic reward R_p) and logical rule (rule reward R_r) levels, and improving the reasoning performance and explainability (explain in sematic and logical rule levels).

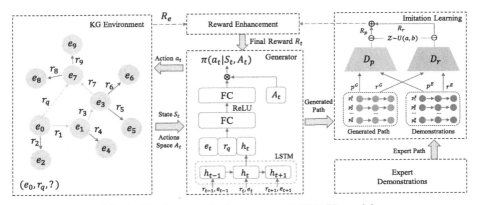

Fig. 1. The framework of our proposed RIGAPR model.

3.2   Path Generator for Reasoning

In order to solve the query answering task, RIGAPR regards the reasoning process as a Markov Decision Process. The generator stepwise selects the most likely action through the interaction with the MDP environment. The interaction process is that the MDP environment feedbacks to the generator a reward according to the action selected by the generator, and update its current state. The generator will select a new action based on the new state. The definition of the MDP component as follows:

State. The state s_t of the generator is formulated as $s_t = (e_s, r_q, e_t)$, which consists of current entity e_t where the generator stays at and global information (e_s, r_q) obtained by query triple.

Action. The set of possible actions A_t at t step are consist of all outgoing edges (relations) and its linked entities of the current entity where the generator stays at. Specifically, $A_t = \{(r, e) | (e_t, r, e) \in \mathcal{G}\}$. In order to give the generator option to recover from mistakes or find short paths (due to the fixed steps of reasoning), we have added self-loop and inverse relations to the KGs, i.e., $(h, NO\_OP, h) \in \mathcal{G}$ and $(t, r^{-1}, h) \in \mathcal{G}$.

Transition. The next state s_{t+1} of the generator is determined by the previous state s_t and the action a_t. Formality, $s_{t+1} = \delta(s_t, a_t) = (e_s, r_q, e_{t+1})$.

Reward. We only consider terminal rewards $R_{e,t} = \{1 | e_T = e_a\}$ and enhance it at the semantic and logical rule levels. In other words, we only enhance the reward for the correct reasoning paths, and the rewards for other paths are still zero. This may lead to the problem of sparse rewards, but we have proved through ablation experiments that it will not affect reasoning performance.

The generator selects the most likely action based on the current state s_t, which can be modeled as a network consisted of an action history encoder and fully connected layers. In particular, the generator learns a reasoning policy $\pi(a_t | S_t, A_t)$ that calculates the probability distribution of the action space A_t and the formula is as follows:

$$h_t = LSTM(h_{t-1}, [r_t; e_t]) \tag{1}$$

$$\pi(a_t | S_t, A_t) = \sigma\left(A_t \times W_{g2} ReLU\left(W_{g1}[e_t; r_q; h_t]\right)\right) \tag{2}$$

Where σ is the Softmax function. $r_t, e_t, r_q \in \mathbb{R}^d$ is the embedding representation of the corresponding entity or relation and [;] denotes the concatenation operator. W_{g1} and W_{g2} are parameters that need to be trained in the generator. To make the reasoning strategy learned by the generator more correct, we enhance the rewards obtained by the generator through imitation learning. Specifically, the demonstration extracted from the KGs is used as the background truth to guide the generator.

3.3 Demonstration Extraction

Whether imitation learning can improve performance mainly depends on the quality of expert demonstrations. A high-quality expert demonstration will guide the generator well in reasoning, otherwise it will be misleading. In this paper, we will use a more direct method to extract expert demonstrations. More specifically, for each query triple in the training dataset, we use a bi-directional breadth-first search (Bi-BFS) to explore the shortest path between the head entity and the tail entity. The reason we extract in this way is that the shortest path can more directly indicate the connection between two entities, and is easier to understand and explain. And the longer path may not always contain richer information, it may be repetition or noise information.

3.4 Rule Injection Adversarial Imitation Learning

To make the path reasoned by the generator similar to the expert demonstration, we utilize Generative Adversarial Imitation Learning [20] to model this process. Different from DIVINE [4], the discriminators in RIGAPR are composed of path discriminator D_p and logical rule discriminator D_r, and distinguish the generated path and expert demonstration from the semantic and logical rule levels respectively. To get more rewards from the discriminator, the generator will reason a path to fool the discriminator by imitating the expert demonstration.

Path Discriminator. D_p distinguishes the generated path and expert demonstration from the semantic feature of the path. For a path $e_0 \xrightarrow{r_1} e_1 \xrightarrow{r_2} \cdots \xrightarrow{r_n} e_n$, we first encode it to a path real value matrix as follow:

$$p = e_0 \oplus r_1 \oplus e_1 \oplus \cdots \oplus r_n \oplus e_n \tag{3}$$

Where $e, r \in \mathbb{R}^d$ is the embedding vectors and $\oplus$ denotes the concatenation operator. And $p \in \mathbb{R}^{(2n+1) \times d}$ is the path truth matrix obtained by vector concatenation. Note that we consider both entities and relations in path embedding, not just relations in DIVINE. Then, we use a CNN to extract the semantic feature s of the path, and it will be further analyzed by full-connection layers. The goal of the path discriminator D_p is to distinguish the path through the further semantic feature and it can be modeled as:

$$s = ReLU\big(conv(p, \textbf{\textit{w}}) + b_p\big) \tag{4}$$

$$D(p) = \sigma\big(W_{p2}ReLU\big(W_{p1}s\big)\big) \tag{5}$$

Here, $\textbf{\textit{w}} \in \mathbb{R}^{h \times l}$is the convolution kernel and b_p denotes the bias. Then, we use the ReLU activation function in both the convolution layer and the full-connection layer. And we use the sigmoid function σ in the output layer to make the path semantic feature outputted by path discriminator D_p on the interval $(0, 1)$.

Logical Rule Discriminator. D_r further distinguishes the path from the point of the first-order predicate logic rule, so that the generator can imitate the logic rule implied in the expert demonstration. To achieve this, we design a path logical rule extraction method. For path $e_0 \xrightarrow{r_1} e_1 \xrightarrow{r_2} \cdots \xrightarrow{r_n} e_n$, we extract the logical rule r_p contained in it as:

$$r_p \triangleq \forall a_0, \ldots, a_n \in E, (a_h, r, a_t) \in G$$
$$(a_0, r_1, a_1) \wedge (a_1, r_2, a_2) \wedge \cdots \wedge (a_{n-1}, r_n, a_n) \Rightarrow \big(a_0, r_q, a_n\big) \tag{6}$$

$$f = (e_0, r_1, e_1) \wedge (e_1, r_2, e_2) \wedge \cdots \wedge (e_{n-1}, r_n, e_n) \Rightarrow \big(e_0, r_q, e_n\big) \tag{7}$$

Where r_p is the first-order predicate logic rule (FOL) contained in the path and f denotes an instance of the rule which consists of the reasoning path. More specifically, we use f_1 to f_n to represent atomic triple in the rule body, and f_h to denote the rule head consisting of the query triple $(e_0, r_q, ?)$ and the reasoned answer entity e_n.

To identify the rationality of the rule contained in the path, we introduce the t-norm fuzzy logic [6], which defines the truth value of rules as a composition of the truth values of its constituent atomic triples (e.g. $f_1, \ldots, f_n, f_h$). We follow the definition [5] of compositions associated with conjunction ($\wedge$), disjunction ($\vee$), and negation ($\neg$) as:

$$I(f_1 \wedge f_2) = I(f_1) \cdot I(f_2)$$

$$I(f_1 \vee f_2) = I(f_1) + I(f_2) - I(f_1) \cdot I(f_2)$$

$$I(\neg f_1) = 1 - I(f_1)$$

$$I(f_1 \Rightarrow f_2) = I(\neg f_1 \vee f_2) = I(f_1) \cdot I(f_2) - I(f_1) + 1$$

We use the feedforward neural network to calculate the truth value $I(f_i)$ of the atomic triples $(f_i : (h_i, r_i, t_i))$, and use the sigmoid function σ to make the rule truth value on the interval $(0, 1)$. It can be formulated as follows:

$$c_i = \tanh([h_i; r_i; t_i]) \tag{8}$$

$$I(f_i) = \sigma(W_{r2} ReLU(W_{r1} c_i)) \tag{9}$$

Where W_{r1} and W_{r2} are the parameters needed to be trained in the rule discriminator. Then we calculate the truth value $I(f)$ of the rule instance f through t-norm fuzzy logical and the larger the truth value is, the more reasonable the rule is. Specifically, the output (3-hop paths) of the rule discriminator is as follows:

$$D(r) = I(f) = I(f_1) \cdot I(f_2) \cdot I(f_3) \cdot I(f_h) - I(f_1) \cdot I(f_2) \cdot I(f_3) + 1 \tag{10}$$

3.5 Reward Enhancement and Model Training

Reward Enhancement. To make the generator imitates learning the demonstrations in terms of semantics and logical rules, we enhance the rewards it receives as follow:

$$R(s_t) = \big(R_{e,t} + \alpha R_p + (1 - \alpha)R_r\big)R_{e,t} \tag{11}$$

Where $R_{e,t}$ is the terminal reward given by the MDP environment, depending on whether the generator finds the correct answer entity. R_p and R_r represent the semantic and the logical rule rewards respectively, obtained from the path discriminator D_p and the logical rule discriminator D_r. And α is a weight to balance semantics and logical rule rewards.

Note that only when the generator generates a path, in other words, when the number of reasoning step reaches the maximum path length T, the path will be assessed in terms of semantic and logical rule, and the generator will be rewarded. Here only the correct reasoning path (i.e. the path obtained the terminal reward) is rewarded and enhanced. Although this may lead to sparse rewards, our experiment has proved that it does not affect reasoning performance.

Similar to the DIVINE, we introduce random noise $\ddagger \sim U(a, b)$ into the reward to discard the generation paths whose quality is not as good as the noise. It can be formulated as follows. p^N and r^N respectively denote path noise and rule noise embeddings composed of random noise with continuous uniform distribution.

$$R_p = \max\big(D(p^G) - D(p^N), 0\big) \tag{12}$$

$$R_r = \max\big(D(r^G) - D(r^N), 0\big) \tag{13}$$

Model Training. We train the path and logical rule discriminators by minimizing its loss so that they can distinguish generation paths and expert demonstrations from semantic and logical rule levels respectively. For path discriminator D_p, we adopt WGAN-GP [21] algorithm to train and its loss $\mathcal{L}_p$ as follows. For logical rule discriminator D_r, we define a classification loss $\mathcal{L}_r$.

$$\mathcal{L}_p = D\left(p^G\right) - D\left(p^E\right) + \lambda(\|\nabla_{\hat{p}}D(\hat{p})\|_2 - 1)^2 \tag{14}$$

$$\mathcal{L}_r = -\left(logD\left(r^E\right) + \log\left(1 - D\left(r^G\right)\right)\right) \tag{15}$$

Where $\mathcal{L}_p$ consists of original critic loss and gradient penalty. p^G and p^E denote the generated path and the demonstration respectively. λ is the gradient penalty coefficient and $\hat{p}$ is sampled uniformly along straight lines between p^G and p^E. Similarly, r^G and r^E is the instance of FOL contained in the generated path and the demonstration respectively. We first pre-train the discriminator with demonstrations and random paths and then use the generated paths and demonstrations for adversarial training.

For all query triples in the training dataset, we will update the parameter θ in the generator by REINFORCE [22] algorithm once it is rewarded. More specifically, the object of generator $J(\theta)$ is to maximize expected rewards and updates parameter $\theta = \{W_{g1}, W_{g2}\}$ with the stochastic gradient as following:

$$J(\theta) = \mathbb{E}_{(e_s, r_q, e_a)\sim D}\mathbb{E}_{A_1...A_T\sim\pi}[R(s_t|e_s, r)] \tag{16}$$

$$\nabla_\theta J(\theta) = \nabla_\theta \sum_t R(s_t)log\pi(a_t|s_t, A_t) \tag{17}$$

4 Experiments

In this section, we introduce the benchmarks for our experiments and the state-of-the-art baselines we compared, as well as the evaluation and experiment details (Sect. 4.1). Then we quantitatively compare the reasoning performance of our model with other baselines (Sect. 4.2). Ablation studies and parameter sensitivity analysis also show the effectiveness of our model components (Sect. 4.3). Finally, some case analysis illustrated the explainability of our model at the semantic and logical rule levels (Sect. 4.4).

Table 1. The statistics of WN18RR and NELL-995 datasets

Dataset	#Entitie	#Relations	#Facts	#Queries
WN18RR	40,945	11	86,835	3,134
NELL-995	75,492	200	154,213	3,992

4.1 Experiment Setup

Dataset. We adopt WN18RR [23] and NELL-995 [15] datasets in our experiments, which are created from the WN18 and NELL datasets respectively. Same as the MINERVA, we combine all the graphs and removed all test triples (including the triples with inverse relations) from the graph for the query answering task. The details of the datasets are shown in Table 1.

Baseline and Evaluation. We compare our method with the state-of-the-art path reasoning method, including MINERVA [1], MARLPaR [17], DIVINE [4], and the model proposed in [18]. And the detailed introduction of these baselines has been discussed in related work. All methods are evaluated by some widely used metrics, including Hits@k and mean reciprocal rank (MRR).

Implementation Details. The default settings of the model in the experiment are as follows. For embedding representations, we set the embedding dimension to 100 and use TransR [9] to generate embedding vectors. For the generator, we use one-layer LSTM as the path history encoder and set its hidden dimension to 200. And the maximum reasoning hop (i.e. the maximum path length) is set to 3. Meanwhile, the size of the convolution kernel is set to 3×5. Through parameter sensitivity analysis, we set the reward enhancement weight α to 0.5 for WN18RR, and 0.6 for NELL-995. The gradient penalty coefficient λ is set to 5 in both datasets. For the model training, we set the batch size to 256 for WN18RR and 512 for NELL-995. And we adopt Adam optimization [24] for generator training and SGD for discriminator training both with the learning rate 0.001.

Table 2. Comparison of reasoning performance to other baselines on two large KG datasets. The results are reported in percentage and the best baseline results are marked with a star (*).

Dataset	WN18RR				NELL-995			
Metrics (%)	@1	@3	@10	MRR	@1	@3	@10	MRR
MINERVA	41.3	45.6	51.3	44.8	66.3*	77.3*	83.1	72.5
MARLPaR	42.1	47.2	52.0	–	– –	–	–	–
[18] (ComplEx)	43.7	–	54.2*	47.2*	65.5	–	83.6	72.2
[18] (ConvE)	41.8	–	51.7	45.0	65.6	–	84.4*	72.7*
DIVINE	43.8*	48.0*	53.8	46.8	65.0	75.4	81.4	71.1
RIGAPR -PD	43.0	47.6	53.3	46.3	62.9	75.1	81.5	69.9
RIGAPR -RD	43.2	48.0	53.4	46.6	65.0	76.6	82.6	71.6
RIGAPR -RN	43.6	47.7	53.6	46.7	65.7	77.1	82.7	72.1
RIGAPR -RE	43.6	47.9	54.2	46.8	66.7	77.2	83.0	72.7
RIGAPR	**44.6**	**48.8**	**54.3**	**47.7**	**68.4**	**78.1**	83.7	**73.9**

4.2   Results and Discussion

We compared our proposed RIGAPR with state-of-the-art baselines in terms of reasoning performance, and the results are listed in Table 2 which shows that RIGAPR outperforms other baselines in overall performance. More specifically, we can find that path reasoning by imitating expert demonstrations is effective and can solve the problem of existing RL-based methods (e.g. MINERVA, MARLPaR, and [18]) relying on reward function design. Compared with the DIVINE model, we have enhanced the generator's rewards from both semantics and logical rules, thus achieving performance improvements.

Furthermore, we noticed that RIGAPR does not perform well in the Hits@10 of the NELL-995 dataset. The possible reason is that the quality of the expert demonstrations extracted through the shortest path is not enough, which makes it difficult for the generator to learn the reasoning strategy. Therefore, it is necessary to study how to balance the diversity of expert demonstrations and noise information in the future.

4.3   Ablation Study

We conduct ablation experiments by removing the components in RIGAPR to verify its effectiveness and parameter sensitivity analysis to determine some parameters.

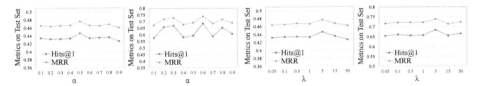

Fig. 2. Parameter sensitivity analysis of α and λ on WN18RR (left) and NEL-995 (right).

Effectiveness of Discriminator. To verify the effectiveness of the discriminator in our model, we removed the path discriminator (RIGAPR-PD) and rule discriminator (RIGAPR-RD) from the complete model RIGAPR respectively. The result of the comparison is shown in Table 2, we can find RIGAPR is better than RIGAPR-PD and RIGAPR-RD in all metrics, which illustrates that the discriminators in our model can distinguish paths at the semantic and logical rule levels and can help the generator to imitate learning expert demonstrations from both levels.

Effectiveness of Random Noise. We introduce random noise when rewarding the generator to discard low-quality generated paths. To verify this, we compare the noise-removed ablation model RIGAPR-RN with RIGAPR. The results show that random noise can improve the model's reasoning performance to a certain extent by improving the quality of the generated path.

Influence of Reward Enhancement. In our model, only when the correct answer entity is found, will the generator be reward enhanced according to the corresponding reasoning path. We compared the ablation model RIGAPR-RE (reward enhancement for all paths) with RIGAPR, and the results show that the reward enhancement method we designed does not reduce the reasoning performance due to sparse rewards.

Influence of Model Weight. We conduct parameter sensitivity analysis to determine the values of the reward enhancement weight α and the gradient penalty coefficient λ, and we set the value of α on the interval $(0, 1)$ and the range of λ is $\{0.05, 0.1, 0.5, 1, 5, 15, 50\}$. As shown in Fig. 2, RIGAPR achieves the best reasoning performance when $\alpha = 0.5$ in WN18RR, and $\alpha = 0.6$ in NELL-995, and the gradient penalty coefficient $\lambda = 5$ in both datasets.

4.4 Case Study

To intuitively understand the reasoning process and explainability of our model, we give some cases in NELL-995 for query answering task. As shown in Table 3, we choose three query relations in NELL-995 and the first two query relations illustrate the explainability of our model at the semantic and rule levels. The latter proves the role of adding self-loop and inverse relations to the KG. For example, it can be formulated as $(Jim\,Murry, works\,for, ?)$ in the first query relation, and the generator finds through stepwise reasoning that both Jim Murry and Michael Hiltzik are writing for the Los Angeles Times newspaper, and then the person Michael Hiltzik works for Los Angeles Times television station. Finally, the generator concludes that Jim Murry also works for the Los Angeles Times television station. The corresponding first-order predicate logic rule can be expressed as: $\forall A, B, C, D \in \mathcal{E}, (A, write\_for\_pub, B) \wedge (B, write\_for\_pub^{-1}, C) \wedge (C, works\_for, D) \Rightarrow (A, works\_for, D)$.

Table 3. Reasoning cases on the NELL-995 dataset and shown at the semantic and rule levels

Query Relation	Reasoning Path & Logic Rule
works_for	$journalist\_jim\_murray \xrightarrow{write\_for\_pub} newspaper\_los\_angeles\_times \xrightarrow{write\_for\_pub^{-1}}$ $writer\_michael\_hiltzik \xrightarrow{works\_for} televisionstation\_los\_angeles\_time$ $(A, write\_for\_pub, B) \wedge (B, write\_for\_pub^{-1}, C) \wedge$ $(C, works\_for, D) \Rightarrow (A, works\_for, D)$
athlete_plays_sport	$craig\_smith \xrightarrow{athlete\_led\_sports\_team} sportsteam\_timberwolves \xrightarrow{team\_plays\_against\_team}$ $sportsteam\_la\_clippers \xrightarrow{team\_plays\_sport} sport\_basketball$ $(A, athlete\_led\_sports\_team, B) \wedge (B, team\_plays\_against\_team, C) \wedge$ $(C, team\_plays\_sport, D) \Rightarrow (A, athlete\_plays\_sport, D)$
organization_headquartered_in_city	$company\_air\_southwest \xrightarrow{has\_office\_in\_city} city\_newquay \xrightarrow{NO\_OP}$ $city\_newquay \xrightarrow{NO\_OP} city\_newquay$ $(A, has\_office\_in\_city, B) \wedge (B, NO\_OP, B) \wedge$ $(B, NO\_OP, B) \Rightarrow (A, org\_headquartered\_in\_city, B)$ $company\_general\_motors \xrightarrow{acquired} automobilemaker\_opel \xrightarrow{acquired^{-1}}$ $company\_general\_motors \xrightarrow{has\_office\_in\_city} city\_lansing$ $(A, acquired, B) \wedge (B, acquired^{-1}, A) \wedge$ $(A, has\_office\_in\_city, C) \Rightarrow (A, org\_headquartered\_in\_city, C)$

5   Conclusion

In this paper, we propose a rule injection-based generative adversarial imitation learning for knowledge graph path reasoning, which solves the issue that the existing GAIL-based model is insufficient in assessing the quality of the path. We first propose a method of expressing the first-order predicate logic rule contained in the path. Then we design the path discriminator and the logical rule discriminator to distinguish the path from the semantic and logical rule levels respectively. And we propose a reward enhancement method to enhance the generator's reward, which can train the generator by imitating learning expert demonstrations at sematic and logical rule levels. Extensively experiments on two benchmarks show that our method outperforms the state-of-the-art baseline. In the future, we will investigate how to extract high-quality and diverse expert demonstrations and avoid introducing noise information.

References

1. Das, R., Dhuliawala, S., Zaheer, M., et al.: Go for a walk and arrive at the answer: reasoning over paths in knowledge bases using reinforcement learning. In: ICLR (2018)
2. Xian, Y., Fu, Z., et al.: Reinforcement knowledge graph reasoning for explainable recommendation. In: SIGIR, pp. 285–294 (2019)
3. Wang, J., Wang, Z., Zhang, D., et al.: Combining knowledge with deep convolutional neural networks for short text classification. In: IJCAI, pp. 2015–2921 (2017)
4. Li, R., Cheng, X.: DIVINE: a generative adversarial imitation learning framework for knowledge graph reasoning. In: EMNLP-IJCNLP, pp. 2642–2651 (2019)
5. Guo, S., Wang, Q., Wang, L., et al.: Jointly embedding knowledge graphs and logical rules. In: EMNLP, pp. 192–202 (2016)
6. Hájek, P.: Metamathematics of Fuzzy Logic. Springer, Dordrecht (2013). https://doi.org/10.1007/978-94-011-5300-3
7. Bordes, A., et al.: Translating embeddings for modeling multi-relational data. In: Advances in Neural Information Processing Systems, pp. 2787–2795 (2013)
8. Wang, Z., Zhang, J., Feng, J., et al.: Knowledge graph embedding by translating on hyperplanes. In: AAAI, pp. 1112–1119 (2014)
9. Lin, Y., et al.: Learning entity and relation embeddings for knowledge graph completion. In: AAAI 2015, pp. 2181–2187 (2015)
10. Nickel, M., Tresp, V., Kriegel, H.P.: A three-way model for collective learning on multi-relational data. In: ICML 2011, pp. 809–816 (2011)
11. Yang, B., Yih, W.T., He, X., Gao, J., Deng, L.: Embedding entities and relations for learning and inference in knowledge bases. arXiv preprint arXiv:1412.6575 (2014)
12. Trouillon, T., Welbl, J., Riedel, S., Gaussier, É., Bouchard, G.: Complex embeddings for simple link prediction. In: ICML, pp. 2071–2080 (2016)
13. Lao, N., Cohen, W.W.: Relational retrieval using a combination of path-constrained random walks. Mach. Learn. 81(1), 53–67 (2010)
14. Gardner, M., Talukdar, P., Krishnamurthy, J., et al.: Incorporating vector space similarity in random walk inference over knowledge bases. In: EMNLP, pp. 397–406 (2014)
15. Xiong, W., Hoang, T., Wang, W.Y.: DeepPath: a reinforcement learning method for knowledge graph reasoning. In: EMNLP, pp. 564–573 (2017)
16. Wang, H., Li, S., Pan, R., Mao, M.: Incorporating graph attention mechanism into knowledge graph reasoning based on deep reinforcement learning. In: EMNLP-IJCNLP, pp. 2623–2631 (2019)

17. Li, Z., Jin, X., Guan, S., et al.: Path reasoning over knowledge graph: a multi-agent and reinforcement learning based method. In: ICDMW, pp. 929–936 (2018)
18. Lin, X.V., Socher, R., Xiong, C.: Multi-hop knowledge graph reasoning with reward shaping. In: EMNLP, pp. 3243–3253 (2018)
19. Qiu, Y., Wang, Y., et al.: Stepwise reasoning for multi-relation question answering over knowledge graph with weak supervision. In: WSDM, pp.474–482 (2020)
20. Ho, J., Ermon, S.: Generative adversarial imitation learning. In: NIPS, pp. 4565–4573 (2016)
21. Gulrajani, I., Ahmed, F., Arjovsky, M., et al.: Improved training of Wasserstein GANs. In: NIPS, pp. 5767–5777 (2017)
22. Williams, R.J.: Simple statistical gradient-following algorithms for connectionist reinforcement learning. Mach. Learn. **8**(3–4), 229–256 (1992)
23. Dettmers, T., Minervini, P., Stenetorp, P., Riedel, S.: Convolutional 2D knowledge graph embeddings. In: AAAI, pp. 1811–1818 (2018)
24. Kingma, D.P., Ba, J.: Adam: a method for stochastic optimization. In: ICLR (2015)

Hierarchical Self Attention Based Autoencoder for Open-Set Human Activity Recognition

M. Tanjid Hasan Tonmoy$^{(\boxtimes)}$, Saif Mahmud, A. K. M. Mahbubur Rahman,
M. Ashraful Amin, and Amin Ahsan Ali

Independent University Bangladesh, Dhaka, Bangladesh
{2015-116-770,2015-116-815}@student.cse.du.ac.bd,
{akmmrahman,aminmdashraful,aminali}@iub.edu.bd

Abstract. Wearable sensor based human activity recognition is a challenging problem due to difficulty in modeling spatial and temporal dependencies of sensor signals. Recognition models in closed-set assumption are forced to yield members of known activity classes as prediction. However, activity recognition models can encounter an unseen activity due to body-worn sensor malfunction or disability of the subject performing the activities. This problem can be addressed through modeling solution according to the assumption of open-set recognition. Hence, the proposed self attention based approach combines data hierarchically from different sensor placements across time to classify closed-set activities and it obtains notable performance improvement over state-of-the-art models on five publicly available datasets. The decoder in this autoencoder architecture incorporates self-attention based feature representations from encoder to detect unseen activity classes in open-set recognition setting. Furthermore, attention maps generated by the hierarchical model demonstrate explainable selection of features in activity recognition. We conduct extensive leave one subject out validation experiments that indicate significantly improved robustness to noise and subject specific variability in body-worn sensor signals. The source code is available at: github.com/saif-mahmud/hierarchical-attention-HAR.

Keywords: Attention mechanism · Human Activity Recognition · Autoencoder · Open-set recognition

1 Introduction

Automated Human Activity Recognition (HAR) has a pivotal role in mobile health, physical activity monitoring, and rehabilitation. Body-worn sensor based HAR can broadly be defined as classification of human physical activity based on signals from multitude of wearable sensors worn at different body locations. Human physical activities include activities of daily living as well as complex

---

M. Tanjid Hasan Tonmoy and Saif Mahmud—Equal contribution.

© Springer Nature Switzerland AG 2021
K. Karlapalem et al. (Eds.): PAKDD 2021, LNAI 12714, pp. 351–363, 2021.
https://doi.org/10.1007/978-3-030-75768-7_28

activities comprised of multiple simpler micro-activities. Increasing usage of smart handheld devices with multi-modal sensors has paved the way to deploy HAR system in applications of elderly activity monitoring, physiotherapy exercise evaluation, and smart home solutions.

HAR techniques rely on spatial information and temporal dynamics of physical activity captured from heterogeneous sensors placed at different human body locations. Activities involve different dominant body parts and thus hierarchical fusion of sensor signals from different sensor placements is able to capture salient information required for the task. Most human activities can be viewed as a session comprising a number of short time windows containing low level actions. Hierarchical fusion of temporal information is able to take this phenomenon into account. Further, HAR systems in real world scenarios are likely to encounter samples different from training classes. An optimal framework should be able to distinguish them from known classes with explainable visualizations.

Though initial works relied on domain knowledge and heuristic based statistical feature representation [19] for activity recognition, recent end-to-end deep learning models utilize convolutional [8] and recurrent [16] architectures for learning representations. Attention mechanism, described as weighted average of feature representations, is adopted to model HAR task like other sequence modeling problem such as NLP in recent works [12,13,15,29,31]. However, such approaches do not utilize hierarchical modelling of spatio-temporal information. Moreover, these approaches follow conventional training method under closed set assumption where unknown samples are forced to be recognised as one of the prior known classes.

Considering the aforementioned requirement towards hierarchical fusion of spatio-temporal features, the methodological approach taken in this paper incorporates hierarchical aggregation of sensor signals from different placements across time to construct representation for a specific window. Feature representations from different windows within the same session are aggregated to yield representation for that session. In the case of predicting label for specific window, we utilize session information guided window feature representation instead. The proposed approach models HAR under open set assumption where test samples are identified as seen or unseen and labeled as one of the known classes from training set simultaneously. Such capabilities are highly desirable in the scenario of a subject performing an activity in a completely unexpected way e.g. doing rehabilitation exercise incorrectly due to physical limitations or malfunctioning of sensor devices. In this regard, we have designed autoencoder architecture along with hierarchical encoder to model the distribution of the known activity classes where unseen activities are supposed to yield higher reconstruction loss. Explainable feature attention maps are obtained from hierarchical self-attention layers to demonstrate dominant sensor placement and temporal importance within session to classify specific physical activity.

We conduct extensive experiments on five publicly available benchmark activity recognition datasets: PAMAP2, Opportunity, USC-HAD, Daphnet and Skoda. Our model outperforms prior methods in several datasets. Furthermore, we evaluate the robustness of the model to noise and subject-specific variability through leave-one-subject-out validation experiments. Moreover, we evaluate open-set recognition performance and generate feature attention maps to demonstrate the activity distinguishing characteristics in the learned representation. In brief, the key contributions of our work are listed below:

1. Proposed hierarchical self attention encoder models spatial and temporal information of raw sensor signals in learned representations which are used for closed-set classification as well as detection of unseen activity class with decoder part of the autoencoder network in open-set problem definition
2. Interpretable visualization of feature attention maps indicate fusion of causal and coherent features for activity recognition
3. Our extensive experiments achieve superior performance in several benchmark datasets and demonstrate robustness to subject-specific variability in sensor readings

2   Related Work

Wearable Sensor Based HAR. The earlier research works on HAR mostly relied on hand-crafted statistical or distribution-based [11,20] features that have been designed based on domain expertise [9]. However, recent works for wearable-based HAR have mostly focused on end to end deep learning systems for modeling effective feature representation. In that regard, various forms of convolutional, recurrent and hybrid architectures such as [7,8,16] were proposed and demonstrated varying levels of success in recognition of the activities under consideration. In recent years, the incorporation of attention mechanism with deep learning-based architectures [12,15,29,31] have demonstrated significant performance improvement. However, most of these works [4,18] do not rely on hierarchical modeling of spatio-temporal information from wearable sensors.

Self Attention Architecture. Recently, self-attention [27] based models have emerged as a popular alternative to recurrent networks for various NLP tasks and has also been proposed for HAR [13]. Using self-attention in a hierarchical manner has been proposed for various tasks such as classifying text documents [6], generating recommendations [10] etc. in order to break up the task into relevant hierarchical parts. However, no such work exists for wearable sensor data to the best of our knowledge though such hierarchy allows for intuitive representation of complex human activities.

Interpretability and Open-Set Recognition. The data from wearable sensor devices are usually high dimensional involving different body placements over some time duration. Furthermore, most of the deep-learning based models used for classification of such data offer little to zero interpretability towards the predicted outcome. Some progress has been made in this regard for video based action recognition task [14,21]. Although attention-based models for wearable HAR offer more interpretability using the attention-scores, it is still scarce in

HAR community. With regards to HAR systems, it is often useful to be able to identify previously unseen activities. Class conditioned [17] or variational [2,26] autoencoder for unseen sample detection has been proposed for image data. Although unseen activity recognition for skeleton data [25] & smart-home environment [1,5] have been proposed, autoencoder architectures for unseen sample detection in wearable based HAR are very scarce.

3   Proposed Method

3.1   Task Definition

We assume that $S = \{S_1, S_2, S_3, ...\}$ is the set of sensors placed at different locations on the body of human subjects. Generally, an Inertial Measurement Unit (IMU) or smart-device contains sensors and records data at a sampling rate of f Hz from multiple axes (e.g. tri-axial accelerometer or gyroscope yields signal along x, y and z-axis). Therefore, there will be record of $m = |S| * \sum_i (a_i)$ signals at any particular time-step where $a_i =$ number of axes at S_i. In a dataset containing sensor signal recording of n time-steps, the readings along time is represented as a multidimensional time-series X as in (1) and reading at particular time-step x_k can be represented as in (2). The human activity recognition problem can be defined as the task to detect physical activity class labels given the multidimensional time-series of sensor signals X of particular duration.

$$X = [x_1, x_2, ..., x_k, ..., x_n] \tag{1}$$

$$x_k = [x_{k1}, x_{k2}, ..., x_{kj}, ..., x_{km}]^T \tag{2}$$

We propose to represent sensor signals hierarchically as an activity session composed of windows representing short segments within the sequence. On the other hand, a window is composed of a fixed number of data-points representing the sensor signal at the corresponding time-stamps. We use the proposed hierarchical encoder to learn representation for a session which is used for both classification and open-set detection. The different components of the model are described in the following subsections.

3.2   Hierarchical Self Attention Encoder

The proposed model incorporates two distinct types of hierarchy - temporal and body location-based. These hierarchies are implemented by the Hierarchical Window Encoder (HWE) and Session Encoder (SE), respectively. Self attention is the core element in both of the aforementioned components and is used in two ways within the components. We refer to them as Modular Self Attention and Aggregator Self Attention, respectively.

Modular Self Attention: Modular self attention consists of N identical blocks of multi-headed self attention and position-wise feed forward layers. For each time-step, three linear transformations referred to as key (K), query (Q), and value (V) are learned. Attention score is obtained by applying softmax function on the scaled dot products of queries and keys and is used to get a weighted version of the values. This operation is performed in matrix form as defined in (3).

$$f_{sa}(Q, K, V) = \text{softmax}(\frac{QK^T}{\sqrt{d_k}})V \tag{3}$$

Here, $Q = XW_Q, K = XW_K, V = XW_V$ where W_Q, W_K, W_V are weight matrices; X is the input and $K \in \mathbb{R}^{t \times d_k}, Q \in \mathbb{R}^{t \times d_k}$ and $V \in \mathbb{R}^{t \times d_v}$. Furthermore, multi-head self attention is defined in (4) where each attention head uses different W_Q, W_K, W_V and the output from different attention heads are combined according to (4).

$$\mathbf{f}_{mhsa}(X) = concat(h^{(1)}, ..., h^{(n)}) \cdot \mathbf{W}_o \tag{4}$$

where, $h^{(j)} = f_{sa}(W_Q^{(j)} X, W_K^{(j)} X, W_V^{(j)} X)$. Position-wise feed forward refers to identical fully connected feed forward network composed of two feed forward layers with ReLU activation in between applied independently to each time-step. In addition, layer normalization is used after self-attention and position-wise feed forward layers and the aforementioned layers contain residual connections.

Aggregator Self Attention: In order to obtain an aggregate representation from all the time-steps in the input sequence, we use Aggregator Self Attention. The primary difference between Modular and Aggregator blocks is in the learned linear representations used in (5). The construction of query and value are the same as the former. The key matrix K_a in (5) is initialized randomly and learned during optimization with the rest of the parameters.

$$f_{agr}(Q_a, V_a) = \text{softmax}(\frac{Q_a K_a^T}{\sqrt{d_{ka}}})V_a \tag{5}$$

Where, $Q_a = XW_{aQ}, V = XW_{aV}; W_{aQ}, W_{aV}$ are weight matrices; X is the input to the layer and $Q_a \in \mathbb{R}^{t \times d_{ka}}, V_a \in \mathbb{R}^{t \times d_{va}}, K_a \in \mathbb{R}^{1 \times d_{ka}}$. In contrast to Modular Self Attention, the position-wise feed forward layer is applied both before and after the self attention operation. Moreover, we use single headed attention in this block in order to simplify the use of the attention scores for interpretability.

Hierarchical Window Encoder (HWE) consists of m modular self attention blocks and an aggregator self attention block where m is the number of sensor placements. First, the values from all sensor modalities placed in a body location are combined using 1-D convolution over single time-step to create a d_{model} sized vector. The position information is incorporated by adding positional encoding based on sine and cosine function. Afterwards, the sequence from each sensor placement goes through modular self attention block. Finally, the transformed time-steps are concatenated along the temporal axis and aggregator self attention is used to obtain a representation for the window as defined in (6) and (7).

$$Z_{window} = concat(f_{mhsa}(X_w^{(i)}), ... , f_{mhsa}(X_w^{(m)})) \tag{6}$$

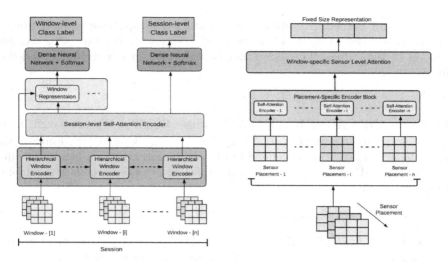

(a) Hierarchical Self Attention Encoder   (b) Hierarchical Window Encoder

Fig. 1. Overview of the (a) **Hierarchical Self Attention Encoder**, consisting of Hierarchical window encoders and session encoder, is used to obtain representation for classification and open set detection (b) **Hierarchical Window Encoder**, where sensor signals from different body locations within short time span are separately transformed and fused later using self attention

$$\tilde{Z}_{window} = f_{agr}(Z_{window}) \tag{7}$$

Session Encoder (SE) comprises n identical HWEs where n is the number of windows within the session. All of the HWE within the session have shared parameters. The input to the session encoder is the output from n temporally ordered window encoders where n is the number of windows within the session. Similar to the HWE, the input goes through Modular Self Attention and Aggregator Self Attention as defined in (8) and (9) to obtain a representation for the session.

$$Z_{session} = f_{mhsa}(X_s) \tag{8}$$

$$\tilde{Z}_{session} = f_{agr}(Z_{session}) \tag{9}$$

Window and Session Classification: For session-level classification, the output from SE is passed to dense and softmax layers to obtain the class label. However, for widow-level classification we concatenate the output from SE with each window representation and pass that to dense and softmax layers. Therefore, we utilize the hierarchical structure to augment the window representation with session information to guide the window-level classification.

3.3   Open Set Human Activity Recognition

Autoencoder, constructed upon the proposed hierarchical self-attention encoder, models the relationship between random variable z representing low-dimensional latent space and random variable x denoting learned representation vector to be reconstructed. We have designed the decoder as multi-layer feed-forward neural network estimating the approximation of distribution $p_\theta(x|z)$ where θ is the learned decoder parameters. On the other hand, the encoder is trained to model the posterior distribution $q_\phi(z|x)$ where ϕ indicates encoder network parameters.

As illustrated in Fig. 2, the objective of proposed autoencoder is to approximate the intractable true posterior $p_\theta(z|x)$ with $q_\phi(z|x)$. The approximation depends on the network parameters and they are tuned based on reconstruction loss and Kullback–Leibler (KL) divergence $D_{KL}(q_\phi(z|x)||p_\theta(z|x))$. As the KL divergence cannot be computed directly from feature representation x and latent space representation z, the loss is minimized through maximizing summation of Evidence Lower Bound (ELBO). Therefore, the loss of autoencoder is computed as, $\mathcal{L}_{AE} = -\sum_i ELBO_i$, where $ELBO_i$ is defined as below in (10):

$$ELBO_i = \mathbb{E}_{q_\phi(z|x_i)}[log(p_\theta(z|x_i))] - D_{KL}(q_\phi(z|x_i)||p_\theta(z)) \tag{10}$$

Here, $ELBO_i$ is the Evidence Lower Bound on the marginal likelihood of the i-th learned representation, x_i. $p_\theta(z)$ indicates prior probability and modeled as unit Gaussian. The expected value defined in the first term indicates reconstruction loss of learned representations. It is assumed according to the characteristics of autoencoder that known activity classes will demonstrate lower reconstruction loss whereas unseen or novel ones should yield higher. The novel activities are detected based on reconstruction loss threshold which is tuned as hyperparameter. The threshold is set from the range $\mu(\mathcal{L}_{known}) - \alpha \cdot \sigma(\mathcal{L}_{known})$ where $\mathcal{L}_{known}$ is the reconstruction loss of autoencoder on training data containing known activity classes and $\alpha \in [0.0, 0.50]$.

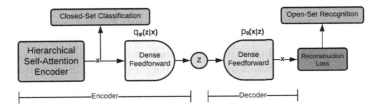

Fig. 2. Overview of the autoencoder architecture where representations from hierarchical self attention encoder is utilized for closed-set classification and reconstructed with decoder for open-set recognition

4   Experiments

Datasets: We use five publicly available benchmark datasets - PAMAP2 [22], Opportunity [23], USC-HAD [30], Daphnet [3] and Skoda [24] for our experiments. A summary of the datasets used is presented in Table 1.

Table 1. Summary of the Datasets used in experiments. For Skoda, we use a 10% split for test and validation since it contains data from a single subject. For the *Sensor Used* column, A = Accelerometer, G = Gyroscope, M = Magnetometer

Dataset	Sampling rate	No. of activity	No. of subject	Validation subject ID	Test subject ID	Sensor used
PAMAP2	100 Hz	12	9	105	106	A, G
Opportunity	30 Hz	18	4	1 (run 2)	2, 3 (run 4, 5)	A, G, M
USC-HAD	100 Hz	12	14	11, 12	13, 14	A, G
Daphnet	64 Hz	2	10	9	2	A
Skoda	98 Hz	11	1	1 (10%)	1 (10%)	A

Construction of Activity Window and Session: Activity sessions are constructed using overlapping sliding-window across the temporal axis. Each activity session comprises a number of non-overlapping short activity segments which we refer to as windows.

Open-Set Experiment: We randomly hold out the data for a fraction of the classes (22% and 27% of the classes in case of Oppotunity & Skoda, 33.33% for rest) as part of the open-set and include the benchmark test data as defined in Table 1 to construct the test set for evaluation. We train the model with the remaining data and report the cross validation results.

Training and Hyperparameters: We implement the proposed model in Tensorflow and train on eight Tesla K80 GPUs. We use Adam optimizer with learning rate set to 10^{-3} with momentum $\beta_1 = 0.9$ and $\beta_2 = 0.999$ and weight decay $\epsilon = 10^{-7}$. The number of identical blocks and head for multi-headed self-attention is set to $N = 2$ and $n = 4$ respectively for our experimental setup. The dropout applied to placement specific encoder block and the size of the representation vector learned from each session is configured to 0.2 and $d_{model} = 64$ respectively.

5   Results and Discussion

Evaluation Metric: For the evaluation of activity recognition performance, we use the macro average F1-score defined in (11) as metric where $|C|$ and

$i = 1, ..., C$ in (11) indicate number of classes and the set of classes respectively.

$$\text{Macro F1} = \frac{1}{|C|} * \sum_{i=1}^{C} \frac{2 * Precision_i * Recall_i}{Precision_i + Recall_i} \tag{11}$$

Baselines and Performance Comparison. We compare the proposed method with a number of baselines which includes most of the prominent feature-based deep learning methods for HAR as well as the recent state-of-the-art models. In particular, we compare our approach with recurrent, convolutional, hybrid and attention-based models. Recurrent network based baselines include LSTM and b-LSTM [8]. For convolutional baselines we compare with simple CNN as well as convolutional autoencoder. Hybrid baselines include DeepConvLSTM (4 CNN and 2 LSTM layers). Attention based baselines include attention augmented DeepConvLSTM as well as SADeepSense [28] (CNN, GRU and sensor & temporal self-attention modules) and AttnSense [12] (attention based modality fusion subnet and GRU subnet). We also compare with self-attention based transformer classifier [13] which does not use any hierarchical modelling.

Table 2. Performance comparison of the proposed method with baselines in terms of window-wise results on benchmark test set

Methods	PAMAP2	Opportunity	USC-HAD	Daphnet	Skoda
LSTM (2014)	0.75	0.63	0.38	0.68	0.89
CNN (2015)	0.82	0.59	0.41	0.59	0.85
b-LSTM [8] (2016)	0.84	0.68	0.39	0.74	0.91
DeepConvLSTM [16] (2016)	0.75	0.67	0.38	0.84	0.91
Conv AE [9] (2017)	0.80	**0.72**	0.46	0.73	0.79
DeepConvLSTM + Attn [15] (2018)	0.88	0.71	0.51	0.76	0.91
SADeepSense [28] (2019)	0.66	0.66	0.49	0.80	0.90
AttnSense [12] (2019)	0.89	0.66	0.49	0.80	0.93
Transformer Encoder [13] (2020)	0.96	0.67	0.55	0.82	0.93
Proposed HSA autoencoder	**0.99**	0.68	**0.55**	**0.85**	**0.95**

Performance on Benchmark Test Set: Table 2 shows that the proposed model outperforms the baseline methods for all of the datasets except Opportunity in terms of window-based results. Specifically, the proposed method outperforms the other methods for PAMAP2, Skoda and Daphnet dataset. On USC-HAD dataset, the performance of the proposed model is on par with the transformer encoder which can be explained by the fact that the dataset contains sensor readings from only one body location (waist) thus not being able to take advantage of sensor location hierarchy. With regards to the Opportunity dataset, some of the mid-level gestures are very short (e.g. less than one second) for which the hierarchical model does not improve much on the existing results. However, the advantage of the proposed hierarchy becomes apparent when we consider the performance on longer and more complex activities. In particular, for the recognition of 5 high level complex activities in the Opportunity dataset, we observe better performance compared to the others (proposed model obtains macro-F1 of 0.91 compared with 0.71, 0.73, 0.791, 0.838 for CNN, LSTM, DeepConvLSTM and AROMA [18] respectively). Therefore, the proposed hierarchical method not only produces better performance in case of longer complex activities (session-wise) but also improves the recognition of shorter activities (window-wise).

Table 3. Performance evaluation in LOSO experiment

Dataset	Proposed model	Transformer encoder	DeepConvLSTM	Conv AE
PAMAP2	**0.94**	0.92	0.61	0.48
Opportunity	0.43	0.42	**0.44**	0.42
USC-HAD	**0.68**	0.60	0.59	0.63
Daphnet	**0.72**	0.71	0.69	0.67

Performance on Leave-One-Subject-Out (LOSO) Experiment: In order to demonstrate the proposed hierarchical model's robustness to subject specific variability in sensor reading, we perform leave-one-subject-out (LOSO) validation experiments. In this regard, we hold data of one subject out for evaluation. Table 3 presents macro F1 score of LOSO experiments on four datasets (Skoda contains single subject) used for experiments. As can be seen from the table, the model demonstrates better performance on LOSO experiments compared to benchmark test data while other models suffer from subject specific variable sensor reading patterns for the same activity.

Attention Maps for Interpretability: We can obtain temporal and sensor-placement specific attention maps based on the attention scores obtained from SE and HWE respectively. The attention maps are useful for understanding the predictions made by the model. With regards to temporal attention maps, a snapshot of which time frames were of more importance for the prediction is useful for understanding both the model output and activity itself in case of unseen activities. Moreover, such attention maps demonstrate good correspondence with mid-level or micro activities that comprise the action. One such example is illustrated in Fig. 3 for a complex activity from the Opportunity dataset using the annotated mid level actions and locomotion. It is evident that more emphasis is given on the relevant actions for recognition of the activity. Furthermore, the sensor-placement based attention maps provide a finer granularity of information regarding which placements played more prominent roles at different times which is in line with the intuitive understanding that distinct micro activities may be dominated by different body parts.

Performance on Open set HAR: Furthermore, the proposed model also produces noteworthy results in the case of open set recognition as shown in Table 4.

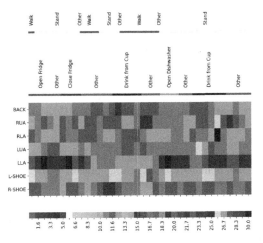

Fig. 3. Attention map for activity 'Cleanup' from Opportunity dataset comprising locomotion and mid-level gestures (top two rows - plotted using ground truth annotation),
bottom x-axis shows temporal attention weight and y-axis indicates body locations of
sensors [(L = Left, R = Right), (L = Lower, U = Upper) & A = Arm] where darker
color indicates higher attention score

Table 4. Open set activity detection performance

Dataset	Total number of classes	Number of novel classes	Accuracy	Macro F1 score
PAMAP2	12	4	0.85	0.69
Opportunity	18	4	0.75	0.58
USC HAD	12	3	0.61	0.52
Skoda	11	3	0.55	0.44
Daphnet	3	1	0.42	0.39

The proposed auto-encoder obtains good performance in terms of accuracy and
macro F1-score on PAMAP2, Opportunity and USC HAD dataset indicating
the capability to distinguish between the activities belonging to the known &
unknown classes. With regards to Daphnet dataset, the scores are lower compared to rest since the unknown class includes transition activities in between
the two known classes resulting in similar distribution for known and unknown
activities.

6   Conclusion

The aim of this work was to design self-attention based model with hierarchical fusion of spatial and temporal features. Our extensive experiments confirmed
that hierarchical aggregation leads to better modelling of spatio-temporal dependency in multimodal time-series sensor signal. These findings have significant

implications for the understanding of how to construct feature representation that leverages better separability for classification and unseen class detection. The findings reported here lays the groundwork of future research into natural language description generation from multimodal sensor signals. However, publicly available benchmark HAR dataset is limited by the lack of complex activity annotations. Notwithstanding these limitations, the hierarchical self-attention model demonstrates interpretable activity recognition as well as robust feature representation.

Acknowledgment. This project is supported by ICT Division, Government of Bangladesh, and Independent University, Bangladesh (IUB).

References

1. Al Machot, F., Elkobaisi, M.R., Kyamakya, K.: Zero-shot human activity recognition using non-visual sensors. Sensors **20**(3), 825 (2020)
2. An, J., Cho, S.: Variational autoencoder based anomaly detection using reconstruction probability. Special Lecture on IE, vol. 2, no. 1, pp. 1–18 (2015)
3. Bachlin, M., et al.: Wearable assistant for Parkinson's disease patients with the freezing of gait symptom. IEEE Trans. Inf Technol. Biomed. **14**(2), 436–443 (2010)
4. Cheng, W., Erfani, S., Zhang, R., Kotagiri, R.: Predicting complex activities from ongoing multivariate time series. In: IJCAI 2018, pp. 3322–3328 (2018)
5. Du, L., Tan, Y.: A novel human activity recognition and prediction in smart home based on interaction. Sensors **19**(20), 4474 (2019)
6. Gao, S., et al.: Classifying cancer pathology reports with hierarchical self-attention networks. Artif. Intell. Med. **101**, 101726 (2019)
7. Guan, Y., Plötz, T.: Ensembles of deep lstm learners for activity recognition using wearables. ACM IMWUT (2017)
8. Hammerla, N.Y., Halloran, S., Plötz, T.: Deep, convolutional, and recurrent models for human activity recognition using wearables. arXiv preprint arXiv:1604.08880 (2016)
9. Haresamudram, H., Anderson, D.V., Plötz, T.: On the role of features in human activity recognition. In: Proceedings of the 23rd International Symposium on Wearable Computers, ISWC 2019. ACM (2019)
10. He, Y., Wang, J., Niu, W., Caverlee, J.: A hierarchical self-attentive model for recommending user-generated item lists. In: Proceedings of the 28th ACM International Conference on Information and Knowledge Management, pp. 1481–1490 (2019)
11. Kwon, H., Abowd, G.D., Plötz, T.: Adding structural characteristics to distribution-based accelerometer representations for activity recognition using wearables. In: Proceedings of the 2018 ACM ISWC. ACM (2018)
12. Ma, H., Li, W., Zhang, X., Gao, S., Lu, S.: AttnSense: multi-level attention mechanism for multimodal human activity recognition. In: Proceedings of the IJCAI 2019, pp. 3109–3115 (2019)
13. Mahmud, S., et al.: Human activity recognition from wearable sensor data using self-attention (ECAI 2020) (2020)
14. Meng, L., et al.: Interpretable spatio-temporal attention for video action recognition. In: Proceedings of the IEEE International Conference on Computer Vision Workshops (2019)

15. Murahari, V.S., Plötz, T.: On attention models for human activity recognition. In: 2018 ACM International Symposium on Wearable Computers, pp. 100–103. ACM (2018)
16. Ordíñez, F., Roggen, D.: Deep convolutional and LSTM recurrent neural networks for multimodal wearable activity recognition. Sensors **16**, 115 (2016)
17. Oza, P., Patel, V.M.: C2ae: class conditioned auto-encoder for open-set recognition. In: Proceedings of the IEEE CVPR, pp. 2307–2316 (2019)
18. Peng, L., Chen, L., Ye, Z., Zhang, Y.: Aroma: a deep multi-task learning based simple and complex human activity recognition method using wearable sensors. In: Proceedings of ACM on Interactive, Mobile, Wearable Ubiquitous Technologies (2018)
19. Plötz, T., Hammerla, N.Y., Olivier, P.L.: Feature learning for activity recognition in ubiquitous computing. In: Twenty-Second International Joint Conference on Artificial Intelligence (2011)
20. Qian, H., Pan, S.J., Da, B., Miao, C.: A novel distribution-embedded neural network for sensor-based activity recognition. In: Proceedings of the IJCAI 2019 (2019)
21. Ramakrishnan, K., et al.: Identifying interpretable action concepts in deep networks. In: CVPR Workshops, pp. 12–15 (2019)
22. Reiss, A., Stricker, D.: Introducing a new benchmarked dataset for activity monitoring. In: Proceedings of the 16th Annual International Symposium on Wearable Computers (ISWC) (2012)
23. Roggen, D., et al.: Collecting complex activity datasets in highly rich networked sensor environments. In: Seventh International Conference on Networked Sensing Systems (INSS) (2010)
24. Stiefmeier, T., Roggen, D., Troster, G., Ogris, G., Lukowicz, P.: Wearable activity tracking in car manufacturing. IEEE Pervas. Comput. **7**, 42–50 (2008)
25. Tian, G., Yin, J., Han, X., Yu, J.: A novel human activity recognition method using joint points information. Jiqiren/Robot **36**, 285–292 (2014)
26. Vasilev, A., et al.: q-space novelty detection with variational autoencoders. In: Bonet-Carne, E., Hutter, J., Palombo, M., Pizzolato, M., Sepehrband, F., Zhang, F. (eds.) Computational Diffusion MRI. MV, pp. 113–124. Springer, Cham (2020). https://doi.org/10.1007/978-3-030-52893-5_10
27. Vaswani, A., et al.: Attention is all you need. In: Advances in Neural Information Processing Systems, vol. 30 (2017)
28. Yao, S., et a.: Sadeepsense: self-attention deep learning framework for heterogeneous on-device sensors in internet of things applications. In: IEEE INFOCOM 2019-IEEE Conference on Computer Communications, pp. 1243–1251. IEEE (2019)
29. Zeng, M., et al.: Understanding and improving recurrent networks for human activity recognition by continuous attention. In: ACM ISWC 2018. ACM (2018)
30. Zhang, M., Sawchuk, A.: USC-had: a daily activity dataset for ubiquitous activity recognition using wearable sensors. In: 2012 ACM Conference on Ubiquitous Computing, pp. 1036–1043 (2012)
31. Zheng, Z., Shi, L., Wang, C., Sun, L., Pan, G.: LSTM with uniqueness attention for human activity recognition. In: Tetko, I.V., Kůrková, V., Karpov, P., Theis, F. (eds.) ICANN 2019. LNCS, vol. 11729, pp. 498–509. Springer, Cham (2019). https://doi.org/10.1007/978-3-030-30508-6_40

Reinforced Natural Language Inference for Distantly Supervised Relation Classification

Bo Xu[1], Xiangsan Zhao[1], Chaofeng Sha[2,3(✉)], Minjun Zhang[1], and Hui Song[1]

[1] School of Computer Science and Technology, Donghua University, Shanghai, China
{xubo,songhui}@dhu.edu.cn,
{2191948,2191725}@mail.dhu.edu.cn
[2] School of Computer Science, Fudan University, Shanghai, China
cfsha@fudan.edu.cn
[3] Shanghai Key Laboratory of Intelligence Processing, Shanghai, China

Abstract. Distant supervision (DS) has the advantage of automatically annotate large amounts of data and has been widely used for relation classification. Despite its efficiency, it often suffers from the label noise problem, which would impair the performance of relation classification. Recently, there are two ways to solve the label noise problem. The first way is to use multi-instance learning to consider the noises of instances, but they do not perform well for sentence-level prediction. The second way is to use reinforcement learning or adversarial learning to directly find noisy label instances but with high computational overhead and poor performance. In this paper, we propose to use the natural language inference (NLI) model to evaluate the quality of the instances directly, and select the high-quality instances as refined training data for sentence-level relation classification. Due to the lack of high-quality supervised data, we use reinforcement learning to train the NLI model. Experimental results on two human re-annotated NYT datasets show the effectiveness and efficiency of our method at the sentence-level relation classification. The source code of this paper can be found in https://github.com/xubodhu/RLRC.

1 Introduction

The task of relation classification is to predict the semantic relations between two entities from a given text. Knowing the relations of entities is essential for many downstream applications, such as knowledge graph completion and question answering.

Typically, each relation classification task requires its own annotated data for training the model, which is expensive and time-consuming. To address this

This paper was supported by the National Natural Science Foundation of China (61906035), Shanghai Sailing Program (19YF1402300) and National Natural Science Foundation of China (61972081).

K. Karlapalem et al. (Eds.): PAKDD 2021, LNAI 12714, pp. 364–376, 2021.
https://doi.org/10.1007/978-3-030-75768-7_29

problem, distant supervision has been proposed to automatically annotate a large number of unlabeled instances from knowledge bases. It assumes that *all sentences mentioning an entity pair from the knowledge base express that relation* [9].

For example, the entity pair (`Steve Jobs, Apple`) has a relation of `Founder` in a given knowledge base. Then a sentence "`Steve Jobs was the co-founder and CEO of Apple and formerly Pixar`" mentions the entity pair (`Steve Jobs, Apple`) will be automatically labeled as relation `Founder` by distant supervision.

Despite its efficiency, distantly supervised relation classification often suffers from the label noise problem. For example, the sentence "`Steve Jobs moved into a house near the Apple office in Cupertino`" also mentions the entity pair (`Steve Jobs, Apple`) but does not express the relation of `Founder`. The label noise problem would impair the performance of relation classification.

Recently, there are two ways to solve the label noise problem. The first way is to use multi-instance learning to consider the noises of instances [2,3,8,10,14]. They divide the training data into many bags, and each bag contains many sentences mentioning the same entity pair. Then the training and test process proceeds at the bag level. However, bag-level relation classification does not perform well for sentence-level prediction [1,18]. The second way is to find the noisy labeling instances directly. They mainly use reinforcement learning [1,7,16] or adversarial learning [6,12] to select high-quality instances or remove noisy instances. However, their computational overhead is high, and the performance of these methods needs to be improved.

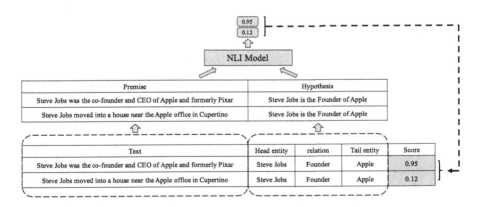

Fig. 1. An example of using the NLI model to evaluate the quality of two instances.

To address these issues, we propose to use the natural language inference (NLI) model [17] to evaluate the quality of the instances directly, and select the high-quality instances as refined training data for sentence-level relation classification. Natural language inference studies whether a hypothesis can be inferred from a premise, where both are a text sequence [17]. In this paper, given an instance of the training data, we transform the triple of the instance into a *hypothesis* and use the text of the instance as a *premise*. Then we use the NLI model to evaluate whether the triple can be inferred from the sentence. For example, in Fig. 1, there are two sentences mentioning the same entity pairs (Steve Jobs, Apple). After transforming the entity pair and their relation Founder into the hypothesis sentence, we use the NLI model to determine whether each sentence can infer the hypothesis, and thus determine whether each sentence express the relation of the entity pair.

Due to the lack of high-quality supervised data, we use reinforcement learning to train the NLI model. Specifically, we first use the NLI model to evaluate the quality of the instances. Then we feed the selected high-quality instances to the relation classification model and use the performance of the relation classification model as a reward to guide the parameter update of the NLI model. Our main contributions can be summarized as follows:

- Firstly, to the best of our knowledge, we are the first to use the natural language inference model to improve distantly supervised relation classification. We evaluate the quality of the instances directly and select the high-quality instances as refined training data for sentence-level relation classification.
- Secondly, due to the lack of high-quality supervised data, we use reinforcement learning to train the NLI model.
- Finally, experimental results on two human re-annotated NYT datasets show our method's effectiveness and efficiency at the sentence-level relation classification.

2   Overview

2.1   Problem Formulation

Our goal is to select some high-quality instances from DS training data to improve sentence-level relation classification, which consists of two subtasks, namely *instance selection* and *sentence-level relation classification*. Sentence-level relation classification has been widely studied in recent years [3,15]. In this paper, we only focus on the instance selection subtask and adopt a CNN architecture proposed by [1] for relation classification in all subsequent experiments.

Let D be the distantly supervised training data, which contains N instances $\{(h_i, t_i, r_i, text_i)\}_{i=1}^{N}$. For each instance D_i, $text_i$ is a text mentioning an entity pair (h_i, t_i). h_i is the head entity, t_i is the tail entity, r_i is one of the relations between them in a given knowledge base. The instance selection subtask's goal is to select a subset of DS training data $D' \subset D$ as refined training data for sentence-level relation classification.

2.2   Framework

The framework of our method is shown in Fig. 2. The general process is as follows: we first select a batch of instances from the DS training data randomly

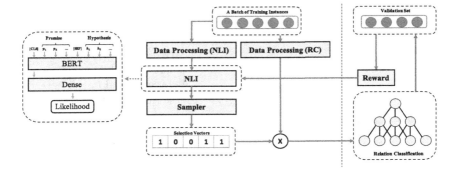

Fig. 2. The framework of our method.

and evaluate the quality of these instances by the NLI model. After that, we use a weighted Bernoulli sampler to select instances based on their NLI scores and obtain a selection vector for the instances, where 1 means the instance is selected, and 0 means not. Due to the lack of high-quality supervised data, we use reinforcement learning to train the NLI model. Specifically, we use the selected instances to train a sentence-level relation classifier (RC) and get a reward according to the performance of the relation classifier on the validation set. Finally, the NLI model updates parameters based on the feedback of the reward.

3   Method

In this section, we will introduce each part of our method in detail. We first introduce how to preprocess the DS training data for the natural language inference model and the relation classification model, then introduce these two models respectively, and finally explain how to update the NLI model's parameters.

3.1   Data Processing

Our framework includes two main models, namely the NLI model and the relation classification model. To train these two models, we need to preprocess the DS training data.

For the NLI model, the input is a premise and a hypothesis. The output is the likelihood between 0 and 1, indicating the probability that the hypothesis can be inferred from the premise. We convert the DS training data into the input of the NLI model as follows: for each instance $D_i = (h_i, t_i, r_i, text_i)$, we regard the text $text_i$ as the premise P_i of the NLI model and use human crafted templates to convert the triples (h_i, t_i, r_i) into the hypothesis H_i of the NLI model. We design a conversion template for each relation. Table 1 shows some examples of the templates.

Table 1. Some examples of human crafted templates to convert the triples (h_i, t_i, r_i) into the hypothesis of the NLI model.

Relations $r_i(h_i, t_i)$	Hypothesis
NA	There's no relationship between t_i and h_i
/location/neighborhood/neighborhood_of	t_i is the neighborhood of h_i
/location/administrative_division/country	h_i is the administrative division of t_i
/people/person/nationality	h_i's nationality is t_i
/business/person/company	h_i works in t_i
/people/person/place_lived	h_i lives in t_i
/location/location/contains	t_i is located in h_i

For the relation classification model, the input is the text mentioning the head entity and tail entity, and the output is the relation between the entity pair. We convert the DS training data into the input, and output of the relation classification model as follows: for each instance $D_i = (h_i, t_i, r_i, text_i)$, we regard the $(h_i, t_i, text_i)$ as the input, and r_i as the output.

3.2   Natural Language Inference Model

Natural language inference model is used for instance selection. As shown in Fig. 2, our NLI model g_ϕ consists of an input layer, a BERT layer and an output layer.

For each input of the NLI model, (P_i, H_i), we represent it as a sequence of words with two special tokens: "[CLS] P_i [SEP] H_i". The first token of the sequence is the [CLS] which contains the information of both the *premise* P_i and the *hypothesis* H_i. The other token [SEP] is used for separating the text of premise P_i and hypothesis H_i. Each token consists of three types of embedding, namely word embedding, position embedding and segment embedding.

BERT is a multi-layer bidirectional Transformer encoder. It has been pre-trained using a combination of masked language modeling objective and next sentence prediction on a large corpus comprising the Book Corpus and Wikipedia. BERT can be used for a variety of downstream tasks. These two pre-training goals allow it to be used for any single sequence and sequence pair task without the need for a large number of task-specific architectural modifications. In this paper, we use BERT to encode the information of both the *premise* P_i and the *hypothesis* H_i. Specifically, it takes an input of a sequence and outputs the representation of the sequence. The final hidden state of the [CLS] token is taken as the output.

Finally, we use a fully-connected dense layer to evaluate the quality for each instance. Specifically, It takes the final hidden state CLS_i of the first token [CLS] as input and outputs the probability that the *hypothesis* H_i can be inferred from the *premise* P_i. The probability is predicted as follows:

$$g_\phi(P_i, H_i) = \sigma(\mathbf{W}_c CLS_i + b_c), \tag{1}$$

where $\mathbf{W}_c$ and b_c are the parameters for the dense layer, σ is the *sigmoid* active function.

3.3   Relation Classification Model

We adopt the CNN architecture proposed by [1] for relation classification. The model consists of an input layer, a convolution layer, a max pooling layer and an output layer.

For each text of the input, $text_i$, we represent it as a sequence of word vector $\mathbf{x} = (\mathbf{w_1}, \mathbf{w_2}, \dots, \mathbf{w_m})$. Each word vector consists of two types of embedding, namely word and position embedding. For word embedding, we use the pre-trained word vectors from word2vec[1]. For position embedding, we follow the previous work [15] to represent the relative distances from the current word respectively to the head entity h_i and tail entity t_i in the text $text_i$. Finally, we concatenate the word and position embedding to form a word vector.

We then use a convolution layer and a max pooling layer to extract features from the input layer. This can be briefly described as follows:

$$\mathbf{C} = CNN(\mathbf{x}) \tag{2}$$

Finally, we feed the output of max pooling layer into a dense layer to predict the relation of the entity pair:

$$p(r|\mathbf{x}) = softmax(\mathbf{W}_o \tanh(\mathbf{C}) + \mathbf{b}_o), \tag{3}$$

where $\mathbf{W}_o$ and $\mathbf{b}_o$ are the parameters of the dense layer. *tanh* and *softmax* are active functions. Given the selected instances $\{(\mathbf{x}_i, r_i)\}_{i=1}^n$ provided by the sampler, we use *cross entropy* to define the objective function of the relation classification model as follows:

$$\mathbf{L}(\theta) = -\frac{1}{n} \sum_{i=1}^{n} \log p(r_i|\mathbf{x}_i) \tag{4}$$

3.4   Model Training

Now we introduce how to train the NLI model and the relation classification model alternately. Inspired by [13], we use the REINFORCE algorithm [11] to optimize the parameters, with the reward obtained from the relation classification model's performance on the validation set. The training procedure is shown in Algorithm 1.

We first initialize parameters θ, ϕ for relation classification model f_θ and natural language inference model g_ϕ (Step 1). Then we iteratively update the

---

[1] https://code.google.com/p/word2vec/.

Algorithm 1. The Training Procedure of the Whole Model

Inputs: Training dataset $D = \{(h_i, t_i, r_i, text_i)\}_{i=1}^N$, validation dataset $D^v = \{(h_k^v, t_k^v, r_k^v, text_k^v)\}_{k=1}^M$, learning rates α, β, batch size B_p, B_s, inner iteration count N_I, moving average window T.

Output: The natural language inference model g_ϕ.

1: Initialize parameters θ, ϕ for f_θ and g_ϕ, moving average $\delta = 0$.
2: **while** until convergence **do**
3:     Sample a batch of data $D^B = (h_j, t_j, r_j, text_j)_{j=1}^{B_s}$ from D
4:     $(\mathbf{x}_j, r_j)_{j=1}^{B_s}$, $(P_j, H_j)_{j=1}^{B_s} = \text{DATAPROCESSING}(D^B)$
5:     **for** $j = 1, ..., B_s$ **do**
6:         Predict selection probability: $w_j = g_\phi(P_j, H_j)$
7:         Bernoulli sample an instance: $s_j = \text{SAMPLE}(w_j)$
8:     **for** $t = 1, ..., N_I$ **do**
9:         Sample a batch of data $(\mathbf{x}_m, r_m)_{m=1}^{B_p}$ from $(\mathbf{x}_j, r_j)_{j=1}^{B_s}$ with $s_j = 1$
10:        Update θ according to Eq. 5
11:    Update ϕ according to Eq. 6
12:    Update the moving average δ according to Eq. 8

---

parameters of the two models (Step 2–12). Specifically, we first sample a batch of data $D^B = (h_j, t_j, r_j, text_j)_{j=1}^{B_s}$ from the training data D (Step 3). Next, we process this data for NLI model and relation classification model, respectively (Step 4). Then, for each instance of batch data, we use the NLI model to evaluate its quality, and perform Bernoulli sampling according to its probability (steps 5–7). $s_j = \{0, 1\}$ is an indicator variable, indicating whether to select this instance for the subsequent relation classification model. After that, we start to update the parameters of the relation classification model N_I times. In each parameter update, we first sample B_p batch size of data from the selected data in the previous step, and then update the parameters according to Eq. 5 (Step 8–10):

$$\theta \leftarrow \theta + \frac{\alpha}{B_p} \sum_{m=1}^{B_p} \nabla_\theta \log p(r_m | \mathbf{x}_m) \tag{5}$$

Then we update the natural language inference model parameters ϕ according to Eq. 6 (Step 11):

$$\phi \leftarrow \phi + \beta \left[F1(f_\theta(\mathbf{x}^v), r^v) - \delta \right] \cdot \nabla_\phi \log \pi_\phi(D^B, (s_1, ..., s_{B_s})), \tag{6}$$

where $F1(f_\theta(\mathbf{x}^v), r^v)$ is the F1 score of relation classification model on validation set D^v, $\pi_\phi(D^B, (s_1, ..., s_{B_s}))$ is the probability that selection vector $(s_1, ..., s_{B_s})$ is selected based on g_ϕ.

$$\pi_\phi(D^B, (s_1, ..., s_{B_s})) = \prod_{i=1}^{B_s} g_\phi(P_j, H_j)^{s_i} \cdot (1 - g_\phi(P_j, H_j))^{1-s_i} \tag{7}$$

To improve the stability of the policy gradient-based learning, we use the moving average of the previous F1 score δ with a window size T as the baseline and update the parameter as follows (Step 12):

$$\delta \leftarrow \frac{T-1}{T}\delta + \frac{1}{T}F1(f_\theta(\mathbf{x}^v), r^v) \qquad (8)$$

4   Experiment

4.1   Experiment Setup

Dataset. We evaluated our method on the widely used DS relation classification dataset, NYT10 [8], which was automatically generated by aligning Freebase with the New York Times corpus. NYT10 contains a total of 53 relations. One of them is the special label NA, which means that there is no relationship between a given entity pair. In our experiment, we follow the previous work [3,14] and use the *accuracy* metric to evaluate the performance of sentence-level relation classification on Non-NA relations. In order to accurately evaluate the performance of sentence-level relationship classification, we used two new test data that were manually relabeled based on the original NYT10 test data. We denote them as HA-Test1 [18] and HA-Test2 [5], respectively. The statistics about these datasets are listed in Table 2.

Table 2. Statistics of the datasets.

Datasets	Relation types	Full size	Size of Non-NA
NYT10 training [8]	53	522,611	136,379
HA-Test1 [18]	22	5,202	5,202
HA-Test2 [5]	28	4,288	4,288

Model and Training Setting. For the NLI model, the input layer consists of two parts: premise and hypothesis. The max size of a premise is 70, and the max size of a hypothesis is 20. For the BERT layer, we use BERT-Base[2] in our model, which contains an encoder with 12 layers (transformer blocks), 12 self-attention heads, and the hidden size of 768. For the output layer, the size of the output is 1.

For the relation classification model, we follow the same settings in [1]. For the input layer, the word and position embedding dimensions are 50 and 5, respectively. For the convolution layer, the convolution operation is performed on 3 consecutive words, and the number of feature maps is set to 230. For the output layer, we set the output size as 53, which is the total number of relations.

For training the whole model, we randomly select 90% of the *NYT10 Training* as the training set, and the remaining 10% as the validation set. Moreover, for Algorfithm 1, the learning rate α is 0.1, β is 0.005. The batch size of B_p is 4 and B_s is 5,120. The inner iteration counts N_I is 10000, and the moving average baseline T is 10.

[2] https://github.com/google-research/bert.

Baselines. Our goal is to build a good instance selector. To verify the effectiveness of our method, we compared different instance selection strategies. Each strategy will select a subset of data from the DS data. This subset of data will be used to train a relation classifier, and the performance of this classifier on the test set is used to evaluate the quality of instance selection strategies. Those strategies are as follows:

- DS. The DS strategy refers to retaining the DS training data without any selection operation.
- RL [1]. The RL strategy refers to using the state-of-the-art reinforcement learning model for instance selection, which is a bag-level instance selector.
- Random. The Random strategy refers to randomly selecting a certain ratio of instances from the DS training data.
- NLIRL. NLIRL is our proposed method, which selects a certain ratio of instances with higher NLI scores from the DS training data.
- NLIRL(Low). NLIRL(Low) is a variant of our proposed method, which selects a certain ratio of instances with lower NLI scores from the DS training data.

4.2   Performance Comparison and Analysis

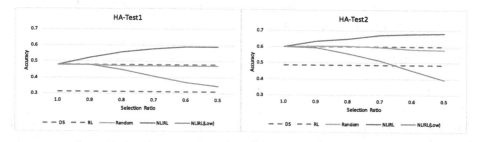

Fig. 3. The overall performance of different selection strategies.

We first compare the overall performance of different selection strategies. As shown in the figure, the DS and RL methods do not need to set a selection ratio threshold, so their performance is straight. Using the RL strategy to filter the training set reduces the performance of the classification method, which shows that this strategy cannot find high-quality instances. The performance of the remaining three strategies varies with the selection ratio threshold. It can be found from the figure that our method NLIRL is better than all other strategies. When the selection ratio is 50%, our method achieves the best results. Compared with NLIRL(Low), if we select instances with low NLI values, namely remove instances with high NLI values, then the classification performance will be reduced. These two aspects prove that our method can find high-quality instances and significantly improve relation classification performance (Fig. 3).

Then we conduct further analysis of some typical relations. We compared the performance of DS, RL and our method (NLIRL with the selection ratio threshold set to 0.5) on these relations. The results are shown in Table 3. Our approach can significantly improve the recall of the relation classifier, thereby improving the overall performance.

Table 3. The performance of four typical relations

Relations	Size	NLIRL			RL			DS		
		Precision	Recall	F1-score	Precision	Recall	F1-score	Precision	Recall	F1-score
HA-Test1										
/location/location/contains	2,754	0.892	**0.822**	**0.856**	**0.934**	0.640	0.760	0.914	0.752	0.825
/people/person/nationality	731	0.946	**0.910**	**0.927**	0.966	0.620	0.755	**0.979**	0.766	0.860
/business/person/company	468	0.965	**0.639**	**0.769**	0.975	0.585	0.732	**0.980**	0.536	0.693
/people/person/place_lived	225	0.791	**0.671**	**0.726**	0.702	0.147	0.243	**0.863**	0.533	0.659
HA-Test2										
/location/location/contains	1,987	0.832	**0.786**	**0.808**	**0.906**	0.444	0.596	0.876	0.636	0.737
/people/person/nationality	568	0.948	**0.900**	**0.923**	0.974	0.526	0.683	**0.979**	0.764	0.858
/business/person/company	277	0.903	**0.570**	**0.699**	0.948	0.523	0.674	**0.963**	0.480	0.641
/people/person/place_lived	270	0.913	**0.659**	**0.766**	0.773	0.063	0.116	**0.927**	0.570	0.706

We also compared the running time of RL and our proposed method NLIRL. Although we all use reinforcement learning methods, our method runs faster. Specifically, we only use 10 h to train our instance selector on a single NVIDIA GeForce RTX2080Ti GPU, while RL needs more than 48 h. That has been proved in [13], our method is not directly proportional to the dataset size, but depends on the number of iterations in Algorithm 1. Besides, we use the pre-trained model Bert for the NLI model to minimize computational overhead.

5   Related Work

Relation extraction based on deep learning requires a lot of training data, but manual labeling of these training data is very time-consuming and expensive. [4] first used distant supervision technology in 2009 to align the sentences in the input text with the triples in the Freebase knowledge graph to solve this problem. At this time, the triples provide supervision information. Despite its efficiency, distantly supervised relation classification often suffers from the label noise problem. Recently, there are two ways to solve the label noise problem.

The first way is to use multi-instance learning to consider the noises of instances [2,3,8,10,14]. They divide the training data into many bags, and each bag contains many sentences mentioning the same entity pair. Then the training and test process proceeds at the bag level. The most important thing is how to find the most relevant sentence from a bag of sentences. For example, [14] considers entity's position when extracting the feature vector representation of the sentence, uses the segmentation max-pooling operation to encode each sentence, and selects the sentence with the highest probability of correctly predicting the relation in a bag for parameter update. Considering the different importance of the expression relation of different sentences in a bag, [3] introduced a sentence-level attention mechanism. A sentence with a larger weight contributes more to the update of the parameter. Conversely, a sentence with a smaller weight contributes to the update of the parameter small, so that all training data can be fully utilized. However, bag-level relation classification does not perform well for sentence-level prediction [1,18]. In this paper, we focus on improving the performance of the sentence-level relation classification task.

The second way is to find the noisy labeling instances directly. There are two main types of methods, namely reinforcement learning and adversarial learning. Many reinforcement learning methods have been proposed in recent years [1,7,16], they decompose the problem of relation extraction into two tasks: instance selection and relation classification. Among them, [1,7] regard the instance selector as a reinforcement learning agent, while [16] regards the relation classification as the agent. Moreover, [7] dynamically identify false positive instances of each relation, and redistribute false positive instances to true negative instances. The first proposed method of adversarial training is to add adversarial noise to word embeddings to extract relations based on CNN and RNN under the framework of multi-instance multi-label learning (MIML) [12]. DSGAN [6] eliminates noisy data in distantly supervised relation classification by learning the generator and discriminator of sentence-level real positive instances. However, their computational overhead is high. In this paper, we use the natural language inference model as the agent, and propose a reinforcement learning method which is not directly proportional to the dataset size, but depends on the number of iterations in the training procedure.

6   Conclusion

In this paper, we propose to use the natural language inference(NLI) model to evaluate the quality of the instances directly and select the high-quality instances as refined training data for sentence-level relation classification. Due to the lack of high-quality supervised data, we use reinforcement learning to train the NLI model to find high-quality instances. Specifically, we use the NLI model to select instances, and then feed the sampled instances to the relation classifier, and use the performance of the relation classifier as a reward to guide the parameter update of the NLI model. Experimental results on two human re-annotated NYT datasets show our method's effectiveness and efficiency at the sentence-level relation classification.

References

1. Feng, J., Huang, M., Zhao, L., Yang, Y., Zhu, X.: Reinforcement learning for relation classification from noisy data. In: Proceedings of the Thirty-Second AAAI Conference on Artificial Intelligence, pp. 5779–5786 (2018)
2. Hoffmann, R., Zhang, C., Ling, X., Zettlemoyer, L., Weld, D.S.: Knowledge-based weak supervision for information extraction of overlapping relations. In: Proceedings of the 49th Annual Meeting of the Association for Computational Linguistics: Human Language Technologies, pp. 541–550 (2011)
3. Lin, Y., Shen, S., Liu, Z., Luan, H., Sun, M.: Neural relation extraction with selective attention over instances. In: Proceedings of the 54th Annual Meeting of the Association for Computational Linguistics, pp. 2124–2133 (2016)
4. Mintz, M., Bills, S., Snow, R., Jurafsky, D.: Distant supervision for relation extraction without labeled data. In: Proceedings of the Joint Conference of the 47th Annual Meeting of the ACL, pp. 1003–1011. Association for Computational Linguistics (2009)
5. Phi, V.T., Santoso, J., Tran, V.H., Shindo, H., Shimbo, M., Matsumoto, Y.: Distant supervision for relation extraction via piecewise attention and bag-level contextual inference. IEEE Access **7**, 103570–103582 (2019)
6. Qin, P., Xu, W., Wang, W.Y.: DSGAN: generative adversarial training for distant supervision relation extraction. In: Proceedings of the 56th Annual Meeting of the Association for Computational Linguistics, pp. 496–505 (2018)
7. Qin, P., Xu, W., Wang, W.Y.: Robust distant supervision relation extraction via deep reinforcement learning. In: Proceedings of the 56th Annual Meeting of the Association for Computational Linguistics, pp. 2137–2147 (2018)
8. Riedel, S., Yao, L., McCallum, A.: Modeling relations and their mentions without labeled text. In: Balcázar, J.L., Bonchi, F., Gionis, A., Sebag, M. (eds.) ECML PKDD 2010. LNCS (LNAI), vol. 6323, pp. 148–163. Springer, Heidelberg (2010). https://doi.org/10.1007/978-3-642-15939-8_10
9. Su, P., Li, G., Wu, C., Vijay-Shanker, K.: Using distant supervision to augment manually annotated data for relation extraction. PloS one **14**(7), 1–17 (2019)
10. Surdeanu, M., Tibshirani, J., Nallapati, R., Manning, C.D.: Multi-instance multi-label learning for relation extraction. In: Proceedings of the 2012 Joint Conference on Empirical Methods in Natural Language Processing and Computational Natural Language Learning, pp. 455–465 (2012)
11. Williams, R.J.: Simple statistical gradient-following algorithms for connectionist reinforcement learning. Machine Learn. **8**(3–4), 229–256 (1992)
12. Wu, Y., Bamman, D., Russell, S.: Adversarial training for relation extraction. In: Proceedings of the 2017 Conference on Empirical Methods in Natural Language Processing, pp. 1778–1783 (2017)
13. Yoon, J., Arik, S., Pfister, T.: Data valuation using reinforcement learning. In: International Conference on Machine Learning, pp. 10842–10851. PMLR (2020)
14. Zeng, D., Liu, K., Chen, Y., Zhao, J.: Distant supervision for relation extraction via piecewise convolutional neural networks. In: Proceedings of the 2015 Conference on Empirical Methods in Natural Language Processing, pp. 1753–1762 (2015)
15. Zeng, D., Liu, K., Lai, S., Zhou, G., Zhao, J.: Relation classification via convolutional deep neural network. In: Proceedings of the 25th International Conference on Computational Linguistics: Technical Papers, pp. 2335–2344 (2014)
16. Zeng, X., He, S., Liu, K., Zhao, J.: Large scaled relation extraction with reinforcement learning. In: Proceedings of the AAAI Conference on Artificial Intelligence, pp. 5658–5665 (2018)

17. Zhang, A., Lipton, Z.C., Li, M., Smola, A.J.: Dive into Deep Learning. Corwin (2020)
18. Zhu, T., Wang, H., Yu, J., Zhou, X., Chen, W., Zhang, W., Zhang, M.: Towards accurate and consistent evaluation: a dataset for distantly-supervised relation extraction. In: Proceedings of the 28th International Conference on Computational Linguistics, pp. 6436–6447 (2020)

SaGCN: Structure-Aware Graph Convolution Network for Document-Level Relation Extraction

Shuangji Yang[1,3], Taolin Zhang[2,3], Danning Su[2], Nan Hu[1,3], Wei Nong[1,3], and Xiaofeng He[1,3(✉)]

[1] School of Computer Science and Technology, East China Normal University, Shanghai, China
{51194501201,51194501160,51194501167}@stu.ecnu.edu.cn
[2] School of Software Engineering, East China Normal University, Shanghai, China
51194501046@stu.ecnu.edu.cn
[3] Shanghai Key Laboratory of Trustworthy Computing, Shanghai, China
hexf@cs.ecnu.edu.cn

Abstract. Document-level Relation Extraction(DocRE) aims at extracting semantic relations among entities in documents. However, current models lack long-range dependency information and the reasoning ability to extract essential structure information from the text. In this paper, we propose SaGCN, a **S**tructure-**a**ware **G**raph **C**onvolution **N**etwork, extracting relation with explicit and implicit dependency structure. Specifically, we generate the implicit graph by sampling from a discrete and continuous distribution, then dynamically fuse the implicit soft structure with the dependent hard structure. Experimental results of SaGCN outperform the performance achieved by current state-of-the-art various baseline models on the DocRED dataset.

Keywords: Document-level relation extraction · Structure-aware information injection · Graph neural network

1 Introduction

The **R**elation **E**xtraction(**RE**) task aims to detect semantic relations between named entities from text. Sentence-level relation extraction tasks have been extensively studied [31,32]. Due to understanding unstructured text in realistic scenarios should extract various entity relations among multiple sentences, document-level relation extraction is necessary.

Prior work [30,33] of document-level relation extraction utilizes position-aware or context-aware attention to combine with LSTM [10] or hierarchical networks [11,22] to learn more accurate entity representations. These approaches only enhance the model's ability to aggregate contextual information, but they are still ineffective for relation extracting cross multiple words or sentences. In Table 1, entity *North American* and *California* are almost 30 words apart.

© Springer Nature Switzerland AG 2021
K. Karlapalem et al. (Eds.): PAKDD 2021, LNAI 12714, pp. 377–389, 2021.
https://doi.org/10.1007/978-3-030-75768-7_30

Long-range relation depend on both local and non-local dependencies. Sahu et al. [19] construct the document-level graph by dependency tree and co-reference links. iDepNN [9] extracts relations within and across sentence boundaries via the shortest and augmented dependency paths. We argue that these approaches are incapable of relation inference. As shown in the Table 1, the first sentence indicates that Antennaria media is a *North American* species, and the second sentence states that it is native to western *Canada*. The relation between the two entities needs to be reasoned. Some recent works [3,17,26] make use of different graph structures to solve reasoning problems. However, these models do not in-depth explore the structural information which is critical to models' understanding of the entire document.

Table 1. An example document from DocRED [30]. All mentions of an entity are represented by same color and only the entities that appear in the example are marked.

Document Text
[1] Antennaria media is a North American species of flowering plants in the daisy family known by the common name Rocky Mountain pussytoes. [2] It is native to western Canada and the Western United States from Alaska and Yukon Territory to **California** to New Mexico. [3] It grows in cold Arctic and alpine regions , either at high latitudes in the Arctic or at high elevations in the mountains (Rocky Mountains, Cascades, Sierra Nevada). ...

Subject: Canada	**Subject: California**
Object: North American	**Object:** North American
Relation: continent, part of	**Relation:** continent
Evidence: 1, 2	**Evidence:** 1, 2

To overcome these drawbacks, we propose a structure-aware model with inferential capabilities. Our primary motivation is to use both explicit and implicit structural information to mine the key information to predict relations between entities. HardKuma distribution [1] is used to obtain implicit structural information which has both continuous and discrete properties. Meanwhile, explicit structural information is introduced by hard connection of dependency tree, then we fuse the implicit and explicit information in a heuristic way to obtain the final fused graph neural network. Additionally, to enable the model to have reasoning ability, we construct a inference graph with mention and entities to propagate information with graph convolution operations.

In summary, our main contributions are following:

- We propose a novel model, SaGCN, to predict relations between entities for document-level relation extraction. SaGCN can handle multiple sentences and long-range information efficiently, and has the ability to reason about entity relations.

- We introduce a structure-aware graph that captures both explicit and implicit structures in a document, with contextual and critical information from dependency tree.
- We conduct experiments on document-level datasets. The results show that SaGCN has superior performance on extracting relations over long distances to other sequence-based, graph-based and BERT-based models.

2 Method

2.1 Model Overview

Given a document D $= [x_1, x_2, ..., x_n]$, where $i \in [1, n]$ and x_i is i^{th} word in document, our goal is to extract the relation r_{ij} between each pair of entities (e_i, e_j) from the document. We work on this task through the following steps. (1) We encode a document by universal context encoder, such as BiLSTM [20] and BERT [5], which can obtain contextual information. (2) Creating a fusion structure aware graph that incorporates explicit and implicit structures between mention and meta dependency path (MDP) nodes. (3) We construct a inference graph with mention and entity nodes to reasoning relation of entities by graph convolution. The overview of our model is shown in Fig. 1.

2.2 Context Encoder

To obtain the contextualized representations of each word, we feed a document D into a contextual encoder. In this paper, we use Bi-directional Long Short-Term Memory (BiLSTM) [20] and BERT [5] as encoder, the word hidden representation calculated by BiLSTM is $h_i^l = [\overrightarrow{h_i^l}; \overleftarrow{h_i^l}]$, specifically

$$\overrightarrow{h_i^l} = \overrightarrow{LSTM}(h_{i-1}^l, h_i^{l-1}) \tag{1}$$

$$\overleftarrow{h_i^l} = \overleftarrow{LSTM}(h_{i+1}^l, h_i^{l-1}) \tag{2}$$

where h_{i-1}^l is i^{th} word hidden vector in l^{th} layer and h_i^{l-1} is i^{th} word hidden vector in $(l-1)^{th}$ layer. Then we can get contextual representation of each token in the document.

2.3 Graph Nodes

We extract three types of nodes: document meta dependency path (DMDP) nodes, mention nodes, entity nodes. Sentence meta dependency path (SMDP) nodes [17] are the tokens on the shortest dependency path between all mention pairs in a sentence. We extend SMDP to DMDP by connecting root nodes of each sentence dependency tree in a document. Concretely, given four mentions m_1, m_2, m_3, m_4 in two sentences s_1, s_2 of a document D, where $m_1, m_2 \in s_1, m_3, m_4 \in s_2$. The DMDP nodes are $MDP_{m_i, m_j}, i, j \in \{1, 2, 3, 4\}$ and $i \neq j$, are different from SMDP nodes which only contain MDP_{m_1,m_2}, MDP_{m_3,m_4}. Mention nodes are the corresponding mention words for all entities in a document. An entity node representation takes the average of all its mention nodes. We generate the structural graph by DMDP nodes and mention nodes. The inference graph is constructed by entity nodes and mention nodes.

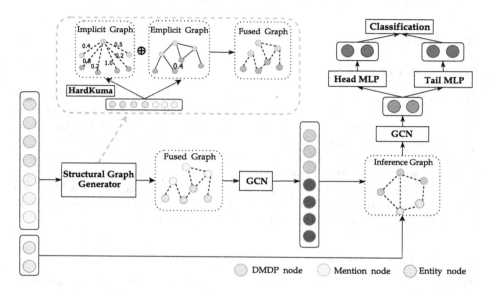

Fig. 1. Overview of SaGCN, the Fused Graph is generated by Structural Graph Generator, Mention Nodes and Entity Nodes construct the Inference Graph. The original DMDP nodes, Mention nodes, Entity nodes are encoded by contextual encoder, such as BiLSTM [20], BERT [5]. The Implict Graph is a fully connected graph with learned edge weight from 0.0 to 1.0, the other nodes' connection is same as the example node.

2.4 Structural Graph

Explicit Graph. We construct explicit graph adjacency matrix $\boldsymbol{A_E}$ by using the dependencies of the syntactic dependency tree generated via spaCy[1]. We define $\boldsymbol{A_{E_{i,j}}} = 1$ when there is a edge connects word i and word j in dependency tree.

Implicit Graph. In order to obtain an implicit graph containing hidden structural information of mention nodes and DMDP nodes, we apply HardKuma distribution [1], which is a hard mode of Kumaraswamy distribution, both continuous and discrete. Samples on this distribution will closely approximate 0 and 1. A brief description is following:

---

[1] https://spacy.io/.

Kumaraswamy Distribution. Like the Beta distribution, the Kumaraswamy [14] is a two-parameters distribution in the open interval $(0, 1)$, we define $K \sim Kuma(\alpha, \beta)$ to represent a Kumaraswamy distributed variable, where $\alpha, \beta \in \mathbb{R}_{>0}$ control the distribution's shape. The Kumaraswamy has a simple Cumulative Distribution Function (CDF) and the inverse of CDF is:

$$C_K^{-1}(u; \alpha, \beta) = \left(1 - (1-u)^{1/\beta}\right)^{1/\alpha} \tag{3}$$

$$C_K^{-1}(U; \alpha, \beta) \sim \text{Kuma}(\alpha, \beta) \tag{4}$$

Then the above formula is used to take samples, and $U \sim \mathcal{U}(0, 1)$ is a uniform distribution, we can exploit this property to perform parameter reconstructions to produce new CDF functions.

HardKuma Distribution. HardKuma is a constraint method that make the value of samples from Kumaraswamy distribution stretched and rectified to 0 and 1. Firstly, stretching the original Kumaraswamy distribution to obtain the variable $V \sim \text{Kuma}(\alpha, \beta, \gamma, \eta)$ which can be sampled in the open interval (γ, η) where $\gamma < 0$ and $\eta > 0$, and the new CDF:

$$C_V(v; \alpha, \beta, \gamma, \eta) = C_V((v - \gamma)/(\eta - \gamma); \alpha, \beta) \tag{5}$$

Second, by applying a hard-sigmoid function, $h = \min(1, \max(0, t))$, the sample $V \sim \text{Kuma}(\alpha, \beta, \gamma, \eta)$ convert to a rectified variable $H \sim \text{HardKuma}(\alpha, \beta, \gamma, \eta)$. Note that the variable H is covered the closed interval $[0, 1]$ now. All negative values can be constrained to 0 and all value over 1 can converge to 1.

Implicit Graph Generator. We generate the implicit graph by sampling the trained parameters $\boldsymbol{\alpha}$ and $\boldsymbol{\beta}$ from the HardKuma distribution. Firstly, calculate the prior $\boldsymbol{c}$ of (α, β) by employing multihead self-attention [23]:

$$\boldsymbol{c}_\alpha = \text{MultiSelfAttention}_\alpha(\boldsymbol{x}) \tag{6}$$

$$\boldsymbol{c}_\beta = \text{MultiSelfAttention}_\beta(\boldsymbol{x}) \tag{7}$$

where $\boldsymbol{x}$ are structural graph nodes. Subsequently, we get $\boldsymbol{\alpha}$ and $\boldsymbol{\beta}$ by:

$$\boldsymbol{\alpha} = \text{LayerNorm}\left(\boldsymbol{c}_\alpha \boldsymbol{c}_\alpha^T\right) \tag{8}$$

$$\boldsymbol{\beta} = \text{LayerNorm}\left(\boldsymbol{c}_\beta \boldsymbol{c}_\beta^T\right) \tag{9}$$

Hence, the implicit graph $\boldsymbol{A_I}$ is sampled via learned parameters $\boldsymbol{\alpha}$ and $\boldsymbol{\beta}$:

$$\boldsymbol{A_{I_{i,j}}} \sim \text{HardKuma}\left(\boldsymbol{\alpha}_{ij}, \boldsymbol{\beta}_{ij}, \gamma, \eta\right) \tag{10}$$

Fusion Graph and Convolution. To benefit from explicit and implicit structural information in learning more plentiful mention representation, we merge the explicit and implicit graph with a fused layer inspired by prior work [2,19]. The adjacency matrix A_F for fused graph with nodes x calculated by:

$$A_F = \sigma(Wx)A_E + (1 - \sigma(Wx))A_I \tag{11}$$

where σ is sigmoid function, $W \in \mathbb{R}^{m \times 1}$ is trainable gate variable. Then we feed mention and MDP nodes into GCN [12] to capture structural information for mention nodes. The GCN iteratively updates the representation of each input graph node i as following:

$$h_i^{l+1} = \text{ReLU}\left(\sum_{j=1}^{n} A_{F_{i,j}} W^l h_j^l + b^l\right) \tag{12}$$

We use ReLU [6] as the activation function, and W^l is the trainable parameter in l^{th} layer, h_i^{l+1} is hidden representation of ith node in $(l+1)^{th}$ convolution layer. The mention nodes contain contextual and cross sentence structural information by dynamically integrating from both explicit and implicit graph. We mark them as m_s.

2.5   Inference Graph

In this section, we generate a fully connect graph with all mention nodes m_s which already captured the backbone information of the document and entity nodes e, and aggregate the information from m_s to e. Meanwhile, reasoning relation from mention-mention, mention-entity, entity-entity levels. The graph convolution calculation is similar to the previous step:

$$h_i^l = \text{ReLU}\left(\sum_{j=1}^{n} A_{ij} W^l h_j^{l-1} + b^{(l)}\right) \tag{13}$$

where A_{ij} is the adjacency matrix of m_s and e, all values of it equal 1. After the operation of mutual reasoning between entities and mentions, we get the final entity representations e, which contains a vast amount of mention information and entities reasoning information.

2.6   Relation Classification

Before classification, we concatenate distance of head entity to tail entity for head entities, and distance of tail entity to head entity for tail entities as same as Yao et al. [30].

$$e_i^{head} = [e_i^K; \textbf{\textit{Dist}}_{head2tail}(e_i^K)] \tag{14}$$

$$e_i^{tail} = [e_i^K; \textbf{\textit{Dist}}_{tail2head}(e_i^K)] \tag{15}$$

where e_i^K corresponds to the ith entity after K blocks of inference graph, and e_i^{head}, e_i^{tail} are resulted head and tail representation for i^{th} entity. Then, relation category scores of entity pairs are calculated by a bi-affine layer.

$$scores(e^{head}, e^{tail}) = \text{ReLU}(e^{head} \boldsymbol{W} e^{tail} + b) \tag{16}$$

where $\boldsymbol{W} \in \mathbb{R}^{d \times r \times d}$ is trainable tensor with r relation types.

3   Experiments

3.1   Dataset

We evaluate SaGCN on DocRED [30], which was built from Wikipedia and Wikidata, covering 96 relation types and $132, 275$ entities. Both manually-annotated and distantly-supervised data are offered. We use the manually-annotated data.

3.2   Baseline Models

We compare the performance of SaGCN with the following models.

- **Sequence-based RE Models.** We select CNN, LSTM and BiLSTM that are adapted by Yao et al. [30]. Context-Aware [30] incorporates the attention mechanism into LSTM. Hierarchical inference network (HIN) [22] aggregates three levels information which are entity, sentence, document to reason relations between entities.
- **Graph-Based RE Models.** GCNN [19] constructs document graph through co-definition, dependency, and adjacency sentence links, and performs relation reasoning on the graph. EoG [3] extracts three types of nodes, entity, mention, and sentence nodes which compose graph and reason with edge-oriented graphs. GAT [24] first leverages attention into graph neural network. AGGCN [8] updates the weight of soft connections between dependency tree nodes through multi-head attention mechanism. LSR [17] uses the maximum tree theory to obtain latent structure information, by iterating multiple GCNs to obtain new entity representations of the information interaction.
- **BERT-Based RE Models.** BERT [5] encodes text using multiple Transformer [23] stacks, has achieved excellent results in many downstream NLP tasks. Two-Phase BERT [27] is very similar to the basic BERT, except that it first predicts whether a relationship exists between two entities, then predicts the relation. GLRE [26] constructs a local and global graph of entities and mentions. This composition gives it the ability to capture both local and global information.

3.3   Results

We show the results of SaGCN for the DocRED in Table 2. From the experimental results, we have following findings. (1) As our proposed model is graph-based, we concentrate on the comparison with graph-based models. SaGCN outperforms the best results among all the graph-based models. Compared with LSR, SaGCN achieves 0.33% and 0.07% higher $F1$ on dev and test set, respectively. It suggests that SaGCN is more effective in capturing long-range dependencies, and the explicit dependency connection is conducive to document-level relation extraction. (2) SaGCN with BERT improves 8.92%/3.1% Ign $F1/F1$ on the dev set compared with BERT, which indicates our model can induce more informative structures for reasoning. (3) Sequence-based models obtained medium performance due to unavailability of structural information.

Table 2. Performance of SaGCN on DocRED. Results with $\star$ are reported in their original papers. Results with $\diamond$ are implemented and published by Nan et al. [17].

Model	Dev				Test	
	Ign $F1$	Ign AUC	$F1$	AUC	Ign $F1$	$F1$
CNN* [30]	41.58	36.85	43.45	39.39	40.33	42.26
LSTM* [30]	48.44	46.62	50.68	49.48	47.71	50.07
BiLSTM* [30]	48.87	47.61	50.94	50.26	48.78	51.06
Contex-Aware* [30]	48.94	47.22	51.09	50.17	48.40	50.70
HIN* [22]	51.06	–	52.95	–	51.15	53.30
GCNN$^\diamond$ [19]	46.22	–	51.52	—	49.59	51.62
EoG$^\diamond$ [3]	45.94	–	52.15	–	49.48	51.82
GAT$^\diamond$ [24]	45.17	–	51.44	–	47.36	49.51
AGGCN$^\diamond$ [8]	46.29	–	52.47	–	48.89	51.45
LSR* [17]	48.82	–	55.17	–	52.15	54.18
SaGCN-GloVe	**52.99**	**51.74**	**55.50**	**55.62**	**52.31**	**54.25**
BERT$^\diamond$ [5]	48.44	50.68	56.57	41.47	47.71	50.0
Two-Phase BERT$^\diamond$ [27]	–	–	54.42	–	–	53.92
HIN-BERT* [22]	54.29	–	56.31	-	53.70	55.60
LSR-BERT* [17]	52.43	—	59.00	–	**56.97**	59.05
GLRE* [26]	56.70	–	58.90	–	–	–
SaGCN-BERT	**56.73**	**55.56**	**59.67**	**58.43**	56.83	**59.37**

3.4   Detail Analysis

Analysis of Intra and Inter Sentence. Since one of the major challenges of document-level relation extraction is that the entity pairs span multiple sentences, we check the performance of the model in intra and inter sentence. The determination of intra- or inter- relation is based on the evidence that have been labeled in the dataset. We consider relation is a intra-relation when the evidence only has one sentence. The results of this experiment show in Table 3. We observe that the performance of SaGCN-GloVe over LSR for extracting relations across sentences is better and can be improved by 0.55%. This indicates the way we extract DMDP nodes and structure-aware graph is able to induce more accurate structures across sentences.

Table 3. Intra and Inter sentence experimental results.

Model	Intra-$F1$	Inter-$F1$
CNN	51.87	37.58
LSTM	56.57	41.47
BiLSTM	57.05	43.49
Context Aware	56.74	42.26
LSR	60.83	48.35
SaGCN-GloVe	**60.99**	**48.52**
BERT	61.61	47.15
Two-Phase BERT	61.80	47.28
LSR-BERT	65.26	52.05
SaGCN-BERT	**65.27**	**52.23**

Ablation Study. To further analyze the effectiveness of each module in SaGCN, we perform ablation experiments. We observe from Table 4 that when the implicit graph is removed, the $F1$ and Inter-$F1$ drop significantly. This result confirms that our implicit structure is valid for this task and implicit information is beneficial to capture long-range dependency relation. There is almost 0.04% $F1$ a slight decrease in the model performance when the explicit graph is removed, and the explicit dependency is helpful in improving the model performance. Accordingly, we remove the inference graph and directly map the mentions that contain the most important information of the entire document to entities in the previous step to extract the relations. There is a 0.4% decrease in $F1$, indicating that information interaction at mention-entity level is necessary. Removal of implicit graph leads to a 0.87% drop in terms of Inter-$F1$ score, which indicates the implicit structure introduction is the most important part of our model.

Table 4. Ablation study of SaGCN on DocRED.

	$F1$	Ign $F1$	Intra-$F1$	Inter-$F1$
Full model	**55.50**	**52.99**	**61.03**	**49.21**
–emplicit graph	55.44	52.70	59.97	49.04
–implicit graph	54.85	51.98	59.45	48.34
–inference graph	55.22	52.03	59.55	48.78

Case Study. We list some examples of DocRED [30] in Table 5 for better understanding. (1) The head entity *Agrippa* and tail entity *Lucius Caesar* in example 1 cross three sentences, which require the model be robust enough to handle long-range cross sentence information. SaGCN predicts the correct relation *sibling* by gathering both explicit and implicit document-level information with graph convolution operation on the structure-aware graph. (2) From example 2, we know Eric Butters Stough is an *American* in sentence 1, he was born in *University of Colorado* in sentence 3. Model needs to reason from the two above relations of entities, then can predict the relation of *University of Colorado* and *American* is *country*. Benefiting from the inference graph in SaGCN, our model correctly predicts the relation. (3) Example 3 needs prior knowledge. Model must knows that *Catholic Church* is a proper noun ahead of time, then extracts the relation *instance of* between *Catholic Church* and *Church*. SaGCN and LSR both lack the knowledge, which we leave to feature work.

Table 5. Case study on the DocRED. Head entities and Tail entities are colored accordingly.

[1] Lucius Caesar was the grandson of Augustus. [2] The son of Marcus Vipsanius Agrippa and Julia the Elder, Augustus' only daughter ...[4] His brother Gaius also died at ... [5] The untimely loss of both heirs compelled Augustus to ... Agrippa Postumus as well as his stepson, Tiberius on 26 June AD 4.

Relation Label: sibling   **SaGCN:** sibling   **LSR:** N/A

[1] Eric Butters Stough (born July 31 , 1972) is an American animator and producer. [2] He is best known as the animation director and a producer on the television series South Park. [3] Born in Evergreen, Colorado, Stough attended the University of Colorado at Boulder and graduated in 1995 with a degree in film.

Relation Label: country   **SaGCN:** country   **LSR:** N/A

[1] Thomas Wolsey (c. March 1473 29 November 1530; sometimes spelled Woolsey or Wulcy) was an English churchman, statesman and a cardinal of the Catholic Church. [2] When Henry VIII became King of England in 1509 ... [3] Wolsey's affairs prospered, and by 1514 he had become ... within the Church, as Archbishop of York, a cleric in England junior only to the Archbishop of Canterbury.

Relation Label: instance of   **SaGCN:** N/A   **LSR:** N/A

4 Related Work

Document-Level Relation Extraction. With the shifting of relation extraction task from sentence-level [7, 28, 31, 33] to document-level, several cross-sentence datasets [30, 33] has been released. To enhance the ability of the model to learn contextual information, Zhang et al. [33] proposed a method for entity location-aware attention that could be combined with LSTM. Tang et al. [22] proposed HIN which is a hierarchical inference network that could aggregate information to entities from entity-level, sentence-level and document-level. EoG [3] considered mention nodes, entity nodes and sentence nodes, verified that the establishment of different nodes improved the accuracy of relation extraction. However, this work only extracted relations based on a walking strategy, did not perform further graph-based information interaction. With the development of graph neural networks, much work [9,18,19,29,32] built document graphs by dependency trees. Nan et al. [17] defined a latent document graph by using Matrix-Tree Theorem [13]. GLRE [26] encoded the document graph into entity global and local representations by using a heterogeneous graph. Different from the previous work, we construct a structural graph to capture context and long-range dependency information. Moreover, we generate an inference graph to reason logical relations between entities.

Dependency Tree for Relation Extraction. The dependency tree represents semantic dependency relations [4,15] in a sentence which is widely utilized in relation extractions [18,19,25,32]. Miwa et al. [16] and Tai et al. [21] combined LSTM [10] and dependency structures for sentence-level relation extraction. Xiong et al. [29] represented the dependency tree as a graph. However, these methods can only handle a single sentence relation extraction. GCNN [19] extended the single sentence dependency tree structure to multiple sentences by connecting the root nodes of each tree. To address the difficulties in parallelizing over different tree structures. Zhang et al. [32] and Guo et al. [8] proposed hard and soft methods to prune dependency tree nodes respectively. Recently, Nan et al. [17] discarded the connectivity present in the dependency tree and generated a latent graph using only the dependency nodes. Unlike previous models that only used explicit or implicit structures based on dependency trees, we argue that both structures should be considered. SaGCN leverages an efficient method to compute implicit graph and dynamically fuses the implicit graph with the dependent explicit graph to obtain abundant structural information.

5   Conclusion

In this paper, we propose a Structure-aware Graph Convolution Network (SaGCN) for document-level relation extraction to alleviate the insufficiency of exploring the structure of documents. SaGCN utilizes HardKuma distribution to capture implicit structure dependency information, and leverages a dynamical fusion mechanism to process both explicit and implicit structures of dependencies. This method is able to capture dependency semantics and process long-range information simultaneously to extract relations end-to-end. In future work, we plan to incorporate the types of dependency paths into document-level RE models.

Acknowledgements. This research is supported by the National Key Research and Development Program of China under Grant No. 2016YFB1000904.

References

1. Bastings, J., Aziz, W., Titov, I.: Interpretable neural predictions with differentiable binary variables
2. Beck, D., Haffari, G., Cohn, T.: Graph-to-sequence learning using gated graph neural networks. In: ACL (2018)
3. Christopoulou, F., Miwa, M., Ananiadou, S.: Connecting the dots: document-level neural relation extraction with edge-oriented graphs. In: EMNLP (2019)
4. Culotta, A., Sorensen, J.: Dependency tree kernels for relation extraction. In: ACL (2004)
5. Devlin, J., Chang, M.W., Lee, K., Toutanova, K.: BERT: pre-training of deep bidirectional transformers for language understanding. In: NAACL-HLT (2019)
6. Glorot, X., Bordes, A., Bengio, Y.: Deep sparse rectifier neural networks. In: AISTATS (2011)
7. Guo, Z., Nan, G., Lu, W., Cohen, S.B.: Learning latent forests for medical relation extraction. In: IJCAI (2020)
8. Guo, Z., Zhang, Y., Lu, W.: Attention guided graph convolutional networks for relation extraction. In: ACL (2019)
9. Gupta, P., Rajaram, S., Schütze, H., Runkler, T.: Neural relation extraction within and across sentence boundaries. In: AAAI
10. Hochreiter, S., Schmidhuber, J.: Long short-term memory. Neural computation
11. Jia, R., Wong, C., Poon, H.: Document-level n-ary relation extraction with multi-scale representation learning. In: NAACL (2019)
12. Kipf, T.N., Welling, M.: Semi-supervised classification with graph convolutional networks. In: ICLR (2017)
13. Koo, T., Globerson, A., Carreras, X., Collins, M.: Structured prediction models via the matrix-tree theorem. In: EMNLP-CoNLL (2007)
14. Kumaraswamy, P.: A generalized probability density function for double-bounded random processes. J. Hydrol. **46**, 79–88 (1980)
15. Liu, Y., Wei, F., Li, S., Ji, H., Zhou, M., Wang, H.: A dependency-based neural network for relation classification. In: ACL-IJCNLP (2015)
16. Miwa, M., Bansal, M.: End-to-end relation extraction using LSTMs on sequences and tree structures. In: ACL (2016)
17. Nan, G., Guo, Z., Sekulić, I., Lu, W.: Reasoning with latent structure refinement for document-level relation extraction. In: ACL (2020)
18. Peng, N., Poon, H., Quirk, C., Toutanova, K., Yih, W.T.: Cross-sentence n-ary relation extraction with graph LSTMs. In: TACL (2017)
19. Sahu, S.K., Christopoulou, F., Miwa, M., Ananiadou, S.: Inter-sentence relation extraction with document-level graph convolutional neural network. In: ACL (2019)
20. Schuster, M., Paliwal, K.K.: Bidirectional recurrent neural networks. IEEE Trans. Signal Process. **45**, 2673–2681 (1997)
21. Tai, K.S., Socher, R., Manning, C.D.: Improved semantic representations from tree-structured long short-term memory networks. In: ACL-IJCNLP (2015)
22. Tang, H., et al.: HIN: hierarchical inference network for document-level relation extraction. In: Lauw, H.W., Wong, R.C.-W., Ntoulas, A., Lim, E.-P., Ng, S.-K., Pan, S.J. (eds.) PAKDD 2020. LNCS (LNAI), vol. 12084, pp. 197–209. Springer, Cham (2020). https://doi.org/10.1007/978-3-030-47426-3_16
23. Vaswani, A., Shazeer, N., Parmar, N., Uszkoreit, J., Jones, L., Gomez, A.N., Kaiser, Ł., Polosukhin, I.: Attention is all you need. In: NIPS (2017)

24. Veličković, P., Cucurull, G., Casanova, A., Romero, A., Liò, P., Bengio, Y.: Graph attention networks. In: ICLR (2018)
25. Veyseh, A.P.B., Dernoncourt, F., Dou, D., Nguyen, T.: Exploiting the syntax-model consistency for neural relation extraction. In: ACL (2020)
26. Wang, D., Hu, W., Cao, E., Sun, W.: Global-to-local neural networks for document-level relation extraction. In: EMNLP (2020)
27. Wang, H., Focke, C., Sylvester, R., Mishra, N., Wang, W.: Fine-tune Bert for DOCred with two-step process. arXiv preprint arXiv:1909.11898 (2019)
28. Xiao, M., Liu, C.: Semantic relation classification via hierarchical recurrent neural network with attention. In: COLING (2016)
29. Xiong, W., Li, F., Cheng, M., Yu, H., Ji, D.: Bacteria biotope relation extraction via lexical chains and dependency graphs. In: BioNLP (2019)
30. Yao, Y., et al.: DocRED: a large-scale document-level relation extraction dataset. In: ACL (2019)
31. Zeng, D., Liu, K., Lai, S., Zhou, G., Zhao, J.: Relation classification via convolutional deep neural network. In: COLING (2014)
32. Zhang, Y., Qi, P., Manning, C.D.: Graph convolution over pruned dependency trees improves relation extraction. In: EMNLP (2018)
33. Zhang, Y., Zhong, V., Chen, D., Angeli, G., Manning, C.D.: Position-aware attention and supervised data improve slot filling. In: EMNLP (2017)

Addressing the Class Imbalance Problem in Medical Image Segmentation via Accelerated Tversky Loss Function

Nikhil Nasalwai[✉], Narinder Singh Punn, Sanjay Kumar Sonbhadra,
and Sonali Agarwal

Indian Institute of Information Technology Allahabad, Prayagraj, India
{mit2019095,pse2017002,rsi2017502,sonali}@iiita.ac.in

Abstract. Image segmentation in the medical domain has gained a lot of research interest in recent years with the advancements in deep learning algorithms and related technologies. Medical image datasets are often imbalanced and to handle the imbalance problem, deep learning models are equipped with modified loss functions to effectively penalize the training weights for false predictions and conduct unbiased learning. Recent works have introduced various loss functions suitable for certain scenarios of segmentation. In this paper, we have explored the existing loss functions that are widely used for medical image segmentation, following which an accelerated Tversky loss (ATL) function is proposed that uses log cosh function to better optimize the gradients. The no-new U-Net (nn-Unet) model is adopted as the base model to validate the behaviour of the loss functions by using the standard benchmark segmentation performance metrics. To establish the robustness and effectiveness of the loss functions, multiple datasets are adopted, where ATL function illustrated better performance with faster convergence and better mask generation.

Keywords: Deep learning · Image segmentation · Medical images · Loss function · Optimization · nn-Unet.

1 Introduction

Image segmentation is widely used in the medical domain to design computer-aided diagnosis (CAD) systems. It is the process of partitioning the image into multiple segments and used to locate regions of interest and associated boundaries. Image segmentation is categorized into semantic segmentation and instance segmentation. In semantic segmentation the multiple objects of the same class are treated as the same entity, whereas instance segmentation extends further to treat multiple objects of the same class as different individual objects.

All authors have contributed equally.

© Springer Nature Switzerland AG 2021
K. Karlapalem et al. (Eds.): PAKDD 2021, LNAI 12714, pp. 390–402, 2021.
https://doi.org/10.1007/978-3-030-75768-7_31

With the advancements in deep convolutional neural networks, many architectures were proposed for image segmentation. The image segmentation models such as SegNet [1], ResNet [6], GoogLeNet [22], proposed for natural images are more efficient and are widely used for various domains and applications. However, for medical image analysis, these models are not suitable and record a poor performance as these models are designed for datasets of huge size. The samples in medical imaging datasets are limited due to limitations in domain expertise and expensive manual delineation of the data [24]. The major role of segmentation in the medical domain is to classify the pixels/voxels concerning targeted regions and localize its position in the modality. Since the number of pixels corresponding to the regions of interest being less (minority class) compared to the background class (majority class), raises th class imbalance problem in the task of medical image segmentation (MIS) which cannot be addressed by the direct implication of architectures used for natural image segmentation [10].

With an uneven distribution of pixels, the model fails to learn the features concerning the minority class, due to which the false negative prediction of the pixel values increases. To address this challenging problem, several deep learning models have been proposed for MIS based on fully convolutional neural networks [12]. Most segmentation models incorporate U-Net model [19] structure that consists of symmetrical encoder-decoder design to capture feature maps efficiently at the multi-scale level and reconstruct the spatial information into desired feature space of imbalanced distribution. In addition to the architecture level solution, the objective functions are also improved due to its direct impact on the learning process of the model. The intention is to improve the performance of the model by enabling the loss function to penalize the training parameters more for false classification as compared to the true classification, thereby making the model to learn the desired features efficiently. The main contribution of the present study can be highlighted as follows:

- An exhaustive analysis for the existing loss functions for MIS by using the nn-UNet [8] as a base model over multiple datasets, where the performance of each loss function is monitored in terms of dice coefficient, Jaccard index, accuracy, precision, recall, specificity and F1-score.
- Accelerated Tversky loss (ATL) is proposed that uses a log cosh function to better optimize the gradients, while resulting in faster convergence and improvement in the segmentation results.

With the objective and contributions highlighted above, the rest of the paper is organized into several sections. The literature review section presents the recent developments in the deep learning segmentation approaches followed by materials and methods section to cover the background knowledge. The later section covers the novel loss function motivation along with the exhaustive experimental trials over multiple datasets. Finally, the results and concluding remarks are presented with future research directions.

2   Literature Review

In recent years, due to the success of deep learning approaches, the development of CAD systems has gained stature in the healthcare sector [5]. Many deep learning frameworks have been proposed for object detection, classification and segmentation in the medical imaging [21], where segmentation of the targeted regions is the initial requirement of any diagnostic procedure. For image segmentation, Badrinarayanan et al. proposed a SegNet [1] model which follows encoder-decoder architecture design. The encoder downsamples the input image to generate feature maps that are utilized by the corresponding decoder blocks to upsample the features into the input spatial resolution, followed from the softmax classifier. The SegNet model was trained using the cross-entropy loss function [27]. In the ResNet model [6], the skip connections were introduced to address the computational overhead that occurs with the increase in depth of layers. The model achieved state-of-the-art results and served as a baseline design for many other models. DenseNet [7] is an architectural advancement over the existing models that contains dense blocks of convolutional layers that are subsequently concatenated for the rich flow of information with less number of parameters.

The U-Net [19] aimed at medical image segmentation with the symmetrical encoder-decoder architecture that generates segmentation mask in a similar spatial resolution as input. Inspired by U-Net architecture, many variants of deep learning models are proposed based on the nature of the dataset covering 2D and 3D segmentation. In one such variant, inception blocks are integrated into the U-Net model to form inception U-Net model [15] that was trained using segmentation loss function proposed as the combination of cross entropy loss, dice loss and Jaccard loss. Similarly, Dense-Inception U-Net [4] is another model which combines the concept of dense blocks and inception modules into the U-Net which outperformed the existing models in terms of dice score. The same models can be extended into three dimensional volumes such as 3D inception U-Net model [16] with the addition of depth channel in the convolution operations. Apart from the architectural improvements, various loss functions were also proposed for MIS [9]. As image segmentation is similar to the classification problem, the most commonly used loss function is pixel-wise cross entropy loss. This loss evaluates the loss for class predictions of each pixel and computes the average loss for all pixels. This approach is suitable for datasets with equal distribution of samples. Medical image datasets are not equally distributed and require advance loss functions to address this imbalance. Following from this context, U-Net introduced a weighting scheme for each pixel such that the pixel at the border is associated with higher weights. The weighting scheme for pixels is further developed to generate various advanced loss functions that improve the segmentation performance. In the next section, we discuss the advanced loss functions proposed for imbalanced datasets in recent years.

3 Materials and Methods

The existing loss functions are categorized based on the mathematical interpretation into the following categories - distribution based loss, region based loss and boundary based loss as mentioned in subsequent sections.

3.1 Distribution Based Loss

These loss functions are focused on minimizing the error between the distributions.

Cross Entropy Loss - It is derived from KL-Divergence and is based on the term entropy, that denotes the number of bits required to differentiate one distribution from another. The formula to minimize the loss is represented in Eq. 1. Cross entropy (CE) used for binary classification is called as binary cross entropy (BCE). The weighted cross entropy (WCE) [14] is another variant of cross entropy, where each class is assigned a weight, w_c, based on the availability of the class samples to address the imbalance problem.

$$\mathcal{L}_{CE} = -\frac{1}{\mathcal{N}} \sum_{i,c=1}^{\mathcal{N},\mathcal{C}} g_i^c log s_i^c \tag{1}$$

where $g \in \{0, 1\}$, if class label c is correct classification for i^{th} pixel and s is the respective predicted probability.

Top-K Loss - It is used when the task is more focused on hard samples instead of easy samples [26]. The loss function shown in Eq. 2 ignores the pixels that are too easy for the model by checking their loss value. If the loss value is below the threshold, $t \in (0, 1]$, it ignores them and focuses only on hard samples.

$$\mathcal{L}_{TopK} = -\frac{1}{\sum_{i,c=1}^{\mathcal{N},\mathcal{C}} 1\{g_i = c \ and \ p_i^c < t\}} (\sum_{i,c=1}^{\mathcal{N},\mathcal{C}} 1\{y_i = c \ and \ s_i^c < t\} log s_i^c) \tag{2}$$

where p is the predicted probability of the pixel.

Focal Loss - This function is based on the standard CE loss, specifically to deal with the case of background and foreground imbalance [13]. This can be formulated as Eq. 3. It reduces the significance of easy samples by down-weighting and trains more on hard samples.

$$\mathcal{L}_F = \begin{cases} -\alpha(1-p)^\gamma log(p), & \text{if y=1} \\ -(1-\alpha)p^\gamma log(1-p), & \text{otherwise.} \end{cases} \tag{3}$$

where α controls the weight of positive and negative samples, while γ handles the easy and hard samples.

Distance Penalized CE Loss - It is an extension to the CE loss to handle the imbalance by calculating the weights based on distance maps of the ground truth masks [3] as shown in Eq. 4. This weight helps the network to focus on hard segment boundary regions.

$$\mathcal{L}_{DPCE} = -\frac{1}{\mathcal{N}}(1+\mathcal{D}) \circ \sum_{i,c=1}^{\mathcal{N},\mathcal{C}} g_i^c logs_i^c \tag{4}$$

where $\mathcal{D}$ represents the distance penalty term, $\circ$ is the Hadamard product, g is the binary output indicator and s is the predicted probability.

3.2   Region Based Loss

These functions calculate the mismatch between the actual truth value and predicted segmentation in terms of regions. The focus is on minimizing the mismatch regions or maximizing the overlapping regions.

Dice Loss - This loss is based on the dice coefficient (DC), a metric used to evaluate the segmentation and the loss function focuses on optimizing the DC. The dice loss can be defined in terms of DC (Eq. 5), as one minus the ratio of area of overlap to total pixel area.

$$\mathcal{L}_{Dice} = 1 - \frac{2y\hat{p} + 1}{y + \hat{p} + 1} \tag{5}$$

Specificity-Sensitivity Loss - Specificity (Sp) and sensitivity (Se) are the evaluation metrics that measure the true negative rate and true positive rate of the model. This loss function as shown in Eq. 7 is based on these metrics [2] to address the imbalance problem by applying weight (w) on each metric.

$$Sp. = \frac{TN}{TN + FP} \, , \, Se. = \frac{TP}{TP + FN} \tag{6}$$

$$\mathcal{L}_{SS} = w * Se + (1 - w) * Sp \tag{7}$$

where, TN is true negative, TP is true positive, FN is false negative and FP is false positive.

IoU Loss - Intersection over union or Jaccard index (JI) is the most widely used metric in segmentation of imbalanced datasets [17]. The IoU index is the ratio of area of overlap to the area of union. The IoU loss is represented in the Eq. 8.

$$JI = \frac{TP}{TP + FP + FN} \quad \mathcal{L}_{IoU} = 1 - JI \tag{8}$$

3.3 Boundary Based Loss

These loss functions work based on the distance between the ground truth region and predicted the segmented region. The aim is to minimize the distance between the ground truth and predicted segment images.

Boundary Loss - This loss function computes the distance mismatch between ground truth and predicted segment image using the integrals over boundary regions [11]. The distance map along with network softmax outputs are used together to form a trainable loss function as shown in Eq. 9.

$$\mathcal{L}_{BD} = \int \Phi_g(p)s_\theta(p)dp \tag{9}$$

where Φ is the representation of boundary distance and s is the softmax probability output.

Hausdorff Distance Loss - The loss function focuses on maximizing the Hausdorff distance (d), an evaluation metric of segmentation, which calculates the error between the distance transforms of ground truth and predicted segmentation images. The Hausdorff metric and loss [18] is given by the Eq. 11.

$$d(X,Y) = \max_{x \in X} \min_{y \in Y} \|x - y\|^2 \tag{10}$$

$$\mathcal{L}_{HD} = \frac{1}{N} \sum_{i=1}^{N}[(s_i - g_i)\dot{(}d_{gi}^2 + d_{si}^2)] \tag{11}$$

3.4 Compound Loss

These loss functions are obtained by combining the above discussed loss functions.

Combo Loss - It is obtained by combining dice loss with weighted cross entropy loss [23].

$$\mathcal{L}_{combo} = w\mathcal{L}_{CE} - (1 - w)\mathcal{L}_{Dice} \tag{12}$$

Exponential Logarithmic Loss - The dice and cross entropy loss are used along with the exponential and logarithmic functions as shown in Eq. 14 such that the network can focus more on areas that are predicted less accurately [25].

$$\mathcal{L}_{Dice} = exp(-ln(DC)^\gamma) \quad \mathcal{L}_{CE} = exp(w_i(-ln(p_t))^\gamma) \tag{13}$$

$$\mathcal{L}_{exp} = w_{Dice}\mathcal{L}_{Dice} + w_{CE}\mathcal{L}_{CE} \tag{14}$$

From the above categories of loss functions, the region based loss functions are more suitable for the medical domain. The boundary based loss functions can be used along with region based loss functions to get better performance. We can also create custom loss functions by combining two or more loss functions based on the dataset and requirement. Hence, the selection of loss function that can be applied for all datasets is difficult. The performance of each of these loss functions is analysed by training the nn-Unet model on multiple datasets.

3.5   nn-UNet

No new U-Net also known as nn-UNet [8] is the same as the basic U-Net in terms of architecture but focused on proposing an automated framework for the entire segmentation process that can configure the non-architectural aspects based on input datasets. The intention to use nn-Unet is that U-Net is the backbone of this architecture which provides optimal results for MIS. Moreover, the nn-UNet framework supports the evaluation of various datasets without the need of manual intervention to tune the parameters. The complicated process of automating segmentation for a dataset is explained using Fig. 1 (from left to right). It shows the steps to be followed in a sequence to setup the nn-UNet architecture and automate the training and tuning process for various datasets. The pipeline of nn-UNet starts with train data, but to handle a wide variety of datasets, nn-UNet proposed a file structure to arrange the input data into train, test and label images along with a JSON file describing the dataset properties. The converted train data is used to extract the dataset fingerprints such as image size, the modality of the image, number of classes etc. Later, the blueprint parameters such as loss functions, optimizer and architecture template are configured. The inferred parameters generate the image resampling, normalization and batch size parameters that are combined with feedback to generate pipeline fingerprints. nn-UNet provides three variants of U-Net namely 2D U-Net, 3D U-Net and 3D-cascade U-Net to handle all possible datasets. The pipeline fingerprints generate the model based on the input, user choice of the training model and the hyper-parameters. After training the model, an ensemble of various networks can be performed to determine the model with the highest dice score which is then used to perform inference on test images.

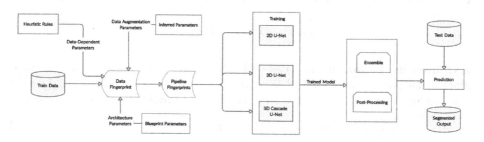

Fig. 1. nnU-Net pipeline to automate the segmentation process.

3.6   Proposed Loss Function

It is evident from the literature review that the class imbalance problem of MIS is addressed using class weights. The class weights modify the loss functions by assigning more weights to penalize the false predictions. In the Tversky loss function, the Tversky index is utilized to evaluate the error as shown in Eq. 15, that assigns different weights to more penalize the training weights for false negative and false positive predictions. But using the Tversky index inside the loss function may result in sub optimal gradients due to its non-convex nature. With this motivation, in the novel accelerated Tversky loss (ATL) function, Tversky index is encapsulated inside log cosh function. The log-cosh function acts as convex surrogate and reduces the non-convexity. This transformation of loss function through encapsulation smoothens the curve thus optimizing the gradients. For $F(x) = log(cosh(x))$, the derivative $F'(x) = tanh(x)$. The $tanh$ function is continuous and differentiable in the range $[-1, 1]$ and due to its nonlinearity, the non-convex nature of the loss function is reduced compared to the Tversky loss function. The ATL function can be formulated as in Eq. 16.

$$\mathcal{L}_{Tv.} = 1 - \frac{p\hat{p}}{p\hat{p} + \beta(1-p)\hat{p} + \alpha p(1-\hat{p})} \tag{15}$$

$$\mathcal{L}_{ATL} = log(cosh(1 - \frac{p\hat{p}}{p\hat{p} + \beta(1-p)\hat{p} + \alpha p(1-\hat{p})})) \tag{16}$$

To fine tune the training parameters of the model the gradient of the $\mathcal{L}_{ATL}$ can be computed with respect to the predicted value during the backward pass using the Eq. 17.

$$\frac{\partial \mathcal{L}_{ATL}}{\partial \hat{p}} = tanh(\mathcal{L}_{Tv.})\frac{\partial \mathcal{L}_{Tv.}}{\partial \hat{p}} \tag{17}$$

$$\frac{\partial \mathcal{L}_{Tv.}}{\partial \hat{p}} = -\frac{\alpha p^2}{(p\hat{p} + \beta(1-p)\hat{p} + \alpha p(1-\hat{p}))^2} \tag{18}$$

4   Experiments and Results

4.1   Experiments

The experiments are conducted to compare the performances of loss functions using the nn-Unet model. To compare the performances, the considered evaluation metrics are accuracy, precision, f1-score, dice score, Jaccard index, specificity and sensitivity. The comparison is done on loss functions that are widely used for MIS such as dice loss, Tversky loss, dice and cross entropy combined, focal loss, specificity-sensitivity loss and the proposed accelerated Tversky loss. The datasets used for the experiment are Hippocampus, Prostate and Heart dataset which contain the MRI images of organs and Hepatic Vessel dataset which contains CT images, under the Medical Segmentation Decathlon challenge [20]. The training and evaluation of the proposed framework is performed in the high performance computing environment having NVidia Titan GPUs in Google Colab.

The 3D-UNet is chosen as the base model and the network topology, batch size and patch size are automatically tuned based on the input dataset dimensions. The model is trained using 5-fold cross validation technique and an early stopping technique is employed to halt the training if there is no improvement in terms of loss for consecutive iterations.

Table 1. Metrics for datasets over various loss functions.

Dataset	Metrics	Dice	DC+CE	Tversky	Focal	SS loss	ATL
Hippocampus	Accuracy	0.994	0.995	0.994	0.994	0.992	**0.995**
	Precision	**0.912**	0.911	0.861	0.902	0.786	0.848
	F1 score	**0.913**	0.912	0.905	0.891	0.869	0.896
	Dice score	**0.913**	0.911	0.904	0.891	0.868	0.895
	Jaccard index	0.813	0.834	0.827	0.801	0.769	**0.842**
	Specificity	0.997	0.997	0.995	0.996	0.992	**0.997**
	Sensitivity	0.918	0.908	0.954	0.881	**0.974**	0.951
Prostate	Accuracy	0.997	0.996	0.997	0.997	0.996	**0.998**
	Precision	0.928	**0.948**	0.849	0.842	0.716	0.833
	F1 score	0.925	**0.934**	0.898	0.819	0.818	0.871
	Dice score	0.918	**0.925**	0.884	0.809	0.799	0.861
	Jaccard index	0.861	**0.862**	0.815	0.704	0.686	0.777
	Pecificity	0.997	0.997	0.998	0.998	0.997	**0.998**
	Sensitivity	0.924	0.921	0.955	0.814	**0.955**	0.924
Heart	Accuracy	0.999	0.998	0.999	0.998	0.999	**0.999**
	Precision	**0.964**	0.939	0.898	0.929	0.821	0.914
	F1 score	**0.964**	0.938	0.936	0.915	0.895	0.941
	Dice score	0.942	0.939	0.936	0.915	0.895	**0.952**
	Jaccard index	**0.908**	0.884	0.881	0.845	0.812	0.888
	Specificity	0.998	0.998	0.998	0.998	0.998	**0.998**
	Sensitivity	0.941	0.937	0.978	0.901	**0.985**	0.969
Hepatic Vessel	Accuracy	0.998	0.998	0.998	0.998	0.998	**0.998**
	Precision	0.758	**0.787**	0.722	0.707	0.545	0.717
	F1 score	0.777	0.778	**0.789**	0.664	0.678	0.768
	Dice score	0.763	0.762	**0.783**	0.644	0.645	0.761
	Jaccard index	**0.834**	0.651	0.654	0.518	0.506	0.633
	Specificity	0.998	0.996	0.998	0.998	0.998	**0.999**
	Sensitivity	0.797	0.769	0.871	0.627	**0.898**	0.829

*Bold values indicate the highest metric value

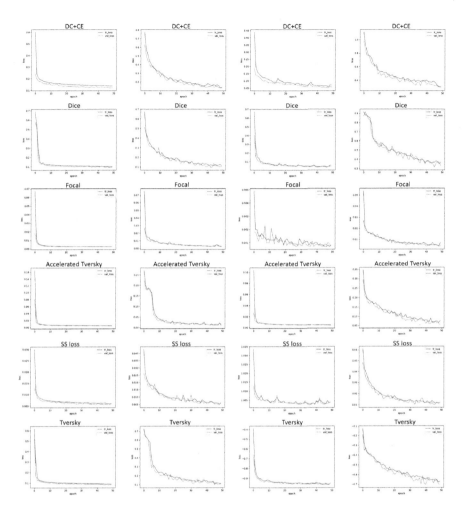

Fig. 2. Loss vs epoch of various loss functions on each datatset. Each columns corresponds to a dataset. The datasets from left to right are - Hippocampus, Prostate, Heart and Hepatic Vessel.

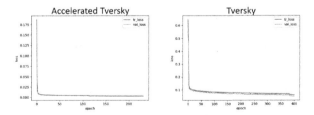

Fig. 3. Convergence of loss functions using early stopping on Hippocampus dataset.

4.2   Results and Discussion

The segmentation metrics for each dataset and corresponding loss functions are stored in a JSON file which contains the metric values for each data item and overall mean values. The training progress for each dataset and the corresponding loss function is presented in a graph, as shown in Fig. 2. Besides, a detailed compilation of the metrics for each loss function is summarized in Table 1. From the table, it is observed that the ATL function provides the best results in specificity, accuracy and is relatively similar in other metrics, indicating that false acceptance of the pixels is minimized. Furthermore, the improvement in the rate of convergence is conveyed through the number of epochs at which the training is halted. From the plots of the loss function in Fig. 3, it is observed that the ATL function converges 46.75% faster than the vanilla Tversky loss function. The improvement in rate of convergence for Prostate, Heart and Hepatic Vessel datasets is 22.3%, 25.7% and 43.2% respectively. With such results, it is believed that the proposed loss function can be applied to any class imbalance problem.

5   Conclusion

In this paper, we discussed the difficulties faced in the medical image segmentation problem and explored various loss functions that are applicable to overcome the class imbalance problem. We have also proposed a novel loss function named accelerated Tversky loss function that tends to better optimize the training gradients. The performance is compared with the widely used segmentation loss functions while using nn-Unet as a base model for training across multiple datasets. With the exhaustive experiments, it is observed that the novel loss function is on par with the best values for most of the evaluation metrics along with the faster rate of convergence towards minimum error. The research can further be extended in the direction to find an optimal loss function that guarantees best results across multiple modalities.

References

1. Badrinarayanan, V., Kendall, A., Cipolla, R.: SegNet: a deep convolutional encoder-decoder architecture for image segmentation. IEEE Trans. Pattern Anal. Machine Intell. **39**(12), 2481–2495 (2017)
2. Brosch, T., Yoo, Y., Tang, L.Y.W., Li, D.K.B., Traboulsee, A., Tam, R.: Deep convolutional encoder networks for multiple sclerosis lesion segmentation. In: Navab, N., Hornegger, J., Wells, W.M., Frangi, A.F. (eds.) MICCAI 2015. LNCS, vol. 9351, pp. 3–11. Springer, Cham (2015). https://doi.org/10.1007/978-3-319-24574-4_1
3. Caliva, F., Iriondo, C., Martinez, A.M., Majumdar, S., Pedoia, V.: Distance map loss penalty term for semantic segmentation. arXiv preprint arXiv:1908.03679 (2019)
4. Zhang, Z., Wu, C., Coleman, S., Kerr, D.: Dense-inception U-Net for medical image segmentation. Comput. Methods Programs Biomed. **192**, 105395 (2020). ISSN 0169-2607

5. Esteva, A., et al.: A guide to deep learning in healthcare. Nature Med. **25**(1), 24–29 (2019)
6. He, K., Zhang, X., Ren, S., Sun, J.: Deep residual learning for image recognition. In: Proceedings of the IEEE Conference on Computer Vision and Pattern Recognition, pp. 770–778 (2016)
7. Huang, G., Liu, Z., van der Maaten, L., Weinberger, K.Q.: Densely connected convolutional networks (2018)
8. Isensee, F., et al.: Automated design of deep learning methods for biomedical image segmentation (2020). arXiv: 1904.08128 [cs.CV]
9. Jadon, S.: A survey of loss functions for semantic segmentation. arXiv preprint arXiv:2006.14822 (2020)
10. Jain, A., Ratnoo, S., Kumar., D.: Addressing class imbalance problem in medical diagnosis: a genetic algorithm approach. In: International Conference on Information, Communication, Instrumentation and Control, pp. 1–8 (2017)
11. Kervadec, H., Bouchtiba, J., Desrosiers, C., Granger, E., Dolz, J., Ben Ayed, I.: Boundary loss for highly unbalanced segmentation. Medical Image Anal. **67**, 101851 (2021). ISSN 1361–8415
12. Lei, T., Wang, R., Wan, Y., Du, X., Meng, H., Nandi, A.K.: Medical image segmentation using deep learning: a survey. arXiv preprint arXiv:2009.13120 (2020)
13. Lin, T., et al.: Focal loss for dense object detection. In: 2017 IEEE International Conference on Computer Vision, pp. 2999–3007 (2017)
14. Sankaran, P., et al.: Multi-task learning and weighted cross-entropy for DNN-based keyword spotting. In: Interspeech, vol. 9, pp. 760–764 (2016)
15. Punn, N.S., Agarwal, S.: Inception u-net architecture for semantic segmentation to identify nuclei in microscopy cell images. ACM Trans. Multimedia Comput. Commun. Appl. (TOMM) **16**(1), 1–15 (2020)
16. Punn, N.S., Agarwal. S.: Multi-modality encoded fusion with 3d inception U-Net and decoder model for brain tumor segmentation. In: Multimedia Tools and Applications, pp. 1–16 (2020)
17. Rahman, M.A., Wang, Y.: Optimizing intersection-over-union in deep neural networks for image segmentation. In: ISVC (2016)
18. Ribera, J., Güera, D., Chen, Y., Delp, E.: Weighted hausdorff distance: a loss function for object localization. ArXiv, abs/1806.07564 (2018)
19. Ronneberger, O., Fischer, P., Brox, T.: U-Net: convolutional networks for biomedical image segmentation. In: Navab, N., Hornegger, J., Wells, W.M., Frangi, A.F. (eds.) MICCAI 2015. LNCS, vol. 9351, pp. 234–241. Springer, Cham (2015). https://doi.org/10.1007/978-3-319-24574-4_28
20. Simpson, A.L., et al.: A large annotated medical image dataset for the development and evaluation of segmentation algorithms (2019). arXiv: 1902.09063 [cs.CV]
21. Suzuki, K.: Overview of deep learning in medical imaging. Radiol. Phys. Technol. **10**(3), 257–273 (2017)
22. Szegedy C., et al.: Going deeper with convolutions. In: 2015 IEEE Conference on Computer Vision and Pattern Recognition, pp. 1–9 (2015)
23. Taghanaki, S.A., et al.: Combo loss: Handling input and output imbalance in multi-organ segmentation (2018). arXiv: 1805.02798 [cs.CV]
24. Tajbakhsh, N., Jeyaseelan, L., Li, Q., Chiang, J., Wu, Z., Ding, X.: Embracing imperfect datasets: a review of deep learning solutions for medical image segmentation. Medical Image Anal. **63**, 101693 (2020)

25. Wong, K.C.L., Moradi, M., Tang, H., Syeda-Mahmood, T.: 3D segmentation with exponential logarithmic loss for highly unbalanced object sizes. In: Frangi, A.F., Schnabel, J.A., Davatzikos, C., Alberola-López, C., Fichtinger, G. (eds.) 3d segmentation with exponential logarithmic loss for highly unbalanced object sizes. LNCS, vol. 11072, pp. 612–619. Springer, Cham (2018). https://doi.org/10.1007/978-3-030-00931-1_70

26. Wu, Z., Shen, C., van den Hengel, A.: Bridging category-level and instance-level semantic image segmentation (2016). arXiv: 1605.06885 [cs.CV]

27. Zhang, Z., Sabuncu, M.: Generalized cross entropy loss for training deep neural networks with noisy labels. Adv. Neural. Inf. Process. Syst. **31**, 8778–8788 (2018)

Incorporating Relational Knowledge in Explainable Fake News Detection

Kun Wu[1(✉)], Xu Yuan[2], and Yue Ning[1]

[1] Stevens Institute of Technology, Hoboken, NJ, USA
kwu14@stevens.edu, xu.yuan@louisiana.edu
[2] University of Louisiana at Lafayette, Lafayette, LA, USA
yue.ning@stevens.edu

Abstract. The greater public has become aware of the rising prevalence of untrustworthy information in online media. Extensive adaptive detection methods have been proposed for mitigating the adverse effect of fake news. Computational methods for detecting fake news based on the news content have several limitations, such as: 1) Encoding semantics from original texts is limited to the structure of the language in the text, making both bag-of-words and embedding-based features deceptive in the representation of a fake news, and 2) Explainable methods often neglect relational contexts in fake news detection. In this paper, we design a knowledge graph enhanced framework for effectively detecting fake news while providing relational explanation. We first build a credential-based multi-relation knowledge graph by extracting entity relation tuples from our training data and then apply a compositional graph convolutional network to learn the node and relation embeddings accordingly. The pre-trained graph embeddings are then incorporated into a graph convolutional network for fake news detection. Through extensive experiments on three real-world datasets, we demonstrate the proposed knowledge graph enhanced framework has significant improvement in terms of fake news detection as well as structured explainability.

Keywords: Fake news detection · Knowledge graphs · Explainable machine learning

1 Introduction

Misinformation in online media has become a menace, from being a public concern [7,12] to causing major financial loss and security risks. Existing work on content-based fake news detection focuses on semantic content using statistical or deep learning models [22] while neglecting rich relational information among entities (names, organizations, etc.). In this paper, we propose to investigate a self-discovered knowledge graph method to enhance the representation learning of entities and relations in fake news detection. While knowledge-based fact checking approaches [4,15,28] have been studied, they often suffer from issues

© Springer Nature Switzerland AG 2021
K. Karlapalem et al. (Eds.): PAKDD 2021, LNAI 12714, pp. 403–415, 2021.
https://doi.org/10.1007/978-3-030-75768-7_32

such as reliability or incompleteness of web knowledge. In contrast to these previous approaches, we extract a credential based multi-relation knowledge graph from our corpus without external domain knowledge. We do not involve external data considering 1) accessibility: extracting knowledge from given news corpora is more flexible and scalable compared to external knowledge base; 2) dynamic context: knowledge is dynamic and updates over time. Keeping external data up to date requires excessive human labor efforts; and 3) relevance: external knowledge often contain global noise rather than useful information.

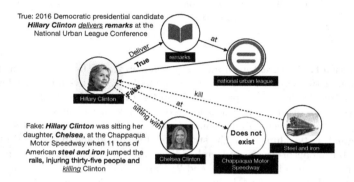

Fig. 1. An example of a knowledge graph extracted from news articles.

In this work, we follow the broad definition of fake news [31] as "false news" where news includes false information related to public figures and organizations in articles, statements, and speeches. The veracity of news articles can be discovered from multiple aspects such as writing styles, languages, and focused stories. From the presented example in Fig. 1, we observe one important factor that distinguishes fake news from real news is the involved entities and their relations. Given the significant roles of entities and their relations in news content, we create a knowledge graph of entities and relations that appear in existing data to represent a few aspects of structured knowledge: credentials, relations, and contexts. In which, each node represents an entity (e.g., persons, organizations, locations) and each edge/relation between a pair of nodes indicates the action (e.g., predicate) among them.

However, several challenges are encountered in learning KG-based representations. First, multiple relations may exist between pairs of entities when two entities appear in different news articles. For instance, [Obama, approves, nuclear deals] (*fake*) and [Obama, plans, nuclear policy changes] (*real*) are two relations between the same pair of entities from different contexts. Thus, relational information embeds credential values while most existing works ignore these structured knowledge. Second, relations between entities can be complex and changing over time. Figure 1 shows an example in the PolitiFact dataset [23], where same entities appear in both fake news and true news. However, the relation between entity "Hilary Clinton" and entity "Steel and iron" is not trustworthy given that they only appear in fake news. Third, integrating and fusing heterogeneous information is challenging. Relational representations into semantic encoding has been proven effective in several natural language processing tasks [30]. However, discovering relational indicators for fake news remains open.

To address aforementioned challenges, we summarize the main contributions of this work as below:

- We build a credential-based multi-relation knowledge graph from existing fake news corpora. Each link between a pair of entities indicates the relation/action from the source entity to the target entity.
- We apply a compositional graph convolutional network to pre-train relational representations of entities and relations simultaneously from the discovered knowledge graph. Thus, the representations of new entities or relations can be inferred and updated.
- We design a new framework to enhance semantic embeddings with structured knowledge in order to predict the trustworthiness of a news article (fake or real). The knowledge embeddings include both relations and entities information. In addition, the proposed framework is able to provide explainable relational evidence for predictions of fake news.

2   Related Work

This section reviews the state-of-the-art methods in the context of fake new detection and knowledge graph learning, and discuss the advantage of our work over them.

Content-Based Fake News Detection. Content-based solutions have attracted wide attention, which mainly focus on extracting the semantics or writing styles of the news articles [9,22,27]. For example, an attention-based deep learning approach (i.e., dEFEND [22]) was proposed for jointly capturing explainable top-k sentences and user comments for fake news detection using a sentence-comment co-attention sub-network. Wang *et al.* [27] presented an event adversarial network in multi-task learning to derive event-invariant features, which can benefit the detection of fake news on newly arrived events. They considered event types along with an adversarial network to better learn the representation of news. Additionally, the images in news articles are also encoded with a CNN model for combining the image features with text features. Levi *et al.* [13] designed a machine learning model using semantic and linguistic features to distinguish fake news from satire stories. Recently, Nguyen *et al.* [16] presented a Markov random field (MRF) model to study the correlation association among documents to assist fake news detection. The most recent approach [14] takes into account short texts (e.g., tweets, users' credits, and propagation patterns) in a Graph-aware Co-Attention Network (GCAN) where the representations of the corresponding source text, user features, and propagation graphs are learned first. Then, a dual co-attention model is developed for prediction. However, most of the content-based detection methods face a few limitations: 1) the leveraged auxiliary features (e.g. user comments, images, etc.) are tailored to the specific domains, which thus cannot be scaled or generalized to a different domain; 2) explainability is limited: attention mechanisms focus on existing features (e.g., words), failing to capture relational dependencies.

Knowledge-Based Fact Checking. Information retrieval methods based on knowledge have been proposed to determine the veracity of news articles. For instance, Magdy *et al.* [15] identified the trustworthiness of a claim by using query results from the web. Wu *et al.* [28] presented a method through "perturbing" a claim from querying knowledge bases and using the result variations as an indicator for fact checking. In addition, Ciampaglia *et al.* [4] considered the shortest path between concepts in a knowledge graph and [21] employed a link prediction algorithm with discriminative meta paths for fact checking. However, these approaches encounter problems of determining the trustworthiness and reliability of the external knowledge (web or knowledge base). In addition, they are deficient when the corresponding entries do not exist in a knowledge base or the knowledge base is compromised.

Knowledge Graphs (KG) that organize relations of entities in directed graphs are widely used in many fields, such as link prediction and question answering. KGs are constructed from triples, e.g., (head, relation, tail) or (subject, predicate, object), to provide rich and strong facts to enhance the understanding of natural languages [26]. Knowledge graph embedding focuses on learning hidden representations of nodes and/or relations. A few state-of-the-art approaches include: TransE [2], DistMult [29], and ConvE [6]. A recent development for multi-relation representation learning in KGs, CompGCN [25], jointly learns the embeddings of nodes and relations using a graph convolutional network. A knowledge graph based fake news detection [18] utilizes the ability of link prediction in KGs to detect fake news. It extracts triples from news and employs TransE to present entities and relations into a vector space. By measuring the distances between subjects combined with relations and objects extracted from news, they can predict the veracity of the news. However, this approach only considers the features of knowledge graphs, omitting global semantic features of news which also provide critical information for fake news detection.

To conclude, our approach will explore semantic content of news articles and enrich the semantic features with structural embeddings from knowledge graphs. Our developed KG embedding model can be compatible with other models and offer relation level explainability beyond keywords' contributions.

3   The Proposed Method

In this section, we present our design of a novel **knowledge graph** enhanced framework for **f**ake news detection, abbreviated as **KGF**, that can be applied in a variety of deep learning models to jointly predict if a news article is fake while providing explainable structured knowledge for the prediction. The overall framework is present in Fig. 2. We first introduce the notations and the problem formulation and then we discuss the details of the proposed framework.

Notations and Problem Formulation. Given a collection of news articles $\mathcal{D}$, each one contains a sequence of words $\{w_1, ..., w_k, ..., w_W\}$ and its corresponding label $y \in \{0, 1\}$ indicating if the news is fake ($y = 1$) or not ($y = 0$). We extract a knowledge graph from this corpus and denote it by $\mathbf{G} = (\mathcal{V}, \mathcal{E}, \mathcal{X}, \mathcal{Z})$ where $\mathcal{V} = \{v_1, v_2, ...v_{|\mathcal{V}|}\}$ denotes the set of vertices (i.e., entities) such as person names or locations. $\mathcal{E}$ denotes the edges between pairs of nodes where each entry $\mathcal{E}[i, j] = \{r_1, r_2, ...r_{|\mathcal{E}[i,j]|}\}$ is a set of relations between node i and node j given that entity i and j may appear in different contexts. $\mathcal{X} \in \mathbb{R}^{|\mathcal{V}| \times d_0}$ denotes the d_0-dimensional input features of each node. $\mathcal{Z} \in \mathbb{R}^{|\mathcal{R}| \times d_0'}$ denotes the d_0'-dimensional input features of each relation.

Knowledge Graph Extraction. We apply the Stanford NLP tool, OpenIE [1], to extract triples from sentences. Each triple (u, r, v) consists of a source entity which is the subject in a clause, a target entity v which is an object in a clause, and a relation r between them. The subject and object entities are usually persons, places, organizations, or general nouns. The relation, also called predicate, is a directed action (e.g., verb) from a subject to an object. As such, we get a set of triples from each news article. During this process, we notice that the OpenIE tool generates some noisy triples. For instance, the triples extracted from the sentence: *"the American people must be able to trust that the American people government is looking out for all of us"* are: *('american people', 'must', 'must able')*. This kind of triples is noise we want to avoid. Hence, we investigate a few techniques to improve the quality of extracted entities and relations. First, We adopt a coreference resolution approach, NeuralCoref, to avoid the ambiguity of pronouns.[1] Next, we use Spacy to lemmatize the verbs in a relation given multiple tenses.[2] To reduce the number of duplicated relations, we remove the adverb in the predicate and only keep the lemmatized verb. As shown in the previous examples, we find that subject or object entities may be invalid. We assume that entities have to contain at least one noun. Therefore, we filter out all the entities that do not contain a noun.

Learning Relational Representations. After cleaning the triples, we organize all the entities into nodes and construct a multi-relational knowledge graph **G** based on all the triples in our training corpus. In this multi-relational graph, we assume each node and relation is encoded by an embedding vector. We adapt the compositional graph convolutional network (CompGCN) [25] to jointly embed both nodes and relations in a relational graph. Assuming node v is an object entity node, $N(v)$ is a set of its immediate neighbors for its incoming edges, and each edge corresponds to a specific relation, CompGCN updates the object node embedding vector as below:

$$\mathbf{h}_v^{(l+1)} = f\Big(\sum_{(u,r) \in N(v)} \mathbf{W}_q^{(l)} \phi\big(\mathbf{h}_u^{(l)}, \mathbf{o}_r^{(l)}\big)\Big) \in \mathbb{R}^{d_{l+1}}, \tag{1}$$

---

[1] https://github.com/huggingface/neuralcoref.
[2] https://spacy.io/.

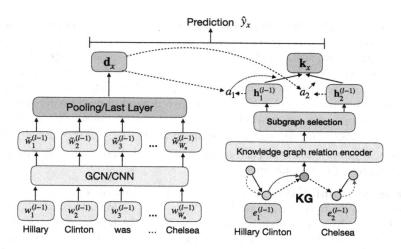

Fig. 2. The overall framework of knowledge graph enhanced fake news detection. The left module is a standard deep neural network on word embedding features. The right module learns the embeddings of entities and their relations using the self-discovered knowledge graph from the corpus. In the aggregation part, the semantic document embeddings learned from the left module are combined with the knowledge embeddings of the document.

where $\phi : \mathbb{R}^d \times \mathbb{R}^d \to \mathbb{R}^d$ is a composition operation [17] between the subject vector $\mathbf{h}_u^{(l)}$ and the relation vector $\mathbf{o}_r^{(l)}$. Layer-wise parameter matrix $\mathbf{W}_q^{(l)} \in \mathbb{R}^{d_{l+1} \times d_l}$ maps the dimension of hidden features from layer l to layer $l+1$. The first layer embedding matrix $\mathbf{H}^{(0)}$ is initialized by $\mathcal{X}$. After the node embedding update, the relation embeddings are also transformed as follows:

$$\mathbf{o}_r^{(l+1)} = \mathbf{W}_{\text{rel}}^{(l)}\mathbf{o}_r^{(l)} \in \mathbb{R}^{d_{l+1}}, \tag{2}$$

where $\mathbf{W}_{\text{rel}}$ is a learnable transformation matrix which projects edges to the same embedding space as nodes. The relation embedding matrix is initialized by $\mathbf{O}^{(0)} = \mathcal{Z}$.

Given a subject entity (u), a relation (r), and their anticipated object entity (v), we design a link prediction task as in ConvE [6] to estimate the embedding parameters. Both subject and relation embeddings are passed through a convolutional layer and several fully connected layers to get an estimated vector for the object (v). The score function s estimates the similarity between the estimated vector and the anticipated object entity. The loss function for the link prediction task is defined as below:

$$L_G = -\sum_{(u,r)} \sum_{e \in \mathcal{V}} \Big[y_e \log \sigma(s(u,r,e)) + (1 - y_e) \log(1 - \sigma(s(u,r,e))) \Big], \tag{3}$$

where e is a randomly sampled object entity. When e is equal to the ground truth object (v), $y_e = 1$. Otherwise, $y_e = 0$. $\mathcal{V}$ is the set of nodes. We adopt ConvE [6] as the score function $s(u, r, e) = s(\phi(\mathbf{h}_u, \mathbf{h}_r), \mathbf{h}_e)$ where ϕ denotes a composition operator for estimating the embedding vector of the object given a subject u and a relation r. All model parameters can be trained via back-propagation and optimized using the Adam algorithm.

Integration of Relational Knowledge in Detection. After learning the embedding vectors of all triples in the knowledge graph, we incorporate the pre-trained embeddings in a global vector to improve prediction performance. For each node, its embedding vector is learned from its neighbors and their corresponding multi-type relations. Therefore, we utilize node embeddings instead of triple embeddings. Given a news article x, assuming there are N_x entity extracted from the article. The corresponding entity/node embedding are denoted as $N = \{n_0, n_1, ..., n_{N_x}\}$. We apply a global attention mechanism to capture the contributions of the nodes embeddings to the global semantic vector of an article. We first introduce how we learn the global vector $\mathbf{d}_x$ for each news article.

Learning Global Semantics. Assuming the word embedding matrix for a document x is represented by $\mathbf{E}_x \in \mathbb{R}^{W \times d}$, where W is the number of unique words in this document. We take advantage of a multi-layer Graph Convolutional Network to learn hidden representations of words given its effectiveness [11]. We build an adjacency matrix $\mathbf{A}$ to represent the context frequency between pairs of words. Following the setup in the Dynamic GCN model for event predictions [5], the edge weight between two words in a document is calculated as below:

$$\mathbf{A}[i,j] = \begin{cases} \text{PMI}(i,j) & \text{PMI}(i,j) > 0 \\ 0 & \text{otherwise.} \end{cases} \tag{4}$$

The PMI value of a word pair i, j is computed as $\text{PMI}(i, j) = \log \frac{s(i,j)}{s(i)s(j)/S}$, where $s(i)$ and $s(j)$ are the total number of sentences in the document containing at least one occurrence of i and j, respectively. S is the total number of sentences in the document. The message passing process on the graph of words is denoted as $\mathbf{H} = f(\hat{\mathbf{A}}\mathbf{E}\mathbf{W}_g)$ where $\hat{\mathbf{A}}$ is the normalized symmetric adjacency matrix $\hat{\mathbf{A}} = \tilde{\mathbf{D}}^{-\frac{1}{2}}(\mathbf{A} + \mathbf{I}_W)\tilde{\mathbf{D}}^{-\frac{1}{2}}$. $\tilde{\mathbf{D}}$ is the degree matrix. $\mathbf{I}_W$ is an identity matrix with dimensions of W. Eventually the hidden features of words are updated by their neighboring word vectors. Assuming the final layer output is $\mathbf{H}^{(L)} \in \mathbb{R}^{W \times d_w}$, we adapt a pooling strategy to get the semantic embedding of the document: $\mathbf{d}_x = \text{pooling}(\mathbf{H}_i) \in \mathbb{R}^{d_w}, i \in \{1, ..., W\}$.

Bridging Relational Knowledge with Global Semantics. We next learn relational representation of news based on the relational embeddings of entities. Assuming there are N_x entities in news x and each entity embedding $\mathbf{h}_i \in \mathbb{R}^{d_L}$ ($i = 1, ...N_x$) is learned from the previous CompGCN, the relational embedding vector of this news is calculated as $\mathbf{k}_x = \sum_i^{N_x} \alpha_i \mathbf{h}_i \in \mathbb{R}^{d_L}$ where α_i is the attention weight of each entity and is computed as follows:

$$\alpha_i = \frac{\exp(\mathbf{t}_i^\top \mathbf{d}_x)}{\sum_{n=0}^{N_x} \exp(\mathbf{t}_n^\top \mathbf{d}_x)}. \tag{5}$$

Here, $\mathbf{t}_i \in \mathbb{R}^{d_w}$ is the entity embedding after projected into the same vector space of semantic embedding by:

$$\mathbf{t}_i = \sigma(\mathbf{W}\mathbf{h_i} + \mathbf{b}) \in \mathbb{R}^{d_w}, \tag{6}$$

where σ is activation function, $\mathbf{W} \in \mathbb{R}^{d_w \times d_L}$ and $\mathbf{b} \in \mathbb{R}^{d_w}$ are trainable parameters, and $\mathbf{h}_i$ is the representation of entity i from the last layer of our multi-relational graph.

Optimization. After obtaining the semantic embedding and the knowledge embedding, we apply a single layer MLP on the concatenation of these two vectors to predict the label of the document $\hat{y}_x = \sigma(\mathbf{w}^\top[\mathbf{k}_x \oplus \mathbf{d}_x])$. The ground truth labels of the news articles are binary. Thus, we adopt binary cross-entropy loss to optimize the model parameters:

$$L = -\sum_{x=1}^{D} \big(y_x \log \hat{y}_x + (1 - y_x) \log(1 - \hat{y}_x)\big), \tag{7}$$

where y is the ground truth and $\hat{y}$ is the model prediction. All model parameters can be trained via back-propagation and optimized using the Adam algorithm given its efficiency and ability to avoid overfitting.

4   Experiment Setup

In this section, we introduce the datasets, the baseline methods for comparison, and the evaluation metrics for measurement in our experiments.

Datasets. To fairly evaluate the performance of our model, we conduct the experiments on three datasets corresponding to different topics: 1) *Celebrity* dataset [19] was collected from web sources targeting rumors, hoaxes, and fake reports on celebrities. We sampled 250 fake news and 250 real news with 1670 relations, 19978 entities, and 31857 triples in total. 2) *PolitiFact* dataset [23, 24] was collected from "politifact.com" and most news are related to political campaigns. We sampled 474 real news and 369 fake news with 51918 entities, 3251 relations, and 91366 triples in total. 3) *GossipCop* dataset [23,24] was collected from "E!Online (eonline.com)" and "GossipCop.com". We sampled 500 real news and 500 fake news with 43371 entities, 2438 relations, and 71842 triples in total.

Comparison Methods. We compare the proposed model with some common NLP models and several state-of-the-art fake news detection methods as baselines including: 1) logistic regression models with news style features by mapping the frequencies of rhetorical relations to a vector space (**RST**) [20]; 2) Recurrent neural networks (RNN) including **vanilla RNN**, Long Short-Term Memory (**LSTM**) [8], Gated Recurrent Units (**GRU**) [3]; 3) Text Convolutional Neural Networks (**Text-CNN**) [10]; 4) Graph based models such as Graph Convolutional Networks (**GCN**) [11], Compositional Graph Covluiontal Networks

(**CompGCN**) [25] using pre-trained knowledge graph features; 5) Attention-based approaches such as **dEFEND**◇ [22], and 6) the Hierarchical Discourse-level Structure (**HDSF**) [9] model. We implement the **dEFEND**◇ model without news comments and use the source code of **HDSF** from its paper directly in our k-fold cross validation.

Hyperparameter Setup. In the pretraining model, the dimensions of the initial and output embeddings (for both nodes and relations) are 100 and 200, respectively. We use the combination of circular-correlation and ConvE as the operator during the training process. We introduce a 30% sparsity dropout into the ConvE layer and utilize the Adam method as the optimizer with the 0.001 learning rate. In the detection framework, we take advantage of Glove as the pretrained word embeddings with the dimension of 100. We use one layer GCN model with 64 hidden units. Afterward, an average pooling layer on the output of GCN is applied to get a context vector of each news.

Evaluation. We apply 5-fold cross-validation on the datasets and compare our approach with the selected baseline methods. In each test set, we make sure the number of fake and real news are the same. Moreover, for each fold, we run all the models 10 times and average the results. To measure the performance of fake news detection, we utilize the commonly used evaluation metrics for classification problems: Accuracy and F1 score, given that our test sets are balanced over the two classes.

Table 1. Performance Comparison of Fake News Prediction using Accuracy (Acc) and F1 score (%). Bold numbers are the best results and underline indicates the second best.

	Celebrity		PolitiFact		GossipCop	
	Acc	F1	Acc	F1	Acc	F1
LR+RST	54.2(±0.035)	54.7(±0.034)	57.8(±0.038)	49.3(±0.059)	53.4(±0.034)	51.6(±0.055)
RNN	53.0(±0.012)	57.1(±0.055)	68.6(±0.016)	68.1(±0.026)	63.9(±0.026)	63.1(±0.035)
LSTM	57.6(±0.047)	63.5(±0.080)	78.8(±0.024)	77.0(±0.025)	66.5(±0.045)	66.9(±0.035)
GRU	59.0(±0.081)	64.9(±0.050)	79.0(±0.027)	77.3(±0.038)	69.7(±0.025)	69.7(±0.039)
HDSF	50.0(±0.009)	66.7(±0.008)	50.4(±0.005)	66.8(±0.003)	50.7(±0.005)	67.1(±0.004)
dEFEND◇	53.2(±0.041)	63.1(±0.056)	70.4(±0.053)	73.9(±0.039)	52.1(±0.025)	65.1(±0.025)
CompGCN	51.8(±0.053)	62.1(±0.050)	63.6(±0.035)	54.8(±0.083)	61.1(±0.067)	65.9(±0.036)
Text-CNN	64.4(±0.060)	65.0(±0.087)	77.5(±0.041)	75.3(±0.046)	69.9(±0.049)	68.4(±0.038)
GCN	62.0(±0.056)	69.1(±0.033)	79.9(±0.020)	76.7(±0.038)	65.4(±0.062)	70.0(±0.046)
KGF-CNN	<u>68.4(±0.083)</u>	<u>71.7(±0.044)</u>	<u>81.6(±0.027)</u>	<u>81.1(±0.028)</u>	<u>71.2(±0.060)</u>	<u>70.8(±0.028)</u>
KGF	**71.4(±0.047)**	**72.1(±0.074)**	**86.0(±0.031)**	**85.3(±0.034)**	**73.3(±0.031)**	**72.3(±0.041)**

5   Results

Fake News Detection Performance. Table 1 exhibits the experimental results of **KGF** and other baseline methods on three datasets in terms of accuracy and F1 score. Overall, our approach outperforms all the baseline models.

When comparing to the Logistic Regression model with Rhetorical Structure Theory(RST) features, we observe our proposed **KGF** model can improve accuracy and F1 score both by 17% on Celebrity. For PolitiFact, the proposed model outperforms LR by 30% in Accuracy and by 35% in F1. CompGCN with only KG features achieves the inferior performance compared to other baselines. From which, we believe the semantic features learned from the original text provide rich information of contexts and backgrounds in detecting fake news. Our **KGF** can also beat both HDSF and dEFEND$^\diamond$, in both performance metrics. But notably, HDSF and dEFEND$^\diamond$ models do not perform well as the reported results in the original papers. For dEFEND$^\diamond$, we think there are three reasons: (1) we used different data sampling strategies; (2) we applied K-fold cross validation for averaged results; (3) our experiments do not consider the comments of the news. For HDSF, the datasets we used in this paper cover different varieties of topics, which differs from the original paper. The GCN model achieves the best performance among all the baseline models on Celebrity and GossipCop regarding F1 scores. However, our **KGF** model can still beat GCN in both accuracy and F1 scores across all datasets, due to the use of encoded KG embeddings.

Ablation Study. In order to investigate the effectiveness of our framework, we define a variant of **KGF**: **KGF**-cnn. We substitute GCN by Text-CNN to obtain the global semantic embedding and combine it with knowledge embedding. From Table 1, we can see **KGF** outperforms **KGF**-cnn and GCN. Meanwhile, **KGF**-cnn outperforms CNN. The results show the effectiveness of relational representations in detecting fake news.

Table 2. Examples of selected KG entities for a fake news prediction. Lime color denotes selected entities by our model and yellow color denotes the detected relations. Cyan denotes the keywords selected by dEFEND$^\diamond$ attention scores.

Police Discover Meth Lab In Back Room of Alabama Walmart DECATUR, Alabama – Police were recently tipped off to a reported meth lab that was being run by Walmart employees in what they are calling one of the biggest busts in decades. Police Chief Robert Garner said that an anonymous tip was left on their drug hotline, expressing concern about a horrible burning smell that was coming from the back of the Decatur WalMart facility. When an officer was sent to investigate, the store was instantly shut down as he discovered a meth lab that took up the entire back room. "The thing was massive, and contained enough materials to make hundreds, if not thousands, of pounds of crystal meth," said Chief Garner. "Apparently, every employee in the store was a part of it, from working with and gathering materials, to cooking, to selling it outside of the store. It was a full, massive operation."

Explainability Evaluation and Case Study. We select an example from the correctly predicted fake news in the test set of PolitiFact. In Table 2, we highlight the entities and relations which received high attention weights obtained .by

Eq. 5. Meanwhile, we highlight the keywords which received high word attention scores from dEFEND with a different color. In this example, we can see that the entities (e.g., "meth lab", "Back Room of Alabama Walmart DECATUR", "Police") with relations (e.g., "tripped", "discovered") chosen by our method are the essential components to the news. Since entities and relations represent facts that the news tries to express, our **KGF** provides the facts in the news that contribute most to the predictions. We can utilize the triples with the highest attention scores to provide explanations of why the news is classified as real or fake. It is worth mentioning that the facts represented by triples in our case can be real or fake.

Model Complexity. The computational complexity of pretraining mainly relies on the number of layers of GCN, i.e., K, the dimension of entity d, the total number of relations $|R|$, and the number of basis vectors $\mathcal{B}$. CompGCN uses the basis vectors $\{v_0, v_1, ..., v_{\mathcal{B}}\}$ to initialize the relation embeddings. Thus, the computational complexity of pretraining is $O(Kd^2 + \mathcal{B}d + \mathcal{B}|R|)$.

6   Conclusion

This paper proposed a new representation learning framework for explainable fake news detection using knowledge graph enhanced embeddings. Without external databases, we first extracted and organized a knowledge graph from accessible and reliable training corpora. Then we adapt a compositional graph neural network to pre-train structured features for entities and relations. Lastly, the pre-trained relational features are incorporated with semantic features for fake news recognition. The extensive experiments on two real-world datasets demonstrated the strengths of our proposed approach in fake news detection tasks, measured by standard classification evaluation metrics. We also exhibit case studies to provide structured explanations for the prediction results. In the future, we plan to investigate meta learning approaches to extract relations from text and examine other types of news including rumor and satire news.

Acknowledgements. This work is supported in part by the US National Science Foundation under grants 1948432, 1763620 and 1948374.

References

1. Angeli, G., Johnson Premkumar, M.J., Manning, C.D.: Leveraging linguistic structure for open domain information extraction. In: ACL, pp. 344–354, July 2015
2. Bordes, A., Usunier, N., Garcia-Duran, A., Weston, J., Yakhnenko, O.: Translating embeddings for modeling multi-relational data. In: NeurIPS, pp. 2787–2795 (2013)
3. Cho, K., et al.: Learning phrase representations using RNN encoder-decoder for statistical machine translation. In: EMNLP, pp. 1724–1734, October 2014
4. Ciampaglia, G.L., Shiralkar, P., Rocha, L.M., Bollen, J., Menczer, F., Flammini, A.: Computational fact checking from knowledge networks. PLoS ONE **10**(6), 1–13 (2015)

5. Deng, S., Rangwala, H., Ning, Y.: Learning dynamic context graphs for predicting social events. In: KDD 2019, pp. 1007–1016. ACM, New York (2019)
6. Dettmers, T., Minervini, P., Stenetorp, P., Riedel, S.: Convolutional 2d knowledge graph embeddings. In: Proceedings of the 32th AAAI Conference on Artificial Intelligence, pp. 1811–1818. AAAI (2018)
7. Grinberg, N., Joseph, K., Friedland, L., Swire-Thompson, B., Lazer, D.: Fake news on twitter during the 2016 U.S. presidential election. Science **363**(6425), 374–378 (2019)
8. Hochreiter, S., Schmidhuber, J.: Long short-term memory. Neural Comput. **9**(8), 1735–1780 (1997)
9. Karimi, H., Tang, J.: Learning hierarchical discourse-level structure for fake news detection. In: NAACL-HLT, pp. 3432–3442, June 2019
10. Kim, Y.: Convolutional neural networks for sentence classification. In: EMNLP, pp. 1746–1751, October 2014
11. Kipf, T.N., Welling, M.: Semi-supervised classification with graph convolutional networks. In: ICLR (2017)
12. Lazer, D.M.J., et al.: The science of fake news. Science **359**(6380), 1094–1096 (2018)
13. Levi, O., Hosseini, P., Diab, M., Broniatowski, D.: Identifying nuances in fake news vs. satire: using semantic and linguistic cues. In: Proceedings of the Second Workshop on Natural Language Processing for Internet Freedom: Censorship, Disinformation, and Propaganda, pp. 31–35, November 2019
14. Lu, Y.-J., Li, C.-T.: GCAN: graph-aware co-attention networks for explainable fake news detection on social media (2020)
15. Magdy, A., Wanas, N.: Web-based statistical fact checking of textual documents. SMUC **2010**, 103–110 (2010)
16. Nguyen, D.M., Do, T.H., Calderbank, R., Deligiannis, N.: Fake news detection using deep Markov random fields. ACL-HLT **2019**, 1391–1400 (2019)
17. Nickel, M., Rosasco, L., Poggio, T.: Holographic embeddings of knowledge graphs. AAAI **2016**, 1955–1961 (2016)
18. Pan, J.Z., Pavlova, S., Li, C., Li, N., Li, Y., Liu, J.: Content Based Fake News Detection Using Knowledge Graphs. In: Vrandevcić, D., Bontcheva, K., Suárez-Figueroa, M.C., Presutti, V., Celino, I., Sabou, M., Kaffee, L.-A., Simperl, E. (eds.) ISWC 2018. LNCS, vol. 11136, pp. 669–683. Springer, Cham (2018). https://doi.org/10.1007/978-3-030-00671-6_39
19. Pérez-Rosas, V., Kleinberg, B., Lefevre, A., Mihalcea, R.: Automatic detection of fake news. In: COLING, vol. 18, pp. 3391–3401 (2018)
20. Rubin, V., Conroy, N., Chen, Y.: Towards news verification: Deception detection methods for news discourse, January 2015
21. Shi, B., Weninger, T.: Fact checking in heterogeneous information networks. In: WWW 2016 Companion, pp. 101–102 (2016)
22. Shu, K., Cui, L., Wang, S., Lee, D., Liu, H.: dEFEND: explainable fake news detection. In: KDD 2019, pp. 395–405 (2019)
23. Shu, K., Mahudeswaran, D., Wang, S., Lee, D., Liu, H.: FakeNewsNet: a data repository with news content, social context and spatialtemporal information for studying fake news on social media (2018)
24. Shu, K., Sliva, A., Wang, S., Tang, J., Liu, H.: Fake news detection on social media: a data mining perspective. SIGKDD Explor. Newsl. **19**(1), 22–36 (2017)
25. Vashishth, S., Sanyal, S., Nitin, V., Talukdar, P.: Composition-based multi-relational graph convolutional networks. In: ICLR (2020)

26. Wang, Q., Mao, Z., Wang, B., Guo, L.: Knowledge graph embedding: a survey of approaches and applications. IEEE Trans. Knowl. Data Eng. **29**(12), 2724–2743 (2017)
27. Wang, Y., Ma, F., Jin, Z., Yuan, Y., Xun, G., Jha, K., Su, L., Gao, J.: EANN: event adversarial neural networks for multi-modal fake news detection. KDD **18**, 849–857 (2018)
28. Wu, Y., Agarwal, P.K., Li, C., Yang, J., Yu, C.: Toward computational fact-checking. Proc. VLDB Endow. **7**(7), 589–600 (2014)
29. Yang, B., tau Yih, W., He, X., Gao, J., Deng, L.: Embedding entities and relations for learning and inference in knowledge bases. CoRR, abs/1412.6575 (2015)
30. Zhang, Z., Han, X., Liu, Z., Jiang, X., Sun, M., Liu, Q.: ERNIE: enhanced language representation with informative entities. In: ACL (2019)
31. Zhou, X., Zafarani, R.: Fake news: a survey of research, detection methods, and opportunities. ArXiv, abs/1812.00315 (2018)

Incorporating Syntactic Information into Relation Representations for Enhanced Relation Extraction

Li Cui, Deqing Yang[✉], Jiayang Cheng, and Yanghua Xiao

Fudan University, Shanghai 200433, China
{lcui18,yangdeqing,chengjy17,shawyh}@fudan.edu.cn

Abstract. Relation Extraction (RE) is a premier task of information extraction (IE) and crucial to many applications including knowledge graph completion (KGC). In recent years, some RE models have employed the topic knowledge of relations through topic words to enrich relation representations, demonstrating better performance than traditional distantly supervised paradigms. However, these models have not taken different syntactic information of relations into account, which have been proven significant in many NLP tasks. In this paper, we propose a novel RE pipeline which incorporates syntactic information into relation representations to enhance RE performance. Representations of sentence and relation in our pipeline are generated by a modified multi-head self-attention structure respectively, where the sentence is represented based on its words and the relation is represented based on the relation-specific embeddings of its topic words. Furthermore, all sentences labeled with the input relation are used to construct an entire weighted directed graph based on their dependency trees. Then, the relation-specific embeddings of words (nodes) in the graph are learned by a GCN-based model. Our extensive experiments have justified that our pipeline significantly outperforms other RE models thanks to the incorporation of syntactic information.

Keywords: Relation Extraction · Relation representation · Syntactic information · Graph convolutional networks

1 Introduction

As one of the premier issues in information extraction (IE) and knowledge graph completion (KGC), relation extraction (RE) has received extensive attention in recent years. The goal of RE is to recognize a relation predefined in knowledge graphs (KGs) for two named entities existing in plain texts. For example, given the entity pair [*Steve Jobs, Apple*] in the sentence*Steve Jobs and Wozniak co-founded Apple in 1976 to sell Wozniak's Apple I personal computer.*, the relation *the-founder-of* can be recognized by a RE model precisely.

This work is supported by Shanghai Science and technology innovation action plan (No. 19511120400).

K. Karlapalem et al. (Eds.): PAKDD 2021, LNAI 12714, pp. 416–428, 2021.
https://doi.org/10.1007/978-3-030-75768-7_33

In recent years, some researchers have employed deep neural networks (DNNs) to achieve RE. For example, convolutional neural networks (CNNs) [26], recurrent neural networks (RNNs) [28] and sentence-level attention mechanism [10,24] has already been proven effective in RE tasks. Rather than only focusing on sentence-level features, the RE pipeline proposed by [11] accomplishes RE task through evaluating the matching degree between a given sentence containing a target entity pair and a candidate relation, and further exploits relation representation learning to improve matching precision. The basic assumption of their solution is that, *the sentence collection of a specific relation contains several latent topics and these topics are semantically related to the relation.* Based on this assumption, a relation is represented by the embeddings of some topic words which are distilled by a topic model from the sentence collection labeled with this relation. However, they only utilized the co-occurrences of topic words to generate relation representations without taking the syntactic context of this relation into account. The syntactic information has been proven to be a class of significant features for many NLP tasks [1,6,18], inspiring us to leverage syntactic information to enhance RE performance.

In order to fully exploit syntactic information of different relations, we propose a novel RE pipeline which also follows the basic principle of sentence-relation matching. In our pipeline, a relation's representation is generated based on the embeddings of its topic words, but the embeddings of topic words are learned by a method different to the embeddings of sentence words. This is because that a topic word may have different syntactic contexts when representing different relations. We illustrate an example in Fig. 1 to explain it, where the word *Apple* acts as the direct object (dobj) of the verb *co-founded* in sentence 1 labeled with the relation *founder*. While it acts as the passive nominal subject (nsubjpass) in sentence 2 labeled with the relation *location*. If each word occurring in different syntactic contexts is represented by a fixed embedding, the performance of the following sentence-relation matching network will likely be limited. Based on this intuition, we should learn an adaptive embedding for a topic word in terms of the relation it represents.

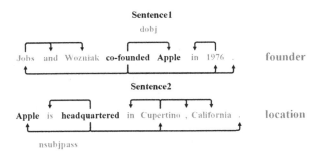

Fig. 1. Example of different syntactic contexts for 'Apple' when occurring in different relations.

In summary, we have the following contributions in this paper.

- We propose a method to learn relation-specific and syntax-enhanced embeddings for the topic words of a given relation, which are beneficial to generate better relation representations, resulting in better RE performance.
- We further construct a modified multi-head self-attention networks to effectively generate the representations of sentences and relations based on the syntax-enhanced word embeddings.
- Our extensive experiments have justified that the rationality of considering different syntactic information of different relations for relation representations.

The rest of this paper is organized as follows. We first introduce some research works related to our work in Sect. 2. Then, we introduce our pipeline briefly in Sect. 3, followed by the detailed descriptions in Sect. 4. We present and analyze our experiment results in Sect. 5 and conclude our work in Sect. 6.

2 Related Works

2.1 Relation Extraction Models

Many early RE models based on supervised paradigms can be categorized mainly into two classes, i.e., feature-based [2] and kernel-based models [5]. These models have demonstrated good performance, but require huge amount of labeled data. Collecting these data is an arduous and labor-intensive task. To address this problem, distant supervision was first introduced by [16] to generate training data automatically. It gives the basic assumption that *any sentence that contains a pair of entities participating in a known Freebase relation is likely to express that relation in some way.* However, this solution not only increases the amount of labeled sentences, but also introduces undesirable noises caused by wrong labeling, which greatly limits RE model's performance. Some related methods were proposed to alleviate this problem. For example, [24] proposed selective attention mechanism to de-emphasize noisy instances introduced by distant supervision, and to highlight the informative sentences. These works have shown promising results.

More recently, many researchers have employed deep-learning models to achieve RE task. For example, [27] proposed an end-to-end CNN model to extract lexical and sentence-level features. [26] further introduced the piece-wise CNNs (PCNNs), and incorporated multi-instance learning into PCNNs. [28] proposed an RNN-based framework to learn long-distance relation patterns which is capable of handling complicated expressions in real-world applications. By transforming extraction problem into a tagging task, [29] proposed a novel tagging scheme to jointly extracted entities and relations. Furthermore, reinforcement learning method was also proposed by [8], which is constituted by an instance selector and a relation classifier, and the instance selection is modeled as a reinforcement learning problem.

Besides exploiting the inherent merits of deep learning models, there have been many works trying to leverage external knowledge to further optimize the performance of RE. For example, [9] incorporated extra information including the separated lexical, syntactic and semantic knowledge into a feature-based RE model. [23] clustered relations into a set of relation topics, and modeled the relationship among relations. In addition, logic knowledge (in form of first-order logic formulae) was used in [20] by matrix factorization to generate enhanced embeddings of relations and entity-pairs. [10] proposed a sentence-level attention model which introduces entity description as external knowledge. Unlike above deep RE models, [11] modeled RE as a matching problem between a sentence containing entity pairs and a candidate relation, which is similar to our pipeline. Their proposed RE framework utilizes semantic knowledge of relation labels through topic word embeddings to improve RE performance.

2.2 Syntax-Based Word Embeddings and GCNs

Many previous works have exploited syntactic information for learning powerful word embeddings which are the basis of many NLP tasks. For instance, [18] first introduced dependency parse based embeddings and demonstrated its advantage on some word-function specific tasks. Furthermore, higher order dependencies were included to enhance embedding performance [1,6]. [21] successfully incorporated syntactic and semantic relationships into word embeddings without expanding the vocabulary using GCNs. These works inspire us to incorporate syntactic information to solve RE problem effectively. In addition, due to the success of GCNs on encoding structural information of graphs [14,17], some researchers also employed GCNs for the tasks of machine translation [4], semantic role labeling [12], document dating [22] and text classification [25]. These pioneer works inspire us to utilize GCNs to design an effective RE pipeline.

3 Framework Overview

As stated before, we model RE as a sentence-relation matching problem, where the given sentence is represented based on the embeddings of its words and the candidate relation is represented by its topic knowledge. Formally, we denote the set of relations as R. A training sample is denoted as $< s, t, r, y >$, where s is a sentence containing an entity pair t, r is a candidate relation and y is the ground-truth label. Specifically, a positive sample (i.e., $y = 1$) indicates the relation between entity pair t under sentence s is r, otherwise it is a negative sample (i.e., $y = 0$). The training goal of our model is to efficiently learn the matching function by which the probability $P(y = 1|s, t, r)$ is calculated, so that the matching probability (score) of any test samples can be predicted. In our pipeline, the matching function is modeled based on the representations of s and r.

The workflow of our RE pipeline is depicted in Fig. 2, where ovals and rectangles denote data and operations, respectively. The overall workflow can be divided into the following three steps.

Fig. 2. The overall workflow of our RE pipeline.

- **STEP1**: For each relation r collected from our training set, we first retrieve all the sentences labeled with r as its *sentence collection*, denoted as C_r. Then, we construct the dependency parsing graph of each sentence s in C_r using Stanford CoreNLP parser [7], and merge them into a weighted directed labeled graph G_r in which the nodes represent the words once appear in C_r. Each edge in G_r represents a weighted directed dependency relation, of which the weight is quantified as occurrence frequency of the edge between its two ends(words) in dependency parsing graphs generated form C_r. Next, we apply another GCN-based model on G_r to obtain the embedding of each node in G_r, the initialization of each node follows the strategy in [21]. Given the construction principle of G_r, such learned embeddings are syntax-enhanced and *relation-specific*, which will be used to generate better r's representation.
- **STEP2**: From the sentence collection C_r collected in STEP1, we extract top-c topic words for the relation r by topic modeling. Therefore, r's topic knowledge can be represented by the embeddings of these topic words which have been learned in STEP1.
- **STEP3**: The semantic distance $dis(s, r)$ between s and r is measured by the Word Movers Distance (WMD) [13]. For each sentence s, we choose some negative relation r' with smaller distance $dis(s, r')$. A deep sentence-relation matching network is constructed in our pipeline, to compute the final matching score between s and r, i.e., $P(y = 1|s, t, r)$.

In the next section, we will introduce the details of **STEP1** and **STEP3** in our RE pipeline.

4   Details of Our Relation Extraction System

In this section, we will first introduce how to learn the relation-specific embeddings for topic words. Then, we will introduce the architecture of our sentence-relation matching network. In the following section, we use a bold uppercase to denote a matrix, and a bold lowercase to denote a vector.

4.1   Relation-Specific Word Embeddings for Relation Representation

Inspired by the merits of syntactic information in many other NLP works [1,6,18], we believe that syntactic information is also helpful to improve the embeddings of relation topic words, which is however ignored in previous works [11]. Therefore, the relation-specific word embeddings for topic words are learned at first in our RE pipeline, in order to generate a better relation representation for sentence-relation matching network.

Weighted Directed Graphs for Relation. For each relation r and its weighted directed labeled graph $G_r = (V_r, E_r)$, V_r is the set of nodes and E_r is the collection of edges in G_r. All edges in E_r are represented in the form of (w_i, w_j, p_{ij}) where p_{ij} is the weight of the dependency relation $w_i \rightarrow w_j$. Specifically, p_{ij} is computed as

$$p_{ij} = \sigma\big(log(freq[w_i \rightarrow w_j])\big) \tag{1}$$

where σ is sigmoid function and $freq[w_i \rightarrow w_j]$ is the frequency of dependency relation $w_i \rightarrow w_j$ occurring in the dependency parsing graphs of the sentences in C_r. In these graphs, the representation of each node in G_r is initialized by pre-trained word embeddings learned by SynGCN [21].

GCN on Weighted Directed Graph. As shown in Fig. 3, for a given relation r, its weighted directed graph G_r is delivered to GCNs of K layers to learn the embedding of each node in G_r. This GCN's propagation rule for its k-th layer is defined as

$$h_i^{k+1} = f\left(\sum_{j \in \mathcal{N}_+(i)} p_{ij} \times \big(W^k h_j^k + b^k\big)\right) \tag{2}$$

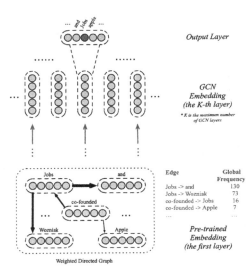

Fig. 3. The GCN-based model utilizing corpus-level syntactic information to learn relation-specific word embeddings for a relation.

where $\boldsymbol{h}_i^{k+1} \in \mathbb{R}^{d_1}$ is the embedding of word w_i after the k-th layer, $\mathcal{N}_+(i) = \{i\} \bigcup \mathcal{N}(i)$ and $\mathcal{N}(i)$ is the direct neighbor set of node, p_{ij} is the weight of the dependency relation $w_i \to w_j$, $\boldsymbol{W}^k \in \mathbb{R}^{d_1 \times d_1}$ and $\boldsymbol{b}^k \in \mathbb{R}^{d_1}$ are both trainable parameters and d_1 is the embedding dimension of each node. Note that by adding p_{ij} to the propagation rule, we leverage the importance of different dependency relations in different relations' syntactic contexts.

GCN's Training Objective. Formally, for each node (word) in graph G_r, the training objective of our GCN-based model is to predict the target word given its direct neighbors in G_r. That is, for the target word w_t, we maximize the following objective:

$$\mathcal{O} = \sum_{t=1}^{|V_r|} log P(w_t | w_1^t, w_2^t, \ldots, w_{|\mathcal{N}(t)|}^t) \tag{3}$$

where $\mathcal{N}(t) = \{w_1^t, w_2^t, \ldots, w_{|\mathcal{N}(t)|}^t\}$ are w_t's direct neighbors.

Specifically, the probability is defined in the form of softmax function:

$$P(w_t | w_1^t, w_2^t, \ldots, w_{|\mathcal{N}(t)|}^t) = \frac{exp(\boldsymbol{v}_t^T \boldsymbol{h}_t)}{\sum_{i=1}^{|V_r|} exp(\boldsymbol{v}_i^T \boldsymbol{h}_t)} \tag{4}$$

where $\boldsymbol{h}_t \in \mathbb{R}^{d_1}$ is w_t's embedding in the final layer of our SynGCN, and $\boldsymbol{v}_t \in \mathbb{R}^{d_1}$ is w_t's target embedding which contains the information of w_t's syntactic contexts and will be used to generate relation r's representation afterwards.

As a result, the objective can be expressed as

$$\mathcal{O} = \sum_{t=1}^{|V_r|} \left(\boldsymbol{v}_t^T \boldsymbol{h}_t - log \sum_{i=1}^{|V_r|} exp(\boldsymbol{v}_i^T \boldsymbol{h}_t) \right) \tag{5}$$

Thanks to GCN's excellent ability to integrate graph structure information, relation-specific syntactic information of different relations is incorporated into their topic words embeddings.

4.2   Sentence-Relation Matching Network

Inspired by [11], we propose a module based on modified multi-head self-attention layers to generate the final representation of sentences and relations.

Sentence Initialization. Suppose an input sentence s consists of m words, denoted as $\{w_1, w_2, \ldots, w_m\}$. Inspired by the basic assumption in [26] that *each block in a sentence has different importance for its relation inference*, we first divide s into three blocks based on the two entities it contains. For example, the sentence in Fig. 4 can be divided by the entity pair *[Jobs, Apple]* into three blocks.

Fig. 4. Sentence-relation Matching Network of our RE pipeline.

In details, the embeddings of a certain word w in s is defined in the concatenation form $[\boldsymbol{x}^w; \boldsymbol{x}^p] \in \mathbb{R}^{d_1+d_2}$, where $\boldsymbol{x}^w \in \mathbb{R}^{d_1}$ is w's word embedding pre-trained by [21] and $\boldsymbol{x}^p \in \mathbb{R}^{d_2}$ is w's position embedding. The method of generating position embeddings we adopted is the same as the one in [24]. We fix the length of each block as l by *truncation* and *zero-padding*. As a result, the initial representation of the i-th block is denoted as $\boldsymbol{I}_i \in \mathbb{R}^{(d_1+d_2)\times l}$. Besides, dividing a sentence into some blocks will significantly reduce the space complexity of our RE pipeline.

Sentence Representation. In recent years, multi-head self-attention mechanism has been proven effective in encoding sentences [3]. Before feeding each block into our multi-head self-attention layers which contains P parallel attention heads. The self-attention operation for each head is defined as below:

$$Attention^s(\boldsymbol{Q}^s, \boldsymbol{K}^s, \boldsymbol{V}^s) = softmax\left(\frac{\boldsymbol{Q}^s \boldsymbol{K}^{sT}}{\sqrt{d}}\right)\boldsymbol{V}^s \qquad (6)$$

where d is the dimension of the self-attention heads, $\boldsymbol{Q}^s, \boldsymbol{K}^s, \boldsymbol{V}^s$ are generated by linear transformation upon the initial representation matrix of each block, i.e., $\boldsymbol{I}_i$. The output of the attention layer for each block $\boldsymbol{I}_i$ is $\boldsymbol{H}_i \in \mathbb{R}^{d\times l}$. Furthermore, we apply a max pooling layer to above self-attention layer. The output of this layer is $\boldsymbol{H}_i^s \in \mathbb{R}^{d\times p}$.

At last, final representation of each sentence s is the concatenation of its block representations $\boldsymbol{H}_i^s$, denoted as $\boldsymbol{O}^s \in \mathbb{R}^{d\times b}$, where $b = 3p$.

Relation Representation. This module aims to generate the final representation of a candidate relation. Topic words for a relation r are generated by topic modeling (e.g., LDA) over C_r.

In this paper, we regard a given relation r's topic words as a weighted bag of words (WBoW), denoted as $\mathcal{A}^r = \{w_1(tw_1), \ldots, w_c(tw_c)\}, 1 \leq i \leq c$, where w_i is the i-th topic word of r with weight tw_i. w_i's relation-specific embedding is $\boldsymbol{x}_i^t \in \mathbb{R}^{d_1}$. To this end, the relation-specific embeddings of r's topics words constitute its initial representation $\boldsymbol{I}^r \in \mathbb{R}^{d_1 \times c}$. In addition, the weight vector $\{tw_1, \ldots, tw_c\}$ is transformed into a diagonal matrix $\boldsymbol{W}^c \in \mathbb{R}^{c \times c}$ where $\boldsymbol{W}_{i,i}^c = tw_i$.

Note that different topic words in r's WBoW have different weights. Thus we propose a modified multi-head self-attention layer which takes the priori weights of r's topic words into account. Then, the definition of this self-attention operation for each relation is as below:

$$Attention^r(\boldsymbol{Q}^r, \boldsymbol{K}^r, \boldsymbol{V}^r, \boldsymbol{W}^c) = softmax\left(\frac{\boldsymbol{Q}^r \boldsymbol{W}^c \boldsymbol{K}^{rT}}{\sqrt{d}}\right)\boldsymbol{V}^r \tag{7}$$

where $\boldsymbol{Q}^r, \boldsymbol{K}^r, \boldsymbol{V}^r$ are generated by linear transformation upon r's initial representation $\boldsymbol{I}^r$, $\boldsymbol{W}^c$ is a diagonal matrix generated by priori weights.

The output of this layer is $\boldsymbol{H}^r \in \mathbb{R}^{d \times c}$. Finally, we further feed $\boldsymbol{H}^r$ into standard multi-head self-attention layer and linear transformation, to obtain r's final representation $\boldsymbol{O}^r \in \mathbb{R}^{d \times b'}$.

Sentence-Relation Interaction and Training Objective. The sentence-relation interaction layer is defined as below:

$$sim(s, r|t) = \boldsymbol{w}^T tanh\big(sum(\boldsymbol{W}_1 \boldsymbol{O}^s) + sum(\boldsymbol{W}_2 \boldsymbol{O}^r) + \boldsymbol{b}_1\big) \tag{8}$$

$$P(y = 1| <s, t, r>) = \frac{1}{1 + e^{-sim(s,r|t)}} \tag{9}$$

where the function $sum(\cdot)$ transforms a matrix into a single column vector by summing all elements in a row. $\boldsymbol{W}_1 \in \mathbb{R}^{d' \times d}$, $\boldsymbol{W}_2 \in \mathbb{R}^{d' \times d}$, $\boldsymbol{w}$ and $\boldsymbol{b}_1 \in \mathbb{R}^{d'}$ are the trainable parameters.

The definition of the loss function is shown as below:

$$\mathcal{L}(\theta) = \sum_{i=1}^{|\mathcal{D}|} \mathcal{L}_B[p(y_i| <s_i, t_i, r_i>; \theta)] \tag{10}$$

where $\mathcal{L}_B$ is the binary cross entropy, $\mathcal{D} = \{<s_i, t_i, r_i, y_i>\}, i = 1, 2, \ldots, |\mathcal{D}|$ is the training set where y_i is the ground-truth label of the sample $<s_i, t_i, r_i>$, θ represents all trainable parameters.

5   Experiments

In this section, we will try to answer the following question: Can our proposed relation-specific embeddings of topic words improve relation representation, resulting in better RE performance? To this end, our proposed RE pipeline was evaluated from two aspects: overall performance and performance of relation-specific topic representation.

5.1 Experiment Settings

Dataset. In our experiments, our RE pipeline and its competitors were evaluated on a widely used distant supervision dataset, which was generated by aligning Freebase relations with New York Times corpus (NYT for short) [19].

Baselines. We compared our RE pipeline with several RE models as below.

- *Logic-MF* [20]: a matrix factorization-based model adopting distant supervision and injects first order logical knowledge into the entity and relation representations.
- *PCNN+ATT* [24]: a CNN-based model applying selective attention mechanism to address the problem of wrong labeling in distant supervision.
- *APCNN+D* [10]: a sentence-level attention model with description-enhanced entity representation for RE.
- *CNN+ATT+RL* [8]: an advanced revision of *PCNN+ATT* with an instance selector based on reinforcement learning.
- *TopicRE* [11]: a RE framework similar to our pipeline with topic word based relation representation.

Evaluation Metrics. We used precision, recall and F1 as the metrics to evaluate the performance of our pipeline and baselines. Specifically, we adopted held-out evaluation [8,24,26] in our experiments.

Table 1. The precision scores under different recall of all models in terms of RE.

Model	Recall							
	0.05	0.1	0.15	0.2	0.25	0.3	0.35	$maxF1$
Logic-MF	0.79	0.75	0.68	0.66	0.63	0.53	0.49	0.408
PCNN+ATT	0.83	0.78	0.71	0.66	0.62	0.59	0.53	0.422
APCNN+D	0.78	0.76	0.72	0.65	0.62	0.58	0.51	0.415
CNN+ATT+RL	0.85	0.73	0.68	0.67	0.59	0.57	0.52	0.418
TopicRE	0.94	0.84	0.75	0.71	0.69	0.65	0.63	0.445
Ours	**0.96**	**0.93**	**0.88**	**0.81**	**0.75**	**0.70**	**0.66**	**0.485**

5.2 Experiment Results

In this subsection, we not only display the performance comparisons between our pipeline and the baselines, but also justify the merits of relation-specific topic-word representation through ablation studies. All results were obtained through held-out evaluation.

Overall Performance Comparison. We compared our pipeline with five baselines in terms of RE performance. Table 1 displays $maxF1$ of each model and their precision scores under different recalls. The results show that our pipeline outperforms the baselines apparently. Even compared with TopicRE which has the best $maxF1$ in all baselines, our pipeline still gains 9% improvement of $maxF1$. Next, we further investigate the effectiveness of relation-specific topic-words' representations which take different syntactic contexts of different relations into consideration.

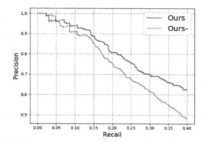
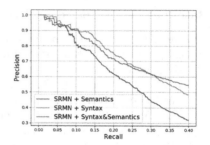

Fig. 5. PR curve comparison of incorporating relation-specific word embeddings or not.

Fig. 6. PR curve comparison of incorporating syntax-based word embeddings or not.

Evaluation of Relation-Specific Topic Word Representation. To this end, we first designed an ablated variant of our pipeline, namely $Ours^-$, by removing relation-specific word representation from relation representation. In other words, the relation's topic words are directly represented by the pre-trained word embeddings learned by SynGCN [21]. Figure 5 displays the Precision-Recall curves of our pipeline and $Ours^-$. Our pipeline's superiority over $Ours^-$ justifies the rationality of learning relation-specific embeddings for relation's topic words, which help our pipeline better recognize a relation in terms of the syntactic contexts involving the relation's topic words.

Also, We are interested in whether syntax or semantics is more important in RE tasks, we conduct another experiment. We use our sentence-relation matching network, which is denoted as SRMN, to accomplish RE tasks with different word embeddings. In fact, the baseline TopicRE uses the pre-trained word embeddings pre-trained on New York Times Annotated Corpus (LDC Data LDC2008T19) [24] by Word2Vec [15], which focus mainly on semantics. In contrast, the proposed SynGCN focus more on syntax. We adopt three word embeddings strategies: *Semantics* follows the strategy in TopicRE, *Syntax* directly uses SynGCN and *Syntax&Semantics* initializes SynGCN with embeddings learned in TopicRE. It is taken for granted that *Syntax&Semantics* should outperform *Syntax* since semantics are also considered. However, it is unexpected that these two strategies are inseparable according to the curves shown in Fig. 6. A possible reason for this observation is that, semantic information and syntactic information are of different scales in sentence representation, as semantic information is extracted mainly in corpus-level while syntactic information is extracted mainly in sentence-level. Furthermore, the available syntactic information is much more than the semantic information [21]. However, both of these two strategies outperform *Semantics* apparently. It also verifies that our RE pipeline's perfect performance is mainly attributed to the syntax-enhanced word embeddings. We try to guess that syntax is more important than semantics for RE tasks, but this requires proof of further works.

6   Conclusion

In this paper, we propose a novel RE pipeline which incorporates syntactic information into relation representation. Specifically, we model RE as a matching problem between a given sentence containing entity pair and a candidate relation. Specifically, we incorporate a modified multi-head self-attention network in our pipeline to compute the final matching scores based on sentence and relation representations.

According to the results of our experiments, our pipeline clearly outperforms all the baselines. The extensive experiments further prove that the improvement of our pipeline over similar baselines is mainly attributed to the relation-specific word embeddings which make full use of the advantages of GCNs in integrating syntactic information. We also found an interesting phenomenon that initialing GCNs with the pre-trained embeddings fused with semantic information will not significantly improve the model's performance, we will further explore whether syntax is more important that semantics for RE tasks in future works.

References

1. Alexandros, K., Suresh, M.: Dependency based embeddings for sentence classification tasks. In: Proceedings of NAACL (2016)
2. Anita, A., Anna, C.: Barrier features for classification of semantic relations. In: Proceedings of RANLP (2011)
3. Ashish, V., Noam, S., Niki, P., et al.: Attention is all you need. In: Proceedings of NIPS (2017)
4. Bastings, J., Titov, I., Aziz, W., et al.: Graph convolutional encoders for syntax-aware neural machine translation. In: Proceedings of EMNLP (2017)
5. Bunescu, R.C., Mooney, R.J.: A shortest path dependency kernel for relation extraction. In: Proceedings of HLT/EMNLP (2005)
6. Chen, L., Jianxin, L., Yangqiu, S., Ziwei, L.: Training and evaluating improved dependency-based word embeddings. In: Proceedings of AAAI (2018)
7. Christopher, M., Mihai, S., John, B., Jenny, F., Steven, B., David, M.: The stanford coreNLP natural language processing toolkit. In: Proceedings of ACL (2014)
8. Feng, J., Huang, M., Zhao, L., Yang, Y., Zhu, X.: Reinforcement learning for relation classification from noisy data. In: Proceedings of AAAI (2018)
9. GuoDong, Z., Jian, S., Jie, Z., Min, Z.: Exploring various knowledge in relation extraction. In: Proceedings of ACL (2005)

10. Ji, G., Liu, K., He, S., Zhao, J.: Distant supervision for relation extraction with sentence-level attention and entity descriptions. In: Proceedings of AAAI (2017)
11. Jiang, H., Cui, L., Xu, Z., et al.: Relation extraction using supervision from topic knowledge of relation labels (2019)
12. Marcheggiani, D., Titov, I.: Encoding sentences with graph convolutional networks for semantic role labeling. In: Proceedings of EMNLP (2017)
13. Matt, K., Yu, S., Nicholas, K., Kilian, W.: From word embeddings to document distances. In: Proceedings of ICML (2015)
14. Michaël, D., Xavier, B., Pierre, V.: Convolutional neural networks on graphs with fast localized spectral filtering. In: Proceedings of NIPS (2016)
15. Mikolov, T., Chen, K., Corrado, G., Dean, J.: Efficient estimation of word representations in vector space (2013)
16. Mintz, M., Bills, S., Snow, R., Jurafsky, D.: Distant supervision for relation extraction without labeled data. In: Proceedings of ACL/AFNLP (2009)
17. N, K.T., Max, W.: Semi-supervised classification with graph convolutional networks. arXiv preprint arXiv:1609.02907 (2016)
18. Omer, L., Yoav, G.: Dependency-based word embeddings. In: Proceedings of ACL (2014)
19. Riedel, S., Yao, L., McCallum, A.: Modeling relations and their mentions without labeled text. In: Proceedings of ECML/PKDD (2010)
20. Rocktäschel, T., Singh, S., Riedel, S.: Injecting logical background knowledge into embeddings for relation extraction. In: Proceedings of NAACL (2015)
21. Shikhar, V., Manik, B., et al.: Incorporating syntactic and semantic information in word embeddings using graph convolutional networks. In: Proceedings of ACL (2019)
22. Vashishth, S., Dasgupta, S.S., Ray, S.N., Talukdar, P.: Dating documents using graph convolution networks. In: Proceedings of ACL (2018)
23. Wang, C., Fan, J., Kalyanpur, A., Gondek, D.: Relation extraction with relation topics. In: Proceedings of EMNLP (2011)
24. Yankai, L., Shiqi, S., et al.: Neural relation extraction with selective attention over instances. In: Proceedings of ACL, pp. 2124–2133 (2016)
25. Yao, L., Mao, C., Luo, Y.: Graph convolutional networks for text classification. In: Proceedings of AAAI (2019)
26. Zeng, D., Liu, K., Chen, Y., Zhao, J.: Distant supervision for relation extraction via piecewise convolutional neural networks. In: Proceedings of EMNLP (2015)
27. Zeng, D., Liu, K., Lai, S., Zhou, G., Zhao, J.: Relation classification via convolutional deep neural network, pp. 2335–2344 (2014)
28. Zhang, D., Wang, D.: Relation classification via recurrent neural network. arXiv preprint arXiv:1508.01006 (2015)
29. Zheng, S., Wang, F., Bao, H., Hao, Y., Zhou, P., Xu, B.: Joint extraction of entities and relations based on a novel tagging scheme (2017)

Correction to: Rule Injection-Based Generative Adversarial Imitation Learning for Knowledge Graph Reasoning

Sheng Wang, Xiaoying Chen, and Shengwu Xiong

Correction to:
Chapter "Rule Injection-Based Generative Adversarial Imitation Learning for Knowledge Graph Reasoning"
in: K. Karlapalem et al. (Eds.): *Advances in Knowledge Discovery and Data Mining*, **LNAI 12714,**
https://doi.org/10.1007/978-3-030-75768-7_27

In the originally published version of chapter 27, the name of the author Xiaoying Chen was spelled incorrectly. This has been corrected.

The updated version of this chapter can be found at
https://doi.org/10.1007/978-3-030-75768-7_27

Author Index

Printed in the United States
by Baker & Taylor Publisher Services